Springer Series in Advanced Manufacturing

Series Editor

Duc Truong Pham, University of Birmingham, Birmingham, UK

The **Springer Series in Advanced Manufacturing** includes advanced textbooks, research monographs, edited works and conference proceedings covering all major subjects in the field of advanced manufacturing.

The following is a non-exclusive list of subjects relevant to the series:

1. Manufacturing processes and operations (material processing; assembly; test and inspection; packaging and shipping).
2. Manufacturing product and process design (product design; product data management; product development; manufacturing system planning).
3. Enterprise management (product life cycle management; production planning and control; quality management).

Emphasis will be placed on novel material of topical interest (for example, books on nanomanufacturing) as well as new treatments of more traditional areas.

As advanced manufacturing usually involves extensive use of information and communication technology (ICT), books dealing with advanced ICT tools for advanced manufacturing are also of interest to the Series.

Springer and Professor Pham welcome book ideas from authors. Potential authors who wish to submit a book proposal should contact Anthony Doyle, Executive Editor, Springer, e-mail: anthony.doyle@springer.com.

More information about this series at https://link.springer.com/bookseries/7113

Alp Ustundag · Emre Cevikcan · Omer Faruk Beyca
Editors

Business Analytics for Professionals

Springer

Editors
Alp Ustundag
Isletme Fakultesi
Istanbul Technical University
Istanbul, Turkey

Emre Cevikcan
Isletme Fakultesi
Istanbul Technical University
Istanbul, Turkey

Omer Faruk Beyca
Isletme Fakultesi
Istanbul Technical University
Istanbul, Turkey

ISSN 1860-5168 ISSN 2196-1735 (electronic)
Springer Series in Advanced Manufacturing
ISBN 978-3-030-93822-2 ISBN 978-3-030-93823-9 (eBook)
https://doi.org/10.1007/978-3-030-93823-9

This Springer imprint is published by the registered company Springer Nature Switzerland AG
The registered company address is: Gewerbestrasse 11, 6330 Cham, Switzerland

Preface

As one of the most valuable assets for industrial systems, data has an enormous impact on sustainable success of businesses to obtain competitive advantage. The increasing size and variety of data has revealed new technologies that enable the analysis of big data. Especially, after real-time data collection and storage are facilitated via smart business and product concepts using digital transformation technologies, business analytics provides a new dimension to enterprises when revealing new insights for decision-making process by using the array of machine learning techniques on a variety of structured and unstructured data under interdisciplinary environment. As a pillar of business analytics, intelligent automated systems have come of age not only because technologies have matured, but also in response to business needs. The new generation of decision making is directly embedded in main business process flows that move from insight to action in a rapid and optimal manner while preventing the limitations of earlier decision support techniques in terms of human intervention and the time to analyze and maintain them.

In this context, this book not only introduces implementation aspects of business analytics, but also provides theoretical background for its management-related approaches and technologies with application cases. This book is divided into two main parts. In the first part, methods, tools and technologies which are needed for business data analytics will be introduced. Techniques used in descriptive, predictive and prescriptive analytics are explained from the business perspective. In the second part, the book addresses giving the comprehension of how businesses receive application-oriented advantages of intelligent automation as an industrial strength. To satisfy this objective, applications of the analytical tools to different areas of the business functions are given. Detailed problems and their solutions for business problems in various areas such as supply chain, marketing and CRM, financial, manufacturing and human resources are presented with their Python codes.

Please note that all dataset and Google Colab Jupyter notebooks for applications and case studies given in the chapters of the book can be reached from the following links or QR Code: https://bit.ly/3E6RKWs, https://github.com/ITU-Business-Analytics-Team.

When the solutions for data-related problems about the management of business functions in real-life industrial systems are concerned, inadequacy in theoretical background can be observed due to challenges and complexities to the current techniques. On the other hand, although a number of algorithms and methods have been developed in scientific manner, very few of them have been validated on real-life data and some obstacles may be encountered about their applicability. Therefore, the book is triggered by the requirement for enhancing theoretical perspective of professionals when bringing business analytics solutions for decision problems in strategical, tactical and operational levels.

There are numerous books about business analytics, but none of them explain both managerial and technological aspects of business analytics in an integrated framework. In this situation, this book fills the gap between real-world business management problems and solution techniques by providing comprehensive guidance for managers about business analytics.

The intended audience of this book will mainly consist of researchers working on the areas of business analytics, AI and machine learning intelligent automation as well as practitioners such as business analysts, data scientists, IT professionals, system integrators and consultants in various sectors. The book is also of interest to graduates and undergraduates in business analytics, data science, engineering management and industrial engineering.

We would like to thank all the authors for contributing to this book:

- Abdullah Emin Kazdaloglu, Istanbul Technical University
- Alp Ustundag, Istanbul Technical University
- Altan Cakir, Istanbul Technical University
- Arif Gulbiter, Istanbul Technical University
- Aycan Pekpazar, Istanbul Technical University
- Basar Oztaysi, Istanbul Technical University
- Behcet Ugur Toreyin, Istanbul Technical University
- Buse Sibel Korkmaz, Technical University of Munich
- Cigdem Altin Gumussoy, Istanbul Technical University
- Cigdem Kadaifci, Istanbul Technical University
- Dogan Oruc, Istanbul Technical University
- Emre Ari, Istanbul Technical University
- Emre Cevikcan, Istanbul Technical University
- Erkan Isikli, Istanbul Technical University
- Hidayet Beyhan, Istanbul Technical University

- İbrahim Yazici, Istanbul Technical University
- Kenan Menguc, Istanbul Technical University
- Kubra Cetin Yildiz, Istanbul Technical University
- Leyla Temizer, Istanbul University
- Mahmut Sami Sivri, Istanbul Technical University
- Mehmet Ali Ergun, Istanbul Technical University
- Mehmet Yasin Ulukus, Istanbul Technical University
- Mirac Murat, Karadeniz Technical University
- Nevcihan Toraman, Istanbul Technical University
- Nizamettin Bayyurt, Istanbul Technical University
- Nursah Alkan, Istanbul Technical University
- Nurullah Calik, Istanbul Medeniyet University
- Omer Faruk Beyca, Istanbul Technical University
- Ozgur Akarsu, Koc Digital
- Ozgur Kabak, Istanbul Technical University
- Sefer Baday, Istanbul Technical University
- Seval Ata, Istanbul Technical University
- Sezi Cevik Onar, Istanbul Technical University
- Sule Itir Satoglu, Istanbul Technical University
- Sultan Ceren Oner, Huawei Turkey
- Tolga Ahmet Kalayci, Istanbul Technical University
- Umut Asan, Istanbul Technical University
- Yakup Turgut, Istanbul Technical University
- Yildiz Kose, Istanbul Technical University
- Yusuf Isik, Istanbul Technical University
- Zeynep Burcu Kizilkan, Istanbul Technical University

We would also like to thank our colleagues Nursah Alkan, Zeynep Burcu Kizilkan, Yildiz Kose, Mahmut Sami Sivri and Seval Ata for their support to prepare the final format of the book. And finally, we thank our families for their moral support and endless patience.

Istanbul, Turkey
2021

Alp Ustundag
Emre Cevikcan
Omer Faruk Beyca

Contents

Business Applications

Editors and Contributors

About the Editors

Alp Ustundag is Head of Industrial Engineering Department of Istanbul Technical University (ITU) and Coordinator of M.Sc. in Big Data and Business Analytics Program. He is also CEO of Navimod Business Analytics Solutions located in ITU Technopark (http://navimod.com/). He has worked in IT and finance industry from 2000 to 2004. He continued his research studies at the University of Dortmund between 2007 and 2008 and completed his doctorate at ITU in 2008. He has conducted a lot of research and consulting projects in the finance, retail, manufacturing, energy and logistics sectors. His current research interests include artificial intelligence, data science, machine learning, financial and supply chain analytics. He has published many papers in international journals and presented various studies at national and international conferences.

Emre Cevikcan is Professor in Industrial Engineering from Istanbul Technical University (ITU). He received the B.S. degree in Industrial Engineering from Yildiz Technical University and the M.Sc. degree and Ph.D. degree in Industrial Engineering from ITU. He studied the scheduling of production systems for his Ph.D. dissertation. His research has so far focused on digital transformation, planning and design of manufacturing and service systems, lean production and scheduling. He has several research papers in scientific journals. In addition, he contributed several book projects about decision making and digital transformation as an editor and chapter author. He has contributed to several industrial projects about lean transformation, manufacturing system design and process improvement.

Omer Faruk Beyca received his B.S. degree in industrial engineering from Fatih University, Istanbul, Turkey, in 2007, and the Ph.D. degree from the School of Industrial Engineering and Management, Oklahoma State University, Stillwater, OK, USA, in 2013. He is with the Department of Industrial Engineering, Istanbul Technical

University, as Assistant Professor. His research has been published in international peer-reviewed journals, conference proceedings and chapters.

Contributors

Ozgur Akarsu Koc Digital, Unalan, Uskudar, Istanbul, Turkey

Nursah Alkan Department of Industrial Engineering, Faculty of Management, Istanbul Technical University, Macka, Istanbul, Turkey

Emre Ari Department of Industrial Engineering, Faculty of Management, Istanbul Technical University, Istanbul, Turkey

Umut Asan Department of Industrial Engineering, Faculty of Management, Istanbul Technical University, Istanbul, Turkey

Seval Ata Department of Industrial Engineering, Faculty of Management, Istanbul Technical University, Istanbul, Turkey

Sefer Baday Applied Informatics Department, Informatics Institute, Istanbul Technical University, Istanbul, Turkey

Nizamettin Bayyurt Industrial Engineering Department, Management Faculty, Istanbul Technical University, Istanbul, Turkey

Omer Faruk Beyca Department of Industrial Engineering, Faculty of Management, Istanbul Technical University, Macka, Istanbul, Turkey

Hidayet Beyhan Department of Management Engineering, Faculty of Management, Istanbul Technical University, Istanbul, Turkey

Altan Cakir Department of Physics Engineering, Faculty of Science and Letters, Istanbul Technical University, Maslak, Istanbul, Turkey;
Artificial Intelligence and Data Science Research and Application Center, Istanbul Technical University Artificial Intelligence, Istanbul, Turkey

Nurullah Calik Department of Biomedical Engineering, Istanbul Medeniyet University, Istanbul, Turkey

Emre Cevikcan Industrial Engineering Department, Management Faculty, Istanbul Technical University, Istanbul, Turkey

Mehmet Ali Ergun Department of Industrial Engineering, Faculty of Management, Istanbul Technical University, Macka, Istanbul, Turkey

Arif Gulbiter Department of Industrial Engineering, Faculty of Management, Istanbul Technical University, Macka, Istanbul, Turkey

Cigdem Altin Gumussoy Department of Industrial Engineering, Faculty of Management, Istanbul Technical University, Istanbul, Turkey

Yusuf Isik Department of Industrial Engineering, Faculty of Management, Istanbul Technical University, Istanbul, Turkey

Erkan Isikli Department of Industrial Engineering, Faculty of Management, Istanbul Technical University, Istanbul, Turkey

Ozgur Kabak Department of Industrial Engineering, Faculty of Management, Istanbul Technical University, Macka, Istanbul, Turkey

Cigdem Kadaifci Department of Industrial Engineering, Faculty of Management, Istanbul Technical University, Istanbul, Turkey

Tolga Ahmet Kalayci Department of Industrial Engineering, Faculty of Management, Istanbul Technical University, Istanbul, Turkey

Abdullah Emin Kazdaloglu Department of Industrial Engineering, Faculty of Management, Istanbul Technical University, Istanbul, Turkey

Zeynep Burcu Kizilkan Department of Industrial Engineering, Faculty of Management, Istanbul Technical University, Macka, Istanbul, Turkey

Buse Sibel Korkmaz Department of Informatics, Faculty of Engineering and Architecture, Technical University of Munich, Garching, Germany

Yildiz Kose Department of Industrial Engineering, Faculty of Management, Istanbul Technical University, Istanbul, Turkey

Kenan Menguc Department of Industrial Engineering, Faculty of Management, Istanbul Technical University, Macka, Istanbul, Turkey

Mirac Murat Department of Industrial Engineering, Faculty of Management, Istanbul Technical University, Istanbul, Turkey

Sezi Cevik Onar Department of Industrial Engineering, Faculty of Management, Istanbul Technical University, Istanbul, Turkey

Sultan Ceren Oner Umraniye, Istanbul, Turkey

Dogan Oruc Department of Industrial Engineering, Faculty of Management, Istanbul Technical University, Macka, Istanbul, Turkey

Basar Oztaysi Department of Industrial Engineering, Faculty of Management, Istanbul Technical University, Istanbul, Turkey

Aycan Pekpazar Department of Industrial Engineering, Faculty of Management, Istanbul Technical University, Istanbul, Turkey

Sule Itir Satoglu Department of Industrial Engineering, Faculty of Management, Istanbul Technical University, Istanbul, Turkey

Mahmut Sami Sivri Department of Industrial Engineering, Faculty of Management, Istanbul Technical University, Macka, Istanbul, Turkey

Leyla Temizer Department of Industrial Engineering, Faculty of Engineering, Istanbul University Cerrahpasa, Avcilar, Istanbul, Turkey

Nevcihan Toraman Department of Industrial Engineering, Faculty of Management, Istanbul Technical University, Istanbul, Turkey

Behcet Ugur Toreyin Informatics Institute, Istanbul Technical University, Istanbul, Turkey

Yakup Turgut Department of Industrial Engineering, Faculty of Management, Istanbul Technical University, Istanbul, Turkey

Mehmet Yasin Ulukus Department of Industrial Engineering, Faculty of Management, Istanbul Technical University, Istanbul, Turkey

Alp Ustundag Department of Industrial Engineering, Faculty of Management, Istanbul Technical University, Istanbul, Turkey

Ibrahim Yazici Department of Industrial Engineering, Faculty of Management, Istanbul Technical University, Macka, Istanbul, Turkey

Kubra Cetin Yildiz Department of Industrial Engineering, Faculty of Management, Istanbul Technical University, Istanbul, Turkey

Methods and Technologies for Business Analytics

Business Analytics for Managers

Yakup Turgut, Yildiz Kose, Alp Ustundag, and Emre Cevikcan

1 Introduction

Humans are decision-makers. Career and lifestyle decisions such as starting a business, purchasing a product, and investing in the future are all personalized decisions. Managers also in businesses make many decisions every day. If there is one truth that may be agreed upon about all managers, it is that they are overwhelmed by the burden of decisions. Known as prioritizing and differentiating in business, budgeting and scheduling are a few of these decisions. Since so many of these decisions are complicated by incomplete data and poor information, it is often difficult to get an accurate assessment of their impact on the future profits.

Moreover, the current business environment is characterized by constant change and is getting even more difficult. Businesses in the modern world must always be evolving in order to remain competitive. This is calling for an organization to be flexible and fast-moving, making tough decisions at times as well as significant and complicated ones. All of this may require a great deal of relevant data and knowledge. The processing of those data must be done rapidly, often in real time, and usually require computerized support within the context of the necessary decisions [1]. Hence, ill-informed decisions will cause one to fail at business.

Fortunately, businesses are inundated with data at an astounding rate and volume. Companies are recording and collecting enormous amounts of data every day. Big data gathered on customer interactions and purchases will offer a variety of advantages to managers, depending on how they are analyzed. The rise of data in marketing

Y. Turgut · Y. Kose · A. Ustundag (✉) · E. Cevikcan
Industrial Engineering Department, Management Faculty, Istanbul Technical University, 34367 Macka, Istanbul, Turkey
e-mail: ustundaga@itu.edu.tr

Y. Turgut
e-mail: turgut16@itu.edu.tr

A. Ustundag et al. (eds.), *Business Analytics for Professionals*, Springer Series in Advanced Manufacturing, https://doi.org/10.1007/978-3-030-93823-9_1

from small to big sets is the impetus for great management that expands beyond standard techniques.

Companies could vastly enhance their business operations through the use of business analytics, such as data analysis, optimization, simulation, and machine learning [2, 3]. *Business analytics (BA)* is the science and art of incorporating quantitative data into decision making. The goal of business analytics is to help organizations improve their decision making with faster, better analysis. No business analytics activity is ever performed in vain. They all seek to increase business value for the enterprise [4, 5].

Recently, many transaction and management tasks have shifted to the cloud, with computer and mobile-based tools in use for problem and issue analysis and solution in many times when you need them. Growth in hardware, network, software, and data storage resources can certainly be identified as major reasons for the analytics industry's recent surge. However, decision support and analytics have flourished thanks to several key developments:

1. **Enhancement of data management**: Many organizations decide to store data not only in their own databases, but also in external databases because they have cloud storage. Data can be in the form of text, sounds, sounds/video, or texts and graphics/videos. In many cases, sending data from locations over long distances is unavoidable. Systems today can store, collect, search, and transmit the data efficiently, all at low cost, without compromising security, quickly, and cheaply.
2. **Data warehouses and big data**: It is virtually impossible to work with data warehouses which contain a great deal of data because it is enormous and unmanageable. Special methods, including parallel computing and Spark, are made available to assist in the organization, search, and extraction of the data. The costs of storing data and processing are steadily decreasing. Big data technologies have enabled gathering enormous amounts of diverse information from a number of sources and in various shapes and forms.

So more advanced analytics and data management and business intelligence tools, including the use of data warehouses, data mining, and cloud-based systems, are essential to today's management.

Some past components of computerized decision making, formerly branded as computer support, have mostly been supplanted by terms like "*analytics*" in the recent past. The analytics literature states that analytics is the art of developing actionable recommendations based on data derived from past experience and insight [6]. Other organizations, of course, have developed their own viewpoints about analytics, too. According to The Institute for Operations Research and the Management Sciences (INFORMS), analytics is the scientific process of translating data into discoveries that improve the decisions made. Before delving into the details of business analytics, it is necessary to define a few key terms for readers: A *model* is a simplified representation of a real-world system, concept, or object. The representation may be in written or verbal terms, a graphical representation, or a mathematical representation. A model can be used to codify a wide variety of decision problems. Models

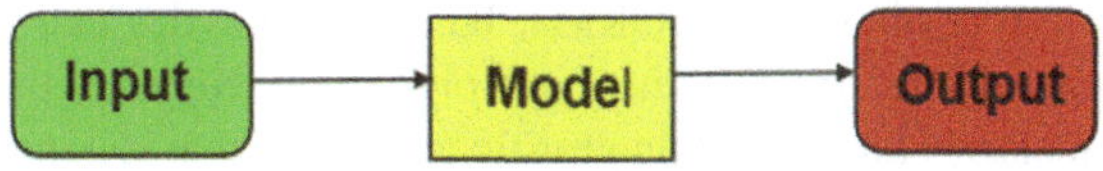

Fig. 1 Simple nature of models

encapsulate the essential characteristics of a problem and portray them in an easily understandable format. It enhances the intuition of decision-makers by providing alternative perspectives and enabling a clearer analysis of a given situation. Models used in business applications are primarily composed of three components, as illustrated in Fig. 1. The term *input* refers to any data or parameter required for the model's computational execution. On the other hand, *output* refers to any data that has been generated as a result of the execution of a model.

Data analytics is the process of looking at massive amounts of datasets to discover and unveil trends and insights that lead to operational and strategic decisions. *Business analytics* is concerned with analyzing various types of data in order to make data-driven business decisions and implementing changes as a result of those decisions. Gaining insight from data is a fundamental aspect of business analytics. The main application of business analytics is the detection and resolution of problems by taking advantage of information obtained from data analysis [5, 7]. Business analytics is not really a separate field; rather, it is an umbrella term that can describe the field because it includes the term "business". Before getting into the analysis of the business, it is essential to have a firm understanding of the business basics. One of the critical components of business analytics is domain knowledge. An effective business analytics solution requires an in-depth understanding of industry functions, industry terminology, industry data handling, and industry obstacles that may arise (Fig. 2).

Business intelligence (BI) is an umbrella term referring to analytical and reporting tools used to determine historical data trends. Insight-based reports, dashboards, and visualizations (all BI presentations) are key distinctions between analytics and BI.

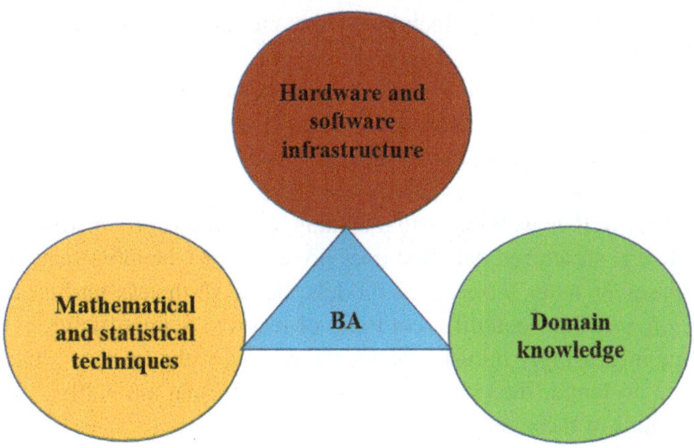

Fig. 2 Main components of BA

According to INFORMs, business analytics can be classified into three types: *descriptive*, *prescriptive*, and *predictive*. Notably, there is a subset of descriptive analytics known as "diagnostic analytics" that is concerned with the question "Why did it happen?" [8]. Four basic analytics application approaches for businesses are detailed as follows:

Descriptive analytics: This term is the fundamental type of business analytics. Most analytics tasks deal with summarizing business activities and therefore are called business reporting. At this level, analytics activities include the creation of reports summarizing business activities in order to answer the question, "What is happening?". The reports provide static snapshots of business activity on a schedule (i.e., daily), as well as real-time performance snapshots with advanced features in dashboards that may be crafted and designed by the decision-maker (e.g., weekly or quarterly).

Diagnostic analytics: Rather than explaining what just occurred, diagnostic analytics asks the question of "Why did it happen?". It utilizes exploratory data analysis techniques such as visualization, business intelligence, and data mining in order to ascertain the underlying causes of a given problem.

Predictive analytics: Once descriptive data has been done, organizations want to learn the answer, "What will happen?". It uses the data to predict what will happen in the future. The concept of prediction involves projecting what some variables might be like in the future, such as market demand, inflation rate, and market trends.

Prescriptive analytics: It is the most advanced level of analytics. While descriptive analysis seeks to give an overview of what happened and prediction analytics helps modeling and predicting what might happen, prescriptive analysis aims to identify, given the known parameters, the best solution or result among various choices. It provides answers to the question of "What should we do?". In other words, prescriptive analytics is seen as the precursor to better business performance and can also be described as proactive decision making ahead of time. Figure 3 illustrates the relationship between time complexity and added value for four different types of business analytics. The four types of analytics are illustrated in Fig. 3 by drawing inspiration from the study [9]. Techniques that can be used in analytical types are indicated in Table 1.

The data processing capabilities used in analytics include structured and unstructured data. The term "structured data" refers to data that is quickly and readily interpreted by computers with data structures.

A relational database stores structured data, while documents, videos, and photographs, as well as web pages that contain all of these, are unstructured. Semi-structured data is structured data that does not adhere to the table structure of relational databases or other table-based models, but nevertheless hashtags or hierarchies to distinguish between different types of records. Semi-structured documents have become increasingly prevalent since the advent of the Internet where full-text databases are no longer the only tool for sharing information. Table 2 summarizes the fundamentals of these data structures.

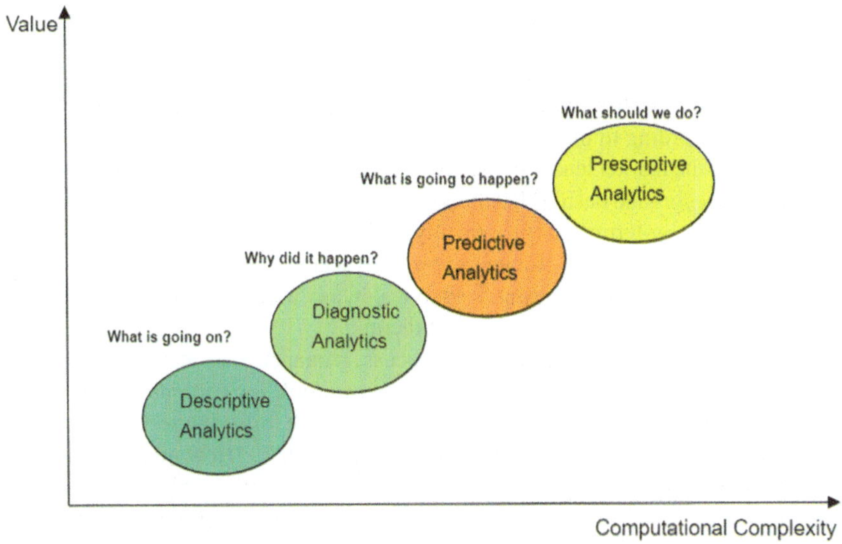

Fig. 3 Classification of business analytics techniques according to computational complexity and value

Table 1 Used techniques for the applications of analytics approaches

Type of analytics	Used techniques/methods
Descriptive	Data aggregation, statistical analysis, data mining, machine learning
Diagnostic	Data mining, machine learning, drill down, feature extraction, feature selection
Predictive	Machine learning, data mining, probabilistic models, statistical analysis
Prescriptive	Mathematical programming, probabilistic models, machine learning, data mining, evolutionary computation, simulation, logic-based models

Table 2 The classification of data structures

	Structure data	Semi-structured data	unstructured data
Definition	Data that is highly organized and in a spreadsheet format	Slightly organized data	Data that lacks a predefined organizational structure and a predefined format
Example data formats	Relational databases, Excel spreadsheets	HTML files, JSON files, XML files	Image files, video files, sound files
Data processing	The data is completely machine-readable, which is why analysis does not require heavy preprocessing	Some preprocessing is necessary before a computer can analyze it	Typically requires significant preprocessing before being analyzed by a computer

Analytics in Practice: Tesco Company

Tesco is an international supermarket based in Welwyn Garden, England [9]. This corporation ranks third in the amount of gross revenue and ninth in terms of revenues and profits, according to market authority. It operates stores in five countries across Europe and is the market leader in the UK's grocery market. Tesco operates a diverse range of businesses and desires that their goods be purchased by all. Tesco's objective is to maintain long-term customer loyalty and offer them better products. Tesco's business operations must be efficient in order to accomplish these objectives. Tesco may utilize various kinds of business analytics in any number of ways to excel their activities. The following are just a few of them:

Descriptive Analytics: Through this process, managers can access standard and custom reports as well as getting into the data to look for patterns and trends. A manager can employ descriptive analytics to address the following sorts of questions: "How much revenue and profit did we make last month?", "How much revenue did we receive in each region?".

Diagnostic Analytics: It can be used to handle the following types of questions: "Why have our sales declined in certain regions?", "What are the possible causes of stock outs during certain seasons?".

Predictive Analytics: It can be used to handle the following types of questions: "Which items are most likely to run out of stock in the coming week?", "How does the price cut affect customer demand?".

Prescriptive Analytics: It can be used to handle the following types of questions: "How should Tesco segment its stores?", "How much stock should Tesco keep for online and in-store sales?".

While professionals recognize the benefits of business analytics in their organizations, there are some significant issues that must be addressed prior to implementing BA. A challenging issue that data professionals will need to confront in the next decade is better understanding how to utilize big data and business analytics in order to create business value [10, 11]. As an organization, to successfully meet these challenges, project management must be used to implement the BA tools.

1.1 Decision-Making Process

Decisions are made at all levels of organizations by people who are managers or technical staff. People wish to be able to justify the decisions so that the reasons why they make are easily understood [12]. Some questions arise in organizations, and some decision making with different levels such as operational, tactical, and strategic is required to answer these questions. Strategic-level decisions are described as "what" the company expectations to achieve and related to the goals of the company are. The strategic-level decisions are oriented to future aspects and have high uncertainty. Tactical-level decisions are efforts to demonstrate how the company will achieve its goals. Operational decisions are daily operations in functional area of the company.

Decision making is simply defined as choosing a plan of action from a set of alternatives to achieve the desired outcome for a particular problem. Several factors affect decision making, including risk, changes in the environment, time pressure, contradictory objectives, many alternatives, and group or individual agents. Decision making is carried out by several agents (such as human agents and software agents) communicated between each other rather than being the task of one person or one group [13]. The characteristics of decision-making situations that emerge under changing environment over time also change. Decision-makers may consent to choose an acceptable decision in place of the optimum decision due to time constraints or cost. Decision making becomes complicated under incomplete information consisting of uncertainty, risk, and ambiguous information. Therefore, the complexity of the decision-making process increases in the case of involving aforementioned factors. However, the ability of effective decision making provides high performance and competitive advantage of the organization. Since management support in decision making supports organizational learning, knowledge management, and technology sourcing in terms of competitive advantage, decision-making process should be a part of the overall organization strategy [14].

Types of decision making are divided into two categories: programmed decision and non-programmed decisions [15]. Programmed decisions develop a guideline to follow. Decisions whose components or variables are known can be measured quantitatively. Non-programmed decisions are unstructured decisions that are generally based on components or variables that are not measured quantitatively. Decision making is performed via information, and a manager's intuition, and judgment. Figure 4 exhibits the relationship between decision levels and decision types. While the strategic decision-making level mostly deals with non-structural decisions, operational-level decision making is related to structural decisions.

After giving main definitions, it will be meaningful to mention about transitions among remarkable breakthrough in decision making. Decision support system (DSS)

Fig. 4 Types of decision making

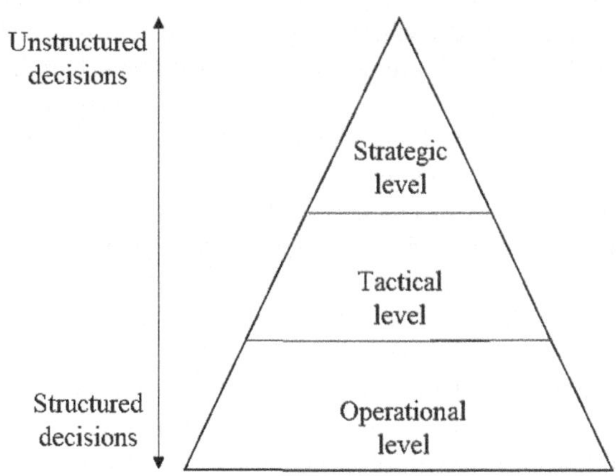

is a computer-based system supporting decision-makers in solving unstructured problems rather than replaces them with automates decision making. DSS emphasized facilitating decision process [16]. Besides, DSS provides value added to the decision-maker and enhances the analytical capacity of firms to evaluate information and create new knowledge [17].

With the integration of AI into traditional decision support systems, efficiency in the decision-making process increases, and higher autonomous levels are achieved. The system, called intelligent DSS or information-oriented DDS, includes the applications of rule-based expert systems or machine learning techniques [18]. Such a system is useful in a complex and volatile environment for managers or decision-makers to make faster decisions and change them quickly as needed [19]. Moreover, the fact that machines have the ability to continuously learn or the "self-learning" trait generates the autonomous decision-making process.

Basically, decision-making process consists of six steps: situation analysis, alternative search, alternative evaluation, objective and criteria setting, making decision, and decision review [15]. Also, in decision-making process, the requirements are determined, various criteria are considered, and the alternatives are evaluated [12].

The development of modern business analytics originated from the requirements of both organizational decision making and complex problem solving. The right decision making is the sign of a successful business analytics. Traditional business decisions are made mainly by people based on available information. Unfortunately, the perception and judgment of the human mind will change when information inputs or psychological factors change. Therefore, business analytics should also be able to deal with uncertainty and risk on behalf of effective decision making. Business analytics and related technologies such as artificial intelligence (AI), cloud-based system, and data mining can improve an organization's business and markets. AI enables modeling human reasoning and learning from examples. Moreover, soft computing, known as the AI method, incorporates imprecise and uncertain information into computer programs [20]. With the aforementioned developments and dynamic conditions, conventional steps in the decision-making process should have been modified comprehensively in order to gain value from the use of business analytics [21]. Moreover, managers spend much of their time in the evaluating alternatives stage. In the business decision, there are not the time or money to search for all alternatives. There is an increasing need for informational assistance to facilitate decision-makers to produce fast and feasible decisions [22]. Therefore, the business decision-making process should be reconfigured for organizations in terms of project management. Table 3 shows the stages of the reconfigured decision-making process based on predictive business analytics.

The decision-making process as given in Table 3 basically includes the structuring of the problem, the construction of the decision model, and the analysis of the problem. These three basic stages are not enabled to realize without cooperation among the data, business, and information technology (IT) departments within the organization. Since the prerequisite for effective business analytics is the availability and sustainability of readily available high-quality data, data analysts provide that relevant and accurate data which is available for analytical use from sources within

Table 3 Stages of decision-making process for predictive business analytics projects

Stage	Action
Business analysis	Define the business problem
	Identify the necessary variables for basic and alternative models
	Determine the estimation horizon and variable frequency
Data integration and extraction	Integrate data from different sources
	Extract inaccurate and inconsistent data
	Identify incomplete information and related methods
	Determine and apply methods for outlier values in data
Variable analysis	Calculate the importance of variables
	Decrease data size and number of variables (PCA, factor analysis, clustering etc.)
	Visualize data
Modeling	Determine forecasting algorithms and methods
	Measure and compare forecasting and estimation performance
	Detect insufficient and excessive compliance situations
	Determine the contribution of variables to forecasting performance
	Construct the final model
Result analysis	Adjust hyper-parameters for performance improvement (fine-tuning)
	Understand the actions implied by the model
	Determine the implications of the action
Implementation of the results	Incorporate the solution into the company
	Monitor the results
	Calibrate the result according to decision-maker requirements

and outside the organization according to the problem determined. In addition, appropriate governance arrangements need to be made to ensure data sustainability [23]. Qualified people in the business unit with the analytical mindset that helps to derive business value from BA ensure that the right model is selected and the model is set up correctly as well as analyzing unstructured data. IT, on the other hand, is complementary to both data and business units, via a platform based on an autonomous infrastructure in order to accelerate the solution of the developed model and ensure data sustainability. Consequently, the most appropriate way of working would be to build collaboration among data, business, and IT on analytical projects in terms of organizational business value. Last but not least, no matter how good the technical infrastructure system is, it will not be possible to benefit from the system unless the employees in the organization are motivated to use the system and have sufficient knowledge of how to use the system effectively [24].

In the business decision-making process, data are compiled from company reports, documents, interviews, or statistical sampling, and their accuracy is ensured. Models developed to solve the problem can be physical, logical, scale, schematic, or mathematical. Mathematical models can be deterministic or probabilistic. In the deterministic models, all values are known. Operational research (OR) as science and source of IT evaluation models includes the use of these models by human decision-makers in organizations [17]. The problems are assumed to be exhaustively described with parameters, data, and possible decision outcomes. Objective functions are generated according to data and parameters to measurements of decision outcomes. It is assumed that these objective functions yielding as a result of performing the model correspond to the best decision [13]. Results should be tested before proceeding to the analysis phase in order to ensure the accuracy of the model. In the analysis phase, sensitivity analysis is performed by changing the input values and parameters.

Critical points of the reconfigured decision-making process are investigated according to stages. Modeling and testing stages are essential due to their complexity, expensive, and time-consuming efforts. After the modeling, even if the model is accurate, its verification should be tested and fine-tuning should be made. Generally, testing a model becomes more difficult as long as the information is non-quantifiable. While trial-and-error experimentation may result in the loss, experimentation with the real system may only occur once. When the model is established, the accuracy of the model should be provided. Assumptions used in the construction of the decision model may underestimate some essential real-life points. The fact that the established model produces a single solution creates an obstacle in front of the development since it will create a limiting effect on the decision-maker. Thereby, the model should have focus on predictability over parametric adjudication. While explaining the analytic approach to decision-makers is time-consuming, understanding the decision-making procedure is important. Although knowledge exists in the organization, it may not be optimally distributed in some cases to the decision-making process. In other words, some bottlenecks occur in the flow of information due to the lack of cognitive insights. As for data collection stage, data integration may take time and money due to gathering information from different sources. The information collected needs to be extremely accurate for the business analytics in order to make the decision-making process reliable. To achieve reliable data output, it is necessary to concentrate on the term "garbage in, garbage out (GIGO)". Generally, GIGO is used to describe failures in the model established and decision making eventually, due to poor quality data. In addition, conflicting perspectives and different expectations from different departments of the organization are some of the challenges in the data collection as well as problem definition stage. However, an accurate model represents real-life conditions in a more sensitive manner saving time and money in the decision-making process. Also, the prescriptive decision model helps not only in problem solving but also in communicating problems and solutions to different agents.

2 Applications

Even if organizations regularly update their business data, there are several limitations like finding the necessary time to review them, determining the data means, and not knowing what to do. In other words, it is difficult to incorporate data-centric approaches into organizations' decision-making processes, and it is not certain that the expected benefits will be achieved [25]. Therefore, recently, business analytics has become a very popular application area for practitioners and researchers in both academic and business communities [26]. Business analytics provides the technologies and methodologies that improve decision making and business data understanding. Especially, competitive companies benefit from programming languages such as Python and R in order to analyze business data and effective decision making. Business analytics is a common tool for applications areas such as planning, scheduling, supply chain analysis, marketing analysis, inventory analysis, project planning, and capital budgeting in different industry sectors such as manufacturing, telecommunications, finance, health care, consumer products, and transport (see Fig. 5).

The evolution and applications of business intelligence and analytics are divided into three eras (see Fig. 5) in terms of their key characteristics and capabilities. The first era, where data is mostly structured, is related to data collection, extraction, and analysis of industries. Statistical analysis and data mining techniques are already included in business intelligence software offered by Oracle, IBM, Microsoft, and SAP for some actions such as dimensionality reduction, clustering, classification, regression analysis, and forecasting modeling. As for the second era, the Internet

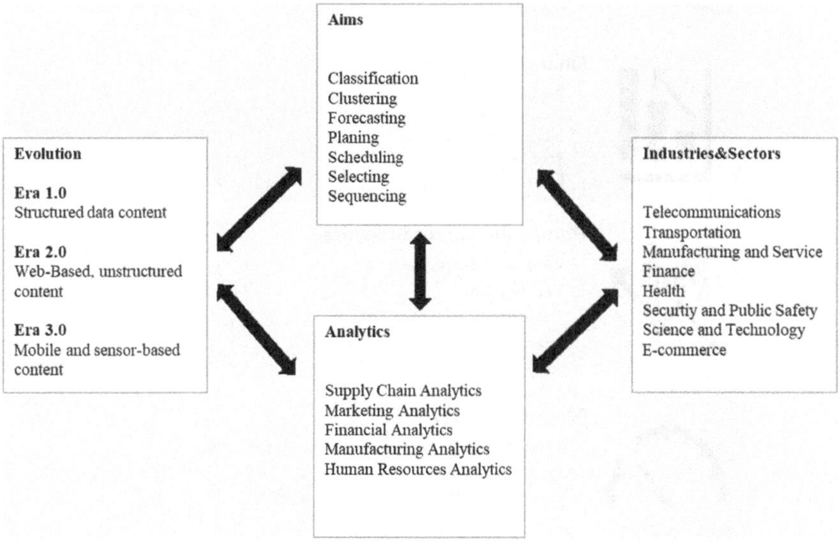

Fig. 5 Characteristics of business analytics

and web are used frequently for data collection and analytical research for business decision making through various text and web mining techniques. In the third era, sensor-based Internet-enabled devices equipped with RFID, barcodes, and radio tags (Internet of things) are addressed to support highly mobile, location-aware, person-centered, and context-relevant applications [26].

The fundamental content of business analytics is stated in Fig. 6. Almost all departments of an organization adopt business analytics for their own purposes, create the process in terms of project management accordingly, and apply it in the required field.

To make the application areas more understandable, sector-oriented applications are mentioned in the following paragraphs.

Companies adopting digitized business models, such as Google, Netflix, Uber, and Airbnb, operate the decision-making process using techniques such as AI, blockchain, cloud system, and data mining. Since the data collected from the web are less structured, text analysis and sentiment analysis are frequently used for these

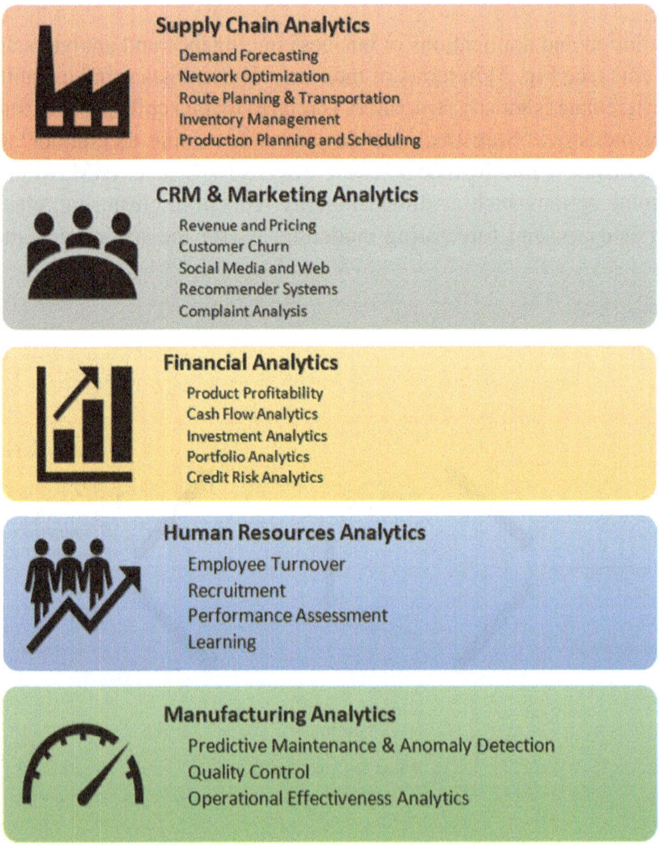

Fig. 6 Content of business analytics

firms. For example, Netflix collects data from its 151 million subscribers and categorizes movie reviews such as liked, loved, and hated to recommend movies to the customer according to their requirements [27]. Meanwhile, in a decision-making process that handles decreasing sales volumes, the media company determines a new pricing strategy with a business analytics mechanism. The pricing strategy generates different pricing strategies, considering regional price differentiation, and estimates sales and subscription volumes.

Since the existing decisions in the healthcare system are generally related to the prediction of an outcome, techniques based on prediction modeling, such as machine learning, neural networks, and deep learning, are among the most used. Thanks to the integration of analytics applications to the information system in health systems, unnecessary hospital stays can be prevented, care delivery models can be improved, and costs can be decreased. The health information system which is compatible with the goals of the method with a business mechanism that can monitor the clinical practices and results of physicians is presented [28]. Kaiser Permanente, which is an American integrated managed care consortium, provided better clinical practice by means of electronic medical record service [29]. Another case analysis in the medical field is the application of a machine learning-based disease prediction system [30].

Data collected from assets and information technologies in the manufacturing industry are analyzed to support business intelligence decisions in the manufacturing industry. Sensor data are stored as a result of high-quality modeling of the current state of physical assets in the workshop, called digital twins, in a virtual environment. With machine learning techniques to be selected in accordance with these data, many decisions can be made that provide lean and more efficient production in the manufacturing industry [18]. In the automobile industry, AUDI collected data from cars driven by 70 AUDI managers in a pilot project, recording 500 data points per second. This data includes parameters such as driving behavior depending on car type, usage time, mileage, and favorite radio station. The analysis of these data provided the basis for the development of services based on automobile data [31]. On the other hand, total quality management technique in the field of manufacturing can be an example of using business analytics [21].

In the financial sector, companies aim to provide investment advice at a low cost and to rebalancing portfolios over time. For example, the investment firm Vanguard established a new "Personal Advisor Services" (PAS) system that combined automated investment advice with the guidance of human advisors via AI. Thus, it has increased customer satisfaction, reduced costs, and quickly raised more than $80 billion in assets under management [32]. Another financial case, CitiBank, can minimize financial risk and pinpoint fraudulent behavior using real-time machine learning and predictive modeling [29].

Companies in the aerospace industry are becoming more competitive and efficient with successful AI applications. At NASA, 86% of human resources applications were carried out without human intervention by means of RPA, which is a business robotic process automation technology based on AI [32]. Airbus used AI to investigate a production problem, analyze a vast amount of data, and make effective decisions [29].

Business analytics can be performed for a company in the transportation sector, considering some parameters such as the changing frequencies, capacities, vehicle types, and fleet structure. The aim is to generate optimal revenues by modeling and simulating the effects of changes on route revenues. Also, transportation companies aiming to determine passenger capacities and prevent passenger overflows by managing the traffic center with a decision support system can benefit from business analytics techniques on the behalf of establishing a simulation system.

Another case is related to multi-mode database analysis. Data compiled from multi-mode databases through voice-controlled interfaces enable the use of voice and sound analytics in business research. New targets are determined in business decisions by analyzing data collected from voice-controlled interfaces such as Google Assistant, Amazon Alexa, or Siri. The R programming language is used for the analysis and processing of audio files collected from such voice-controlled interfaces. Specifically, the analysis of Amazon Alexa app's audio files distinguishes different emotional states of consumers and provides shopping-related implications such as purchasing probability and accuracy as input to machine learning models [33].

At McCann Japan company, which operates in the field of marketing, better advertisements were produced using the information of historical advertisements in the data collected by a robot designed with AI technology [29].

New applications of network analysis may include online virtual communities, criminal and terrorist networks, social and political networks, and trust and reputation networks [26].

Further information on BA applications can be found in [27, 29].

3 Summary

To get ahead in today's rapidly changing business world, managers need new and cutting-edge decision-making tools to aid them in their endeavors. To find answers to all of your business needs, business analytics encompasses multiple analytic tools, some of which are technical and others of which are simply creative. This term "business analytics", originally coined by business world, may be the best metaphor for that specific group of tools geared toward turning raw data into useful insights through a process that is systematic and rigorous. Many studies in the application section demonstrate that BA has already established itself as a business enabler in many businesses. So far, it is safe to say that BA represents an amazing paradigm-shifting opportunity.

To learn about business analytics, this chapter presents a holistic, overarching characterization of the subject as it is found in the real world, an issue for exploration, a methodology for decision making, and a place for businesses. It covers some variety elements that relate to business analytics such as the historical development, the current situation, needs, and issues. In a nutshell, this chapter discussed the following fundamental concepts related to the BA: definitions of basic terms and their interdependencies, decision-making processes and BA relationships, data structures in BA, methodologies used in BA applications, how BA is used in real-world situations, the steps of BA applications, the boundaries of BA, and the essential points for organizations thinking about incorporating BA in their organizations.

It is essential to have a solid understanding of a few essential points before businesses use business analytics to improve their business processes. These points are summarized in the following paragraphs, so read on to learn more.

Although the business analytics decision-making process consists of sequential steps, feedback loops may occur in some cases.

Modeling nonlinear relationships in a high-dimensional space via AI tools, complexity of real-life can be better represented by these models. The main difference of business analytics is that it creates a boundless platform by creating integration between multiple disciplines such as behavioral sciences, psychological factors, and human judgments as well as technical aspects.

As with other competencies in an organization, the successful implementation of the BA approach hinges on several organizational-specific factors other than merely the handling of data and data analysis, data processing, and technological issues. Major aspects are ensuring everyone has a clear understanding of the organization's goals, while developing an analytics-friendly workplace; employing a management philosophy that recognizes and supports the use of business analytics; and implementing methods for avoiding bias due to ignorance, anxiety, or coercion.

Data management within the organization is a major requirement for BA-oriented applications to succeed. Data collected in a variety of formats from a variety of sources must be filtered through the project leader. Additionally, by instituting a discipline to improve the data management process within the organization through BA applications, it is necessary to ensure the continuity of data flow among stakeholders by storing the desired data in the appropriate format and sharing it during the specified time periods, i.e., ensuring data sustainability.

A large part of this book deals with explaining quantitative approaches that can be applied directly to your work. This book is about applying several quantitative methods with your hands-on experience in mind. In the hands-on chapters, a variety of quantitative approaches described and their implementation are examined in a multitude of diverse business environments. This book includes several examples from the fields of businesses, marketing, business, accounting, and other fields of business, as you will see.

References

1. Delen D, Demirkan H (2013) Data, information and analytics as services. Decis Support Syst 55(1):359–363. https://doi.org/10.1016/j.dss.2012.05.044
2. S, Mathew SK (2018) Business analytics and business value: a comparative case study. Inf Manage 55(5):643–666. https://doi.org/10.1016/j.im.2018.01.005
3. Whitelock V (2018) Business analytics and firm performance: role of structured financial statement data. J Bus Anal 1(2):81–92. https://doi.org/10.1080/2573234X.2018.1557020

4. Bayrak T (2015) A review of business analytics: a business enabler or another passing fad. Procedia Soc Behav Sci 195:230–239. https://doi.org/10.1016/j.sbspro.2015.06.354
5. Poornima S, Pushpalatha M (2020) A survey on various applications of prescriptive analytics. Int J Intell Netw 1(July):76–84. https://doi.org/10.1016/j.ijin.2020.07.001
6. Hindle G, Kunc M, Mortensen M, Oztekin A, Vidgen R (2020) Business analytics: defining the field and identifying a research agenda. Eur J Oper Res 281(3):483–490. https://doi.org/10.1016/j.ejor.2019.10.001
7. Holsapple C, Lee-Post A, Pakath R (2014) A unified foundation for business analytics. Decis Support Syst 64:130–141. https://doi.org/10.1016/j.dss.2014.05.013
8. Lepenioti K, Bousdekis A, Apostolou D, Mentzas G (2020) Prescriptive analytics: Literature review and research challenges. Int J Inf Manage 50:57–70.https://doi.org/10.1016/j.ijinfomgt.2019.04.003
9. Griva A, Bardaki C, Pramatari K, Papakiriakopoulos D (2018) Retail business analytics: customer visit segmentation using market basket data. Expert Syst Appl 100:1–16. https://doi.org/10.1016/j.eswa.2018.01.029
10. Delen D, Ram S (2018) Research challenges and opportunities in business analytics. J Bus Anal 1(1):2–12. https://doi.org/10.1080/2573234X.2018.1507324
11. Vidgen R, Shaw S, Grant DB (2017) Management challenges in creating value from business analytics. Eur J Oper Res 261(2):626–639. https://doi.org/10.1016/j.ejor.2017.02.023
12. Montgomery H (1983) Decision rules and the search for a dominance structure: towards a process model of decision making. Adv Psychol 14:343–369. https://doi.org/10.1016/S0166-4115(08)62243-8
13. Barthélemy JP, Bisdorff R, Coppin G (2002) Human centered processes and decision support systems. Eur J Oper Res 136(2):233–252. https://doi.org/10.1016/S0377-2217(01)00112-6
14. Zhu K, Kraemer KL, Xu S (2006) The process of innovation assimilation by firms in different countries: a technology diffusion perspective on e-business. Manage Sci 52:1557–1576. https://doi.org/10.1287/mnsc.1050.0487
15. Simon HA (1960) The new science of management decision. In: Harper Brothers, New York.https://doi.org/10.1037/13978-000
16. Lee SM, Eom HB (1990) Multiple-criteria decision support systems: the powerful tool for attacking complex, unstructured decisions. Syst Pract 3(1):51–65. https://doi.org/10.1007/BF01062821
17. Bernroider EWN, Schmöllerl P (2013) A technological, organisational, and environmental analysis of decision making methodologies and satisfaction in the context of IT induced business transformations. Eur J Oper Res 224(1):141–153. https://doi.org/10.1016/j.ejor.2012.07.025
18. Trakadas P, Simoens P, Gkonis P, Sarakis L, Angelopoulos A, Ramallo-González AP, Skarmeta A, Trochoutsos C, Calvo D, Pariente T, Chintamani K, Fernandez I, Irigaray AA, Parreira JX, Petrali P, Leligou N, Karkazis P (2020) An artificial intelligence-based collaboration approach in industrial iot manufacturing: Key concepts, architectural extensions and potential applications. Sensors (Switzerland) 20(19):5480. https://doi.org/10.3390/s20195480
19. Negulescu OH (2014) Using a decision-making process model in strategic management. Rev General Manage 19(1):111–123
20. Azvine B, Nauck D, Ho C (2003) Intelligent business analytics—a tool to build decision-support systems for eBusinesses. BT Technol J 21(4):65–71. https://doi.org/10.1023/A:1027379403688
21. Sharma R, Mithas S, Kankanhalli A (2014) Transforming decision-making processes: a research agenda for understanding the impact of business analytics on organisations. Eur J Inf Syst 23:433–441. https://doi.org/10.1057/ejis.2014.17
22. Jalal-Karim A (2013) Evaluating the impact of information security on enhancing the business decision-making process. World J Entrepreneurship Manage Sustain Dev 9(1):55–64. https://doi.org/10.1108/20425961311315719
23. Seddon PB, Constantinidis D, Tamm T, Dod H (2017) How does business analytics contribute to business value? Inf Syst J 27:237–269. https://doi.org/10.1111/isj.12101

24. Purvis RL, Sambamurthy V, Zmud RW (2001) The assimilation of knowledge platforms in organizations: an empirical investigation. Organ Sci 12:117–135. https://doi.org/10.1287/orsc.12.2.117.10115
25. Kowalczyk M, Buxmann P (2014) Big data and information processing in organizational decision processes: a multiple case study. Bus Inf Syst Eng 6(5):267–278. https://doi.org/10.1007/s12599-014-0341-5
26. Chen H, Chiang RHL, Storey VC (2012) Business intelligence and analytics: from big data to big impact. MIS Q Manage Inf Syst 36(4):1165–1188. https://doi.org/10.2307/41703503
27. Akter S, Wamba SF, Gunasekaran A, Dubey R, Childe SJ (2016) How to improve firm performance using big data analytics capability and business strategy alignment? Int J Prod Econ 182:113–131. https://doi.org/10.1016/j.ijpe.2016.08.018
28. Kohli R, Kettinger WJ (2004) Informating the clan: controlling physicians' costs and outcomes. MIS Q Manage Inf Syst 28(3):363–394. https://doi.org/10.2307/25148644
29. Akter S, Michael K, Uddin MR, McCarthy G, Rahman M (2020) Transforming business using digital innovations: the application of AI, blockchain, cloud and data analytics. Ann Oper Res, 1–33.https://doi.org/10.1007/s10479-020-03620-w
30. Rongqiang S, Yan C, Chang D, Qingyue G (2020) Research on promoting the application of disease prediction system based on machine learning. In: Proceedings of the 2020 international symposium on artificial intelligence in medical sciences, pp 45–50. https://doi.org/10.1145/3429889.3429898
31. Dremel C, Herterich MM, Wulf J, Waizmann JC, Brenner W (2017) How AUDI AG established big data analytics in its digital transformation. MIS Q Exec 16(2):81–100
32. Davenport TH, Ronanki R (2018) Artificial intelligence for the real world. Harv Bus Rev 96(1):108–116
33. Hildebrand C, Efthymiou F, Busquet F, Hampton WH, Hoffman DL, Novak TP (2020) Voice analytics in business research: conceptual foundations, acoustic feature extraction, and applications. J Bus Res 121:364–374. https://doi.org/10.1016/j.jbusres.2020.09.020

Yakup Turgut was born in Turkey, in 1993. He received a B.E. degree in industrial engineering from Yıldız Technical University, Turkey, in 2015, and a master's degree from the Istanbul Technical University, Turkey, in 2018. He is a Ph.D. candidate in industrial engineering at Istanbul Technical University. His research interests are simulation modeling, mathematical modeling, and artificial intelligence.

Yildiz Kose graduated from Industrial Engineering Department of Gazi University, Ankara, in 2014. She received her M.Sc. degree from Industrial Engineering Department of Karadeniz Technical University, Trabzon, in 2017. Since 2018, she has been a Ph.D candidate and research assistant at Istanbul Technical University. Her current research interests are production planning, assembly line and lean production systems.

Alp Ustundag is the Head of Industrial Engineering Department of Istanbul Technical University (ITU) and the coordinator of M.Sc. in Big Data & Business Analytics Program. He is also the CEO of Navimod Business Analytics Solutions located in ITU Technopark (http://navimod.com/). He has worked in IT and finance industry from 2000 to 2004. He continued his research studies at the University of Dortmund between 2007 and 2008 and completed his doctorate at ITU in 2008. He has conducted a lot of research and consulting projects in the finance, retail, manufacturing, energy, and logistics sectors. His current research interests include artificial intelligence, data science, machine learning, financial and supply chain analytics. He has published many papers in international journals and presented various studies at national and international conferences.

Emre Cevikcan is a professor in Industrial Engineering from Istanbul Technical University (ITU). He received the BS degree in Industrial Engineering from Yildiz Technical University and the MSc degree and Ph.D. degree in Industrial Engineering from ITU. He studied the scheduling of production systems for his Ph.D. dissertation. His research has so far focused on digital transformation, planning and design of manufacturing and service systems, lean production and scheduling. He has several research papers in scientific journals. In addition, he contributed several book projects about decision making and digital transformation as an editor and chapter author. Prof. Cevikcan has contributed to several industrial projects about lean transformation, manufacturing system design and process improvement.

Descriptive Analytics

Nizamettin Bayyurt and Sefer Baday

1 Introduction

In this chapter, descriptive analytics and statistical analyses are introduced. Specifically, data exploration and visualization, probability distributions, statistical inference, and Bayesian statistics are explained. Along with theory, practical applications on a sample data set are provided. Applications are performed using the following Python libraries: Pandas, Seaborn, and Statmodels.

2 Key Definitions

Statistics is the collection of methods for analyzing data. It allows examining the relationships between variables, making predictions and forecasts. In terms of its subjects, statistics is classified into two types: descriptive and inferential statistics. Descriptive statistics describe data by charts and graphs while inferential statistics allows making predictions.

Data is information gathered about a problem. It can be classified into two types: quantitative data and qualitative data. Quantitative data: numerical or countable data. These kinds of data can be subdivided into two: discrete or continuous data.

- Discrete data: This type of data consists of whole numbers (0, 1, 2, 3…), for example, the number of children in a family.

N. Bayyurt (✉)
Industrial Engineering Department, Management Faculty, Istanbul Technical University, Istanbul, Turkey
e-mail: bayyurt@itu.edu.tr

S. Baday
Applied Informatics Department, Informatics Institute, Istanbul Technical University, Istanbul, Turkey

A. Ustundag et al. (eds.), *Business Analytics for Professionals*, Springer Series in Advanced Manufacturing, https://doi.org/10.1007/978-3-030-93823-9_2

Table 1 View of IBM SPSS data

gender	age	educ	jobcat	salary	salbegin	jobtime	prevexp
Male	43	15	Manager	57,000	27,000	98	144
Male	37	16	Clerical	40,200	18,750	98	36
Female	66	12	Clerical	21,450	12,000	98	381
Female	48	8	Clerical	21,900	13,200	98	190
Male	40	15	Clerical	45,000	21,000	98	138

- Continuous data: Data that can take any value within an interval, for example, people's height (between 50 and 210 cm) or weight (between 50 and 150 kg).

Qualitative data: non-numeric or categorical data. It measures qualities and generates categorical answers. (color, gender, and nationality).

A variable is anything that can take different values across data set (e.g., age, salary or gender). Scales of measurement tell us how precisely variables are recorded.

There are four types of scales: nominal, ordinal, interval, and ratio.

- **Nominal Scale**: The data can only be categorized (e.g., gender, ethnicity, marital status, etc.)
- **Ordinal Scale**: The data can be categorized and ranked (e.g., preference level; first, second, language ability; beginner, intermediate, and advanced
- **Interval Scale**: The data can be categorized, ranked, and evenly spaced. The differences of observations are meaningful (e.g., exam scores, salaries)
- **Ratio scale**: The data can be categorized, ranked, evenly spaced, and has a natural zero. The ratio of observations is meaningful. (e.g., height, age, weight, etc.)

Employee data from IBM SPSS data files was used for this chapter's applications, which included the annual salary, starting salary, age, gender, job time, previous experience, and job type of 474 employees.

The information is presented in Table 1.

3 Descriptive Statistics

3.1 Frequency Distributions

The distribution created by finding how many times the observations in a variable repeat is called frequency distribution. When the number of observations is large, data is first divided into certain classes, and the table showing the number of observations falling into these classes is called the classified frequency distribution.

The education frequencies are depicted in Table 2. Employees' educational backgrounds range from 8 to 21 years. There are personnel with 12 years of education

Table 2 Education frequency

Education	Count	Percentage
8	53	0.111814
12	190	0.400844
14	6	0.012658
15	116	0.244726
16	59	0.124473
17	11	0.023207
18	9	0.018987
19	27	0.056962
20	2	0.004219
21	1	0.002110

the most (190 people) and then with 15 years of education (116 people). The relative frequencies of those are given in the last column. 45.08% of employees have a 12-year education, whereas 24.47% have a 15-year education.

Table 3 shows the frequencies of education by gender. There are no women with higher education levels (>17); however, there are more women with lower education levels than men (8, 12).

Salary is a continuous variable. For the frequency table, it must be divided into classes. The number of employees per class is given in Table 4 when it is divided into ten classes. The first class (15,630.749–27,675] has the greatest number of employees (210). The number of employees per class reduces as the salary rises.

Table 3 Education frequency by gender

Education	Female	Male	All
8	30	23	53
12	128	62	190
14	0	6	6
15	33	83	116
16	24	35	59
17	1	10	11
18	0	9	9
19	0	27	27
20	0	2	2
21	0	1	1
All	216	258	474

Table 4 Salary frequency

Salary interval	Frequency
(15,630.749, 27,675.0]	210
(27,675.0, 39,600.0]	159
(39,600.0, 51,525.0]	38
(51,525.0, 63,450.0]	28
(63,450.0, 75,375.0]	22
(75,375.0, 87,300.0]	8
(87,300.0, 99,225.0]	4
(99,225.0, 111,150.0]	4
(111,150.0, 123,075.0]	0
(123,075.0, 135,000.0]	1

3.2 Measures of Central Tendency

Measures of central tendency are the measures that allow us to learn about a series tendency and make various comparisons. These kinds of measures are used to find out which values the data tend toward. The measures are classified as sensitive and insensitive measures. Sensitive measures are arithmetic mean, weighted mean, square mean, harmonic mean, and geometric mean while insensitive measures are mode, median, and quartiles. The reason why they are called sensitive is that all the values in a variable are taken into account in calculation. This method of calculation pulls the mean toward the extremes, making it difficult to make an accurate estimation, especially in the case of extreme values in the variable.

Mean is the average of a variable. Mode is the most frequently observed value in a variable and median is the middle value when the variable is sorted in order of magnitude.

The P-th percentile in the ordered set is that value below which lies P% of the observations in the set. Quartiles are the percentage points that break down the ordered data set into quarters. The first quartile is the 25th percentile. It is the point below which lies 1/4 of the data. The second quartile is the 50th percentile, which is the median, and the third quartile is the 75th percentile.

Table 5 shows summary statistics for numerical variables in the data set. The mean, standard deviation, minimum and maximum values, and first, second, and third quartiles of the variables are shown in the table. The second quartile and the median are the same. The youngest employee is 24 years old, the oldest is 66 years old, and the average age is 38.67 years old, according to the table. 25% of the employees are under the age of 24, 25% are between the ages of 24 and 33.5, 25% are between the ages of 33.5 and 47, and 25% are between the ages of 47 and 66 years.

Table 5 Summary statistics for numerical variables in the data set

	Age	Educ	Salary	Salbegin	Jobtime	prevexp
count	474.00	474.00	474.00	474.00	474.00	474.00
mean	38.68	13.49	34,419.57	17,016.09	81.11	95.89
std	11.77	2.88	17,075.66	7870.64	10.06	104.56
min	24.00	8.00	15,750.00	9000.00	63.00	0.00
25%	30.00	12.00	24,000.00	12,487.50	72.00	19.25
50%	33.50	12.00	28,875.00	15,000.00	81.00	55.00
75%	47.00	15.00	36,937.50	17,490.00	90.00	138.75
max	66.00	21.00	135,000.00	79,980.00	98.00	476.00

3.3 *Measures of Variability*

Various measures of distances of observations from each other or from any mean value in a variable are called measures of variability. Mean value of a variable alone is not sufficient to understand the variable. In addition to mean, measures of variations are also useful. Some of the measures of variability are;

- Range: Difference between maximum and minimum values
- Interquartile Range: Difference between third and first quartile
- Variance: Average of the squared deviations from the mean
- Standard Deviation: Square root of the variance

4 Data Visualization

Data visualization is the process of displaying data with graphs. Graphs show meaningful relationships in data, making the data easier to understand and interpret.

Bar Graphs

Bar charts are one of the most popular ways of visualizing data. Heights of rectangles represent group frequencies. It is used for categorical data. The chart in Fig. 1 depicts the distribution of jobcat variable, which includes clerical, manager, and custodian. The frequencies are represented on the vertical axis.

The average salaries of these three groups are shown in Fig. 2. Managers are paid twice as much as the other two groups, as shown in the graph. Although custodians' incomes appear to be slightly higher than clericals' salaries on average, the salaries of clericals and custodians seem to be equal.

Next graph (Fig. 3) was created to see if there is a wage disparity between men and women in similar roles. As can be shown, men receive more salaries on average than women in all groups. There is not security woman in the data set.

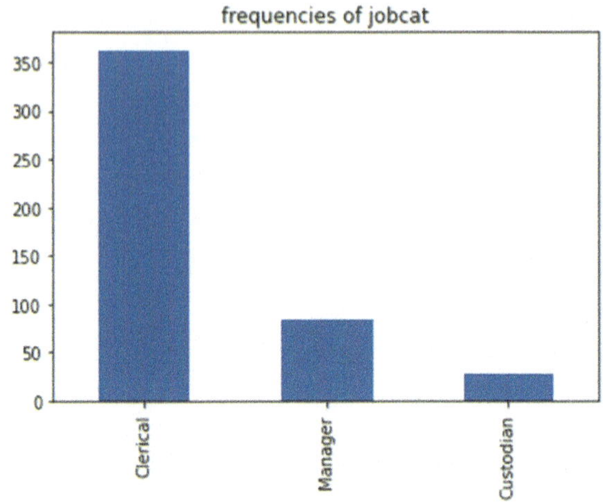

Fig. 1 Bar plot of category distribution

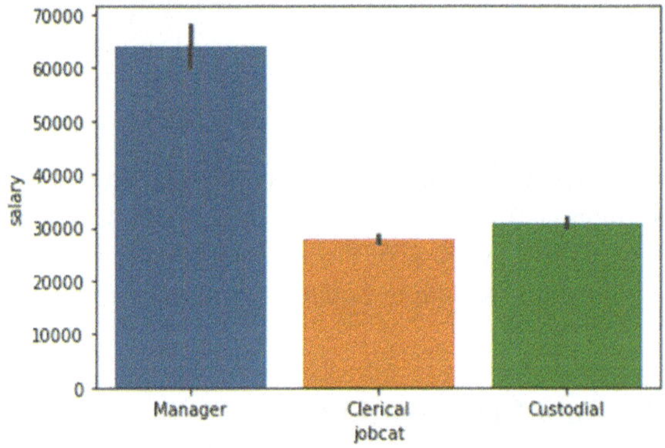

Fig. 2 Bar plot of average salaries for job categories

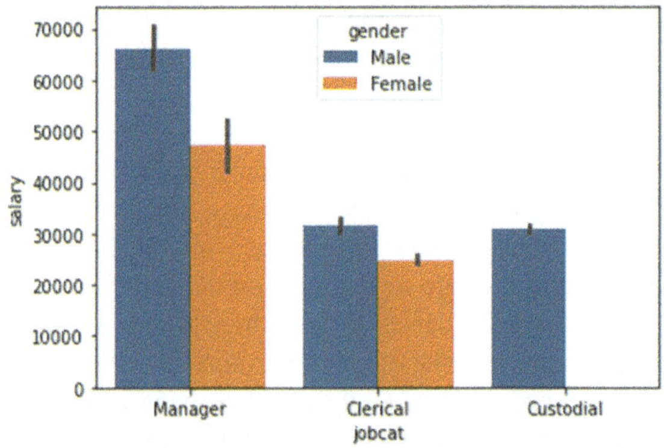

Fig. 3 Bar plot for average salaries of job categories grouped by gender

Histogram

Histograms show frequencies on the vertical axis and class intervals on the horizontal axis. Each box is assigned to a class. It is used for continuous data. The histogram of the variable salary is shown in Fig. 4. Salary seems to be skewed right rather than distributed normally.

Pie Charts

Categories represented as percentages of total. It is used for categorical data. In data set, 76.6% of the employees are clerical, 17.7% are managers, and 5.7% are custodians (Fig. 5).

Box Plot

In this chart type, five statistical summaries are seen together. These are; max, min values, first, second, and third quarters. It is useful for numerical data.

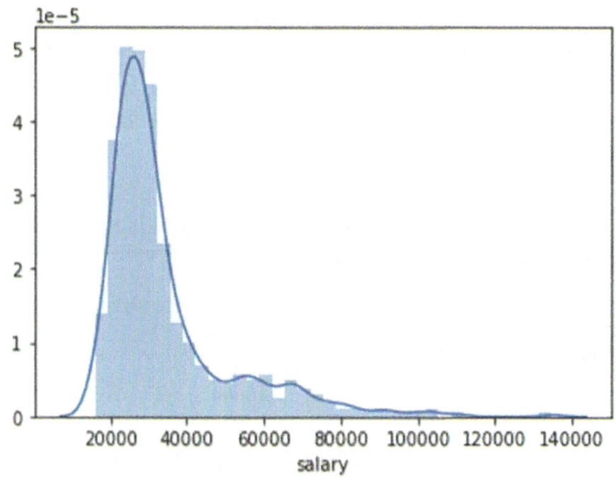

Fig. 4 Histogram for salaries

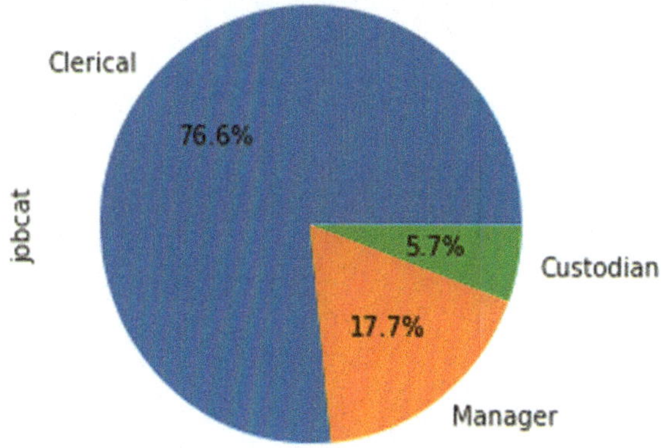

Fig. 5 Pie chart for job category distribution

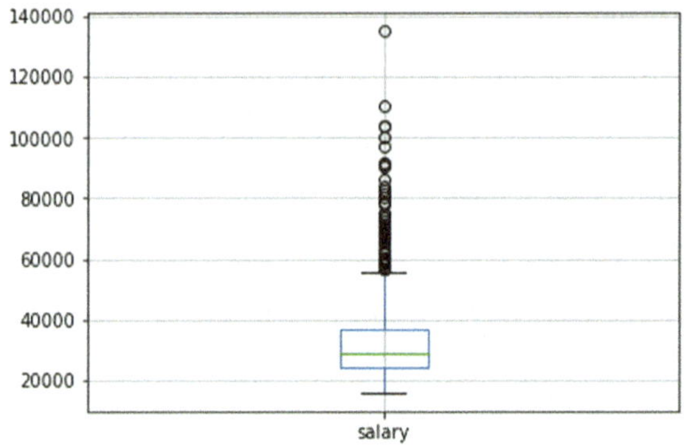

Fig. 6 Box plot for salary

The salary variable's box plot graph is presented in Fig. 6. The minimum wage is $15,000, while the maximum salary without outliers is little less than $60,000. The three lines in the box represent the first quartile value of $24,000, the median value of $28,000, and the third quartile value of $36,000.

A box plot graph of salary and education is shown below to illustrate how salary varies at different levels of education (Fig. 7). As one's education level rises, so does the salary, as well as the variability of the salary.

Line charts

These are used to display changes or trends in data over a period of time. It is useful especially for time series.

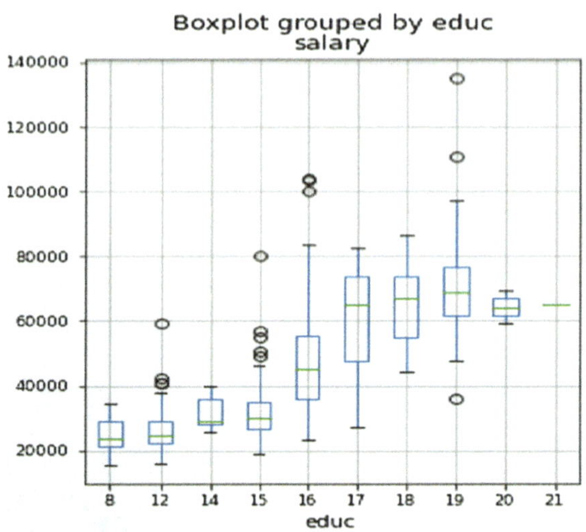

Fig. 7 Box plot for salary grouped by education

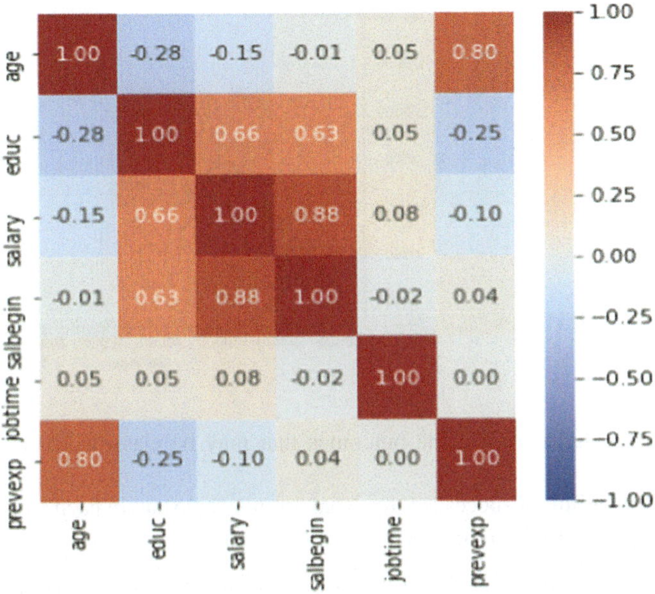

Fig. 8 Heat map for correlation of numerical variables

Heat maps

Heat maps represent individual values from a data set on a matrix using variations in color or color intensity. A heat map for correlation of numerical variables is given in Fig. 8. Thanks to a divergent color map, it is easier to detect highly positively or highly negatively correlated variables.

Scatter Plots

These types of graphs are used to identify and report any underlying relationships among pairs of data sets. Each point represents an observation. It is useful for numerical data. A scatter plot for current salary versus beginning salary is given in Fig. 9.

5 Discrete Probability Distributions

5.1 Binomial Distribution

An experiment named as Bernoulli process if it possesses the following properties [2]:

1. The experiment consists of repeated trials.

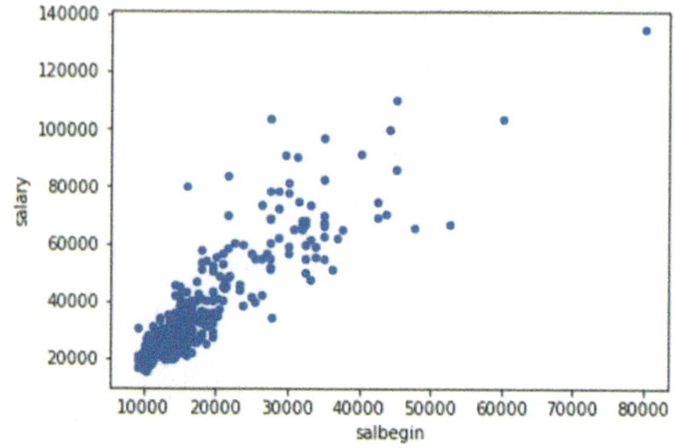

Fig. 9 Scatter plot beginning salary versus salary

2. Each trial has two possible outcomes that may be classified as success and failure.
3. The probability of success, p, in a trial remains constant in each trial.
4. The repeated trials are independent of each other.

The number x of successes in n Bernoulli trials is called a binomial random variable. The probability distribution of this discrete random variable is called the binomial distribution. The number of successes in n independent trials is

$b(x; n, p) = \binom{n}{x} p^x q^{n-x}$ where $x = 0, 1, 2, \ldots, n$ and $p + q = 1$.

Here is an example for the application of binomial distribution. The probability of a person has *H*. pylori infection in a country is 0.73. If 50 people are randomly selected from the population, the probability that 10 of them will have *H*. pylori is a binomial distribution where $n = 50$, $x = 10$, $p = 0.73$, and $q = 0.27$.

5.2 *Poisson Distribution*

Poisson distribution is a discrete probability distribution used to determine the probability of a number of successes per unit of time, when the events are independent, and the average number of successes per unit of time remains constant [2].

$f(x) = \frac{e^{-\lambda}\lambda^x}{x!}$ where $x = 0, 1, 2, \ldots,$ is the number of successes per unit time, and λ is the average number of successes per unit time.

If four accidents occur on average per month at Istanbul's Taksim Square, the probability that at most two accidents will occur in the following month is a Poisson distribution where $\lambda = 4$

6 Continuous Probability Distributions

6.1 The Normal Distribution

One of the most important distributions for continuous random variables is the normal distribution. The reasons of this are that most of the continuous random variables observed in our daily life are showing the normal distribution, and the normal distribution is used as the basic distribution in statistical inferences. For example, the monthly returns of an investment, the weekly sales of a company, the weights, heights, or IQ level of people are approximately normally distributed. The probability density function of a normally distributed X random variable [1].

$$N\left(x;\mu,\sigma^2\right) = f(x) = \frac{1}{\sqrt{2\pi\sigma^2}}e^{-\frac{1}{2}\left(\frac{x-\mu}{\sigma}\right)^2} \quad -\infty < x < \infty \tag{1}$$

μ: the mean of the random variable.
σ^2: the variance of the random variable.
$e = 2.7182818\ldots$ constant, $\pi = 3.1415926\ldots$ constant.

Normal distribution is bell-shaped and symmetric (Fig. 10). Because the distribution is symmetric, one-half (50%) lies on either side of the mean. Because mean divides the distribution into two equal parts and it has the highest frequency, it is median and mode at the same time.

Normal distribution is characterized by a different pair of mean, μ and variance, σ^2. That is: [$X \sim N(\mu, \sigma^2)$]. Normal distribution is asymptotic to the horizontal axis both to left and right. The area under any normal probability density function is 1 and the same for any normal distribution, regardless of the mean and variance.

The standard normal random variable, Z, is the normal random variable with mean $\mu = 0$ and standard deviation 1.

$$Z \sim N\left(0, 1^2\right).$$

The probabilities of being within one standard deviation, two standard deviations, and three standard deviations about the mean are approximately

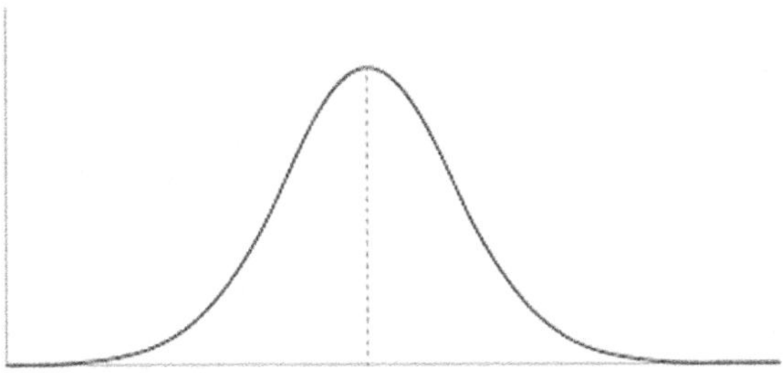

Fig. 10 Shape of normal distribution

$$P(\mu - \sigma \leq x \leq \mu + \sigma) \cong 0.68$$
$$P(\mu - 2\sigma \leq x \leq \mu + 2\sigma) \cong 0.95$$
$$P(\mu - 3\sigma \leq x \leq \mu + 3\sigma) \cong 1$$

Tests for Normality

Almost all parametric statistical methods are based on the assumption that data is drawn from a normal distribution. The hypotheses for the analysis of normality are as follows.

$H_{0:}$ Normal probability density function is suitable for the distribution.

$H_{1:}$ Normal probability density function is not suitable for the distribution.

Histograms (for $n > 25$) can give an idea of whether the data is normally distributed. If the data is approximately normally distributed, the graph will be symmetrical and bell-shaped. The histogram of salary reveals that it is not symmetrically distributed but rather skewed to the right (Fig. 4).

Another method that can be used to test normality is the Q-Q plot (normal quantile plot). Let $x_{(1)}, x_{(2)}, \ldots. x_{(n)}$ show n observations of a variable, X_i, and let us assume that these values are ordered from smallest to largest $x_{(1)} < x_{(2)} < \ldots < x(n)$.

$x_{(j)}$ indicates that j-numbers of data are less than or equal to $x_{(j)}$.

$\frac{j-1/2}{n}$ is the approximation of the ratio of data smaller than $x_{(j)}$ to all, that is, j/n.

Quantiles in the standard normal distribution are defined by

$$P\left[Z \leq q_{(j)}\right] = \int_{-\infty}^{q_{(j)}} \frac{1}{\sqrt{2\pi}} e^{-z^2/2} dz = p_{(j)} = \frac{j - \frac{1}{2}}{n} \quad (2)$$

If data is normally distributed, then $(q_{(j)}, x_{(j)})$ form a line [3].

A theoretical normal is represented by the red line on Fig. 11. The salary variable by blue dots. The data is closer to the red line the closer it is to normal. The salary variable seems to be far from the normal distribution.

Fig. 11 Q-Q plot

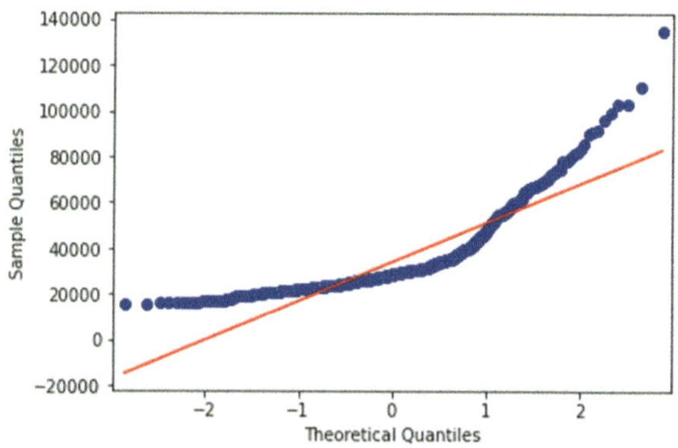

Kolmogorov–Smirnov Test

The test is based on comparison of the cumulative frequency distribution obtained from the sample with the theoretical cumulative distribution. The D statistic to be calculated is the maximum absolute difference between the observed cumulative frequency distribution (g) and the theoretical cumulative frequency distribution (f).

$$D = \max|f - g| \tag{3}$$

D value for salary variable is calculated as 219.3 with a p-value of 2.32e-48 using Python statsmodel library.

The closer the D statistic is to zero in Kolmogorov–Smirnov test, the better the data fits to a normal distribution. In sample data D is 219.3 which is far from zero. Its significance level is 2.3×10^{-48} that rejects the normality of the salary variable.

Skewness and Kurtosis tests

The skewness and kurtosis measures based on the moments give information about the distribution of the data. A measure of skewness of a sample $x_1, x_2 \ldots..x_n$ with n units drawn from a population is

$$\alpha_3 = \frac{\sum_{i=1}^{n} (x_i - \overline{x})^3/n}{s^3} \tag{4}$$

A negative skewness value indicates that the data is skewed to the left, and a positive skewness value indicates that the data is skewed to the right. If the distribution is normal, the skewness value is zero. The skewness of the salary variable is calculated as 2.12.

The kurtosis measure of the distribution is;

$$\alpha_4 = \frac{\sum_{i=1}^{n} (x_i - \overline{x})^4/n}{s^4} \tag{5}$$

In a normal distribution, the kurtosis measure is 3. The kurtosis measure shows the weights of the distribution in the tails. A kurtosis measure greater than 3 indicates that the distribution is narrow, and a kurtosis measure of less than 3 indicates that the distribution is flattened. The kurtosis for salary is calculated to be 5.37.

The skewness measure of the variable salary in data set is 2.12 that shows it is skewed to the right. The distribution's kurtosis value is 5.37. This is still far from 3, which is the normal distribution's kurtosis measure, indicating that the data have a peak distribution.

Shapiro–Wilk Test

It is based on the calculation of the W statistic to test whether a random sample is drawn from a normally distributed population. W statistic is obtained by ratio of the variance estimator obtained as the linear combination of the squares of the

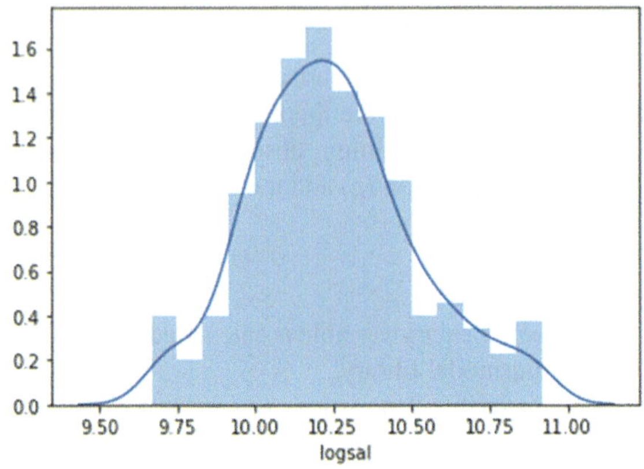

Fig. 12 Log transformed salary

observations ordered from the smallest to the largest, and the variance estimator obtained by the sum of the squares of the differences of observations and the mean [4]. W statistic is calculated from the following formula:

$$W = \frac{\left(\sum a_i x_{(i)}\right)^2}{\sum_{i=1}^{n}(x_i - \overline{x})^2} \tag{6}$$

W is between zero and one. For very small W values, the hypothesis that the population is normally distributed is rejected. The "a" values in the W statistic are obtained from the standard tables prepared according to various sample sizes for Shapiro–Wilk test. Using stats module of the Python SciPy library, W statistic is calculated as 0.77 with a p-value of 3.28e-35.

The Shapiro–Wilk Test, with a W statistic of 0.77 and a p-value of 3.29×10^{-25}, rejects the salary variable's normality. According to the statistics, managers earn significantly more than the rest of the workforce. By splitting managers from clerical and custodians, it may be able to achieve a more normal distribution. Applying a transformation to the data is another way to approximate the normal distribution. Figure 4 presented an income graph that was highly skewed to the right. After applying a logarithmic transformation on the distribution of income less than $55,000, the graph obtained (Fig. 12) shows a distribution that is closer to normal.

7 Statistical Inference: Estimation

7.1 Sampling

A population is a set that consists of all measurements for which the investigator is interested.

A sample is a subset of the measurements selected from the population.

(e.g., the set of all voters in an election in Turkey is population). A sample is a subgroup that can be selected from the population. The size of the sample is always less than the total size of the population. In general, the calculations are done using sample data, and then, the population is estimated by the results of the sample.

- The sample mean, $\overline{x}$, is the estimator of the population mean, μ.
- The sample variance, s^2, is the estimator of the population variance, σ^2.
- The sample standard deviation, s, is the estimator of the population standard deviation,σ.
- The sample proportion, $\widehat{p}$, is the estimator of the population proportion, p.

7.2 *Sampling Distribution Means*

A statistic calculated from all possible samples with certain sizes from a population varies from one sample to another. The probability distribution of these values is called sampling distribution. For instance, the sampling distribution of $\overline{x}$ is the probability distribution of all possible values the random variable, $\overline{x}$, may get when all possible samples of size n are taken from a population.

The expected value of the sample mean is equal to the population mean as shown in Eq. 7.

$$E(\overline{X}) = \mu_{\overline{X}} = \mu_X \tag{7}$$

The variance of the sample mean is equal to the population variance divided by the sample size:

$$\sigma_{\overline{X}}^2 = \frac{\sigma_X^2}{n}\sqrt{\frac{N-n}{N-1}} \tag{8}$$

when N is large enough compared to n, $\sigma_{\overline{X}}^2$ approaches $\sigma_{\underline{X}}^2 = \frac{\sigma_X^2}{n}$

When sampling from a normal population with mean μ and standard deviation σ, the sample mean, $\overline{X}$, has a normal sampling distribution [1].

$$\overline{X} \sim N(\mu, \frac{\sigma^2}{n}) \tag{9}$$

7.3 Central Limit Theorem

If $\overline{X}$ is the mean of a random sample of size n taken from a population with mean μ and variance σ^2, then the distribution of $Z = \frac{\overline{X}-\mu}{\sigma/\sqrt{n}}$ as $n \rightarrow \infty$ (the sample size becomes large, $n > 30$) is the standard normal distribution, $N(Z; 0, 1)$.

7.4 Confidence Intervals of Means and Proportions

It is difficult, sometimes even impossible, to get information about the population by examining the entire population data. For this reason, it is attempted to estimate the parameters of the population by using sample data from the population and sampling methods.

There are two types of parameter estimates;

1. Point estimation: A single numerical value of a statistic that best estimates the parameter is called point estimation. Sample mean, for example, is a point estimator of the population mean.
2. Interval estimation: Its goal is to find the lower and upper boundaries within which the unknown population parameter can be identified with a high degree of certainty (Fig. 13).

$P(-Z_{\frac{\alpha}{2}} < Z < Z_{\frac{\alpha}{2}}) = 1 - \alpha$ and since $Z = \frac{\overline{X}-\mu}{\sigma/\sqrt{n}}$
A $(1 - \alpha)\%$ confidence interval for population mean is

$$P\left(\overline{X} - Z_{\alpha/2}\frac{\sigma}{\sqrt{n}} \leq \mu \leq \overline{X} + Z_{\alpha/2}\frac{\sigma}{\sqrt{n}}\right) = (1 - \alpha) \tag{10}$$

If the population variance is not known than, replace σ with sample standard deviation, s, and using the t distribution, the confidence interval as follows.

$$P\left(\overline{X} - t_{\alpha/2}\frac{s}{\sqrt{n}} \leq \mu \leq \overline{X} + t_{\alpha/2}\frac{s}{\sqrt{n}}\right) = (1 - \alpha) \tag{11}$$

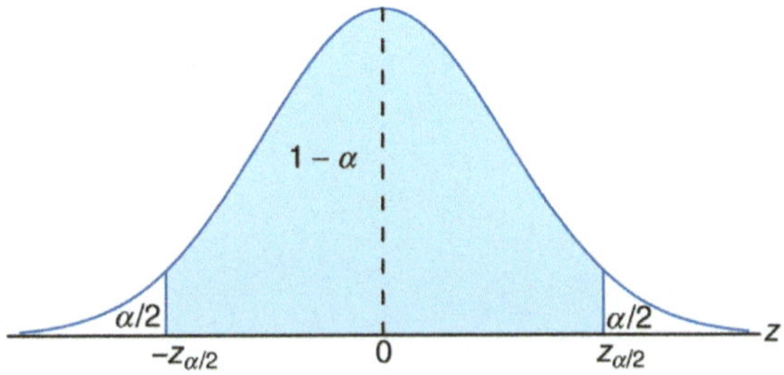

Fig. 13 Confidence interval estimation

Table 6 Confidence interval statistics for salary, education, and previous experience variables

Variable	N	Mean	SD	SE	95% Conf.	Interval
Salary	474.0	34,419.5675	17,075.6615	784.3111	32,878.4025	35,960.7326
Educ	474.0	13.4916	2.8848	0.1325	13.2312	13.7519
Prevexp	474.0	95.8861	104.5644	4.8028	86.4486	105.3235

degrees of freedom: $df = n - 1$.

The 95% confidence interval for education is (13.23 – 13.75). Table 6 provides the 95% confidence intervals for the salary, education, and previous experience variables, as well as their summary statistics.

The estimator of the population proportion, p, is the sample proportion, $\widehat{p}$; if sample size is large, $\widehat{p}$ has an approximately normal distribution with mean and standard deviation, $\widehat{p}$ *and* $\sigma_{\widehat{p}}$, where $\mu_{\widehat{p}} = p$ *and* $\sigma_{\widehat{p}} = \sqrt{\frac{pq}{n}}$, $p + q = 1$.

Then a $(1 - \alpha)\%$ confidence interval can be constructed for population proportion,

$$\widehat{p} \pm Z_{\alpha/2}\sqrt{\frac{\widehat{p}\widehat{q}}{n}} \text{ where } np \geq 5 \text{ and } nq \geq 5$$

The ratio of the number of 1s in the variable to the total number, after coding the managers with 1 and the others with zero, displays the manager ratio in the data, which is also equal to the column average. 95% of confidence interval for the managers proportion is (0.14–0.21).

7.5 *Testing Hypotheses*

A statistical hypothesis is a statement made concerning one or more unknown population parameters and denoted by H_0. Every hypothesis has a contradiction or alternative, denoted by H_1.

H_0: $\mu = 100$ (Null hypothesis).

H_1: $\mu \neq 100$ (Alternative hypothesis).

The null hypothesis is written that the difference between the value of the sample statistic and the population parameter is not significant, it is statistically zero.

A hypothesis is either true or false; one may reject or fail to reject it on the basis of information obtained from sample data.

A test of any statistical hypothesis where the alternative is one-sided, such as

H_0: $\theta = \theta_0$ or perhaps H_0: $\theta = \theta_0$.

H_1: $\theta > \theta_0$ H_1: $\theta < \theta_0$.

is called a one-tailed test. One-tailed tests are given in Fig. 14.

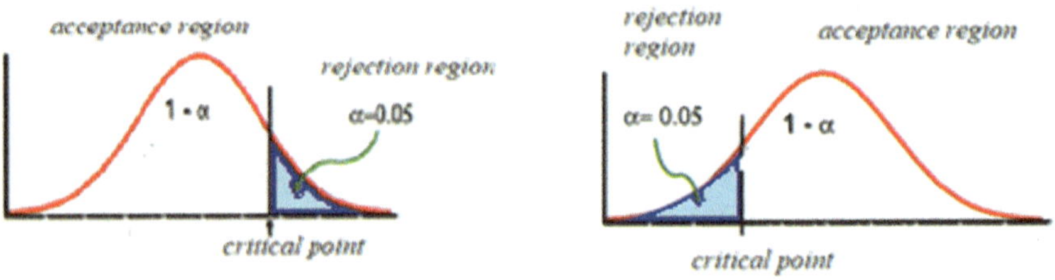

Fig. 14 Representations for one-tailed test

If the hypothesis is two-sided, the level of significance is divided into two equal parts; on the right and left (Fig. 15). If the hypothesis is one-sided, the rejection region is only one side; right or left.

The null hypothesis is frequently used to indicate the status quo or a preexisting view. It is kept, or considered to be true, until it is rejected in favor of the alternative hypothesis by a test.

There are two possible states of nature:

H_0 is true OR H_0 is false.

There are two possible decisions:

Fail to reject H_0 as true OR Reject H_0 as false.

Rejection of null hypothesis when it is true is a wrong decision and called type I error. The probability of a Type I error is denoted by α. α is called the level of significance of the test.

Non-rejection of null hypothesis when it is false is called type II error, and the probability of a type II error is denoted by β. $1 - \beta$ is called the power of the test. α and β are conditional probabilities;

The possible outcomes of a statistical hypothesis test are depicted in a contingency Table 7.

The p-value is the minimum level of significance, α, at which the null hypothesis may be rejected using the test statistic's obtained value. When the p-value is less than α, reject H_0.

Fig. 15 Representation for two-sided test

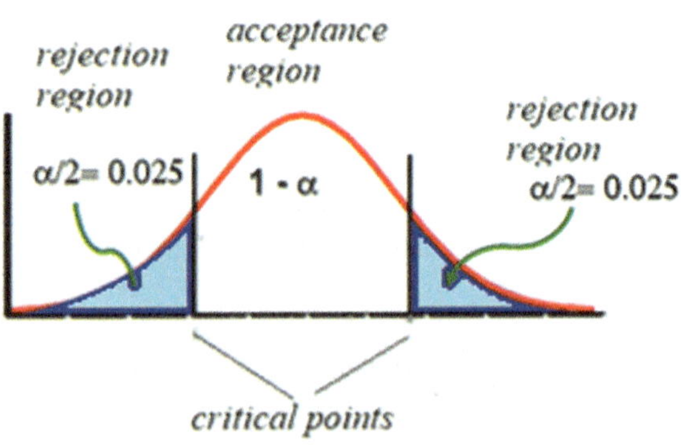

Table 7 Contingency table of a statistical hypothesis

	State of nature	
Decision	H_0 true	H_0 false
Don't reject H_0	Correct	Type II error (β)
Reject H_0	Type I error (α)	Correct

7.6 *Testing Hypotheses About the Population Mean and Proportion*

Suppose that a sample of n observations has a mean $\underline{x}$ and the standard deviation s, selected from a population whose mean is claimed to be equal to a certain value, μ,. Is the difference between μ and $\underline{x}$ statistically significant? Or is it a sampling error? In other words, can the claim that the population mean $\mu = \mu_0$ be rejected by means of the sample statistics at α level of significance?

H_0: $\mu = \mu_0$.

H_1: $\mu \neq \mu_0$

Significance level:α

Test statistics

$$Z = \frac{\overline{x} - \mu_0}{\sigma/\sqrt{n}}, \sigma\ known$$

$$T = \frac{\overline{x} - \mu_0}{s/\sqrt{n}}\ df = n - 1, \sigma\ unknown$$

The null hypothesis is rejected if the test statistic exceeds the absolute critical value. Otherwise, it cannot be rejected at the predetermined α level of significance.

Hypotheses for the population proportion, p

H_0: $p = p_0$.

H_1: $p \neq p_0$.

Test statistic is $Z = \frac{\hat{p} - p_0}{\sqrt{\frac{pq}{n}}}$

The null hypothesis is rejected if the test statistic exceeds the absolute critical value which is 1.96 for: α =0,05. Otherwise, it cannot be rejected at the predetermined: α level of significance.

The following hypotheses were tested in sample data. The first is regarding salary, whether it is 34,000 dollars or not; hence, the null and alternative hypotheses are as follows.

H_0: $\mu = 34{,}000$.

H_1: $\mu \neq 34{,}000$.

The null hypothesis could not be rejected because the T statistic is 0.53 and the p-value is 0.59. If the null hypothesis is rejected, the likelihood of making a wrong judgment is 0.59, which is quite high. The second hypothesis is regarding whether or not the salary is 32,000 dollars. The null hypothesis is rejected since the T statistic is 3.08 with a p-value of 0.002. Thus, the alternative hypothesis was indirectly accepted. Because the sample's average salary is 34,419, it is assumed that the population's average wage is not 32,000 but higher. The third hypothesis tests whether population education is equal to 14 years or not. Thus, the null hypothesis, population education equals 14, is rejected with a T value of 3.83 and a significance level of 0,00,014, implying that true education is less than 14, as the sample average education is 13.49, which is less than 14.

The following hypotheses were developed to test if the proportion of women in the industry is 60%.

H_0: $p = 0.60$.

H_1: $p \neq 0.60$.

The null hypothesis is rejected at the significance level of 2.8×10^{-10} with $T = -6.308$.

7.7 Testing Hypotheses for Differences Between Two Means and Proportions

If there is a need to compare whether there is a significant difference between the means of two independent populations, the hypotheses are as follows:

I: Difference between two population means is 0

$\mu_1 = \mu_2$.

H_0: $\mu_1 - \mu_2 = 0$.

H_1: $\mu_1 - \mu_2 \neq 0$.

II: Difference between two population means is less than 0

$\mu_1 \leq \mu_2$.

H_0: $\mu_1 - \mu_2 \leq 0$.

H_1: $\mu_1 - \mu_2 > 0$.

III: Difference between two population means is less than a number "*d*"

$\mu_1 \leq \mu_2 + d$.

H_0: $\mu_1 - \mu_2 \leq d$.

H_1: $\mu_1 - \mu_2 > d$.

Large-sample (for $n1 \geq 30$ and $n2 \geq 30$ because of Central Limit Theorem) test statistic for the difference between two population means is

$$Z = \frac{(\overline{x}_1 - \overline{x}_2) - (\mu_1 - \mu_2)}{\sqrt{\frac{\sigma_1^2}{n_1} + \frac{\sigma_2^2}{n_2}}} \tag{12}$$

if population variances, σ_1^2andσ_2^2, are known. x_1 and x_2 are sample means and $n1$ and $n2$ are sample sizes. The term in the denominator is the standard deviation of the difference between the two sample means.

A $(1 - \alpha)100\%$ confidence interval for the difference between two population means, $\mu 1 - \mu 2$, using independent random samples will be

$(\overline{x}_1 - \overline{x}_2) \pm z_{\frac{\alpha}{2}}\sqrt{\frac{\sigma_1^2}{n_1} + \frac{\sigma_2^2}{n_2}}$ σ_1^2andσ_2^2 are known.

The test statistics depending on whether $\sigma_1^2 = \sigma_2^2$ and $\sigma_1^2 \neq \sigma_2^2$ when they are unknown are as follows.

σ_1^2andσ_2^2 unknown but if $\sigma_1^2 = \sigma_2^2$

$$T = \frac{(\overline{x}_1 - \overline{x}_2) - (\mu_1 - \mu_2)}{\sqrt{\frac{1}{n_1} + \frac{1}{n_2}}} / \sqrt{\frac{(n_1 - 1)s_1^2 + (n_2 - 1)s_2^2}{n_1 + n_2 - 2}} \quad \text{with } df = n_1 + n_2 - 2$$

σ_1^2andσ_2^2 unknown but if $\sigma_1^2 \neq \sigma_2^2$

$$T = \frac{(\overline{x}_1 - \overline{x}_2) - (\mu_1 - \mu_2)}{\sqrt{\frac{s_1^2}{n_1} + \frac{s_2^2}{n_2}}} \qquad df = \frac{(s_1^2/n_1 + s_2^2/n_2)^2}{[(s_1^2/n_1)^2/(n_1 - 1)] + [(s_2^2/n_2)^2/(n_2 - 1)]}$$

Test statistic for a large-sample test for the difference between two population proportions is

$$z = \frac{(\widehat{p}_1 - \widehat{p}_2) - (p_1 - p_2)}{\sqrt{\left(\frac{\widehat{p}_1(1-\widehat{p}_1)}{n_1} + \frac{\widehat{p}_2(1-\widehat{p}_2)}{n_2}\right)}} \tag{13}$$

And a $(1 - a)$ 100% confidence interval for the difference between two population proportions by means of the test statistic is

$$(\widehat{p}_1 - \widehat{p}_2) \pm z_{\frac{\alpha}{2}}\sqrt{\left(\frac{\widehat{p}_1(1 - \widehat{p}_1)}{n_1} + \frac{\widehat{p}_2(1 - \widehat{p}_2)}{n_2}\right)} \tag{14}$$

Are women and men educated at the same level? The following are the hypotheses to be tested.

H_0: $\mu_m - \mu_w = 0$.

H_1: $\mu_m - \mu_w \neq 0$.

Table 8 Statistical test results

Test	Statistics result	*P*-value
Levene (equal variances)	13.96	0.000209
T-test (different variances)	−8.45	3.4588e-16

Table 9 Mean values for variables

Gender	Age	Educ	Salary	Salbegin	Jobtime	Prevexp
Female	39.3	12.3	26,031.9	13,091.9	80.3	77.1
Male	38.1	14.4	41,441.7	20,301.3	81.7	111.6

For both equal variance and different variance cases, the hypothesis that women and men have similar educational levels is rejected at the 5% significance level. The null hypothesis is rejected and concluded that men and women have different educational levels. When the sample data are examined to better understand the differences between the groups, it can be revealed that men, on average, have more education than women. Men have a mean education of 14.43, while women have a mean education of 12.37 years (Table 9).

7.8 Analysis of Variance

The hypothesis of whether there is a difference between more than two means is tested using analysis of variance (ANOVA). In an analysis of variance, there are r independent samples each of which corresponds to a population subjected to a different treatment. The samples have $n = n_1 + n_2 + n_3 + \cdots + n_r$ total observations and means: $x_1, x_2, x_3, \ldots x_r$ and variances: $s1\hat{}2$, $s2\hat{}2$, $s3\hat{}2$, ..., $sr\hat{}2$. ANOVA has two assumptions: The populations are normally distributed with equal variances.

The hypothesis test of analysis of variance:

H_0: $\mu_1 = \mu_2 = \mu_3 = \mu_4 = \ldots \mu_r$.

H_1: null hypothesis is wrong (at least one of them is different).

The test statistic of analysis of variance is based on the ratio of two estimators of a population variance:

$$F(r-1, n-r) = \frac{\text{Estimate of variance based on means from } r \text{ samples}}{\text{Estimate of variance based on all sample observations}} \tag{15}$$

Grand mean: Grand mean is obtained by dividing all the data obtained by the total number of observations.

Total variation (SST): Total variation is obtained by summing the squares of the difference of each observation and the overall mean. The degrees of freedom of total variation are $(n-1)$.

The total variation is divided into within-group variation and between-group variation. The logic of the analysis of variance is to compare the ratio of the between-group variation to the within-group variation. If the between-group variation is greater than the within-group variation, at least one of the group means is said to be different from the others.

Variation Between Groups (sum of square of treatment, SSTR): The variation between groups is obtained from the difference of each group mean from the overall mean. If the means of the groups is very close to each other, the variation between groups will be small. If the number of groups is r, the degree of freedom for the variation between groups will be $(r - 1)$.

Variation within groups (Sum of square of error, SSE): Error variation obtained by the differences of observations in each group from the group mean to which they belong. If the observations in the groups are close to each other, the variation within the group will also be small. Degrees of freedom for intra-group variation can be represented as sd $= n - r$.

F-test statistic: The F-test statistic is obtained from the ratio of the mean squares of between-group to the mean of squares of within-group with $(r - 1; n - r)$ degrees of freedoms.

Table 10 Key terms for ANOVA analysis

Source of variation	Degrees of freedom	Sum of squares	Mean sum of squares	F
Between groups	$r - 1$	SSTR	SSTR/$(r - 1)$	MSTR/MSE
Within groups	$n - r$	SSE	SSE/$(n - r)$	
Total	$n - 1$	SST		

The null hypothesis is rejected if the F value is greater than the critical table F value. At least one group is interpreted to be distinct from the others. Pairwise comparisons are used to determine which group or groups differ. One of such post hoc tests, the Tukey test is based on the studentized range distribution, q, with r and $(n - r)$ degrees of freedom. The critical point in a Tukey Pairwise Comparisons test is the Tukey Criterion:

$$T = q_{\alpha} \frac{\sqrt{\text{MSE}}}{\sqrt{n_i}} \tag{16}$$

where n_i is the smallest of the r sample sizes. The test statistic is the absolute value of the difference between the appropriate sample means, and if the test statistic is greater than the Tukey Criterion's critical point, the null hypothesis which is the equality of two population means is rejected.

On the variable jobcat, there are three groups: manager, clerical, and custodian. Is the average educational level of these groups the same? The ANOVA test can be used because there are more than two groups (assuming the ANOVA assumptions are met). The hypotheses are as follows:

H_0: $\mu_{\text{cler}} = \mu_{\text{cus}} = \mu_{\text{man}}$.

H_1: at least one of them is different.

$F = 165.2$, p-value $= 4.3 \times 10^{-55}$ are the test findings presented in the table above. As a result, the null hypothesis is rejected, and it is inferred that the mean education

Table 11 Pairwise Tukey analysis

Group1	Group2	MeanDiff	*p*-adj	Lower	upper	Reject
Clerical	Custodial	−2.6826	0.001	−3.7221	−1.6431	True
Clerical	Manager	4.3822	0.001	3.7513	5.0132	True
Custodial	Manager	7.0648	0.001	5.912	8.2176	True

of at least one group differs from that of the others. To figure out which groups were different from the others, the Tukey test was used. As a result, the following results have been obtained. The Tukey test, which compares the group means pairwise, reveals that all of the groups are different from one another (Table 11).

Managers have an average of 7.06 years more education than clericals, 4.38 years more education than custodians, while clericals have an average of 2.68 years more education than custodians.

8 Bayesian Statistics

Bayesian statistical analysis integrates a prior probability distribution and likelihoods of observed data to calculate a posterior probability distribution of occurrences based on Bayes' theorem [5].

Bayes' theorem for a discrete random variable:

$$P(\theta|x) = \frac{P(x|\theta)P(\theta)}{\sum_i P(x|\theta_i)P(\theta_i)} \quad \text{where } P(x) = \sum_i P(x|\theta_i)P(\theta_i) \tag{17}$$

where θ is an unknown population parameter to be estimated using the data. The denominator sums all possible values of the parameter of interest, θi, with x being the observed data set.

$P(\theta)$: The prior, which refers to the knowledge of the parameters before to viewing any data.

$P(x|\theta)$: The likelihood, or the chance of getting the data x from the model.

$P(\theta|x)$: Is known as the posterior in Bayesian jargon.

$P(x)$: The evidence, which is the likelihood of the data given the model we chose, rather than any specific set of parameters.

Bayes' theorem for continuous distributions:

$$f(\theta|x) = \frac{f(x|\theta)f(\theta)}{\int f(x|\theta)f(\theta)\mathrm{d}\theta} \tag{18}$$

Deciding which model to use as a prior is complex. The following are some examples of simple cases:

When analyzing natural events such as employee salary, education or IQ levels of people, the normal distribution is used. When modeling proportions, the beta distribution is used. When modeling the frequency of events, the Poisson distribution is used.

Beta function is defined as;

$B(\alpha, \beta) = \int_0^1 (1-\theta)^{\beta-1}\theta^{\alpha-1} d\theta$ for $\alpha, \beta > 0$

Beta distribution with the parameters $\alpha > 0$and$\beta > 0$

$$f(\theta) = \frac{\Gamma(\alpha+\beta)}{\Gamma(\alpha)\Gamma(\beta)}\theta^{\alpha-1}(1-\theta)^{\beta-1} \tag{19}$$

Γ: gamma function,

$$\Gamma(\alpha) = \int_0^{\infty} \theta^{\alpha-1}e^{-\theta}d\theta \text{ for} \alpha > 0 \tag{20}$$

Beta distribution has two parameters, α and β, that control the distribution. The graphical representation for some α and β values is given in Fig. 16.

Getting the posterior: The posterior is proportional to the product of likelihood and prior, according to Bayes' theorem.

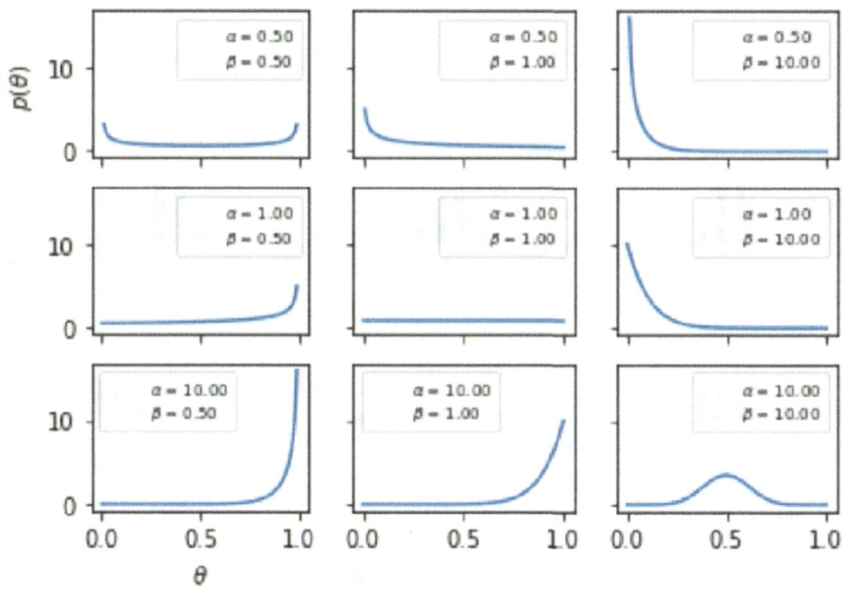

Fig. 16 Beta distributions for various α and β

$$f(x) \propto f(\theta)f(\theta)$$

The posterior will be proportional to the product of the binomial and beta distributions if estimation of the women's proportion is wanted:

$$f(x) \propto \frac{n!}{n!(n-x)!}\theta^{x}(1-\theta)^{n-x}\frac{\Gamma(\alpha+\beta)}{\Gamma(\alpha)\Gamma(\beta)}\theta^{\alpha-1}(1-\theta)^{\beta-1}$$

By simplifying the expression above, the following is got.

$$f(x) \propto \theta^{x+\alpha-1}(1-\theta)^{n-x+\beta-1}$$

The graphical representation of a binomial distribution is given below (Fig. 17). Graph shows the probability distribution of getting × heads out of n (3, 7, 10) coin tosses, given a fixed value of θ(0.25, 0.50, 0.75).

The following graphs are generated by using the analytical expression of the posterior for estimating the women's proportion in the industry for different beta parameters and given number of observations. See for codes and additional information [5].

The most likely value is determined by the posterior mode. The spread of posterior is proportional to the degree of uncertainty regarding a parameter's value; the wider the distribution, the less certain we are. The real value for θ is represented by a black vertical line at 0.50, the posterior by a green curve, and the prior by a red curve.

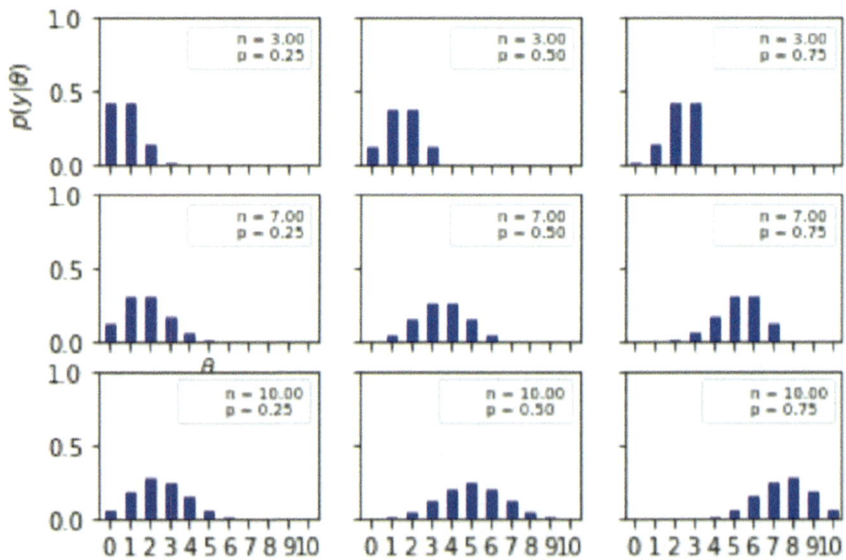

Fig. 17 Probability distributions of getting × number of heads for n-number of coin tosses

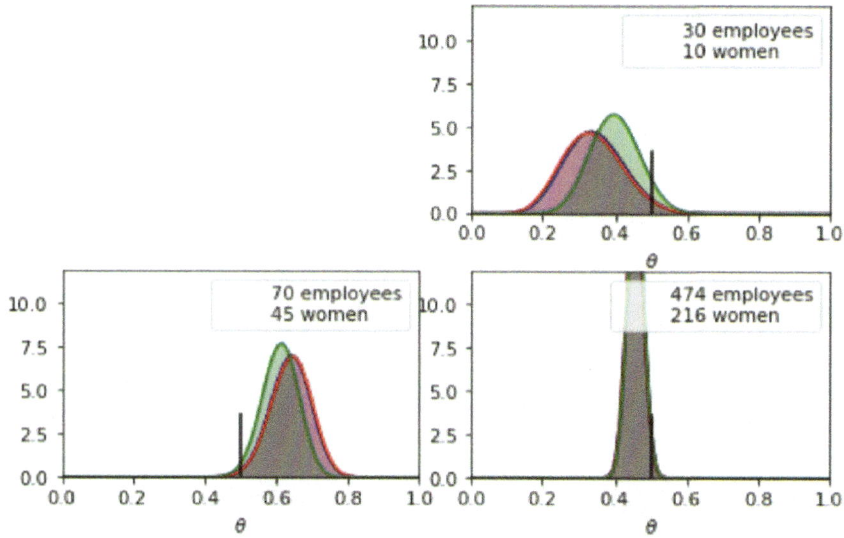

Fig. 18 Posterior for estimating the true women proportion in the industry at $\theta = 0.50$ using employee data

References

1. Walpole RE, Myers RH, Myers SL, Ye K (2012) Probability & statistics for engineers & scientists, 9th edn. Prentice Hall
2. Salvatore D, Reagle D (2002) Theory and problems of statistics and econometrics, 2nd edn. McGraw-Hill
3. Johnson RA, Wichern DW (2002) Applied multivariate statistical analysis. Prentice Hall
4. Bai ZD, Chen L (2003) Weighted W Test for Normality and Asymptotics a Revisit of Chen-Shapiro Test for Normality. J Statis Plann Inference 113
5. Martin O (2016) Bayesian analysis with python. Packt Publishing

Nizamettin Bayyurt is a Quantitative Methods professor. He received an undergraduate degree in Mathematics from Middle East Technical University in Ankara, Turkey and a master's degree in mathematics from Fatih University in Istanbul, Turkey. He obtained his Ph.D. in Quantitative Methods from Istanbul University in 2004. For the past two decades, he has taught undergraduate and graduate statistics and operations research courses. His research has been published in international peer-reviewed journals, conference proceedings, and book chapters. He has recently been teaching at Istanbul Technical University's Faculty of Management.

Sefer Baday is an assistant professor at the Informatics Institute of Istanbul Technical University. He received his B.S degree in Chemical Engineering from Bogazici University in 2006 and M.S. degree in Computational Science and Engineering from Koç University in 2008. He recevied his Ph.D. degree in Biophysics from Biozentrum in the University of Basel (Switzerland) in

2013. After that, he continued his research as a postdoctoral researcher at Biozentrum until 2014. Between 2014 and 2015 he worked at the Chemistry Department in the University of Cambridge (UK) as a postdoctoral research associate. His main research areas are computeraided drug design, biomolecule modeling and simulation, and protein-ligand interactions.

Prediction Modeling

Nursah Alkan, Yakup Turgut, Emre Ari, Seval Ata, Mehmet Yasin Ulukus, Mehmet Ali Ergun, and Omer Faruk Beyca

1 Introduction

Predictive modeling can be defined as modeling the historical data using statistical and machine learning techniques to predict future observations. Prediction modeling tasks can be grouped into three categories: supervised learning, unsupervised learning and reinforcement learning. Supervised learning is the task of mapping input features to labeled outputs. Supervised learning may be utilized for many machine learning problems such as stock price prediction, health care diagnosis, face recognition, spam detection and demand forecasting. Unsupervised learning is used to extract useful information from unlabeled data. Some of its applications are fraud detection, customer segmentation, recommender systems, etc. Reinforcement learning teaches agents to take actions in order to maximize the cumulative reward. In this chapter, several algorithms for supervised learning as well as unsupervised and reinforcement learning techniques will be discussed.

2 Linear Regression

Linear regression is one of the most widely used prediction algorithms in statistics and machine learning in the last several decades. Popularity of linear regression models can be attributed (but not limited) to the following reasons [1]:

- The training phase is often quite fast.
- Model outputs are easily interpretable even for non-technical audiences.
- It is easy to implement in any programming language.

N. Alkan · Y. Turgut · E. Ari · S. Ata · M. Y. Ulukus · M. A. Ergun · O. F. Beyca (✉)
Department of Industrial Engineering, Faculty of Management, Istanbul Technical University, 34367 Macka, Istanbul, Turkey
e-mail: beyca@itu.edu.tr

A. Ustundag et al. (eds.), *Business Analytics for Professionals*, Springer Series in Advanced Manufacturing, https://doi.org/10.1007/978-3-030-93823-9_3

Linear regression assumes a linear relationship between the input (independent[1]) variables x and output variable (dependent[2]) y. In other words, when using a linear regression model, it is assumed that y can be calculated from a linear combination of input variables. Linear regression models can be classified according to the number of input variables used. The version in which there is a single input variable x is often called simple linear regression. Similarly, when there are multiple input variables, then it is referred as a multiple linear regression model. In general, the relationship between input variables x and output variable y are assumed to have the following form:

$$y = \beta_0 + \beta_1 x_1 + \beta_2 x_2 + \ldots \beta_n x_n \tag{1}$$

where x_i is the ith independent variable, β_i is called the coefficient[3] of the independent variable x_i and β_0 is often referred as the intercept. This is actually a typical linear relation equation. Please note that, if it is known the values of β_i, it can be easily made predictions about the value of y given particular x_i just by plugging x_i and β_i values into Eq. (1). In fact, when a linear regression model is trained on the data under consideration, what is done is to calculate and determine the best-fit β_i's based on the trends observed in the data.

For smaller models with small number of independent variables and/or small number of data points, the values of β_i can be calculated analytically. The method used for the analytical solution is called the ordinary least squares (OLS) method. However, as the number of features and data points grow, OLS method requires considerable amount of computation time. This is where gradient descent (GD) method comes in. More on both methods will be discussed later in this chapter.

Once a regression model has been trained, its success must be measured and compared with models trained using other regression algorithms. There are several metrics quantifying the difference between observed and predicted values. Later in this chapter, some metrics commonly used in practice will also be discussed.

2.1 Simple Linear Regression

As discussed earlier, a simple linear regression model is a two-dimensional model that includes a single independent variable x and a dependent variable y. Then the expression in Eq. (1) reduces to:

$$y = \beta_0 + \beta_1 x \tag{2}$$

---

[1] Other terms used for output variable are dependent variable or target value.

[2] See Footnote 1.

[3] Coefficients are referred as weights in some sources.

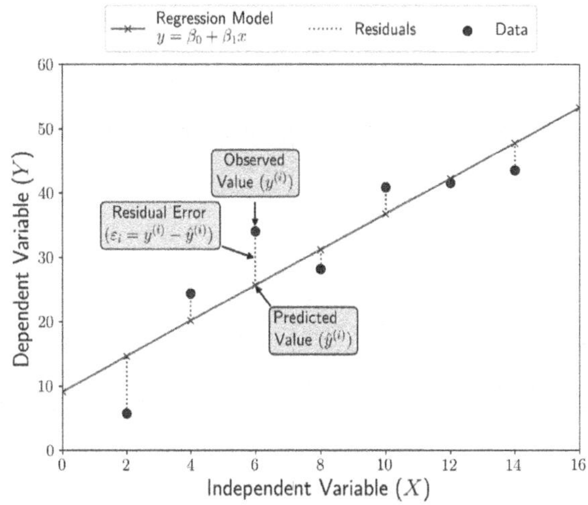

Fig. 1 Graphical representation of a simple linear regression model. Circles represent the observed data, and the line represents the trained linear regression model from that data

There will always be some difference between the observed values of $y^{(i)}$ and the predicted values $\hat{y}^{(i)}$ because it is very unlikely for all $y^{(i)}$ to lay on a straight line. For simple linear regression models, it is used OLS method whose main aim is to minimize the errors between the observed data (y_i) and predictions made for these data points ($\hat{y}_i$) by calculating the optimum β_0 and β_1 values. With the explicit introduction of an error term, the linear regression equation becomes:

$$y = \beta_0 + \beta_1 x + \mathcal{E} \tag{3}$$

In Eq. (3); β_0 is the intercept, β_1 is the slope of Eq. (2) and the $\mathcal{E}$ is the random error (residual) component. An example of a simple linear regression model trained on a small sample data (number of data points $(n) = 7$) is shown in Fig. 1. The observed data ($y^{(i)}$) are marked with circles and the model learned from this data is given as a straight line. The "x" marks along the line shows the predicted values ($\hat{y}^i$) for corresponding $x^{(i)}$ value. Residual errors ($\mathcal{E}_i$) are marked with dotted lines.

In order to analytically calculate optimal β_0 and β_1 using OLS method, it is first expressed Eq. (3) in terms of individual data points:

$$y^{(i)} = \beta_0 + \beta_1 x^{(i)} + \mathcal{E}_i \tag{4}$$

where it is assumed that the dataset contains n pairs of $(x^{(i)}, y^{(i)})$. As discussed earlier, the objective of OLS method is to minimize sum of residual errors across all training data which is often referred as the sum of squared errors[4] (SSE):

[4] Sum of squared errors (SSE) may be referred as "residual sum of squares" (RSS) or "sum of squared residuals" (SSR) in other sources.

$$SSE = \sum_{i=1}^{n} \varepsilon_i^2 \tag{5}$$

Since it is desired that the negative residuals are as close to zero as possible, it is necessary to minimize the square of the errors rather than the error itself. By using Eq. (4), the residual errors can express as:

$$\mathcal{E}_{\text{i}} = (\text{y}^{(\text{i})} - \underbrace{\beta_0 - \beta_1 x^{(i)}}_{\hat{y}_{\text{i}}} \tag{6}$$

Replacing $\mathcal{E}_i$'s in equation of (5) leads to the following form for SSE for a given (β_0, β_1) pair.

$$SSE(\beta_0, \beta_1) = \mathcal{E}_i^2 = \sum_{i=1}^{n} (y^{(i)} - \beta_0 - \beta_1 x^{(i)})^2 \tag{7}$$

To find parameters of β_0, β_1 that minimizes Eq. (7), the partial derivatives of Eqs. (7) with respect to β_0 and β_1 are calculated. The optimal values can be calculated by solving for β_0, β_1 where the following partial derivatives are 0:

$$\frac{\partial SSE}{\partial \beta_0} = -2\sum_{i=1}^{n}\left(y^{(i)} - \beta_0 - \beta_1 x^{(i)}\right) = 0 \tag{8}$$

$$\frac{\partial SSE}{\partial \beta_1} = -2\sum_{i=1}^{n}\left(y^{(i)} - \beta_0 - \beta_1 x^{(i)}\right)x^{(i)} = 0 \tag{9}$$

When it is solved[5] Eqs. (8) and (9) for optimal values $\widehat{\beta_0}$ and $\widehat{\beta_1}$, it is get the following set of formulas to calculate $\widehat{\beta_0}$ and $\widehat{\beta_1}$ from the data: $\widehat{\beta_0} = \overline{y} - \widehat{\beta_1}\overline{x}$ and $\widehat{\beta_1} = \frac{\sum_{i=1}^{n}(y^{(i)} - \overline{y})x^{(i)} - \overline{x})}{\sum_{i=1}^{n}(x^{(i)} - \overline{x})^2}$ where $\overline{y} = \frac{1}{n}\sum_{i=1}^{n} y^{(i)}$ and $\overline{x} = \frac{1}{n}\sum_{i=1}^{n} x^{(i)}$.

Assumptions of Simple Linear Regression Models

When it is trained and used a simple linear regression model, it is assumed the following conditions hold:

1. *Linear relationship:* output variable y has a nearly linear relationship with input variable x,
2. *Homoscedasticity:* For each particular value of x, the residuals (errors) have the same variance. That is, if it is picked two random values for x, the variance of the residual errors should be roughly the same for both values chosen.
3. *Independent Errors:* Residuals (Errors) should be uncorrelated.
4. *Normality:* The residuals at a particular value of x follow a normal distribution.

[5] The math of solving for $\widehat{\beta_0}$ and $\widehat{\beta_1}$ is left as an exercise for the curious readers.

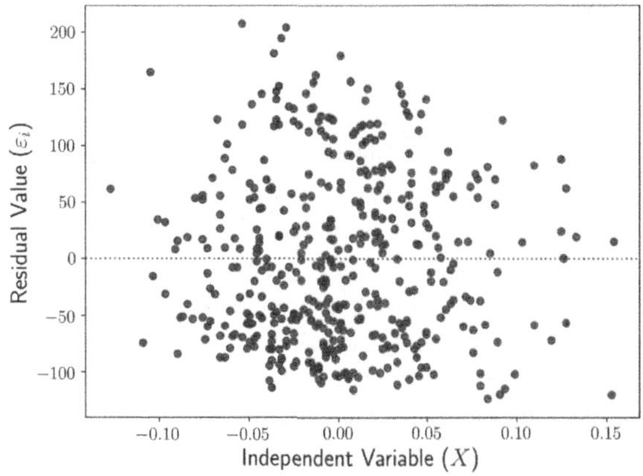

Fig. 2 An example of a residual plot

One way to check if these assumptions hold is to create a residual plot that plots independent variable values $x^{(i)}$ against residual values $\mathcal{E}_i$ (Fig. 2). In ideal world, it should see the following patterns:

- Roughly same number of randomly placed points above and below the residual value of 0. This indicates the assumption of a linear relationship holds well.
- Residuals forming a roughly horizontal band around residual value of 0. That is, the spread of residuals remains roughly the same across all values of independent variable x. This condition is looked to confirm the homoscedasticity assumption.
- There are no residual points standing far away from the rest. This makes sure that it do not have any outliers in data.

It is also important not to over-read residual plots. If there is no obvious patterns deviating just discussed, then the model is in good shape.

For other assumptions that residual plots do not cover, one can use goodness-of-fit tests (Chi-Square [2], Anderson–Darling [3] and so on) to check the normality and a Durbin–Watson test [4] for independence of errors.

2.2 *Multiple Linear Regression*

Multiple linear (multi-linear) regression models differ from the simple linear regression models in the number of independent variables (x_i's) used to make prediction for the output variable y. Specifically, in multi-linear regression, more than one explanatory variable are used. With that in mind, Eq. (2) can extend for multiple explanatory variables:

$$y = \beta_0 + \beta_1 x_1 + \beta_2 x_2 + \cdots + \beta_m x_m + \mathcal{E} \tag{10}$$

where y is the dependent variable, β_0 is intercept of the model, $\beta_{1,}\beta_2 \ldots \beta_m$ are the coefficients of $x_1, x_2, \ldots x_m$ respectively, $\mathcal{E}$ is the residual (error term).

As it can see, many of the terminology carries over from the simple linear regression. One of the major differences is that we are no longer fitting a line to the data. Depending on the number of explanatory variables, the resulting model may be a 2D plane (in case of 2 explanatory variables) or a hyperplane (3 + explanatory variables) which make it very hard or even impossible to visualize in 2-dimension.

When a multi-linear regression model is trained, the objective is the same: finding the optimum values of β_i that minimizes the error terms. To that end, there are two major methods it can use: OLS method and the gradient descent (GD). Similar to simple regression models, OLS method solves for optimal coefficient values analytically. OLS method solves the following system of linear equations for the optimal coefficients:

$$\beta = \left(X^T X\right)^{-1} X^T Y \tag{11}$$

where $\beta = \begin{bmatrix} \beta_1 \\ \beta_2 \\ \vdots \\ \beta_m \end{bmatrix}, X = \begin{bmatrix} X_{1,1} & \cdots & X_{1,m} \\ \vdots & \ddots & \vdots \\ X_{n,1} & \cdots & X_{n,m} \end{bmatrix}, Y = \begin{bmatrix} y_1 \\ y_2 \\ \vdots \\ y_n \end{bmatrix}$.

In words, matrix X includes all data for explanatory variables. Each column of X contains data for a particular explanatory variable so the number of columns is equal to the number of explanatory variables (m). The rows of X include data for a particular data point so there will be a total of n rows. Similarly, the vector Y contains all target values for n data points. Performing the matrix operations given in Eq. (11) will yield a vector (β) that contains all of the optimal β_i's except the intercept (β_0). However, with slight modification of adding a row of 1's at the top of matrix X, OLS can solve for the intercept as well.

On the other hand, as the matrix X gets larger (i.e. as number of explanatory variables and/or data points increases), the matrix operations required to solve Eq. (11) get computationally expensive. In turn, it results in a slow training phase. In such cases, gradient descent (GD) algorithm can be used instead of OLS. Unlike OLS, GD takes an iterative approach to finding optimum coefficients.

Before beginning the iterations, GD algorithm initializes all coefficients to randomly selected values. Then, with these coefficients the predicted values are calculated:

$$\hat{y}^{(i)} = \beta_0 + \beta_1 x_1^{(i)} + \beta_2 x_2^{(i)} + \cdots + \beta_n x_n^{(i)} \forall i \tag{12}$$

Although there are other alternatives, GD most commonly uses a squared loss function which needs to be minimized:

$$L = \sum_{i=1}^{n} \left(y^{(i)} - \hat{y}^{(i)}\right)^2 \tag{13}$$

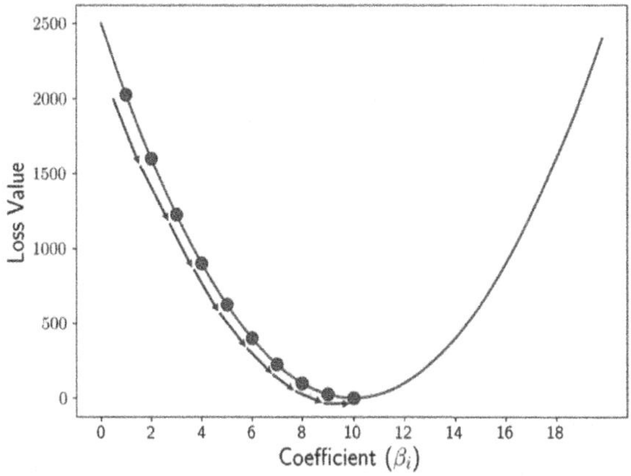

Fig. 3 Evolution of coefficient values over each iteration of GD

Then GD algorithm will update each coefficient according to the gradient at given iteration by using the following formula:

$$\beta_i^{t+1} = \beta_i^t - \eta \frac{\partial L}{\partial \beta_i^{t+1}} \tag{14}$$

where t is the current iteration, β_i^t is the coefficient value at the beginning of iteration t, β_i^{t+1} is the coefficient value after the update at iteration t. Aside from that, η is called the learning rate or the step-size that is determined by the user.

At each iteration, GD moves coefficient values closer to their optimal value, if the learning rate η is chosen appropriately (Fig. 3). Since the gradient is zero at the optimal value, updates will not affect the value of the coefficients once they attain their respective optimal values.

If η is set to a value that is too small, the steps will be very small which may lead the GD algorithm to either run very slow or fail to converge to the optimal solution. On the contrary, if the learning rate is too large, the values of coefficients may oscillate around or diverge from the optimal values.

Assumptions of Multi-Linear Regression

All the assumptions listed for simple linear regression should also hold to multiple linear regression. In addition to these assumptions, the explanatory variables are also required to be uncorrelated. A model in which there are two or more highly correlated explanatory variables is said to have multicollinearity problem. It can investigate if multicollinearity exists by calculating correlations between each pair of explanatory variables. If correlations closer to 1 or -1 are seen, the issue may need attention. Multicollinearity problems may not affect the predicted power of the model significantly but reduces the interpretability of the model. Reducing the effect of multicollinearity becomes more important if one wishes to measure the individual effects of each explanatory variable on the output. Several methods can be seen later in this chapter to use in the case of associated explanatory variables.

2.3 Performance Metrics for Regression Models

So far, how to train a linear regression model and some techniques for validity checking are discussed. After training and validating the model, its performance in the given task should also be evaluated. More specifically, it is necessary to measure how well the model's predictions and the observed data fit. To that end, several metrics can be used in practice. A few of the most popular metrics namely Mean Square Error (MSE), Mean Absolute Error (MAE), Mean Absolute Percentage Error (MAPE) and R-Squared (R^2) are introduced in this chapter. It is important to note that these metrics are not specific to linear regression models and can be used to assess regression models created with other ML algorithms. These metrics are also useful to compare performances of several models and pick the best among them. Also, any performance evaluation strategy should include some form of cross-validation, in which the performance of the model is evaluated on a separate dataset (referred to as validation or test data) apart from the data on which the model is trained (training data). Both train and test data should come from the same distribution, although at least two separate datasets are being mentioned.

Mean Square Error (MSE)

The mean squared error (MSE) for a given dataset can be calculated by using Eq. (15).

$$MSE = \frac{1}{n}\sum_{i=1}^{n}(y^{(i)} - \hat{y}^{(i)})^2 \tag{15}$$

As the name suggests, MSE sums the squares of the residual errors and takes the average over all data instances. The squaring is one of the ways to measure the magnitude of the error. This measure is more sensitive to any outliers as it squares the difference. The lower the MSE value the better the regression model.

Mean Absolute Error (MAE)

The Mean absolute error (MAE) is very similar to MSE as it sums all the residual errors and takes the average over all data instances (n is the number of data instances). However, MAE measures the magnitude of the error with its absolute value rather than squaring it (Eq. (16)).

$$MAE = \frac{1}{n}\sum_{i=1}^{n}|y^{(i)} - \hat{y}^{(i)}| \tag{16}$$

Compared to MSE, MAE is less sensitive to outliers as it uses the absolute value operator. Similar to MSE, MAE needs to be as small as possible.

Mean Absolute Percentage Error

One drawback of both MSE and MAE is that they are not easily interpretable. For instance, in order to judge if a model with MAE value of 5 is a good model or not, it is necessary to know the spread and scale of the values of the output variable y. If the difference between minimum and maximum values[6] of y is small, say 10, then MAE of 5 is an unacceptable error. On the other hand, If the difference between minimum and maximum values of y is large, say 100,000, then MAE of 5 is a very good performance.

It can use Mean Absolute Percent Error (MAPE) to work with an easier-to-interpret metric.

$$MAPE = \frac{100}{n}\sum_{i=1}^{n}\left|\frac{y^{(i)} - \hat{y}^{(i)}}{y^{(i)}}\right| \tag{17}$$

When calculating the MAPE, the errors are normalized to the observed value, so the error is measured as a percentage of the observed value. Thus, MAPE provides an error measure distilled to a percentage which makes it easier to judge if the error seen is an acceptable error or not.

R-Square (R^2)

R-square measure is another metric that is easily interpretable. The coefficient of determination or R-squared represents the proportion of the variance in the dependent variable which is explained by the linear regression model. It is a scale-free score i.e., irrespective of the values being small or large, the value of R-square will be less than one.

$$\mathrm{R}^2 = 1 - \frac{\sum_{i=1}^{n}\left(y^{(i)} - \hat{y}^{(i)}\right)^2}{\sum_{i=1}^{n}\left(y^{(i)} - \overline{y}\right)^2} \tag{18}$$

where $\overline{y}$ is the average of $y^{(i)}$'s. R-square should be as close to 1 as possible. In most cases, R-square gives a measure between 0 and 1. However, in rear cases, it may get values outside of this range. This is an indication that the model is ill-formed and should not be used for predictions. Reasons for an ill-formed model can be various but one can focus on trying different learning rates (if GD is used) or regularization parameters. One drawback of R-square is that addition of extra explanatory variables usually increases R-square (due to increase in variance) even if the new variable has little effect on predictive performance. Because of this, it either need to keep an eye on other metrics (such as MSE, MAE) together with R-square or use adjusted R-square.

---

[6] Looking at minimum and maximum values of y is one way to gauge the spread of the data and may be misleading if there are outliers. Other measures, such as the variance, could be used for this purpose as well.

Adjusted R-Square (R^2)

As the name suggests, Adjusted R-square is a modified version of the R-square metric. More specifically, it is adjusted for the number of independent variables in the model, and it will always be less than or equal to R-squared (Eq. (19)).

$$R^2_{\text{adj}} = 1 - \left[\frac{(1 - R^2)(n - 1)}{n - m - 1}\right] \tag{19}$$

This adjustment makes sure that Adjusted R-square score goes up only if addition of a new explanatory variable provides significant predictive power to the model. That is, Adjusted R-square metric is robust to superficial increases of R-square due to increases in variance.

2.4 Regularization in Linear Regression

As a broader concept, regularization is a process of enforcing constraints on ML models to avoid overfitting and/or deal with insufficient/correlated data. Regularization methods applying to linear regression models usually come in forms of extra penalties added to the loss function. In linear regression applications, it is necessary to consider regularization if:

- There is multicollinearity between two or more independent variables.
- There is not enough data to fit a meaningful model.

Adequate data means that there should be at least as many data instances as the number of independent variables (so $n \geq m$). For example, if there are 2 independent variables, at least 2 data instances are needed to find a meaningful relationship. Otherwise, there will be infinitely many ways to fit a linear model (i.e. we can define infinitely many lines passing from a single point).

However, in some applications, the number of independent variables or features are very large and data collection is very costly (i.e. genetic applications). In such cases, it may need to work with data for which $n \leq m$. If that is the case, it needs to apply some form of regularization.

Ridge Regression

Ridge regression is a version of linear regression in which a penalty term is added to the loss function. The penalty term is the sum of squared coefficients and is also referred as the L2 regularization. The loss function for ridge regression becomes:

$$\text{Ridge Loss} = \sum_{i=1}^{n}\left(y^{(i)} - \hat{y}^{(i)}\right)^2 + \overbrace{\lambda\sum_{i=1}^{n}\beta_i^2}^{L2\text{ Penalty Term}} \tag{20}$$

where λ is the parameter that tunes the magnitude of the regularization. Since the aim is to minimize the loss function, the addition of L2 penalty makes sure that β_i coefficients do not get unnecessarily large. Thus, ridge regression shrinks the coefficients, and it helps to reduce the model complexity and multicollinearity. When $\lambda \to 0$ the ridge regression model turns into a regular linear regression model. On the other hand, a larger value of λ means more shrinkage to the coefficients and results in a more constrained model. In the presence of multicollinearity, ridge regression shrinks the parameters of all correlated variables in equal proportions.

Lasso Regression

In Lasso regression a slightly modified version of the ridge regression is used. Instead of L2 penalty, An L1 penalty term is used, which is the sum of absolute value of the coefficients.

$$\text{Lasso Loss} = \sum_{i=1}^{n} \left(y^{(i)} - \hat{y}^{(i)}\right)^2 + \overbrace{\lambda \sum_{i=1}^{n} |\beta_i|}^{L1 \text{ Penalty Term}} \tag{21}$$

Similar to ridge regression, parameter λ is used to tune the intensity of the regularization. However, unlike ridge regression, the use of L1 penalty quickly drives coefficients of correlated features toward zero. Within a set of correlated features, the coefficient of one of the features will be nonzero and the rest will be set to zero given that λ is sufficiently large. This quality of lasso regression makes it useful for automatic feature selection tasks as well.

3 Logistic Regression

Logistic Regression is a staple statistical learning method and a fundamental tool for machine learning (ML) community. Although name includes the word "regression," it is mostly used for classification tasks in which the target (y) is a categorical variable. Due to the discreet nature of categorical variables, it is no longer directly try to predict the actual value/class label of a given data point. Rather, most classification algorithms aim to learn a decision boundary that separates instances of different classes. This type of classification algorithms is usually referred as the "Discriminative" classification methods. Most of the classification algorithms (logistic regression, ANNs, SVM, decision trees, random forest, etc.) covered in this book fall under discriminative category. There are also "Generative" classification methods that learns the underlying distribution for each class label and then do the classification task using that distribution.

Logistic regression is a discriminative classification algorithm used for binary classification in its basic form. In binary classification, there are two possible outcomes for the target variable y such as negative–positive, true–false, healthy–sick and so on. It can also be extended for use in multi-class problems in which there are more than 2 class labels. Most contemporary logistic regression implementations (such as logistic regression in scikit-learn) handle multi-class problems seamlessly.

For the discussions in this chapter, it is focused on binary classification problems. Without loss of generality, the reference class (i.e. negatives or 0's) and the other class (i.e. positives or 1's) can be labeled as the class of interest. In training phase, it is learned to estimate the probability of actually belonging to the positive class. That is, a logistic regression model predicts $P(Y = \text{Positive}|x_i)$ for each data point x_i. Then, it uses these probabilities to classify data points. If $P(Y = \text{Positive}|x_i) > 0.5$ then x_i is classified as a positive. Similarly, if $P(Y = \text{Positive}|x_i) < 0.5$ (which means $P(Y = \text{Negative}|x_i) > 0.5$), then x_i is classified as belonging to the negative class.

3.1 The Reasons for Using Logistic Function

For the points in the dataset, $P(Y = \text{Positive}|x_i)$ values are equal to 1 for instances of the positive class and 0 for the instances of the negative class because their class is known with certainty. When plotted against a single independent variable X, they are located along two horizontal lines as seen in Fig. 4.

If a linear regression model is used to predict the probability of $P(Y = \text{Positive}|x_i)$, some problems are encountered (Fig. 4a). First of all, it may get values less than 0 or greater than 1 which are not valid values for a probability. This problem could be solved by casting those values to either 0 or 1. However, a more serious problem is that a linear line performs poorly for the intermediate values of X. One of the reasons for this is that the considered data see as a step-function like shape rather than a linear trend.

A logistic (sigmoid) function performs much better when it comes to predicting probabilities. A logistic function asymptotically approaches to 0 for small X values and similarly asymptotically approaches to 1 for larger X values which conforms

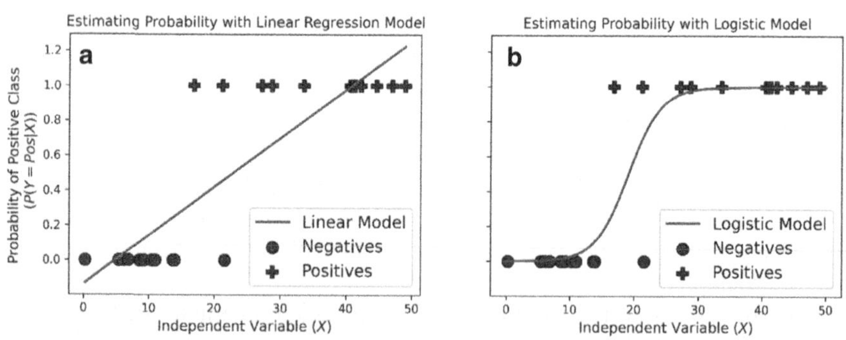

Fig. 4 Linear versus logistic regression for estimating a probability

with values of a probability. It also attains a value between 0 and 1 for intermediate values X and it can fit the probability data much tighter compared to a linear function (Fig. 4b). These qualities, along with many others, make logistic function very useful for probability prediction.

3.2 The Logistic Function and Logistic Regression

As all the classification algorithms, there is also a mathematical formulation behind Logistic regression. A logistic (sigmoid) function has an s-shape form the following mathematical formula:

$$\sigma(z) = \frac{1}{1+e^{-z}} \tag{22}$$

Logistic regression has lots of applications from statistics to sociology in the different scientific fields. The domain of Logistic Function is the real numbers range from $-\infty$ to $+\infty$ and the range of the Logistic regression between two points which upper limit is the constant number, generally 1, when $z$goes to $+\infty$ lower limit is zero when $z$goes to $+\infty$.

Similar to linear regression, logistic regression linearly combines the values of independent variables (X_i) (Eq. 23).

$$z = \beta_0 + \beta_1 X_1 + \beta_2 X_2 + \cdots + \beta_n X_n \tag{24}$$

where β_0 is the intercept of the model and $\beta_{1,}\beta_2 \ldots \beta_n$ are the coefficients of $X_1, X_2, andX_n$ respectively. After calculating the intermediate value z, the resulting value is fed through a logistic function for the predicted probability (Fig. 5).

$$\hat{y} = \sigma(z) \tag{25}$$

3.3 Training a Logistic Regression Model

In training phase, the main aim is, again, to identify the β_i coefficients that minimizes a certain error function for a given training dataset. For logistic regression, this can be achieved in two fundamentally different ways.

First approach is to formulate it as an optimization problem and solve for optimal β_i coefficients. Some ML libraries including scikit-learn employ such algorithms to train a logistic regression model, it is referred to [5] for the details of this approach.

Logistic regression models can also train by using an error-driven approach via gradient descent. Gradient descent methods are usually coupled with the cross-entropy loss function:

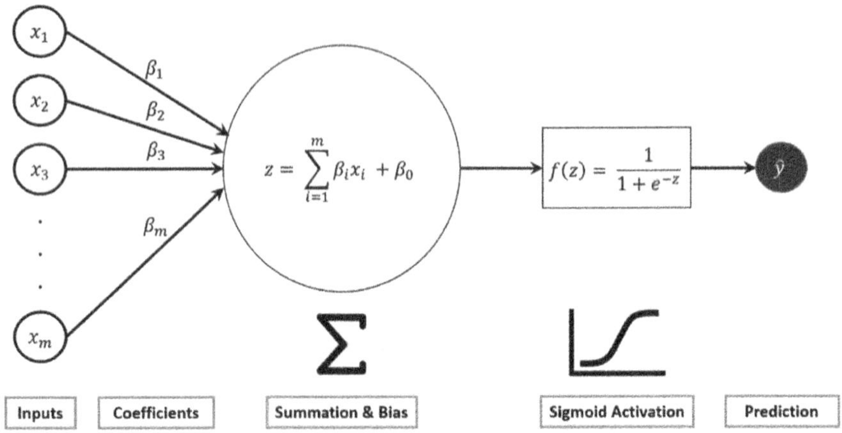

Fig. 5 Graphic representation of logistic regression predictor

$$L(y, \hat{y}) = -\sum_{i=1}^{c} ylog(\hat{y}) \tag{26}$$

In Eq. (26), y is the actual class label, $\hat{y}$ is the predicted probability value and c is the number of possible classes.

For the two-class problems loss function will turn into the following formula:

$$L(y, \hat{y}) = -\left[y\log\hat{y} + (1-y)\log(1-\hat{y})\right] \tag{27}$$

where it is assumed that y attains either value of 1 or 0. The cross-entropy loss function is close to zero if y and $\hat{y}$ values are close to each other. Otherwise, it can take on arbitrarily large positive values.

Since it is wanted to minimize the difference between y and $\hat{y}$, the average of the cross-entropy loss function is tried to be minimized in all training instances by adjusting the coefficients. This average over all instances is known as the cost function ($\mathcal{I}$).

$$\mathcal{J} = \frac{1}{m}\sum_{i=0}^{m} \mathcal{L}(y_i, \hat{y}_i) \tag{28}$$

Coefficients are adjusted by using gradient descent with which the value of each coefficient is moved according to the gradient. This approach is also known as the backpropagation algorithm as the errors made in prediction are propagated back to the coefficients. The coefficients are adjusted in a way that the model, potentially, produces results closer to observed classes. The backpropagation algorithm was first introduced by Rumelhart et al. [6] and rapidly become the go-to algorithm for training,

Artificial Neural Networks. It is referred to [7] for further details in the theoretical background of the backpropagation algorithm.

The backpropagation algorithm will update beta parameters iteratively applying value of gradient as given in the following equation:

$$\widehat{\beta}_{new} = \widehat{\beta}_{current} - \eta \frac{\partial \mathcal{J}}{\partial \widehat{\beta}_{current}} \tag{29}$$

In Eq. (29), η is the learning rate that is determined by the user and $\widehat{\beta_i}$ are the estimation of the coefficients.[7] The coefficients are updated until either $\widehat{\beta_i}$ converge to their respective optimal values or maximum number of iterations reached.

In the gradient descent algorithm, the change in the value of each coefficient depends on three values:

The gradient value: If the absolute value of the gradient is greater, a bigger change in coefficient value is seen.

The learning rate (or step-size): It is used as a multiplier to the gradient value. The learning rate is usually chosen between 0 and 1 and most denoted by $\boldsymbol{\alpha}$. The lower the value of alpha the slower the learning will be.

The actual value of the feature x_i

The learning rate is one of the hyperparameters that needs to be specified by the user and has a substantial effect on the performance of the model. If the learning rate is chosen too small, learning process will be too slow and the model will fail to converge to the optimal coefficients. On the other hand, if it is chosen too large, then the model will be very likely to oscillate around the optimum value or diverge from the optimal solution (i.e., goes in the opposite direction).

It must note that, in order to train and use a logistic regression model, one does not need to know the algorithmic details given here. In most ML libraries, training a logistic regression model is one line of code or press of a button given that your training set is ready. However, these details are useful when you get unexpected results from your model. Knowing the inner works of the implementations, one will have easier experience in debugging the model and make it work.

3.4 Model Output

Given training data, a trained logistic regression model outputs the optimal set of coefficients for the classification task. With those parameters, it can predict the probability of belonging to a positive class. As discussed earlier, if the trained model predicts probability of being positive to be greater than the threshold probability, it is

[7] Coefficients are also referred as "parameters" in some sources.

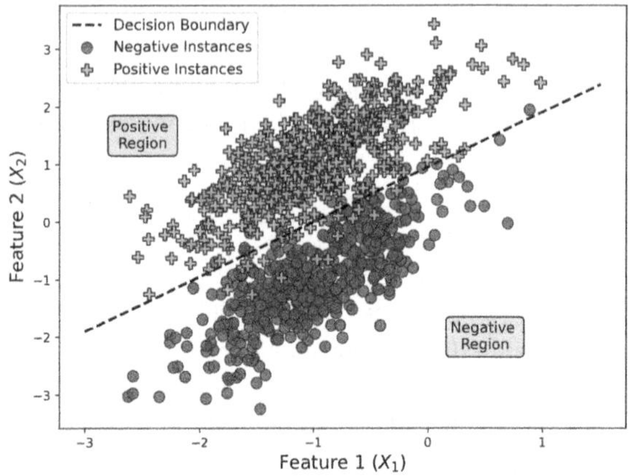

Fig. 6 An example decision boundary obtained from a logistic regression model

classified that instance as positive and vice versa.[8] The threshold for the probability is commonly chosen to be 0.5 but other thresholds could be utilized depending on the application. A given probability threshold corresponds to a line or plane that separates positive instances from negative ones. This separating line is usually referred to as the decision boundary. Although logistic function is a nonlinear function, the resulting decision boundary is linear (Fig. 6) because the features are combined linearly.

This means that the data needs to be linearly separable for logistic regression to perform well. If this is not the case (i.e., data is not linearly separable), several logistic regression learners may need to stack which, in fact, leads to an artificial neural network (ANN) model. For this reason, logistic regression models are sometimes referred to as building blocks of the ANNs [4]. Other algorithms such as tree-based approaches, support vector machines with nonlinear kernels or K-nearest neighbor (K-NN) are also suitable alternatives when a nonlinear decision boundary is needed.

3.5 *Model Evaluation*

There are many metrics that measure the performance of a model on a classification task. Performance evaluation is necessary as it is desired to identify the model with the highest discriminating power. That is, the model should be able to separate the positive instances from the negative ones quite well. The metrics discussed in this section are some of the most commonly used metrics for this task. It is important to note that these metrics are not specific to Logistic regression models and apply to all classification algorithms.

[8] In most ML tools, you do not explicitly need to get the parameters, calculate the probabilities and classify your data. There will be functions that automate this process. For instance, you call *predict(..)* function in scikit-learn.

Before diving into the definitions of these metrics, four possible outcomes from a classification exercise should be identified.

- True Positive: True positive is an outcome that the model correctly classifies an instance as positive when the actual value is also positive.
- True Negative: True negative is an outcome that the model correctly classifies an instance as negative when the actual value is also negative.
- False Positive: False positive is an outcome that the model incorrectly classifies an instance as positive when the actual value is negative. This type of error is also referred as Type-I Error.
- False Negative: False negative is an outcome that the model incorrectly classifies an instance as negative when the actual value is positive. This type of error is also referred as Type-II Error.

For a more concrete example, consider a classification task in which a classification model is predicting if there is a house in the image. An example of each of these four cases in the context of a house picture identification task can be seen in Fig. 4. When the numbers belonging to each category (TP, FP, TN and FN) are summed up and written in the corresponding cells, the table in Fig. 7 is also called a confusion matrix.

Accuracy

The accuracy metric is simply the percentage of correctly labeled instances. It is the default metric for many ML libraries for classification tasks. The accuracy of a given model can be calculated as:

$$\text{Accuracy} = \frac{\text{TP} + \text{TN}}{\text{TP} + \text{TN} + \text{FP} + \text{FN}} \tag{30}$$

It should be noted that changing the class referred as positive cases does not change the accuracy score.

In many cases, a high accuracy score is a sign of a satisfactory model. This is specifically true when there is a balanced dataset in which proportion of each class label is comparable. However, for imbalanced datasets, accuracy score may be misleading.

Consider an extreme case in which it is desired to identify a rare disease seen in 1 case per 100,000 people. Now consider a dummy classifier that classifies everybody as healthy (as opposed to a carrier of the disease). What is the accuracy of this dummy classifier? It is, in fact, very high (≈ 0.999). However, it is not doing the job we want it to do which is catching positive cases for the disease. To overcome this, it can use metrics specific to a class label which will be discussed later.

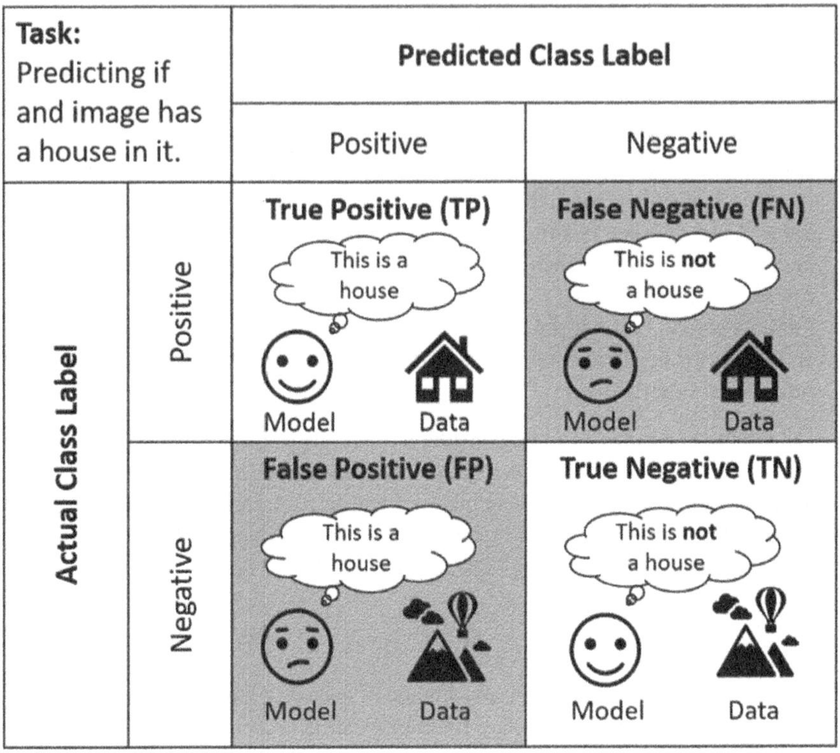

Fig. 7 Four possible outcomes in classification for an image classification task

Precision and Recall

Precision[9] and Recall[10] metrics are two of the widely used class-specific metrics. Formulas for these metrics are as follows:

$$\text{Precision} = \frac{\text{TP}}{\text{TP} + \text{FP}}, \text{Recall} = \frac{\text{TP}}{\text{TP} + \text{FN}} \tag{31}$$

$$\text{Recall} = \frac{\text{TP}}{\text{TP} + \text{FN}} \tag{32}$$

Precision is a measure that reflects the percentage of correctly classified instances when the model predicts as a positive. That is, a high precision means that when the model predicts as positive, it is very likely to be correct.

[9] Sometimes referred as positive predictive value (PPV).

[10] Recall is also referred as "sensitivity" or "true-positive rate (TPR)" in some sources.

Recall measures the percentage of the actual positive cases that are correctly labeled as positive. A high recall rate means that the model catches most of the positive cases.

A careful reader will realize that precision and recall values will be different if he/she changes what is referred as the positive class. Also, these metrics allow us to catch model imperfections in case of an imbalanced dataset. For instance, both precision and recall values will be zero for the dummy classifier for the rare disease when positive cases are identified as the people with the disease.

In some applications, recall is favored over precision and vice versa. For instance, for a first-step medical test, catching as many positive cases as possible can be more desirable so a high recall rate is preferred. This can come at the expense of increasing false positives and in turn, hurts precision value. However, those cases may be eliminated with a more detailed test.

On the other hand, for a spam filter a higher precision would prefer because emails labeled as spam are often not seen and read. If the model is not precise when it says the email is spam then it could cause user to miss an important email (because it goes to the spam filter). However, making a spam email into the inbox is less of a problem since user can quickly delete and mark it as spam.

F1 Score

If the specific application being worked does not warrant to favor precision to recall or vice versa, a balanced score for both is preferred. This is where F1 score comes in. F1 score is one the best alternative metrics to accuracy and is essentially the harmonic mean of the precision and recall scores. F1 score can be calculated by using the following formula:

$$\text{F1 Score} = \frac{2 * \text{Precision} * \text{Recall}}{\text{Precision} + \text{Recall}} \tag{33}$$

ROC-AUC

A Receiver Operating Characteristics (ROC) curve is a graphical representation of discriminative ability of a binary classifier as the decision threshold is varied. An ROC curve is constructed by calculating false-positive rate (FPR) and true-positive rate (TPR, also known as recall) for different decision thresholds.

For logistic regression, it may think these different thresholds as the thresholds for $P(Y = \text{Positive}|x_i)$ value. Remember that most implementations classify an instance as positive if $P(Y = \text{Positive}|x_i) > 0.5$. When building ROC curve, $P(Y = \text{Positive}|x_i)$ is ranged from 0 (for which everything is classified as positive) to 1 (everything is classified as negative). Then FPR and TPR rates are plotted against each other to form the ROC curve. It usually has FPR on x-axis and TPR on y-axis at the coordinate system. An example ROC curve can be seen in Fig. 8.

$$\text{TPR(Recall)} = \frac{\text{TP}}{\text{TP} + \text{FN}} \tag{34}$$

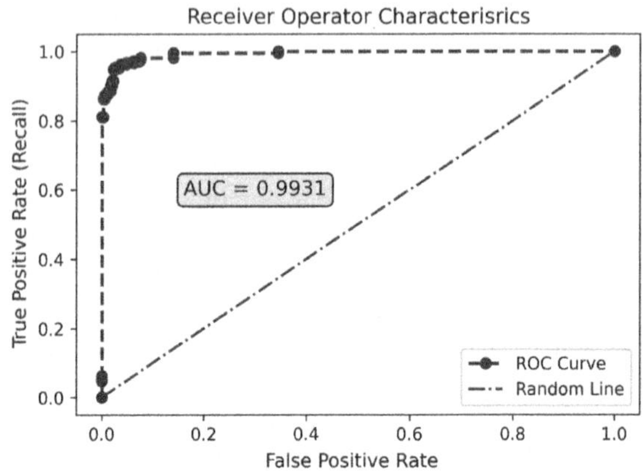

Fig. 8 An example: ROC curve

$$\mathrm{FPR} = (1 - \mathrm{Specivity}) = \frac{\mathrm{FP}}{\mathrm{FP} + \mathrm{TN}} \tag{35}$$

The area under the ROC curve (ROC-AUC) is a good measure of discriminative power of the ML model. Both TPR and FPR have the maximum value of 1. In turn the maximum value of ROC-AUC metric to attain is also 1 (Area of a 1 $\times$ 1 square). Thus, a ROC-AUC value close to 1 means that the model has high discriminative power at various decision thresholds. The worst case is that ROC-AUC value to be close to 0.5 which means the models performance is very similar to a dummy classifier that picks a random class.

4 K-Nearest Neighbor (K-NN)

K-nearest neighbor (K-NN) is an algorithm that is used as a machine learning algorithm in many fields. K-NN algorithm is a supervised learning algorithm used for classification and regression problems. However, it is mainly used for classification problems in the industry.

K-NN algorithm can be implemented in binary (two-class) problems and multi-class problems. K-NN is one of the simplest machine learning algorithms based on supervised learning technique. K-NN algorithm assumes the similarity between the new case/data and available cases and puts the new case into the category that is most like the available categories. K-NN algorithm stores all the available data and classifies a new data point based on the similarity. This means when new data appears then it can be easily classified into a well suited category by using K-NN algorithm. K-NN is a nonparametric algorithm, which means it does not make any assumption on underlying data. It is also called a lazy learner algorithm because it does not learn from the training set immediately instead it stores the dataset and at the time of classification, it performs an action on the dataset. K-NN algorithm at the training phase just stores the dataset and when it gets new data, then it classifies that data into a category that is much like the new data.

4.1 K-NN Algorithm

First a new data point enters to dataset. Then, it is necessary to find this new data which class belongs to. Second, the number of neighbors (K) is going to be determined. Choosing the right value of K is called parameter tuning and it's necessary for better results. By choosing the K value, the square root of the total number of data points available in the dataset is taken. [8].

K value should be chosen as an odd number, because the region with the most neighbors to the new point should be selected, and in case of an even number, an equal number of neighbors can be obtained.

Third, with the Euclidian distance or Euclidian metrics K-nearest neighbors are found. According to new data point belongs to which class points has the highest neighbors. Finally, new data point assigned to that class.

K-NN algorithm steps can be shown as follows.

- Select the number K of the neighbors
- Calculate the Euclidean distance of K number of neighbors
- Take the K-nearest neighbors as per the calculated Euclidean distance.
- Among these k neighbors, count the number of the data points in each category.
- Assign the new data points to that category for which the number of the neighbor is maximum.

4.2 Advantages and Disadvantages of the K-NN Algorithm

Advantages

- Very Simple
- Training is trivial
- Works with any number of classes
- Easy to add more data
- It has few parameters such as K and distance matrix.

Disadvantages

- The computation cost is high because of calculating the distance between the data points for all the training samples.
- Categorical features don't work well.
- Not good with the high dimensional data.

5 Naive Bayes

Naive Bayes is a classification technique that utilizes maximum likelihood. Although Naive Bayes has assumptions that generally do not hold in real life problems it works quite well for complex classification tasks such as document classification and spam classification [9].

Let assume y is the categorical response variable with n classes and $x_1, x_2 \cdots x_k$ are the input variables. The conditional probability of y is the class i can be written as:

$$p(y = i|x_1, x_2 \cdots x_k) = \frac{p(y = i)p(x_1, x_2 \cdots x_k|y = i)}{p(x_1, x_2 \cdots x_k)} \tag{36}$$

Naive Bayes algorithm assumes that predictors $x_1, x_2 \cdots x_k$ are independent of each other. Thus the probability $p(x_1, x_2 \cdots x_k|y = i)$ can be defined as follows:

$$\begin{aligned} p(x_1, x_2 \cdots x_k|y = i) &= p(x_1|y = i) * \ldots * p(x_k|y = i) \\ &= \prod_{j=1}^{k} p(x_j|y = i) \end{aligned} \tag{37}$$

Since the term $p(x_1, x_2 \cdots x_k)$ is the normalization constant and is the same for all classes the term write likelihood function can omit as follows:

$$p(y = i|x_1, x_2 \cdots x_k) = p(y = i)\prod_{j=1}^{k} p(x_j|y = i) \tag{38}$$

Lastly prediction is made by finding the class i that will maximize the above equation.

$$\begin{aligned} \hat{y} &= \underset{i}{\operatorname{argmax}}\, p(y = i|x_1, x_2 \cdots x_k) \\ &= p(y = i)\prod_{j=1}^{k} p(x_j|y = i) \end{aligned} \tag{39}$$

The variations of Naive Bayes are based on different assumptions made on the distribution $p(x_j|y = i)$.

Gaussian Naive Bayes assumes $\boldsymbol{p(x_j|y = i)}$ is distributed normally.

$$p(x_j|y = i) \propto \mathcal{N}(\mu_{ji}, \sigma_{ji}^2) \tag{40}$$

where μ_{ji} and the σ_{ji} are the mean and standard deviation of the observations for the variable j that belongs to class i.

Multinomial Naive Bayes assumes data is distributed multinomially especially used for document classification.

$$p(x_j|y = i) \propto \frac{N_{i,j} + \alpha}{N_i + \alpha n} \tag{41}$$

For the document classification example $N_{i,j}$ is the frequency of feature j that belongs to class i. N_i is the number of all features that belongs the class i. Smoothing parameter α prevents the occurrence of zero probabilities which may result in inaccurate predictions.

Bernoulli Naive Bayes assumes each variable only consists of 0 and 1 values.

$$p(x_j|y = i) \propto p(x_j|y = i)x_j + (1 - p(x_j|y = i))(1 - x_j) \tag{42}$$

6 Support Vector Machine

Support vector machines (SVMs) are a popular machine learning algorithm, due to their superior predictive power and highly flexible modeling techniques. In SVMs, the function between the input and output vectors can be defined as a classification problem used to assign the observations to predefined classes, or as a regression problem used to estimate a continuous value of the desired output [10]. In SVM algorithms, after each data item is specified as a point in n-dimensional space, the value of each feature expresses the value of a particular coordinate. Here, SVM uses new learning techniques that realize structural risk minimization. In this way, the SVM algorithm for classification aims to find a maximum hyperplane to divide the input data representing highly complex nonlinear relationships into a predefined discrete number of classes where it becomes linearly separable. The misclassifications obtained in the training step are minimized thanks to the optimal separating hyperplane, and a maximum margin classifier with the optimal decision limit is found [11]. For this, the learning problem of SVM is considered as a constrained nonlinear optimization problem, and a model with a quadratic cost function and linear constraints is obtained [12].

A decision surface that is maximally distant from any data point forms the boundary of positive and negative hyperplanes. In SVM, which also allows data to be divided into more than two different classes using multiple hyperplanes, the larger the distance or margin between the positive and negative hyperplanes, the greater the predictive power of the classifier is. The possible cases of hyperplanes on a sample space and an example of a linear support vector machine are shown in Fig. 9.

Data points in the training data are given as followed:

$$\langle(\vec{x}_1, y_1), (\vec{x}_2, y_2), \ldots., (\vec{x}_k, y_k)\rangle \tag{43}$$

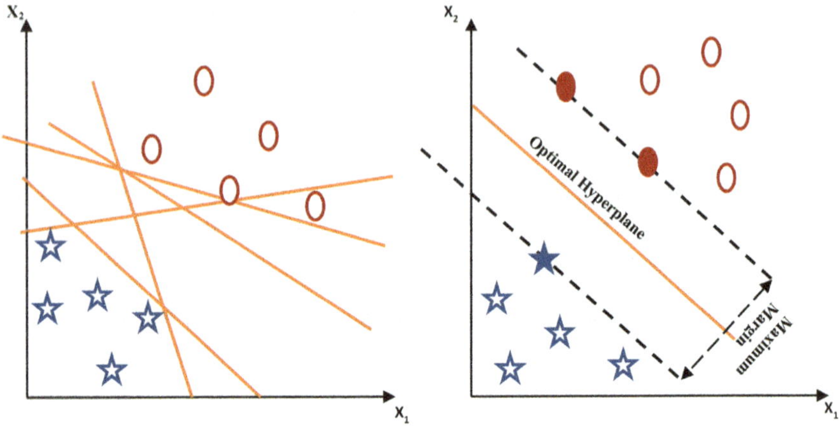

Fig. 9 Possible hyperplanes and an example of a linear support vector machine

where each member is a pair of the *ith* input variable vector $\vec{x}_i$ and a class label y_i corresponding to it. In SVMs, each data point is scaled [0, 1] or [−1, +1] values. In the case of classifying linearly separable data, the one with the largest margin is found among all hyperplanes that minimizes the training error. For this, the bias term b is determined to select among all hyperplanes perpendicular to the normal vector $\vec{w}$. In the absence of an intersection term, a hyperplane that is forced to pass through the origin is encountered and this leads to the limitation of the solution. Correct classification is expressed by means of a dividing hyperplane in mathematical form as given in Eq. (44) [10].

$$\vec{w} \cdot \vec{x} + b = 0 \tag{44}$$

In a given training dataset, the *ith* input vector $\vec{x}_i$ which lie on the hyperplane satisfy $\vec{w} \cdot \vec{x_i} + b = 0$, where $\vec{w}$ is the normal vector of hyperplane, b is a constant term called bias, $\frac{|b|}{\|\vec{w}\|}$ is the perpendicular distance from the hyperplane to the origin, and $\|\vec{w}\|$ is the Euclidean norm of $\vec{w}$.

Since the dividing line is the midline between two support lines, the parallel hyperplanes expressing these two support lines can be defined by Eq. (45) [13].

$$\hat{y}_i = \begin{cases} -1, & (\vec{w} \cdot \vec{x}_i + b) \leq -1 \\ 1, & (\vec{w} \cdot \vec{x}_i + b) \geq +1 \end{cases} \tag{45}$$

This can be rewritten by combining into one set as given in Eq. (46).

$$y_i(\vec{w} \cdot \vec{x}_i + b) \geq 1 \tag{46}$$

Parallel hyperplanes are determined with no points between them in case of linear separation of the training data [14]. In order to increase the confidence in the prediction results to be obtained, the maximum distance between the decision boundary and the data points on both sides of the line should be considered. The surface area with the maximum distance is called the margin so that an optimal classifier is obtained by maximizing the margin size. Hence, the margin value is measured by distance between the vectors $\overrightarrow{x}_+$ (vector of positive x data examples) and $\overrightarrow{x}_-$ (vector of negative x data examples) the closest to each other as shown in Fig. 10. The vector $\overrightarrow{x}_+$ is on the positive hyperplane with normal vector $\overrightarrow{w}$ and perpendicular distance from the origin $\frac{(1-b)}{\|\overrightarrow{w}\|}$. The vector $\overrightarrow{x}_-$ is on the negative hyperplane with normal vector $\overrightarrow{w}$ and perpendicular distance from the origin $\frac{(-1-b)}{\|\overrightarrow{w}\|}$. If the distance between the hyperplanes is denoted by d, the margin is calculated geometrically by Eqs. (46) and (47) [13].

$$d = \left(\overrightarrow{x}_+ - \overrightarrow{x}_-\right) \tag{47}$$

$$d = \frac{(1-b) - (-1-b)}{\|\overrightarrow{w}\|} = \frac{2}{\|\overrightarrow{w}\|} \tag{48}$$

Since the objective is to maximize the margin, the expression $\frac{2}{\|w\|}$ needs to be maximized. However, such an optimization problem with $y_i\left(\overrightarrow{w} \cdot \overrightarrow{x}_i + b\right) \geq 1$ constraints are difficult to solve. In this situation, realizing the fraction maximization carries the

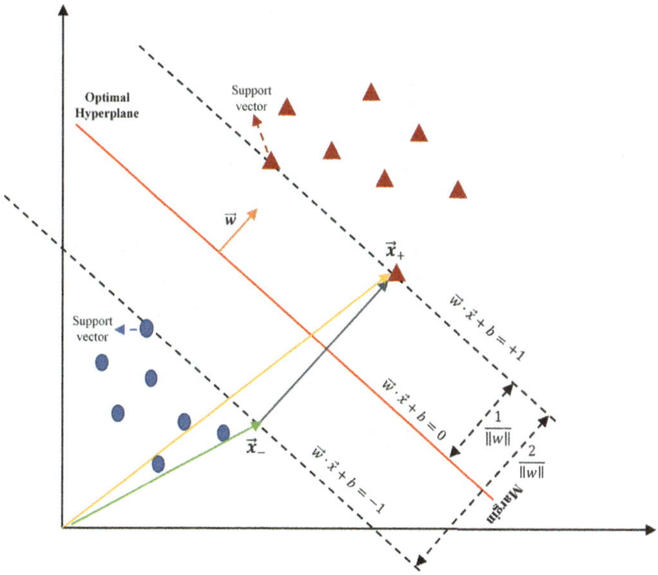

Fig. 10 Linearly separable SVM

same expression as minimizing the denominator. By focusing on minimizing $\| \vec{w} \|$, the optimization problem is transformed into a quadratic programming problem that is relatively simple to solve [15].

$$\text{minimize } \frac{1}{2} \| \vec{w} \|^2 \tag{49}$$

$$\begin{gathered}\text{Subject to} \\ y_i \cdot (\vec{w} \cdot \vec{x}_i + b) \geq 1, \quad i = 1, 2, \ldots, k\end{gathered} \tag{50}$$

This quadratic optimization problem is solved by transforming a Lagrange factor α_i into a dual problem by relating it to each constraint in the primal problem. The constraint, which is difficult to solve in the optimization problem given in above, becomes easier by changing with Lagrange multipliers. In this way, the constrained optimization problem is made unconstrained. The Lagrange function L_p used in the solution of the problem is as given in Eq. (51) [13].

$$L_p = \frac{1}{2} \| \vec{w} \|^2 - \sum_{i=1}^{k} \alpha_i \left[y_i \cdot \left(\vec{w} \cdot \vec{x}_i + b \right) - 1 \right] \tag{51}$$

where α_i are non-negative Lagrange multipliers. L_p is a function that minimizes the weight vector $\vec{w}$ and the constant b and maximizes the non-negative dual variable α_i. By taking the partial derivatives of the Lagrange function with respect to w and b, the Karush Kuhn–Tucker conditions in Eqs. (52) and (53) are obtained [13].

$$\frac{dL_p}{d\vec{w}} = y - \sum_{i=1}^{k} \alpha_i y_i x_i = 0 \rightarrow \vec{w} = \sum_{i}^{k} \alpha_i y_i x_i \tag{52}$$

$$\frac{dL_p}{db} = \sum_i \alpha_i y_i = 0 \tag{53}$$

The obtained equations are put in the relevant places in the Lagrange function, and the problem turns into a maximization type dual Lagrange problem. Writing the optimization problem in dual form reveals that the classification is only a function of the support vectors [10].

$$\text{maximize } \sum_{i=1}^{k} \alpha_i - \sum_{i,j} \alpha_i \alpha_j c_i c_j \vec{x}_i^{\,T} \vec{x}_j \tag{54}$$

$$\begin{gathered}\text{Subject to} \\ \textstyle\sum_i \alpha_i y_i = 0 \quad i = 1, 2, \ldots, k\end{gathered} \tag{55}$$

$$\alpha_i \geq 0 \quad i = 1, 2, \ldots., k \tag{56}$$

where, i and j represent the observations in the training dataset. The α_i values are used to calculate the vector $\overrightarrow{w}$ in terms of the training set.

$$\overrightarrow{w} = \sum_i \alpha_i y_i \overrightarrow{x}_i \tag{57}$$

After the optimization problem is solved, the training points located on the positive and negative hyperplanes parallel to the decision boundary create the vectors corresponding to the nonzero α's and provide all the necessary support in finding the optimal values for $\overrightarrow{w}$. These points are called support vector and the generated algorithm is called support vector machines [10].

6.1 *Soft Margin Classifier*

The learning procedure presented above is valid for cases where the data can be separated linearly, but in general the data are not linearly separated, that is, completely. The quadratic programming presented above cannot be used in case of data whose solutions are not linearly separable, since constraints $y_i \cdot \left(\overrightarrow{w} \cdot \overrightarrow{x}_i + b\right) \geq 1$ cannot be met. As shown in Fig. 11, loose variables ξ_i are added to the SVM objective function to allow the hard decision margin to make a few errors if the data is misclassified. In this case, SVM is not looking for the hard margin that would classify all data perfectly. Instead, SVM allows most of the data to be classified correctly while allowing a few data to be misclassified around the decision boundary, and SVM is

Fig. 11 Soft margin SVM

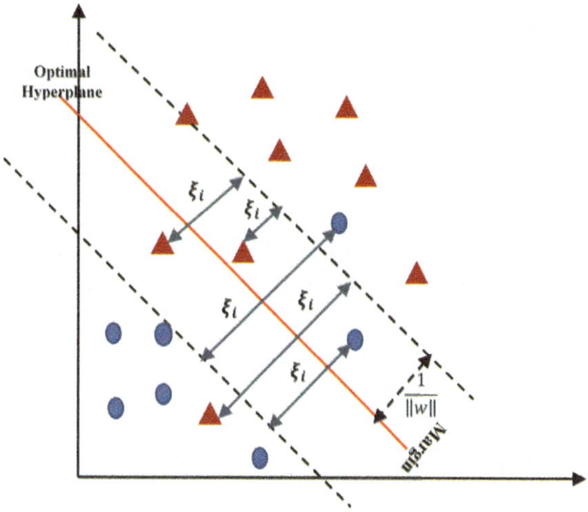

now characterized as a soft margin classifier [14]. The soft margin SVM optimization problem with slack variables is given as follows.

$$\text{Minimize} \quad \frac{1}{2}\| \vec{w} \|^2 + C\sum_{i=1}^{k} \xi_i \tag{58}$$

$$\text{Subject to} \\ y_i \cdot (\vec{w} \cdot \vec{x}_i + b) \geq 1 - \xi_i \quad i = 1, 2, \ldots, k \tag{59}$$

$$\xi_i \geq 0 \quad i = 1, 2, ..., k \tag{60}$$

By increasing the objective function by a function that penalizes nonzero ξ_i, the optimization problem becomes a trade-off between a large margin and a small error penalty. For $\xi_i > 0$ and a point $\vec{x}_i$ the margin may be less than 1, but then the penalty C is paid for this in minimization. The regulation term C, which qualifies as a penalty parameter, provides control of how the SVM will handle the errors. The larger the parameter C, which varies depending on the optimization goal, the tighter the margin is achieved, and thus more emphasis is placed on minimizing the number of errors resulting from misclassification. As C decreases, it is aimed to maximize the margin between the two classes, resulting in more errors [15]. The dual problem for soft margin SVM is given as follows.

$$\text{maximize} \sum_{i=1}^{k} \alpha_i - \frac{1}{2}\sum_{i,j} \alpha_i \alpha_j c_i c_j \vec{x}_i^{\,T} \vec{x}_j \tag{61}$$

$$\text{Subject to} \\ \sum_i \alpha_i y_i = 0 \quad i = 1, 2, \ldots, k \tag{62}$$

$$0 \leq \alpha_i \leq Ci = 1, 2, ..., k \tag{63}$$

6.2 *Kernel Trick for Nonlinear Data*

The SVM linear classifier is based on a dot product between vectors of training data points. In the optimization problem given in Eq. (10), the training data is in the form of the scalar products $\vec{x}_i^{\,T} \vec{x}_j$. Each data point is moved to a higher-dimensional space through the$\phi : \vec{x} \rightarrow \phi(\vec{x})$ transform to make finding higher-dimensional relationships easier. In this case the dot products of the training data will be replaced by scalar products $\phi(\vec{x}_i)^T \phi(\vec{x}_j)$ in a higher-dimensional space. Then, it will be expressed using a kernel function $K(\vec{x}_i, \vec{x}_j) = \phi(\vec{x}_i)^T \phi(\vec{x}_j)$ to provide a simpler calculation [3].

When a problem is not linearly separable in the input space, kernel tricks aim to transform the original observations into a higher-dimensional space, called the kernel space, where it can be linearly separable using nonlinear functions. For this, using the Mercer method, any continuous, symmetric and positive semi-definite kernel function $K(\vec{x}, \vec{y})$ are expressed as a dot product in a high-dimensional space, such that

$$\iint K(\vec{x}, \vec{y}) f(\vec{x}) f(\vec{y}) d\vec{x} d\vec{y} \geq 0 \quad \text{for all } f \tag{64}$$

$$\int f^2(\vec{x}) d\vec{x} < +\infty \tag{65}$$

There are many possible kernels that meet Mercer's requirements, and the most popular ones are given as follows [10].

Linear Kernel:

$$K(\vec{x}, \vec{y}) = \vec{x}^T \cdot \vec{y} \tag{66}$$

Polynomial Kernel:

$$K(\vec{x}, \vec{y}) = \left(1 + \vec{x}^T \cdot \vec{y}\right)^d d > 0 \tag{67}$$

Gaussian radial basis function:

$$K(\vec{x}, \vec{y}) = e^{\left(-\frac{(\vec{x}-\vec{y})^2}{2\sigma^2}\right)} \tag{68}$$

Sigmoid Kernel:

$$K(\vec{x}, \vec{y}) = \tanh\left(\alpha \vec{x} \cdot y + \beta\right) \tag{69}$$

The kernel trick, which relies only on the dot product between two vectors, converts a linear algorithm to a nonlinear algorithm using any algorithm, which is equivalent to a linear algorithm operating in the w interval space. Because kernels are used, high-dimensional space can be infinitely dimensional, and so the vector $\vec{w}$ can never be calculated explicitly [16].

An Example: Airline Customer Satisfaction

Airline A, which has a large customer pool, appeals to customers from large masses. Since the airline's economy is largely dependent on customer satisfaction, the company wishes to identify the conditions and measures it can provide to increase the satisfaction of customers. For this, some actionable information has been provided

Table 1 Performance metrics of SVM models

Model type		Confusion matrix		Accuracy (%)	Recall score (%)	Precision score (%)	F1 score (%)	AUC
		T	F					
SVM (linear)	**T**	14,544	3010	83.4	83.9	85.6	84.7	0.9
	F	3453	17,957					
SVM (radial)	**T**	16,348	1206	94.1	94.9	94.4	94.6	0.99
	F	1092	20,318					
SVM (poly.)	**T**	15,909	1645	92.2	93.5	92.4	92.9	0.98
	F	1397	20,013					
SVM (sigmoid)	**T**	12,248	5306	72.2	74.2	75	74.6	0.78
	F	5531	15,879					

from the airline to help increase customer satisfaction. The data have been obtained through surveys that will enable the airline company to follow the satisfaction of customers on each flight. Customer satisfaction can vary depending on many variables that a customer and airline experience during travel. Therefore, the customer survey, which evaluates different aspects of airline service, consists of a combination of customer-specific information, such as age, and information provided by the airline, such as flight distance. In the study, it is aimed to predict whether a future customer will be satisfied with the service by considering the data provided through the variables included in the survey. The prediction model is developed using support vector machines technique. In this way, it is tried to help the airline company in which aspects of its services should be emphasized more in order to increase the satisfaction of its customers.

The dataset includes 129,880 responses received through surveys conducted by the airline on its customers over 3 months. In the dataset to be used to estimate customer satisfaction, there are categories that measure the relative satisfaction of customers with services such as seat comfort, leg room, inflight entertainment and cleanliness, as well as personal information of customers such as age, gender, customer type. Hence, the dataset contains a mix of both numeric and non-numeric features [17]. A summary of the performances of the SVM method based on accuracy, recall score, precision score and F1 score as well as confusion matrices is presented in Table 1. The results show that the SVM method classifies airline customer satisfaction well, with a high accuracy score.

7 Decision Trees

A decision tree is a type of diagram that uses nodes to represent the variables and branches springing from these nodes depend on possible values for the variables [18]. In other words, a decision tree is an illustration of decision-making by representing it

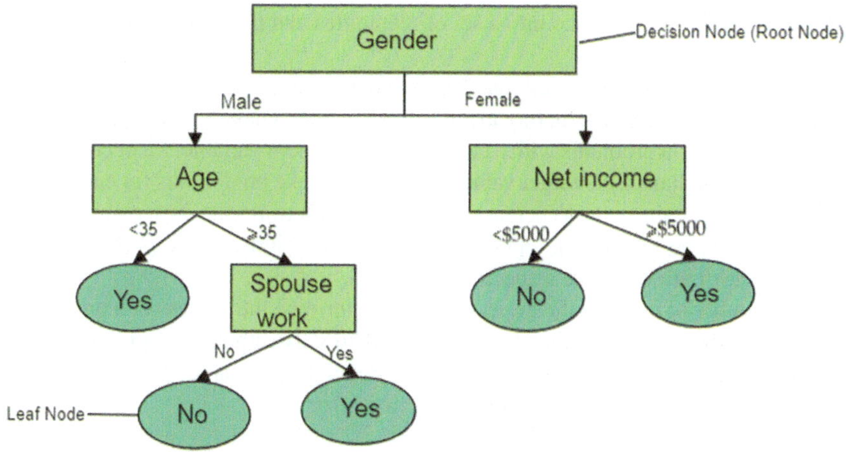

Fig. 12 A simple representation of a decision tree (all green boxes are decision nodes, and blue ovals are leaf nodes)

as a branching path from one or more possible options to multiple possible branches. Decision trees can be used for both classification and regression. Figure 12 illustrates a decision tree used for classification.

A decision node is a node in a decision tree at which the flow branches into multiple alternative flows. Terminal nodes, also known as leaf nodes, represent the final subsets. Each internal node in a decision tree is assigned a discrete function of the values of one or more attributes and then divides the feature space into two or more subspaces based on that function (Fig. 13 [19]). Tree-based methods involve

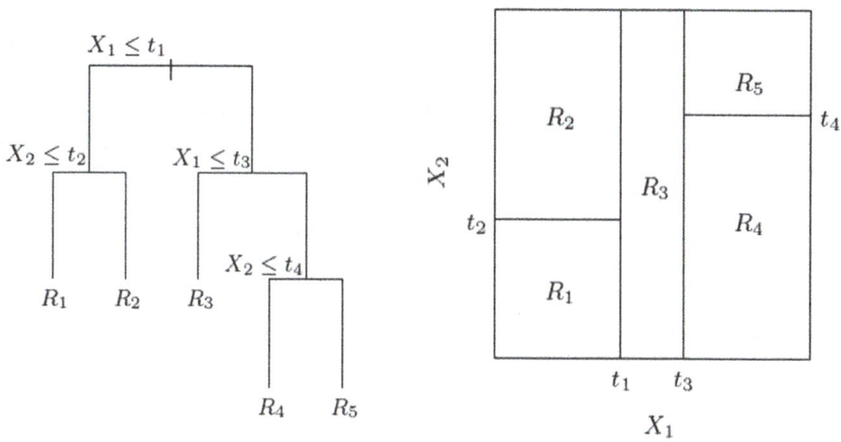

Fig. 13 Left: the decision tree, right: partition of feature space

segmenting the feature space into a set of rectangles and then predict a constant for each region.

Note that target values are assigned to each leaf node. Regression or classification models are built as a tree structure using a decision tree. Classification trees are used to solve prediction problems with a categorical dependent variable. Most commonly, the mean or median of the target values are used to make predictions for each region in the setting of regression. Whereas in classification settings the most common class is returned as the prediction. Optionally, the leaf may indicate the probability that the target attribute will have a specific value [20].

The number of misclassifications is used to determine the success of a prediction. Regression trees, on the other hand, are used to solve prediction problems with continuous or ordered discrete values as the dependent variable. The MSE (Mean Square Error) is used to measure the prediction error [21].

7.1 Mathematical Formulation of Decision Trees

The mathematical formulation of decision trees is as follows [5]. Given $x_i \in \mathbb{R}^n$, $i = 1, 2, \ldots, l$ as the training vector and $y \in \mathbb{R}^l$ as the label vector. A decision tree performs recursive partitioning of the feature space, grouping samples with similar labels or target values. Let node m represent data with N_m samples using the Q_m graphical representation. Split the data into subsets by constructing an arbitrary feature and a threshold set for each candidate split value of $\theta = (j, t_m)$, resulting in sets of the form $Q_m^{\text{left}}(\theta)$ (θ) and $Q_m^{\text{right}}(\theta)$.

$$Q_m^{\text{left}}(\theta) = \{(x, y) | x_j \leq t_m\} \tag{70}$$

$$Q_m^{\text{right}}(\theta) = Q_m \backslash Q_m^{left}(\theta) \tag{71}$$

An impurity function or loss function $H()$ is then used to determine the quality of a candidate split of node m. The particular impurity function or loss function used depends on the task to be solved (classification or regression).

$$G(Q_m, \theta) = \frac{N_m^{\text{left}}}{N_m} H\left(Q_m^{\text{left}}(\theta)\right) + \frac{N_m^{\text{right}}}{N_m} H\left(Q_m^{\text{right}}(\theta)\right) \tag{72}$$

The next step is to minimize the impurity by selecting the parameters.

$$\theta^* = \text{argmin}_\theta G(Q_m, \theta) \tag{73}$$

Continue searching for subsets in $Q_m^{\text{left}}(\theta))$ and $Q_m^{\text{right}}(\theta)$ until the maximum allowable depth is reached, and if necessary, search at a maximum depth of $N_m < min_{samples}$ or at a depth of $N_m = 1$..

Classification Criteria: Suppose there is a classification outcome taking on values $0, 1, \ldots, \boldsymbol{K} - 1$, for node m. The probability of this classification outcome can be calculated as follows: $\boldsymbol{p_{mk}} = \frac{1}{N_m}\sum_{y\in Q_m}\boldsymbol{I}(\boldsymbol{y} = \boldsymbol{k})$ is the proportion of class k observations in node m. Prediction probability for this region is set to $\boldsymbol{p_{mk}}$ if $\boldsymbol{m}$ is a terminal node. Generally, impurities are measured using the following criteria:

$$H(Q_m) = \sum_k p_{mk}(1 - p_{mk}) - \text{Gini} \tag{74}$$

$$H(Q_m) = -\sum_k p_{mk}\log(p_{mk}) - \text{Entropy} \tag{75}$$

$$H(Q_m) = \sum_k 1 - \max(p_{mk}) - \text{Misclassification} \tag{76}$$

Regression Criteria: If the target is a continuous value, then the mean squared error (MSE or L2 error), Poisson deviance and mean absolute error (MAE or L1 error) are commonly used to determine future split locations. To estimate the mean of terminal nodes, the MSE and Poisson deviance utilize the mean value $\overline{\boldsymbol{y}}_{\boldsymbol{m}}$ of the node whereas the MAE utilizes the median.

$$\overline{y}_m = \frac{1}{N_m}\sum_{y\in Q_m} y \tag{77}$$

$$H(Q_m) = \frac{1}{N_m}\sum_{y\in Q_m}(y - \overline{y}_m)^2 - \text{Mean Squared Error} \tag{78}$$

$$H(Q_m) = \frac{1}{N_m}\sum_{y\in Q_m}\left(y\log\frac{y}{\overline{y}_m} - y + \overline{y}_m\right) - \text{Half poisson deviance} \tag{79}$$

$$\text{median}(y)_m = \underbrace{\text{median}(y)}_{y\in Q_m} \tag{80}$$

$$H(Q_m) = \frac{1}{N_m}\sum_{y\in Q_m}\left|y - \text{median}(y)_m\right| - \text{Mean absolute error} \tag{81}$$

Before the most popular decision tree algorithms are discussed, it is necessary to understand the fundamental components of decision trees.

Splitting: Splitting means dividing a node into several sub-nodes so that relatively pure nodes are created. This procedure is repeated until a predetermined level of homogeneity or stopping criteria is achieved. The splitting rules are also known as attribute selection rules, as they define how tuples should be split at a given node. The most popular attribute selection approaches are: (I) reduction in variance, (II) Gini index, (III) information gain and (IV) chi-square. When the target variable is continuous, or when there is a regression problem, the reduction in variance approach is used. In contrast, when the target variable is categorical, i.e., when there is a classification problem, other approaches are used. Another critical point is that the modeler should define a split point for categorical features prior to beginning splitting. The most frequently used method for determining the split point is to sort the values and then evaluate the midpoints between adjacent values using some metric, typically Information gain or Gini impurity. The following are the most frequently used attribute selection criteria:

Information Gain: The amount of information contained in each attribute is tried to be estimated by using the information gain criterion. Entropy is a quantity that measures randomness and uncertainty of a random variable.

$$E = -\sum_{i=1}^{n} p_i \log_2(p_i) \tag{82}$$

where p_i denotes the probability that an object will be classified into a specific class and n denotes the total number of classes. The following formula is used to calculate the Information Gain for an attribute X:

$$IG(T, X) = E(T) - E(T, X) \tag{83}$$

where T represents the target variable.

GINI Index is a metric that indicates how often a randomly selected element will be incorrectly identified. This implies that you should favor attributes with a lower Gini index. It is calculated as follows:

$$GI = \sum_{i=0}^{n} p_i(1 - p_i) \tag{84}$$

where p_i denotes the probability that an object will be classified into a specific class and n denotes the total number of classes.

Stopping: When the decision tree's size and complexity increase, the generalizability and robustness of the model will diminish. In other words, the larger and more complex the decision tree, the less generalizable and robust the resulting model will be. To avoid this, the decision tree building process should employ stopping rules to prevent it from becoming too complex. Stopping rules could be determined using common parameters such as minimum record counts in leafs, minimum node counts prior to splitting and leaf depth (steps) from the root node.

Pruning: Sometimes stopping rules are inefficient. Instead of growing a complex decision tree, one may choose to prune a large one to get it to a size that is optimal. Pruning is classified into two types: pre-pruning and post-pruning. Pre-pruning uses Chi-square or multiple-comparison tests to identify non-significant branches. After creating a full decision tree, a post-pruning procedure is used to remove branches that help increase overall classification accuracy. Pruning can be accomplished in a variety of ways. Two common ways are as follows:

1. Information gain: The most straightforward approach is to remove branches that result in the least amount of information gain.
2. Classification/Prediction performance on validation set: Another option is to prune the tree to obtain the highest validation set classification performance.

A tree can be grown using several different algorithms. These algorithms can differ on several fronts, such as how many splits exist per node, how they are found and when to stop splitting. The most frequently used decision tree algorithms are summarized in Table 2 by the study [22].

The most used tree induction approach is undoubtedly the classification and regression trees (CART) algorithm. The CART algorithm is described as follows:

7.2 Classification and Regression Trees (CART) Algorithm

CART is a classification algorithm based on Gini's impurity index as the criterion for splitting the data into smaller pieces. The CART is a binary tree structure in which node division occurs iteratively [7]. This algorithm has three steps:

1. Determine the optimal split for each feature. There are N-1 possible splits for each feature having N different values. To maximize the splitting criterion, find the split that maximizes it.
2. Determine the optimal split for the node. Choose the split that maximizes the criteria for splitting, as indicated by Step 1.
3. Start over from Step 1 and continue splitting nodes until the stopping condition is met.

7.3 The Advantages and Disadvantages of Decision Trees

Decision trees are widely used in the area of machine learning because of their success as well as the fact that they are straightforward. The following are some of the advantages of decision trees:

1. It is simple to understand and interpret.
2. It is capable of dealing with both numeric and categorical data.
3. It does not require a lot of processing like normalization or encoding.

On the other hand, it comes with the following drawbacks:

1. It has a relatively high spatial and temporal complexity.
2. Decision tree model reproducibility is extremely sensitive to changes in the data, and even a small alteration in the structure of the data can produce a significant change in the tree structure.

Table 2 Classification of popular decision tree algorithms [22]

Algorithm	CART	C4.5	CHAID	QUEST
Attribute selection measure	Gini index, towing criteria	Information gain	Chi-square	Chi-square for categorical variables; J-way ANOVA for continuous/ordinal variables
Target variable	Categorical/continuous	Categorical/continuous	Categorical	Categorical
Input variables	Categorical/continuous	Categorical/continuous	Categorical/continuous	Categorical/continuous
Pruning approach	Pre-prunning	Pre-prunning	Pre-prunning	Post-prunning
Split type at each node	Binary	Multiple	Multiple	Binary

Table 3 Hyperparameters of decision trees

Parameter	Explanation
Max_depth	The overall complexity of decision tree is controlled by this parameter
Min_impurity_split	Tree growth stopping threshold
Min_samples_leaf	Minimum number of leaf node samples required
Min_impurity_decrease	If a split decreases the impurity value more than or equal to this value, a node will be split
Max_leaf_nodes	In relative terms, best nodes are nodes where impurity is less
Max_features	The number of features to consider when seeking the best split
Min_samples_split	The smallest sample size required to split an internal node

3. Overfitting is a common problem with decision trees. it is necessary to carefully tune the hyperparameters of decision trees to overcome this problem. Table 3 presents the hyperparameters of the decision tree models. These parameters are set in the Scikit-Learn classifier model that implements the decision tree algorithm [23]. More information about the parameters can be found in the Scikit-Learn library, which is available to readers.

An Example: Predicting Customer Churn Through the Use of Decision Trees

In this section, a decision tree application to a classification problem is demonstrated. The customer data from a telecom has been used to derive customer churn using a decision tree. Table 4 contains a summary of the characteristics of customer churn data. There are 18 independent variables, some of which pertain to the customer's characteristics and the majority of which pertain to the company's services to the customer. Customer churn decisions are the target variable. Specifically, it is necessary to learn which features are important in a customer's churn decision, as well as develop a good predictive model that can accurately predict a customer's churn

Table 4 Summary of the features

Feature no.	Feature ID	Type	Feature no.	Feature ID	Type
1	Gender	Binary	11	Device protection	Binary
2	Senior citizen	Binary	12	Tech support	Categorical
3	Partner	Binary	13	Streaming TV	Categorical
4	Dependents	Binary	14	Streaming movies	Categorical
5	Tenure	Discrete	15	Contract	Categorical
6	Phone service	Binary	16	Paperless billing	Binary
7	Multiple lines	Categorical	16	Payment method	Categorical
8	Internet service	Categorical	17	Monthly charges	Continuous
9	Online security	Categorical	18	Total charges	Continuous
10	Online backup	Categorical	19	Churn	Binary

decision. For these goals, decision tree implementation provided by Scikit-Learn library in Python is used [23]. The details of the algorithm's implementation can be found in the supplementary code library. 80% of the data samples are used for training and the remaining 20% for testing. The findings indicate that Contract and Tenure are the first two most significant variables influencing a customer's churn decision. 80% prediction accuracy on test data is obtained. Additionally, decision rules for predicting a customer's churn decision are obtained such as the following:

> if (Contract_Month-to-month < = 0.5) and (Monthly Charges < = 93.675) and (Online Security_No < = 0.5) and (Contract_Two year < = 0.5) and (Monthly Charges < = 72.4) and (Total Charges > 41.8) then class: 0 (proba: 97.27%) | based on 440 samples.

8 Ensemble Learning

Ensemble learning is a machine learning paradigm where the outputs of multiple learning models (weak learners) are combined to train and produce better results [24]. Ensemble learning is also called committee-based learning or learning multiple classifier systems [25]. Ensemble methods use a set of classifiers to make a better prediction so that an ensemble's ability to generalize is often much stronger and has better overall accuracy than any individual ensemble member. Although predictive modeling is dominant in the use of ensembles, it can also be used for other analytical modeling capabilities such as clustering and association rule mining. That is, ensemble models can take on both supervised and unsupervised machine learning tasks [10].

One of the most important features of ensemble methods is that it elevates weak learners (or base learners) to learners with strong generalization skills [26]. Weak learners or base learners may have a high bias (such as low degree of freedom models) or have too many variances to be robust (such as high degree of freedom models). This is why these weak learners cannot perform very well on their own. To decrease the variance or bias in such weak learners, ensemble methods create a strong learner (or ensemble model) which enables better performances [27]. Therefore, researchers and practitioners have generally preferred to use ensemble learning models to obtain better accuracy and more stable/robust/consistent/reliable results. Many studies conducted by researchers in recent years have proven that ensemble learning models almost always improve prediction accuracy and rarely predict worse than single models for any given problem [10].

In 1979, in the study on ensemble learning systems of Dasarathy and Sheela [28], the issue of segmentation of the feature space using multiple classifiers was discussed. Then, ensemble learning methods, which have become a major learning paradigm since the 1990s, have achieved significant development with the contributions of two pioneering studies. One of them is the study conducted by Hansen and Salamon [29] in 1990 showing that a set of structured neural networks can be used to develop classification performance. The other was the study conducted by Schapire [30], which proved that by combining weak classifiers through boosting, a properly strong classifier could be produced. Later, this study has provided the basis for the development of AdaBoost algorithms. After these pioneering studies, ensemble learning studies have expanded rapidly by many researchers and took place in the literature [31–33]. Ensemble methods, whose popularity has increased since the 2000s, which started to be seen in the 1990s, have become a necessary method to win the modeling competitions based on data mining and analytics. KDD-Cup, one of the most famous data mining competitions, has been held annually since 1997 and has attracted great interest from data mining teams around the world. The competition includes many problems in every field, from applications in the field of medicine such as pulmonary embolism detection to application areas in the service sector such as customer relationship management. Among the various techniques used in solving the problems considered in KDD-Cup competitions, ensemble learning models have been the most remarkable methods. It has also been included as the method used in the winning team of the competition many times [25]. Another famous data mining competition is the Netflix Prize, which aims to increase the accuracy of the predictions about the user preferences of movies, that is, how much they will enjoy a movie. In the $1 million prize, the winner, second and nearly all of the other teams that placed at the top of the ranking have offered a solution based on ensemble learning models. As a result, the winning team has presented an ensemble learning model that is a combination of various predictive models. Moreover, the team, which achieved the desired performance but lost because it presented the result 20 min later, has used its name as the Ensemble [10, 25]. In addition to impressive results in competitions, ensemble learning methods have achieved great success by applying in a variety of real-world tasks such as object detection, recognition, tracking, recommendation systems, medical diagnostics, text categorization [25].

The diversity of classifiers that make up the ensemble indicates the success of an ensemble learning model, in other words, its ability to correct the errors of some elements in the model. Individual classifiers in an ensemble system must obtain different errors in different situations since it will not be possible to correct a potential error if the same output is obtained from all the classifiers. The variety of classifiers to include in the model can be obtained in different ways. One way is to train individual classifiers, such as using different training datasets. Such datasets are obtained through resampling techniques such as random bootstrapping or bagging, in which subsets of training data are obtained from all training data, usually by replacement. In the ensemble models, a single-based learning algorithm is used to ensure that homogeneous weak learners are trained in different ways, thus making the model "homogeneous".

Using different training parameters for different classifiers is another approach that can be used to achieve diversity. Another factor that creates diversity is to ensure the use of models that are different from each other and look at the data from a different perspective. Because of combining the results of different types of weak learning algorithms, they are called heterogeneous ensemble models. The completely different types of classifiers such as multilayer artificial neural networks, support vector machines and decision trees are used in such models. A final factor of achieving diversity is to use different features or different subsets of existing features [10, 27,

34]. In general, ensemble methods make up a broad class of algorithms based on sound learning theoretical principles. The most attention algorithms in the literature have been considered here.

8.1 Bagging

The bagging method, one of the oldest and simplest but most effective ensemble learning algorithms, was first proposed by Breiman [32] in 1996 as an abbreviation of bootstrap aggregation. In the bagging method, instead of sampling a new, independent training dataset each time, different subsets of training data are randomly generated by replacing the entire training dataset. Diversity in bagging is provided by using bootstrapped copies of the training data so that a subset of training data is used to train a different base learner in the same type [24]. The sub-datasets created by using resampling are different from each other, but they are not independent as they are all derived by a single dataset. The bagging method performs significantly better than the outputs obtained by a single model created using the original training data but creates a combined model that is never significantly worse [26]. The bagging method can be used for classification and regression models. The results from each of the models considered in classification-type prediction problems are combined by using the simple or weighted majority voting technique, and then the ensemble prediction for that sample or record is determined by the class label with the highest votes. In regression type prediction models where the target variable is a number/continuous variable, each output of the considered models is combined using the simple or weighted average technique [10]. A graphical representation of the bagging algorithm is given in Fig. 14.

Since the basic learner combination strategy used in the bagging method enables the predictions to behave more decisively on new data, the bagging method is a technique that provides a lower variance. The bagging method is more attractive especially for available data with limited size. In order to have enough training samples in each subset of data, relatively large portions of the samples considered are drawn into each data subset. This results in significant overlapping of individual training data subsets, that is, sharing the same data. Thus, in bagging, like many ensemble methods, a variation on training data is achieved by using a relatively unstable base learner that produces different generalization behavior allowing different decision limits to be obtained against small perturbations in different training datasets [24]. The point to be noted here is that some learning methods, such as decision stumps and K-nearest neighbors, are insensitive to perturbations in training samples because they have a very stable structure. Since the basic learners formed by these methods will be similar to each other, combining them will not improve the generalization ability. Therefore, bagging should be used with methods such as decision trees with unstable learners [26]. In this way, the trained classifiers are then combined using a voting technique, resulting in a more stable structure with lower variance [24]. As a result, the higher the instability of the base learners, the greater the performance

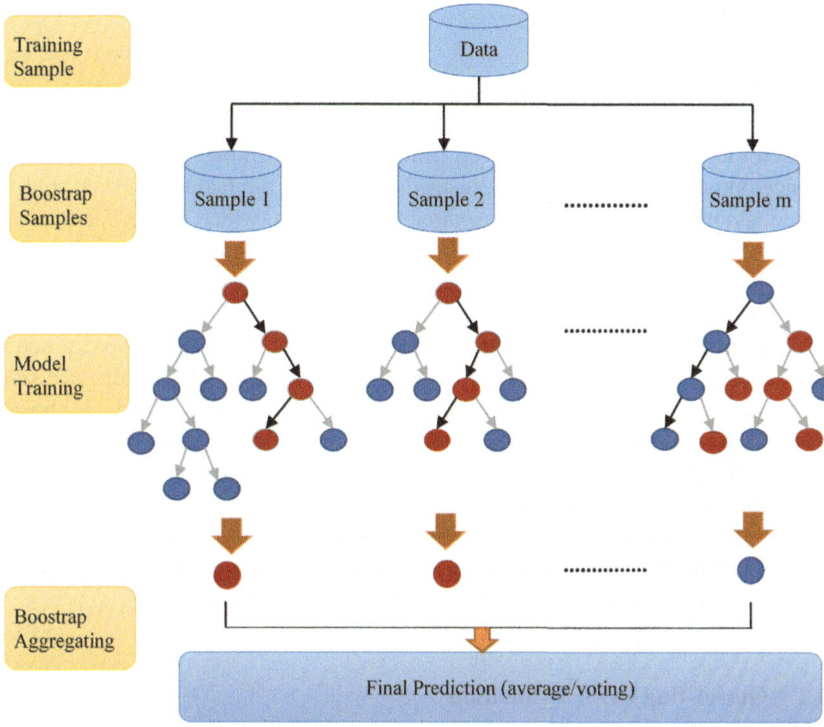

Fig. 14 Bagging method based on decision tree ensembles

improvement [26]. For a classification problem consisting of m instances, where each x_i instance in the training set is labeled as $y_i \in \{-1, +1\}$, the bagging algorithm is defined as follows.

Input: Dataset $B = \{(x_1, y_1), (x_2, y_2), \ldots, (x_m, y_m)\}$;
Base the learning algorithm K;
Number of learning rounds N;
Process:
1. $For\ n = 1, 2, \ldots, N$:
2. $B_n = Boostsrap(B)$; % Generate a bootstrap sample from B
3. $h_n = K(B_n)$ % Train a base learner h_n from the bootstrap sample
4. ***End***
Output: $H(x) = argmax_{y \in Y} \sum_{n=1}^{N} 1(y = h_n(x))$ % the value of $1(\alpha)$ is 1 if α is true % and 0 otherwise

---

8.2 *Random Forest*

The random forest algorithm was first introduced by Breiman [35] in 2001 as an extension of the simple bagging algorithm. Random forests are a bagging method in which multiple trees are combined using bootstrap samples to achieve a lower variance output. In random forests, instead of using all the features in each split in the tree, a random subset of features/variables is chosen, which further randomizes the tree. As a result, the bias of the resulting model is slightly increased, but due to the combination of less related trees, its variance is reduced, resulting in a better model [10]. The output of the random forest model for classification problems is a class chosen by the majority of trees, while the output of the random forest model for regression problems is obtained by averaging each tree. In addition, the decision at each node in the tree structure uses random node optimization, where it is chosen by a random procedure rather than a deterministic optimization. Thus, random forest models often produce more accurate prediction results than simple bagging and simple boosting (i.e. AdaBoost) [25].

Random forest models take less training time than other methods and can maintain accuracy in data that is largely missing. Random forest ensembles are also used for density estimation, an unsupervised task and detection anomalies [25].

8.2.1 Out-of-Bag Error Estimation

In the bagging method, each new model is trained using bootstrap aggregation, where the training sampling is fitted using bootstrapping samplings. In bagging method, cross-validation or validation set approach can be used when estimating the test error of a model. However, these approaches can lead to big computational efforts, especially in big datasets [5, 36]. Instead of these approaches, Out-of-bag (OOB) error, also called out-of-the-bag estimation, is an extremely useful and simpler way to estimate generalization error and variable significance. When a bootstrap sample is selected from the dataset, some observations are not included in the tree building phase and these observations, called OOB, are used to make an internal estimate of the generalization error. To obtain the OOB error rate, each tree estimates a class value for the OOB dataset, and these estimates are recorded. At any point, the generalization error can be calculated by averaging (for regression) or majority vote (for classification) of the error rate estimates in trees where it is OOB for each observation. A general error rate can be calculated by including all observations using the mean-squared error for regression or using classification error for classification [5].

8.2.2 Variable Importance

Although decision trees are a simpler and easier-to-interpret method, when many trees are packed, it is not possible to represent the statistical learning procedure that occurs with a single tree. This also raises the problem of showing which variable is more important. In this case, the bagging method is preferred as it improves the prediction accuracy against the more difficult interpretability feature.

The predictiveness of a variable is measured by the variable importance. The first step in measuring variable importance is to fit a random forest to the data, as the random forest performs direct variable selection when generating the classification rules. Variable importance can be calculated in two ways: Gini significance and permutation-based variable importance.

Gini importance is obtained by the Gini index used when constructing direct random forest trees. When a node is split over the *jth* variable, the Gini index value of the obtained two new nodes will be smaller than the Gini value of the split node. The Gini importance value of the *jth* variable for each tree is obtained by calculating the difference between these two values. After all the trees in the forest are created, the importance of the *jth* variable is determined by summing the Gini importance of the trees in which the *jth* variable takes place [37].

The process for obtaining permutation-based variable importance is as follows. When a random forest model is fit, the accuracy estimate from the OOB error for each data point is recorded and averaged over the forest. To measure the importance of the *jth* feature, the values of the *jth* feature are permuted among the training data and the prediction accuracy of the OOB samples is recalculated on this permuted dataset. As a result of this process, two predictive values are obtained for each *jth* feature. The accuracy estimates from the Original OOB before the permutation and the modified OOB after the permutation over all trees are subtracted and the importance score for the *jth* feature is calculated by averaging the difference obtained. Then the score value is obtained by normalizing the differences with the standard deviation. According to the obtained score values, the degree of importance of the features is ranked from the features that produce large values to the features that produce small values [38].

8.3 Boosting

Boosting is a family of models in which weak learners are brought together to obtain a stronger learner with better performance [26]. However, unlike bagging, which basically aims to reduce variance, boosting uses an iterative approach that combines a number of weak learners. In this method, each new model is created by giving more weight to the training examples that were incorrectly predicted by the previous model. Then the same process is applied for the training samples that were predicted incorrectly at each iteration, resulting in a strong learner with lower variance through a weighted combination of the models obtained at the end of the process [26, 39]. The simple boosting process in constructing ensemble models is as shown in Fig. 15.

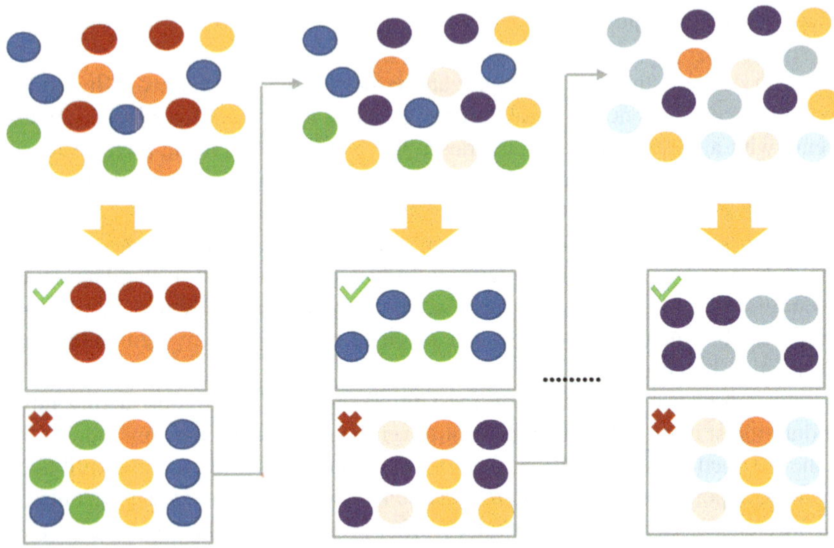

Fig. 15 Boosting-based ensembles

Like the bagging method, the boosting method uses voting for classifications or averaging for numerical estimates to combine the outputs of individual models. Also, like bagging, the boosting method combines models of the same type by using algorithms that are unstable estimators, such as decision trees. Although these two methods seem similar to each other in terms of structure and purpose, these methods have different strategies in using the training dataset and creating the best estimation model ensemble. Boosting uses the entire training dataset unlike the bagging method, which uses bootstrapping copies of the training dataset to create decision trees. Bagging combines the independent trees while boosting uses the interdependent trees to form the final ensemble [10].

Many algorithms based on the boosting method such as AdaBoost, gradient tree boosting, LP Boost, XGBoost, LightGBM have been developed by many researchers.

8.3.1 AdaBoost

The AdaBoost algorithm, first introduced by Freund and Schapire [33] is one of the most effective and well-known boosting methods. The general boosting procedure described isn't a real algorithm because there are some specific parts in an AdaBoost algorithm, such as "adjust distribution" and "combine outputs" [25]. The considered samples are drawn into datasets after a sample distribution, which allows the training data to be updated iteratively. The classifiers are combined through weighted majority voting, where the voting weights weighted by the sample distribution are based on the training errors of the classifiers.

Considering a binary classification problem, the AdaBoost algorithm has a process that proceeds as follows. Initially, all observations have equal weights. Then, a classifier is created for these observations through the learning algorithm and each sample is reweighted based on the output of the classifier. In each iteration, the weights of correctly classified samples are reduced while the weights of misclassified samples are increased. These result in a series of "easy" instances with low weight and a series of "hard" instances with a high weight. At each iteration, more classifiers are created for the reweighted data, thus forcing weak learners to focus on hard instances of the training set. Furthermore, the overall accuracy of the classifier is measured by giving a weight to each classifier. This weight is also a function of the total weight of correctly classified models. Then, the weights of the instances are increased or decreased based on the output of this novel classifier. This process is repeated until the function remains unchanged or the number of estimators reaches the maximum limit. As long as each sample in a model performs slightly better than random guessing, it can be proven that the final model is learned and converges to a strong learner [39]. For a classification problem consisting of m instances, where each x_i instance in the training set is labeled as $y_i \in \{-1, +1\}$, the AdaBoost algorithm is defined as follows.

Input: Dataset $B = \{(x_1,y_1), (x_2,y_2), \ldots, (x_m, y_m)\}$;
Base the learning algorithm K;
Number of learning rounds N;
Process:
1. $B_1(i) = 1/m$ % Initialize the weight distribution
2. $For\ n = 1, 2, \ldots, N$:
3. $h_n = L(B, B_n)$; % Train a base learner h_n from B using distribution B_n
4. $\varepsilon_n = P_{r_i \sim B_i}[h_n(x_i \neq y_i)]$; %Measure the error of h_n
5. $if\ \varepsilon_n > 0.5\ then\ break$
6. $\alpha_n = \frac{1}{2} ln \frac{1-\varepsilon_n}{\varepsilon_n}$; % Determine the weight of h_n
7. $B_{n+1}(i) = \frac{B_n(i)}{Z_n} \times \begin{cases} exp(-\alpha_n) & if\ h_n(x_i) = y_i \\ exp(\alpha_n) & if\ h_n(x_i) \neq y_i \end{cases}$
$= \frac{B_n(i) x exp(-\alpha_n y_i h_n(x_i))}{Z_n}$ % Update the distribution, where Z_n is normalization factor with enables D_{n+1} to be a distribution
End
Output: $H(x) = sign(f(x)) = sign \sum_{n=1}^{N} \alpha_n h_n(x)$

8.4 Gradient Boosting

AdaBoost algorithm, one of the many boosting algorithms in the literature, was the most widely used boosting technique during the 1990s [33]. After the development of this algorithm, the gradient boosting idea has emerged based on Loe Breiman's observation that boosting can be considered as an optimization algorithm on an appropriate cost function [40]. After the explicit regression gradient boosting algorithms were developed by Friedman, a more general concept of functional gradient boosting has been introduced simultaneously by Llew Mason, Jonathan Baxter, Peter Bartlett and Marcus Frean [41].

Like other boosting methods, the gradient boosting method is defined as a machine learning algorithm for regression and classification models that iteratively generates the ensemble form of weak predictive models, such as decision trees, in a prediction model. The basic idea behind this algorithm is to establish novel base learners in such a way that they are maximally correlated with the negative gradient of the loss function associated with the entire ensemble. Here, it can be applied to the popular loss functions such as least squares, least-absolute-deviation, Huber, logistic binomial log-likelihood [38, 42]. The gradient boosting algorithm has a process that proceeds as follows

- Initially, a decision tree is built from the first dataset and the error between the prediction and the output is found.
- This error value is used as new output values for instances of the dataset.
- With the errors in the dataset, a new decision tree is created as the label of the sample and the tree is trained to reproduce the error created by the previous tree.
- The tree continues to be added to the process until the error value between the previous output and the current prediction reaches the desired level.

For a regression problem consisting of n instances, where each x_i instance in the training set is labeled as y_i, the gradient boosting algorithm is defined as follows [40].

Input: Training set $\{(x_i, y)\}_{i=1}^{n}$;
Differentiable loss function $L(y, F(x))$;
Number of iterations M;
Process:
1. $F_0(x) = argmin_\rho(\sum_{i=1}^{n} L(y, \rho))$ % Initialize model with a constant value
2. $For\ m = 1, \ldots, M$:
3. $\tilde{y}_i = -\left[\frac{dL(y_i, F(x_i))}{dF(x_i)}\right]_{F(x)=F_{m-1}(x)}$ $for\ i = 1, \ldots, n$ % Compute so- called pseudo-residuals
4. $a_m = argmin_{a,\beta} \sum_{i=1}^{n} [\tilde{y}_i - \beta h(x_i; a)]^2$ % Fit a base learner (weak learner) closed under scaling $h(x_i; a_m)$ to pseudo-residuals
5. $\rho_m = argmin_\rho \sum_{i=1}^{n} L(y_i, F_{m-1}(x_i) + \rho h(x_i; a_m))$ % Find the best gradient descent step-size ρ_m
6. $F_m(x) = F_{m-1}(x) + \rho_m h(x; a_m)$ % update the function estimate
End
Output: $F_m(x)$

An Example: Human Resource Analytic

Company X, which provides services to the large-scale market in the field of information technologies and automation, has campuses in different locations and has a large number of employees. One of the most important concerns of the company's human resources department is the gradual departure of employees over time based on employee attrition. Employee turnover is very costly for the company, given the time it takes to interview and find a replacement, placement fees and lost productivity for several months. Therefore, understanding why and when employees are most likely to leave will enable the development of a variety of human resources policies for employee retention as well as planning new hires in advance. In this example, it is aimed to predict when employees will leave and which employees are likely to leave, taking into account the dataset containing information about the company's employees. For this, the model is developed using ensemble methods. Thus, an attempt is made to help the company design a better human resources policy and to consider in advance a serious issue such as the cost that may arise from employee turnover. The dataset, which includes 14,999 employees, includes variables such as average monthly working hours, time spent in the company on a yearly basis, salary level, number of projects and satisfaction level. The data in the dataset are supervised and consist of a mixture of numerical and non-numerical variables. In addition, in the given dataset, the target variable consists of two types of employees, those who remain with the company and those who leave the company [43]. A summary of the performances of the ensemble models based on accuracy, recall score, precision score and F1 score as well as confusion matrices are presented in Table 5. As the results show, Bagging has the best classification accuracy and the greatest AUC among the models. Besides, it is seen with the high values of the performance evaluation criteria and the AUC value that other ensemble methods give similar results with the Bagging method.

Table 5 Performance metric values of the models

Model type		Confusion matrix		Accuracy (%)	Recall score (%)	Precision score (%)	F1 score (%)	AUC
		T	F					
Bagging	T	3410	18	97.3	90.3	98.3	94.1	0.99
	F	104	968					
Random forest	T	3409	19	97.1	89.6	98.1	93.6	0.99
	F	112	960					
Gradient boosting	T	3388	40	97.2	92.2	96.1	94.1	0.99
	F	84	988					
XG Boost	T	3388	40	97	91.1	96	93.5	0.99
	F	95	977					
AdaBoost	T	3363	65	91.3	69.6	92	79.2	0.97
	F	326	746					

9 Unsupervised Methods

In unsupervised learning, as the name suggests, there is no specific information like parameters or labels that could supervise the system to learn. Instead, the system learns only from input data by identifying the underlying structure of it. Unsupervised learning provides useful and practical insights especially when there is no sufficient data to train a model. It makes previously intractable problems more solvable and is much nimbler at finding hidden patterns both in the historical data that is available for training and in future data [44]. In this manner, it is better in drawing conclusions and similar to human learning process for more open-ended problems even though the performance of the model cannot be clearly measured. On the other hand, unsupervised learning is also used to improve the performance of supervised learning methods in some cases like missing labels, overfitting, outlier detection, dimensionality reduction by eliminating insignificant features, or selecting the most important features [44].

Unsupervised learning problems mostly tackle the clustering problem for machine learning tasks. In this section several clustering techniques will be introduced.

9.1 Clustering

Clustering is a method to group the data in a meaningful way. The main idea is bringing the objects with higher similarities together and separating them from the objects with higher dissimilarities as much as possible. It is useful in finding internal structures, patterns and underlying rules in the data and categorizing an unlabeled set of data. The real-world applications of clustering vary widely. The most common examples of clustering applications are in market and customer segmentation, document segmentation, image compression and bioinformatics fields.

9.2 K-means Algorithm

K-means algorithm is one of the commonly used and simple distance-based clustering methods. It depends on the parameter k which is the predetermined number of clusters. At the end of the algorithm, each data point is assigned to exactly one of these k clusters. Optimal grouping is achieved by minimizing the within-cluster sum of squares (WCSS), also known as inertia, such that the sum of the within-cluster variations across all k clusters is as small as possible [44]. Hence, every data point is allocated to the nearest cluster.

$$WCSS = \sum_{i=1}^{k} \sum_{x \in C_i} \|\boldsymbol{x} - \mu_i\|^2 \tag{85}$$

where the k is the number of clusters, x are the observations, C_i is the ith cluster and μ_i is the center of the ith cluster.

The steps of the algorithm are as following:

Initialization: K-many randomly selected points are introduced as centroids, which are used as the beginning points for every cluster. Centroids are not necessarily one of the input data points. If the centroids are required to be one of the input data points, this method is called the k-medoids algorithm.

Recursion: Each data point is assigned to a cluster based on the minimum distance between the data point and the centroids of the clusters. Distance between is computed as Euclidean distance which is based on the sum of squares. It is also possible to use other distance types such as Manhattan distance. Johnston [45] suggests that Manhattan distance may outperform Euclidean distance when it comes to higher-dimensional data. After the assignment of clusters, the new centroids of the clusters are computed for each cluster by averaging the data points in that cluster. This process iteratively assigns each data point to a cluster with updated centroids until a stopping criterion is met.

Termination: The algorithm ends if the centroids are the same with the centroids in the previous iteration or maximum iteration number is reached.

The optimum number of clusters can be decided with Elbow method. In the elbow method, the k-means algorithm is run with a set of different k parameters. For each k, WCSS is calculated and plotted where the x-axis shows the number of clusters and the y-axis shows the WCSS. WCSS is highest when $k = 1$, yet decreases as the number of clusters increases and therefore, the plot looks like an arm with a noticeable elbow point. The optimum k value is chosen as this elbow point. For example, in Fig. 16, the optimal number of clusters is 3.

Fig. 16 Elbow method

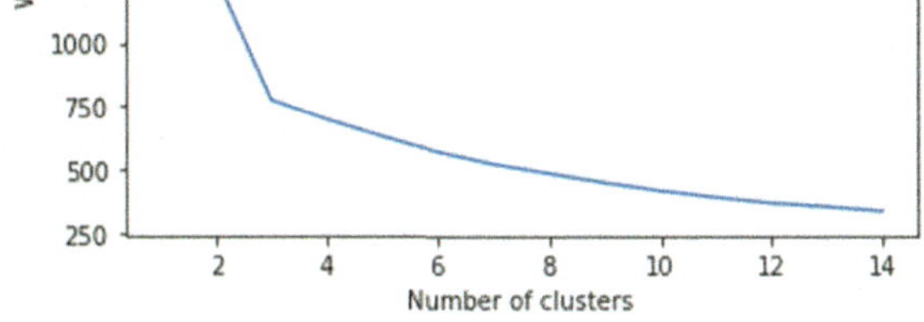

9.3 DBSCAN

DBSCAN (density-based spatial clustering of applications with noise) is considered as one of the powerful clustering algorithms. Its similarity criteria is based on density estimation [46]. DBSCAN groups the data points that are close to each other by considering the following two parameters that are chosen by user:

"eps": It specifies the maximum distance between two data points to be considered as neighbors and fall into the same cluster.

"min_samples": It specifies the minimum number of points in an eps-neighborhood.

The steps of the algorithm are as following [45]:

- Move through each unvisited point in a loop and mark as visited.
- From each point, look at the distance to every other point in the dataset.
- Connect all points within the eps-neighborhood as neighbors.
- Check to see whether the number of neighbors is at least as many as the minimum points required.
- If the minimum point threshold is reached, group together as a cluster. If not, mark the point as noise.
- Repeat until all data points are categorized in clusters or as noise.

The parameters can be tuned by examining the dataset first. If the dataset is large, a small "min_samples" is better; if the dataset is noisier, a large "min_samples" is better and generally it should be greater than or equal to the dimension of the dataset. As a rule of thumb, "min_samples" can be chosen as 2 times of the dataset dimension [47]. As a useful way of deciding on "eps" parameter, K-nearest neighbors (K-NN) algorithm can be utilized as suggested in [48]. In this method, after deciding on "min_samples," *k*-NN algorithm is run with its k-neighbor parameter equal to "min_samples." When the average distance between each point and its *k*-neighbors is sorted and plotted as in Fig. 17, the value with highest curvature, as in elbow method, can be chosen as the eps parameter which is chosen 0.3 here.

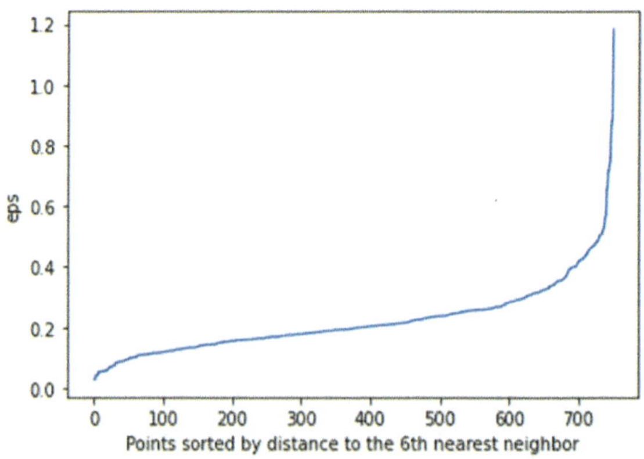

Fig. 17 Specification of eps parameter

On the contrary of k-means which tends to work well with circular shaped clusters, DBSCAN gives better results with clusters of arbitrary shape from noisy datasets and requires only one scan of the dataset [49]. Also, only the data points that have close neighbors will be seen as members within the same cluster and those that are farther away can be left as not clustered outliers [45]. Therefore, DBSCAN is more robust with the presence of outliers.

9.4 Mean Shift Algorithm

Mean shift algorithm does not require specifying the number of clusters in advance. The number of clusters is determined by the algorithm with respect to the data.

The downside to Mean Shift is that it is computationally expensive $O(n^2)$. Mean shift algorithm updates the centroids of the candidate clusters by moving toward mode of the cluster region using following formula

$$c_i^{t+1} = \frac{\sum_{x \in c_i^t} K\left(x - c_i^t\right)x}{\sum_{x \in c_i^t} K\left(x - c_i^t\right)} \tag{86}$$

where c_i^{t+1} is the centroid of the candidate cluster at iteration $t + 1$. K is the kernel function:

1. Mean shift clustering is done with the following steps:
2. Select a random observation as centroid of the candidate cluster.
3. Use $c_i^{t+1} = \frac{\sum_{x \in c_i^t} K\left(x - c_i^t\right)x}{\sum_{x \in c_i^t} K\left(x - c_i^t\right)}$ formula to update centroid till convergence
4. Merge with existing clusters if converged centroid is close to existing cluster's centroid.

9.5 Gaussian Mixture

Gaussian mixture models probabilistic clustering algorithm that assumes observations are coming from a finite mixture model as follows:

$$p(x) = \sum_{i=1}^{k} w_i \mathcal{N}(\mu_i, \Sigma_i) \tag{87}$$

where observations x is coming from a distribution that is the linear combination of k multivariate normal distribution ($\mathcal{N}(\mu_i, \Sigma_i)$) with mean of μ_i and the covariance matrix of Σ_i. In order to calculate the parameters of multivariate normal distributions, Expectation Maximization (EM) algorithm is used which is similar to k-means algorithm. The EM algorithm is as follows:

1. Initialize k random clusters with random parameters of mean and covariance matrix
2. Assign observations to clusters that will maximize the likelihood.
3. Update the mean and covariance parameters using the observations in each cluster
4. Repeat 2–3 steps until convergence.

9.6 Hierarchical Clustering

Hierarchical Clustering as the name suggests constructs clusters by merging and splitting them from top to bottom or bottom to top. In agglomerative hierarchical clustering, every observation starts as individual clusters and similar clusters are merged as you move up. In divisive hierarchical clustering, at the start all observations are assumed to be in the same cluster, as you move down the hierarchy, splitting operations are applied to clusters.

Clusters are merged or split using the dissimilarity measure between sets of observations. Euclidean, Manhattan, max distance or Mahalanobis distances can be used as dissimilarity measures.

Hierarchical Clustering algorithms can vary according to merging criteria they used.

1. Ward minimizes sum of squared distances within all clusters same as k-means algorithm
2. Complete Linkage minimizes the maximum distance between two observations in a cluster
3. Average Linkage minimizes the average distance between all pair of observations in cluster
4. Single Linkage minimizes the distance between two observations that are closest to each other.

10 Reinforcement Learning

Reinforcement learning (RL) is a subfield of machine learning that focuses on the broad and difficult subject of optimizing behavior in complex environments. RL is entirely different paradigm than many other machine learning approaches such as supervised and unsupervised learning. The learning process is entirely motivated by reward value and environmental observations. RL problems involve learning what to do, how to map situations to actions so as to maximize a numerical reward. The learner is not told which actions to take, as in many forms of machine learning, but instead must discover which actions yield the most reward by trying them out. This model is versatile and can be applied to a wide range of real-world applications, including gaming, self-driving cars, algorithmic trading, improving manufacturing

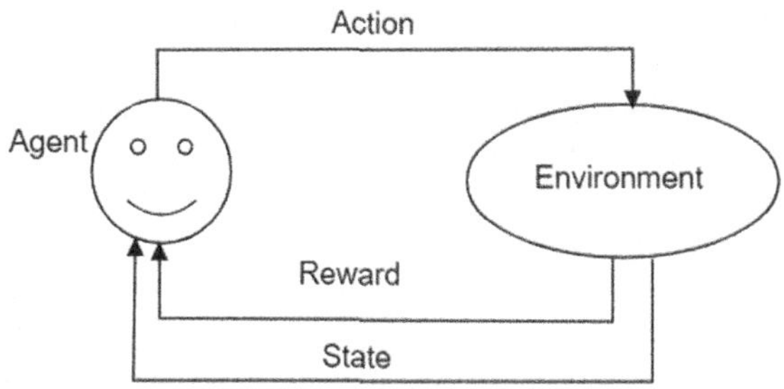

Fig. 18 Basic components of reinforcement learning

process efficiencies etc. The topic of reinforcement learning is developing swiftly, due to the flexible and generative nature of it. It draws attention from researchers who are seeking to enhance the methods they currently use, as well as those who are interested in solving their problems the most efficiently [50, 51]. The fundamental building blocks of reinforcement learning are illustrated in Fig. 18.

Before delving into the specifics of the technique, let's quickly define these terms.

Agent is the decision-making entity that has been trained to discover optimal behaviors.

Environment is everything with which the Agent has the potential to interact, either directly or indirectly.

Action: When an agent engages in an action, it makes use of various means to interact with and change its environment, thus allowing it to transition between states.

State represents the world or the environment that is currently involved in the task.

Reward is an immediate and direct response to the agent's actions.

Policy: A policy determines the behavior of the learning agent at a given point in time. A policy, in a general sense, is a set of rules which maps from observed environmental conditions to behaviors which are enacted when in those conditions.

These terms will be discussed in detail in the following section in relation to the Markov decision process.

10.1 Markov Decision Process

Markov decision process (MDP) is used to formalize the RL problem elements. States, actions, transitions between states and a reward function are the building blocks of MDPs [52]. Each of them is explained in the following manner:

State represents the world or the environment that is currently involved in the task. A finite set $\{s_1, \ldots, s_N\}$ that contains the collection of all possible states is known to be the set of environmental states. This set, also referred to as a state space, has a size of N.

Actions: It includes all activities carried out by the agent at any point in time while in a state. When an agent engages in an action, it makes use of various means to interact with and change its environment, thus allowing it to transition between states. Action set A is defined as a finite set $\{\boldsymbol{a}_1, ..\boldsymbol{a}_M\}$ where the size of action space is M.

Transition between states: To create a transition between two states, an agent employs action $\boldsymbol{a} \in \boldsymbol{A}$ in a state $\boldsymbol{s} \in \boldsymbol{S}$. In that way, the system goes from $\boldsymbol{s}$ to a new state $\boldsymbol{s}' \in \boldsymbol{S}$, using a transition probability distribution over the set of all possible transitions. These quantities completely specify the dynamics of the MDP, or in other words how the system evolves through states in time. Transition probability function, $\boldsymbol{T} : \boldsymbol{SxAxS} - > [0, 1]$ denotes the probability of being in state $\boldsymbol{s}'$ after doing action $\boldsymbol{a}$ in state $\boldsymbol{s}$. For every action $\boldsymbol{a}$, and every state $\boldsymbol{s}$ and every state $\boldsymbol{s}'$, $\boldsymbol{T}(\boldsymbol{s}, \boldsymbol{a}, \boldsymbol{s}')$ must be greater than or equal to zero and $\boldsymbol{T}(\boldsymbol{s}, \boldsymbol{a}, \boldsymbol{s}')$ must be less than or equal to one. It follows that for all states, and all actions, $\sum_{\boldsymbol{s}' \in \boldsymbol{S}} \boldsymbol{T}(\boldsymbol{s}, \boldsymbol{a}, \boldsymbol{s}') = 1$. While the transition function determines the state in which the agent will be at time $\boldsymbol{t} + 1$, the function only considers the state in which the agent is at time $\boldsymbol{t}$ and the action it takes in that state. This is referred to as the Markovian property or condition. Without the Markov condition, previous states and action taken will be instrumental in predicting the next state, which will subsequently necessitate the use of a memory in order to determine the best behavior. It is defined as follows:

$$P(s_{t+1}|s_t, a_t, s_{t-1}, a_{t-1}, ...) = P(s_{t+1}|s_t, a_t) = T(s_t, a_t, s_{t+1}) \quad (88)$$

A reward function: Another critical component of the formulation of RL problems is the reward function. It is simply a scalar value that the agent obtains from the environment on a periodic basis. This scalar value can be either positive or negative. The purpose of rewards is to assist the agent in evaluating how well they behaved. The reward function can be defined in two ways: It can be defined as either $\boldsymbol{R} : \boldsymbol{S} \times \boldsymbol{A} \rightarrow \mathbb{R}$ or $\boldsymbol{R} : \boldsymbol{S} \times \boldsymbol{A} \times \boldsymbol{S}' \rightarrow \mathbb{R}$. The first of these two options provides reward for performing an action in a particular state, and the second offers reward for specific transitions between those states.

Finally, MDP can be defined as follow:

MDP: A Markov decision process is a tuple of four items: $\boldsymbol{S}, \boldsymbol{A}, \boldsymbol{T}$ and $\boldsymbol{R}$. The item labeled $\boldsymbol{S}$ describes the state of the system, $\boldsymbol{A}$ describes the available actions, and $\boldsymbol{T}$ describes the transition function. The item labeled $\boldsymbol{R}$ describes the reward function. This definition of an MDP can be used to model a variety of different types of systems.

Tasks in a system modeled using the MDP structure can take one of two forms: (i).

Episodic tasks: In episodic tasks, the agent aims to reach a final or goal state, and an episode is completed when the agent begins from an initial state and reaches the final state. (ii) Continuing tasks: There is no desired final state for the agent, and the agent's interaction with the environment can continue indefinitely.

Policies: In an MDP-modeled system, the policies define how the agent should behave. A policy specifies the course of action that the agent should take at any state. It can be either deterministic or stochastic in nature. A deterministic policy, in formal terms, is a function defined as $\boldsymbol{\pi} : \boldsymbol{S} \rightarrow \boldsymbol{A}$. On the other hand, a stochastic policy does not advise a specific action outright. It defines the possible actions in each state using a probability distribution. It is defined as $\pi : S \times A \rightarrow [0, 1]$ such that it holds true for each state $s \in S$ that $\pi(s, a) = 0$ and $\sum_{a \in A} \pi(s, a) = 1$.

The objective or goal in MDP-modeled systems is to ensure that the agent accomplishes its objective, that is, to determine the agent's actions so that it collects the maximum reward possible until the task is completed. It is necessary to add up the immediate rewards collected by the agent throughout the process until it reaches its goal in order to determine how well the agent can accomplish its objective. The cumulative reward can be calculated in three ways: (i) $E[\sum_{t=0}^{h} R_t]$, (ii) $E[\sum_{t=0}^{\infty} \gamma^t R_t]$, and (iii) $\underbrace{\lim}_{h \rightarrow \infty} E\left[\frac{1}{h} R_t\right]$ These three distinct calculations vary according to the planning horizon used and the weight assigned to rewards received at various points in time. For instance, $E[\sum_{t=0}^{\infty} \gamma^t R_t]$ calculates the cumulative reward over an infinite horizon and utilizes a discounting factor γ to weight rewards collected over time.

10.2 Value Functions and Bellman Equations

Value functions provide a mechanism for coupling optimality criteria and policies. A value function indicates how well the agent is doing or how well the agent performs given the state they are in. Optimality, i.e. expected return, is a convenient and effective way to express how good something is. Policies are specified with respect to particular policies. Value function $V^{\pi}(s)$ is the expected return when starting in s and following π thereafter. More formally, the value function for policy π is as follows:

$$\begin{aligned} V^{\pi}(s) &= E_{\pi}\left[\sum_{k=0}^{\infty} \gamma^{t+k} R_{t+k} | s_t = s\right] = E_{\pi}\left[R_t + \gamma R_{t+1} + \gamma^2 R_{t+2} + \ldots | s_t = s\right] \\ &= E_{\pi}\left[R_t + \gamma V^{\pi}(s_{t+1}) | s_t = s\right] \\ &= \sum_{s'} T(s, \pi(s), s')\left(R(s, a, s') + \gamma V^{\pi}(s')\right) \end{aligned} \quad (89)$$

The above equation is also known as Bellman Equation. Similarly, it is defined the value of taking action a in states under a policy π, denoted $Q^{\pi}(s, a)$, as the expected return starting from s, taking the action a, and thereafter following policy π, which can be formally defined as

$$Q^{\pi}(s, a) = E_{\pi}\left[\sum_{k=0}^{\infty} \gamma^{t+k} R_{t+k} | s_t = s, a_t = a\right] \quad (90)$$

$Q^{\pi}(s, a)$ is also referred as the state-action value function. A primary objective of any MDP is to find the best policy. That is, find the policy that generates the greatest return, which is in essence finding the policy that optimizes the value function. An optimal policy, denoted by π^*, is one that ensures that $V^{\pi^*}(s) \geq V^{\pi}(s)$ for all $s \in S$ and all policies π. The following equation, dubbed the Bellman optimality equation, expresses the value of state s when an optimal policy is used.

$$V^*(s) = \underbrace{\max}_{a \in A} \sum_{s' \in S} T(s, a, s')(R(s, a, s' + \gamma V^*(s')) \tag{91}$$

The following rule is useful in determining an optimal action given an optimal state value function:

$$\pi^*(s) = \underbrace{\arg\max}_{a} \sum_{s' \in S} T(s, a, s')(R(s, a, s') + \gamma V^*(s')) \tag{92}$$

The state-action value corresponding to an optimal situation is equivalent to:

$$Q^*(s, a) = \sum_{s' \in S} T(s, a, s')(R\left(s, a, s' + \gamma \underbrace{\max}_{a} Q^*(s', a')\right) \tag{93}$$

The relationship between Q^* and V^* is defined by following equations:

$$Q^*(s, a) = \sum_{s' \in S} T(s, a, s')(R(s, a, s' + \gamma V^*(s')) \tag{94}$$

$$V^*(s) = \underbrace{\max}_{a} Q^*(s, a) \tag{95}$$

$$\pi^*(s) = \underbrace{\arg\max}_{a} Q^*(s, a) \tag{96}$$

After discussing the fundamentals of MDP, it can be moved on to discussing how to solve problems that have been formulated using the MDP setting. Algorithms for solving MDP problems can be broadly classified into two types: (i) model-based algorithms and (ii) model-free algorithms. Algorithms in the first category, also known as DP methods, assume that all elements of an MDP are known. In contrast, algorithms in the second category assume that the transition probability function and reward function of the MDP are unknown. The agent attempts to collect transition samples through interaction with the environment. Reinforcement learning problems are the type of problems that model-free algorithms attempt to solve. An explanation of the algorithms involved in both categories of algorithms is presented below.

Model-based algorithms: Policy iteration and value iteration are two of the more common DP methods used to address model-based problems. Policy iteration alternates between the two phases (policy evaluation and policy improvement). The first phase calculates the value function for the current policy, and the second phase modifies the current strategy to increase its value function. These processes are repeated until the optimal policy is found. The policy iteration algorithm performs the evaluation and improvement phases entirely apart from each other. Computing the value function in the limit is a mandatory step in the evaluation process. However, you don't have to wait for the model to converge, but you can evaluate the model and policy up to this point and use that to improve the policy. The value-based algorithms are based on this concept and terminate evaluation after a single step. To keep things simple, these algorithms will not describe in detail; readers can find these details and more about model-based algorithms in [53].

Model-free algorithms: There are two approaches to RL problems. The first is to acquire an understanding of the transition and reward models through interactions with the environment. After learning the missing pieces of the MDP structure through interaction, the problem can be solved using any DP algorithm. This is referred to as model-based reinforcement learning or indirect reinforcement learning. The second option, dubbed direct RL, entails directly estimating values for actions without first estimating the MDP model. There is a great deal of effort being made in the literature to develop more efficient model-free algorithms. It is not possible to discuss these algorithms in detail here; instead, the most well-known reinforcement learning algorithm, Q-learning will be explain using a simple problem. The details of these algorithms may be found in [51].

10.3 Methods for Solving RL Problems

Solving RL problems requires two steps: 1-policy evaluation 2-policy improvement. Policy evaluation is the step to estimate the value function for a given policy. This is also referred as the prediction problem. Once the value of the policy is determined the next step is to seek for better policies that provide better values. This step is called policy improvement. Almost all RL methods cycle between policy evaluation and policy improvement until no further improvement can be made. Figure 19 summarizes the general framework of RL methods also called generalized policy iteration [2].

Three general methods are available for estimating value functions and then determining optimal policies in reinforcement learning problems.

10.3.1 Dynamic Programming

Dynamic programming (DP) is a collection of algorithms for solving problems modeled by Markov decision processes in which the environment model is fully known. The solutions of DP techniques are exact. While classical dynamic programming algorithms are not appropriate for solving RL problems because they require that the environment model is known in advance. Dynamic programming, on the other hand, provides a theoretical foundation for RL algorithms. The common characteristic between DP and RL algorithms is that they both employ the value function

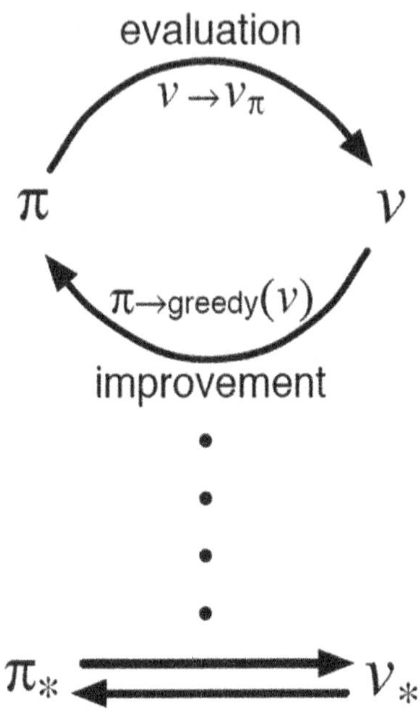

Fig. 19 Generalized policy iteration

to determine the optimal policies. The pseudo-code of policy iteration algorithm is given as follows (see [51] for details):

1. Initialization

Initialize $V(s)$ and $\pi(s)$ arbitrarily for all s

2. Policy Evaluation

Repeat

$\Delta \leftarrow 0$

For each s:

$v \leftarrow V(s)$

$V(s) \leftarrow \sum_{s',r} P(s',r|s,\pi(s))[r + \gamma V(s')]$

$\Delta \leftarrow \max(\Delta, |v - V(s)|)$

Until $\Delta < \theta$ (a small positive number)

3. Policy Improvement

policy-stable$\leftarrow$ *true*

For each s:

$a \leftarrow \pi(s)$

$\pi(s) \leftarrow argmax_a \sum_{s',r} P(s',r|s,\pi(s))[r + \gamma V(s')]$

If $a \neq \pi(s)$, then policy-stable$\leftarrow$ *false*

If policy-stable, then stop and return V and π,

Else go to step 2

The policy iteration algorithm exploits the bellman equations to evaluate the policy. Recall that Bellman equations is a system of $|S|$ simultaneous linear equations in $|S|$ unknowns and can be solved easily with a cost of computation time. However, the algorithm above utilizes an iterative approach to solve this system of equation. One can show that the iterative approach will converge to the solution of the equation system. In policy improvement step the algorithm now uses the estimated values and selects the actions that provides the maximum value. The algorithm stops when two consecutive policies are identical.

10.3.2 Monte Carlo (MC) Methods

The agent interacts with its environment and generates experienced samples (i.e., sequences of states, actions and rewards), and then calculates the value of each state based on the average return from those samples. Mathematically speaking, the value of s is the expected return starting from s and following the policy π afterward, hence the gist of the technique is to create multiple sequences and take the average of returns to obtain and approximation for the expectation,

$$\begin{aligned} V^{\pi}(s) &= E[G_t|S_t = s] \\ &= E\left(R_{t+1} + \gamma R_{t+2} + \ldots + \gamma^{n-1} R_{t+n}|S_t = s\right) \qquad (97) \\ &\approx \frac{1}{N}\sum_{i=1} G_{i,s}. \end{aligned}$$

The critical aspect of MC techniques is that they are applied to episodic tasks (tasks that have a terminal state). The solutions provided by MC methods do not have guaranteed optimality. The pseudo-code of the MC method is provided as follows [51]:

a. Initialization

Initialize $Q(s, a)$ and $\pi(s)$ arbitrarily for all s, and Returns (s, a) empty list

b. Repeat

Choose a random state s and random action a, all having positive probability
Generate a sequence of states and actions following policy π

2.1. Policy Evaluation

For each state-action pair (s, a) appearing in the sequence

Record the return G after the first occurrence of (s, a)
Append G to Returns (s, a)

$$Q(s, a) \leftarrow \text{Average(Returns(s, a))}$$

2.2. Policy Improvement

For each s appearing in the episode:

$\pi(s) \leftarrow \text{argmax}_a Q(s, a)$

Note that the algorithm starts from a random state and follows policy π and computes the returns. Namely, it simulates the policy until the episode is over. The algorithm learns the value of state-action pairs through experience and updates the estimates by averaging. Policy improvement is applied after each episode with the updated values.

10.3.3 Temporal Difference (TD) Learning

These techniques combine Monte Carlo and dynamic programming concepts. TD methods are similar to the Monte Carlo approach, which does not use an environment model and instead learns solely through the agent's interaction with the environment. Unlike MC, TD methods update estimates using previous estimates which have been discovered, rather than requiring the estimation to be completed before the updates are applied, as is the case with DP methods. This strategy is also referred to as bootstrapping. Q-learning is a TD learning algorithm. Following are the steps in the algorithm [54]:

1. Initialize $Q(s, a)$ for all $s \in S^+, a \in A(s)$, arbitrarily except that $Q(terminal, .) = 0$
2. Loop for each episode:

   Initialize S
   Loop for each step of episode:
   Choose A from S using policy derived from Q ($\epsilon-$ greedy)
   Take action A, observe R, S'

   $$Q(S, A) \leftarrow Q(S, A) + \alpha\left[R + \gamma \max_a Q(S', a) - Q(S, A)\right]$$

   Set $S \leftarrow S'$,

Until S is terminal

Application of Reinforcement Learning: Grid Maze Problem

Let's say a mouse is trying to learn the shortest path from the origin to the target location within a $m \times n$ square matrix, where the origin and target location is specified. The problem is illustrated in Table 6, where the black cells denote cells that contain obstacles. The mouse can move in four directions: up, down, left and right. The mouse is asked to reach the last cell as fast as possible. When the mouse moves from one cell to another, it gets -1 reward, and if it reaches the end it gets a reward of 10. If the mouse tries to move out of the grid, it stays in the same position, but gets -1 reward. This problem is solved by implementing policy iteration algorithm, Monte Carlo

Table 6 Simple illustration of the maze problem

(0,0)	(0,1)	(0,2)	(0,3)
(1,0)	(1,1)	(1,2)	(1,3)
(2,0)	(2,1)		(2,3)
(3,0)	(3,1)	(3,2)	End

algorithm and Q-learning algorithm. The details of the algorithms' implementations can be found in the supplementary code library.

Figure 20 illustrates the optimal values and optimal policy generated by Q-learning algorithm. The results of policy iteration and Monte Carlo algorithms are very similar and can also be found in the supplementary code files.

Figure 21 depicts the cumulative reward through episodes of the Q-learning algorithm. At first, the agent is inept at learning appropriate actions due to random action selection. After a while, the agent begins to learn how to perform good actions and earn good rewards. At the end of the learning process, the algorithm converges and the agents learn what action to take for each state.

```
---------------------------
 1.81| 3.12| 4.58| 6.20|
---------------------------
 3.12| 4.58| 6.20| 8.00|
---------------------------
 4.58| 6.20| 0.00| 10.00|
---------------------------
 6.20| 8.00| 10.00| 0.00|
```

```
---------------------------------------
 Right | Right | Right | Down |
---------------------------------------
 Right | Right | Right | Down |
---------------------------------------
 Right | Down |       | Down |
---------------------------------------
 Right | Right | Right |      |
```

Fig. 20 Optimal values and optimal policy

Fig. 21 Cumulative reward through the episode

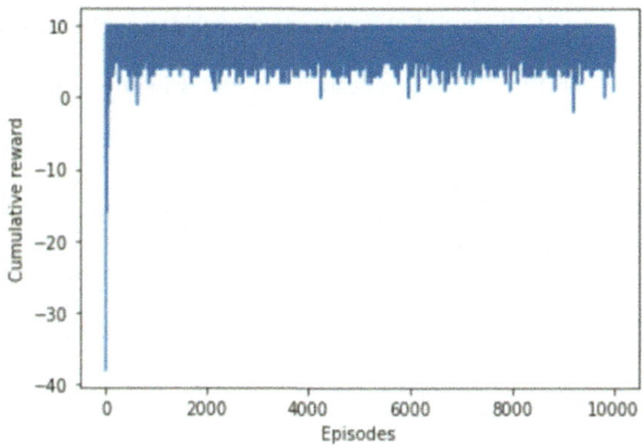

References

1. Mueller JP, Massaron L (2021) Machine learning for dummies. Wiley
2. Pearson KX (1900) On the criterion that a given system of deviations from the probable in the case of a correlated system of variables is such that it can be reasonably supposed to have arisen from random sampling. London, Edinburgh, Dublin Philos Mag J Sci 50(302):157–175
3. Anderson TW, Darling DA (1952) Asymptotic theory of certain goodness of fit criteria based on stochastic processes. Ann Math Stat 193–212
4. Durbin J, Watson GS (1950) Testing for serial correlation in least squares regression: I. Biometrika 37(3/4):409–428
5. James G, Witten D, Hastie T, Tibshirani R (2013) An introduction to statistical learning. 103
6. Rumelhart DE, Hinton GE, Williams RJ (1986) Learning representations by back-propagating errors. Nature 323(6088):533–536
7. Robert HN (1989) Theory of the backpropagation neural network. Proc 1989 IEEE IJCNN 1: 593–605
8. Zhang ML, Zhou ZH (2007) ML-KNN: a lazy learning approach to multi-label learning. Pattern Recognit 40(7):2038–2048
9. Zhang H (2004) The optimality of naive Bayes. AA 1(2):3
10. Sharda R, Delen D, Turban E (2020) Analytics, data science, and artificial intelligence. Pearson Educ
11. Mountrakis G, Im J, Ogole C (2011) Support vector machines in remote sensing: a review. ISPRS J Photogramm Remote Sens 66(3):247–259
12. Kecman V (2005) Support vector machines–an introduction. In: Support vector machines: theory and applications. Springer p 1–47
13. Burges CJC (1998) A tutorial on support vector machines for pattern recognition. Data Min Knowl Discov 2(2):121–167
14. Cortes C, Vapnik V (1995) Support-vector networks. Mach Learn 20(3):273–297
15. Manning CD, Raghavan P, Schütze H (2008) Support vector machines and machine learning on documents. Introd Inf Retr 319–348
16. Cristianini N, Shawe-Taylor J (2000) An introduction to support vector machines and other kernel-based learning methods. Cambridge University Press
17. Kaggle (2021) https://www.kaggle.com/prachi15gupta98/airline-passenger-satisfaction Accessed 15 Oct 2021
18. de Ville B (2013) Decision trees Wiley interdiscip. Rev Comput Stat 5(6):448–455
19. Friedman JH (2017) The elements of statistical learning: data mining, inference, and prediction. Springer open
20. Apté C, Weiss S (1997) Data mining with decision trees and decision rules. Futur Gener Comput Syst 13(2–3):197–210
21. Loh W (2011) Classification and regression trees. Wiley Interdiscip Rev Data Min Knowl Discov 1(1):14–23
22. Song YY, Ying LU (2015) Decision tree methods: applications for classification and prediction. Shanghai Arch Psychiatry 27(2):130
23. Pedregosa F (2011) Scikit-learn: machine learning in Python. J Mach Learn Res 12:2825–2830
24. Wang G, Hao J, Ma J, Huang L (2010) Empirical evaluation of ensemble learning for credit scoring. Business intelligence in economic forecasting: technologies and techniques, IGI global pp 118–137
25. Zhou Z (2012) Ensemble methods foundations and algorithms. Chapman and Hall/CRC
26. Liu X, Zhou Z (2013) Ensemble methods for class imbalance learning. Imbalanced learning: foundations, algorithms, and applications. Wiley pp 61–82
27. Polikar R (2012) Ensemble learning. ensemble machine learning methods and applications. Springer pp 1–34
28. Dasarathy B, Sheela B (1979) Composite classifier system design: concepts and. Proc IEEE 67(5):708–713

29. Hansen L, Salamon P (1990) Neural network ensembles. IEEE Trans Pattern 12(10):993–1001
30. Schapire R (1990) The strength of weak learnability. Mach Learn 5(2):197–227
31. Opitz D, Shavlik J (1996) Actively searching for an effective neural-network ensemble. Connect Sci 8:337–353
32. Breiman L (1996) Bagging predictors. Mach Learn 24(2):123–140
33. Freund Y, Schapire R (1996) Experiments with a new boosting algorithm. In Proceedings of the thirteenth international conference on machine learning
34. Opitz D, Maclin R (1999) Popular ensemble methods: an empirical study. J Artif Intell Res 11:169–198
35. Breiman L (2001) Random forests. Mach Learn 45(1):5–32
36. Breiman L (1996) Out-of-bag estimation
37. Cutler A, Cutler DR, Stevens JR (2012) Random forest. Ensemble machine learning. Springer
38. Hastie T, Tibshirani R, Friedman J (2009) The elements of statistical learning data mining, inference, and prediction. Springer
39. Rokach L (2019) Ensemble learning: pattern classification using ensemble methods. World scientific publishing
40. Friedman J (1999) Greedy function approximation: a gradient boosting machine
41. Mason L, Baxter J, Bartlett PL, Frean M (1999) Boosting alghoritms as gradient descent. Advances in neural information processing systems 12, MIT Press pp 512–518
42. Natekin A, Knoll A (2013) Gradient boosting machines, a tutorial. Front Neurorobotics 7(21)
43. Kaggle (2021) https://www.kaggle.com/markxie/hra-data Accessed 15 Oct 2021
44. Patel AA (2019) Hands-on unsupervised learning using Python: how to build applied machine learning solutions from unlabeled data. O'Reilly Media
45. Johnston B, Jones A, Kruger C (2019) Applied unsupervised learning with Python: discover hidden patterns and relationships in unstructured data with Python. Packt Publishing Ltd
46. Garcia Marquez FP (2019) Handbook of research on big data clustering and machine learning. IGI Global
47. Sander J, Ester M, Kriegel HP, Xu X (1998) Density-based clustering in spatial databases: the algorithm gdbscan and its applications. Data Min Knowl Discov 2(2):169–194
48. Rahmah N, Sitanggang IS (2016) Determination of optimal epsilon (eps) value on dbscan algorithm to clustering data on peatland hotspots in sumatra. IOP Conf Ser: Earth Environ Sci 31(1):12012
49. Mehrotra KG, Mohan CK, Huang H (2017) Anomaly detection principles and algorithms. Springer
50. Busoniu L, Babuska R, De Schutter B, Ernst D (2017) Reinforcement learning and dynamic programming using function approximators. CRC press
51. Sutton RS, Barto AG (2018) Reinforcement learning: an introduction. MIT press
52. Van Otterlo M, Wiering M (2012) Reinforcement learning and markov decision processes. Reinforcement learning. Berlin, Heidelberg, Springer pp 3–42
53. Wiering MA, Van Otterlo M (2012) Reinforcement learning. Adaptation, learning, and optimization 12(3)
54. Watkins CJCH (1989) Learning from delayed rewards

Nursah Alkan is a Ph.D. candidate and a Research Assistant in the Department of Industrial Engineering at Istanbul Technical University since 2019. She received the MSc degree in Department of Industrial Engineering from Yildiz Technical University, Turkey, in 2019. Her research interests include fuzzy sets, multi-criteria/objective decision making, data analysis, machine learning and digital transformation.

Yakup Turgut was born in Turkey, in 1993. He received a B.E. degree in industrial engineering from Yıldız Technical University, Turkey, in 2015, and a master's degree from the Istanbul Technical University, Turkey, in 2018. He is a Ph.D. candidate in industrial engineering at Istanbul Technical University. His research interests are simulation modeling, mathematical modeling, and artificial intelligence. His email address is turgut16@itu.edu.tr.

Emre Ari is a Research Assistant and a Ph.D. student in the Department of Industrial Engineering at Istanbul Technical University, Turkey. In 2016, he received his MSc Degree in Statistics from Queen Mary University of London in UK and he received his BSc degree in Mathematics from Kahramanmaraş Sütçü İmam University in 2009. His current research interests lie in the area Machine Learning, Deep Learning, Reinforcement Learning, Data Analysis, Financial Approaches and Statistics. He plans to continue his academic career by trying to finding new approaches in his interest areas.

Seval Ata is a Research Assistant in the Department of Industrial Engineering at Istanbul Technical University and a PhD Candidate in the Industrial Engineering department at Boğaziçi University. She received her BS and MS degrees in Industrial Engineering from Boğaziçi University in 2015 and 2018, respectively. Her research interests include stochastic modeling of hybrid manufacturing systems, optimal production decision mechanisms and currently, she is mainly focusing on adaptive production control mechanisms with the aim of increasing environmental sustainability.

Mehmet Yasin Ulukus is an assistant professor at İstanbul Technical University Department of Industrial Engineering. He got his Ph.D. from the Department of Industrial Engineering at the University of Pittsburgh. He earned his MS and BS in Industrial Engineering from Bogazici University. His research interests span predictive health and healthcare operations. His methodological interests include machine learning, deep learning, reinforcement learning/approximate dynamic programming, stochastic processes, and queueing theory. His current research focuses on data driven management of intensive care units.

Mehmet Ali Ergun is an academician and researcher currently working at Industrial Engineering Department of Istanbul Technical University as an Assistant Professor. He holds a Ph.D. degree in Industrial Engineering from University of Wisconsin—Madison and a BS degree in Computer Engineering from Bogazici University. His research focuses on applying stochastic optimization, simulation and machine learning techniques to help making data-driven decisions in healthcare related challenges. In one of his work, he was a major contributor to the National Cancer Institute-funded Cancer Intervention and Surveillance Modeling Network (CISNET) project that was used to develop the breast cancer screening guidelines published in the U.S.

Omer Faruk Beyca received his BS degree in industrial engineering from Fatih University, Istanbul, Turkey, in 2007, and the Ph.D. degree from the School of Industrial Engineering and Management, Oklahoma State University, Stillwater, OK, USA, in 2013. He is with the Department of Industrial Engineering, Istanbul Technical University, as an Assistant Professor. His research has been published in international peer-reviewed journals, conference proceedings, and book chapters.

Time Series Analysis

Erkan Isikli, Leyla Temizer, Abdullah Emin Kazdaloglu, and Emre Ari

1 Introduction

The emergence of digital technologies has been changing how things are done in the workplace, in society, and even at home. Recent technological advancements enable the instantaneous recording, processing, and dissemination of information and therefore decision-making processes become more efficient and effective. Time series forecasting has been a key mechanism for modeling various phenomena to support strategic decisions in areas as diverse as air quality monitoring, health informatics, power consumption, and financial engineering and it seems to expand its sphere of influence in the era of big data. Recent developments in the field of computational intelligence have led to a renewed interest in time series analysis. There is a tendency on modifying the traditional time series techniques or creating hybrid models that combine them with machine learning algorithms to improve forecast accuracy when the data is complex, voluminous, or volatile. Since the related literature is massive in terms of both theory and practice, this study provides a concise introduction to time series analysis defining the key and relevant concepts along with the time series forecasting techniques that have been most widely used in various fields.

A time series is a collection of data points that follow a chronological order. Mathematically, it can be denoted as $\{Y_t\}_{t \in T}$, where Y_t is a random variable and T, the index set, usually contains consecutive and evenly spaced integer values. Y_t can be discrete (e.g., emergency department visits, medical tourism demand) or continuous (e.g., electricity price, infant mortality rate) and T is often finite (i.e., $|T| < \infty$).

E. Isikli · A. E. Kazdaloglu · E. Ari
Department of Industrial Engineering, Faculty of Management, Istanbul Technical University, 34367 Macka, Istanbul, Turkey

L. Temizer (✉)
Department of Industrial Engineering, Faculty of Engineering, Istanbul University Cerrahpasa, 34320 Avcilar, Istanbul, Turkey
e-mail: leylatemizer@gmail.com

A. Ustundag et al. (eds.), *Business Analytics for Professionals*, Springer Series in Advanced Manufacturing, https://doi.org/10.1007/978-3-030-93823-9_4

Let $\{Y_1, Y_2, \ldots, Y_n\}$ be a time series of length n. A forecasting model uses the first m observations in this realization and aims to generate values, often denoted by $\hat{y}_t$, that do not greatly vary from y_t (actual values). $\hat{y}_t$ is called the *fitted value* for y_t and this process is called *in-sample forecasting*. The difference between y_t and $\hat{y}_t$ is called a *residual* and generally denoted by ε_t. The values generated by this model for y_{m+1} through y_n are called *forecasts* or *predicted values* and $n - m$ is called the *forecast horizon*. This process is called the *out-of-sample forecasting* and the difference between the actual values and forecasts are called *forecast errors*. The forecast horizon specifies how far ahead we are forecasting. When $n - m = 1$, one step ahead is forecasted; for $n - m > 1$, multi-step ahead. The forecast horizon is generally classified into three categories: short term, medium term, long term. However, the definitions of these categories depend on the context. When forecasting power consumption, short term may correspond to one week to one month (very short term may correspond to one hour to one day), medium term and long term may correspond to one month to one year and 1–10 years, respectively.

In some cases, the time series in hand consists of two or more time-dependent random variables. That is, there is a random vector instead of a single random variable. Forecasting models for such vectors are called *multivariate time series models*, which are especially useful to study the dynamic relationships between several time series while maintaining a satisfactory level of prediction accuracy. In this chapter, univariate models are handled, but for details on multivariate time series models, the interested readers are referred to [1].

A time series may include four basic components: trend, seasonality, cycle, and random variation. *Trend* is a persistent and long-term upward or downward movement or tendency over time that can be classified as linear or nonlinear, or deterministic or stochastic. Basic models that can be used in the presence of a trend are given in Eqs. 1–3:

$$\text{Linear \& Deterministic Trend: } y_t = \beta_0 + \beta_1 t + \varepsilon_t \tag{1}$$

$$\text{Non - Linear \& Deterministic Trend: } y_t = \beta_0 + \beta_1 t + \beta_2 t^2 + \varepsilon_t \tag{2}$$

$$\text{Stochastic Trend: } y_t = \beta_0 + Y_{t-1} + \varepsilon_t \tag{3}$$

Seasonality is a repeated behavior that is tied to the calendar (e.g., weekly pattern observed in daily visits to the emergency department) [2]. *Cycle* is a repeated behavior due to economic growth or downsizing that is not tied to the calendar. *Random variation (noise)* is irregular and unpredictable fluctuations over time. A forecasting model may have integrated these components either in an additive or multiplicative manner.

The procedure to find an appropriate forecasting model is iterative and often tedious, with a series of model alternatives being built to fit data in hand and then compared to each other. It mainly involves the following steps:

Model identification: To identify tentative models of different types, the data in hand is first plotted against time to detect potential outliers (i.e., unusual events, shocks), missing values, or structural breaks (i.e., abrupt changes that cause the model parameters change permanently) and to see what type of trend and/or seasonality it exhibits (if it does). A comprehensive review on detecting univariate and multivariate *outliers* in a time series is provided in [3], and [4] presented a strategy to deal with outliers which influenced several other researchers [5, 6]. As a significant stream of research, outlier detection and correction has significantly grown with the proliferation of computationally intelligent methods [7]. A *structural break* refers to a change point that causes an increase or decrease in the mean and variance of the time series. Note that for a time series technique to perform well, the behavior of the time series should not vary significantly throughout time. In other words, the data in hand is expected to repeat its past behavior given a sufficiently long period of time in the future. To understand how the time series behaves, its pattern can be decomposed into sub-patterns using a time series decomposition technique such as STL, a seasonal trend decomposition procedure based on local regression smoothing [8]. If the trend is stochastic, taking the first difference of the original time series usually produces a mean-stationary series. In the presence of a time series with changing variance (i.e., heteroscedasticity), taking the logarithm of the series or performing a Box-Cox transformation [9] on it usually produces a variance-stationary series. Care should be taken when taking the logarithm of a differenced time series since it is highly likely to contain negative values. A proper constant should be thus added to the differenced series to avoid missing values. In some cases, where the first difference of the original series is still non-stationary in terms of mean, second-order differencing may be attempted; however, the interpretation for a higher order of difference may be complicated and impractical. Examining the autocorrelations and partial autocorrelations of the residuals or testing for a unit root are also suggested.

Parameter estimation: Once a tentative model is specified, the time series should be first split into an estimation (training) data set (to develop the models) and a validation (post-sample forecasting) data set (to validate the models). Recent research often adopted an expanding-window approach [10] in which the estimation data set is extended to include the very first observation in the validation data set and then a one-step-ahead forecast is generated for the second observation in the original validation data set. This process is repeated until the estimation data set contains all the observations in hand. Model parameters are estimated using a proper estimation method. Ordinary least squares (OLS) and its extensions, and maximum likelihood estimation (MLE) are widely used.

Diagnostic check: For each model, various model selection criteria such as Akaike Information Criterion (AIC), Root Mean Squared Error (RMSE), Mean Absolute Percentage Error (MAPE), Mean Absolute Scaled Error (MASE), and Theil's Inequality Coefficient (TIC) are calculated and the residuals and forecast errors are investigated through visual analysis to see how well the models fit the data. The model with the best fit may be selected immediately or the process returns to the first step generating the alternatives of this model.

Note that model building is not a one-time-task and even the best performing model may need to be refined and improved over time. Moreover, the ability of a time series model to produce accurate forecasts does not depend on how sophisticated it is. In general, simpler models perform well enough to use as a forecasting tool. As claimed by [11], "statistically sophisticated or complex methods do not necessarily produce more accurate forecasts than simpler ones." It was also agreed by [12] stating that "complex models usually give a better fit than simpler models but the resulting forecasts need not be more accurate." Nevertheless, the need for more complex models in some cases such as energy price forecasting [13] should not be overlooked.

2 Smoothing Methods

Research into smoothing methods has a long history. *Simple exponential smoothing (SES)* was introduced by R. G. Brown in the late 1950s. The value at the last observed time point and a correction constant are necessary to implement SES. Basically, a weighted moving average technique, SES has shown effectiveness especially in cases where forecasts should be made frequently and quickly. Since it requires only three values to make a prediction, SES is pretty simple and easy to update. It is particularly useful for generating forecasts in the short or medium term. SES assumes that the data is stationary. If the data exhibits trend or cycle, more advanced exponential smoothing techniques are required. SES was modified by C. C. Holt to work with data exhibiting trend, and then Winter modified Holt's version to model data with both trend and seasonality. These smoothing methods and several other versions of them were included in a standardized framework [14], which allows for the MLE of model parameters and the construction of prediction intervals, using the state space approach.

2.1 *Moving Average Methods*

In moving average methods, the next value in a time series x_t is simply assumed to be an average of the k previous observations. F_{t+1}, the forecasted value for time $t+1$, can thus be found using Eq. 4:

$$F_{t+1} = \frac{x_t + \cdots + x_{i-k}}{k} \tag{4}$$

The model in Eq. 4 is modified by assigning weights to each of these observations instead of taking their arithmetic mean. This Weighted Moving Average Model assumes that the next value in the time series is proportional to the weighted average of the k previous observations. The contribution of each previous observation to forecasting the value of the next observation can thus be different. F_{t+1}, the forecasted value for time $t + 1$ in this case is given as in Eq. 5:

$$F_{t+1} = w_1 x_1 + w_2 x_2 + \cdots + w_k x_{t-k} \quad (5)$$

where $\sum_{i=1}^{k} w_i = 1$.

Moving Average Models are not only used for smoothing time series but also in regression models, where moving averages are taken for error terms instead of past values.

2.2 *Simple Exponential Smoothing*

Simple Exponential Smoothing Method (SES) is a univariate time series forecasting method that is used when there is no trend and seasonality in the data. SES has been also called Single Exponential Smoothing, Brown's Simple Exponential Smoothing, or Exponentially Weighted Moving Average. Since the weights of past values decrease exponentially, the model uses the word exponential. This is a rather simple method that only uses past values. SES is not preferred for long-term forecasting since after a certain number of steps, forecasts approach to a constant value. The method can be mathematically expressed as in Eq. 6:

$$\hat{y}_{t+1} = \alpha y_t + (1 - \alpha)\hat{y}_t \quad (6)$$

where α, the *correction factor*, is a real number between 0 and 1, $t \in Z^+$ and $\hat{y}_0 = y_0$. When the number of observations is large, small values (e.g., as low as 0.05 or 0.10) for the correction factor lead to higher correction. Larger values for the correction factor, on the other hand, provide lower correction, but they respond to changes in the data more quickly. As an extreme example, if the constant is zero, the value of the next period will be equal to the value of the last period; if it is one, the value of the next period will be equal to the value of the current period.

2.3 *Double Exponential Smoothing*

SES is not efficient when the data exhibits trend. In such cases, the Double Exponential Smoothing method (DES), which has also been called Brown's Linear Exponential Smoothing, Brown's Double Exponential Smoothing, or Second-Order Exponential Smoothing, is adopted. In DES, aiming to eliminate the trend, the SES approach is applied to the data twice. The mathematical formulation of DES is given in Eqs. 7–8:

$$\hat{y}_{t+1} = \alpha y_t + (1 - \alpha)\big(\hat{y}_t - B_{t-1}\big) \quad (7)$$

$$B_t = \beta\big(\hat{y}_{t+1} - \hat{y}_t\big) + (1 - \beta)B_{t-1} \quad (8)$$

where B_t is the best estimate of trend at time t and β is the trend smoothing factor taking values between 0 and 1. For $t > 1$, $\hat{y}_1 = y_1$ and $B_1 = y_1 - y_0$.

2.4 *Triple Exponential Smoothing*

The Triple Exponential Smoothing (TES), which is also known as Holt-Winter's seasonal additive model, deals with the case where data exhibits both trend and seasonality. In this model, the seasonality effect does not change over time. TES consists of three equations as shown below. The seasonal effect is introduced to the model as in Eq. 11.

$$L_t = \alpha(y_t - S_{t-s}) + (1 - \alpha)(L_{t-1} + B_{t-1}) \quad (9)$$

$$B_t = \beta(L_t - L_{t-1}) + (1 - \beta)B_{t-1} \quad (10)$$

$$S_t = \gamma(y_t - L_t) + (1 - \gamma)S_{t-s} \quad (11)$$

where γ is the seasonality smoothing factor and S_{t-s} is the seasonal effect at time $t - s$. The k-step-ahead forecast $\hat{y}_{t+k}$ equals $L_t + kB_t + S_{t-s+k}$.

3 Box–Jenkins Models

Box–Jenkins models are basically divided into two main classes as linear stationary stochastic models and nonlinear stochastic models, depending on whether the examined time series is stationary or not. Linear stationary stochastic forecasting models are autoregressive (AR), moving average (MA) and a combination of these two: autoregressive moving average (ARMA). Non-stationary stochastic models are known as ARIMA. If the non-stationary model includes seasonal effects, it is called seasonal ARIMA (SARIMA).

The majority of real-world time series used in practice are non-stationary. The stationarity of these series deteriorates due to various reasons (e.g., trend), and ensuring stationarity in these models is a preliminary step for modeling. To ensure stationarity, an appropriate number of differences is considered. In ARIMA models,

Table 1 ARIMA processes

White noise	ARIMA (0, 0, 0)
Random walk	ARIMA (0, 1, 0) with no constant
Random walk with drift	ARIMA (0, 1, 0) with a constant
Autoregression	ARIMA (p, 0, 0)
Moving average	ARIMA (0, 0, q)

the degree of differencing is indicated by the parameter "d" and often the first difference ($d = 1$) is sufficient to eliminate the stochastic trend in the time series. In practice, it is not common for d to be greater than 2.

In ARIMA models, the "p" parameter from the AR model, the "q" parameter from the MA model, and the "d" parameter that expresses the differencing process are used and the model is written as ARIMA(p, d, q). ARIMA models can be the same as AR, MA or ARMA processes depending on the values of p, d and q. Special ARIMA processes are shown in Table 1.

A non-stationary time series can become stationary taking the dth order difference. The stationary time series can be expressed as ARMA(p, q) process as follows:

$$y_t = \sum_{i=1}^{p} \varphi_i y_{t-i} + \sum_{j=1}^{q} \theta_j \epsilon_{t-i} + \epsilon_t \tag{12}$$

where ϵ_t is white noise with zero mean and constant variance.

The difference operation is defined as generating a new time series by taking the difference of two consecutive data points as follows:

$$y_t' = y_t - y_{t-1}(13)$$

As in AR(p) and MA(q) models, the ideal model for the time series of interest can be decided by analyzing the ACF and PACF plot in ARIMA models. Yet, in practice, it is often not possible to make direct inferences from these plots. Therefore, the most suitable model should be selected by testing the compatibility of various alternatives based on the AIC (Akaike Information Criterion), SIC (Schwarz Information Criterion) and HQIC (Hannan–Quinn Information Criterion) criteria.

4 Time Series Forecasting Using Explanatory Variables

Time series techniques are intrinsically endogenous; they only consider the historical data patterns which are identified and projected into the future to obtain forecasts. However, for instance, when forecasting demand for a brand-new product, alternative approaches are required since historical data is not available. In such cases, regression-based forecasting is usually adopted.

Regression-based forecasting attempts to define a function that can relate the behavior of a time series (e.g., demand of a product) to a single or multiple time series (e.g., advertising expenditure, price, promotions) as correctly as possible. It requires several assumptions to generate accurate forecasts; however, the functional form (e.g., log-linear, curvilinear) can be often easily adapted to cases where violations to these assumptions are present. The formula for regression-based forecasting can be adopted as:

$$y_t = f(y_t, y_{t-1}, \ldots, y_{t-l}) \tag{14}$$

where $f(\cdot)$ is the regression model, such as linear regression, support vector regression, neural networks, etc., and l is the lag value.

5 Multi-frequency Modeling

Traditional time series methods such as TES can only capture simple seasonal patterns. On the other hand, most the time series in real life exhibits more complex seasonal patterns. Some time series may even have non-integer periods and some others may have multiple seasonal periods. In this section, two methods, namely, TBATS and Prophet that can help overcome seasonal complexities in time series modeling are explained [15, 16].

5.1 *TBATS*

TBATS is the acronym for Trigonometric Box–Cox transform, ARMA errors, Trend, and Seasonal components. It was developed to model multiple, complex seasonal patterns for time series. The traditional approach to incorporate two seasonal components into time series model is developed using the following state space model [17]:

$$F_t = l_{t-1} + b_{t-1} + s_t^1 + s_t^2 + \varepsilon_t \tag{15}$$

$$l_t = l_{t-1} + b_{t-1} + \alpha\varepsilon_t \tag{16}$$

$$b_t = b_{t-1} + \beta\varepsilon_t \tag{17}$$

$$s_t^1 = s_{t-m_1}^1 + \gamma_1\varepsilon_t \tag{18}$$

$$s_t^2 = s_{t-m_2}^2 + \gamma_2 \varepsilon_t \tag{19}$$

where F_t is the forecasted value, l_t and b_t correspond to level and trend at time t, respectively. The ith seasonal component at time t is represented by s_t^i. The notations for level, trend and seasonal smoothing parameters are α, β, γ_1, γ_2. Seasonal cycles are denoted with m_1 and m_2, and ε_t represents the white noise prediction error.

The model described above cannot handle nonlinearity and auto correlated noise in the time series. An extended model including Box-Cox transformation, ARMA errors and multiple seasonal periods is defined as follows:

First Box-Cox transformation is applied according to the following formula:

$$y_t = \begin{cases} \frac{y_t^{\omega}-1}{\omega}, \ \omega \neq 0 \\ \log y_t, \ \omega = 0 \end{cases} \tag{20}$$

Then, forecasted values for multiple seasonal components with ARMA errors are calculated as follows:

$$F_t = l_{t-1} + \phi b_{t-1} + \sum_{i=1}^{T} s_{t-m_i}^i + \varepsilon_t \tag{21}$$

$$l_t = l_{t-1} + \phi b_{t-1} + \alpha \varepsilon_t \tag{22}$$

$$b_t = (1-\phi)b + \phi b_{t-1} + \beta \varepsilon_t \tag{23}$$

$$s_t^i = s_{t-m_i}^i + \gamma_i \varepsilon_t \tag{24}$$

$$\varepsilon_t = \sum_{i=1}^{p} \varphi_i \varepsilon_{t-i} + \sum_{j=1}^{q} \theta_j \epsilon_{t-i} + \epsilon_t \tag{25}$$

where m_i is the ith seasonal period. While the local level and trend are denoted as l_t and b_t, respectively, b denotes the trend in the long-run. The ith seasonal component at time t is represented by s_t^i. The prediction error ε_t is defined by an ARMA(p, q) process with Gaussian white noise of ϵ_t. α, β and γ_i are smoothing parameters for level, trend and the ith seasonal component.

The model above cannot accommodate non-integer seasonal components. To overcome this challenge, a more flexible approach that is based on Fourier Series is defined as in Eqs. 26–28:

$$s_t^i = \sum_{j=1}^{k_i} s_{j,t}^i \tag{26}$$

$$s_{j,t}^{i} = s_{j,t-1}^{i} \cos \lambda_j^i + s_{j,t-1}^{*i} \sin \lambda_j^i + \gamma_1^i \varepsilon_t \tag{27}$$

$$s_{j,t}^{*i} = -s_{j,t-1}^{i} \sin \lambda_j^i + s_{j,t-1}^{*i} \cos \lambda_j^i + \gamma_2^i \varepsilon_t \tag{28}$$

k_i is the number of harmonics needed for the seasonal component i. The stochastic level for the ith seasonal component is denoted by $s_{j,t}^{i}$ and the change in the seasonal component i is represented by $s_{j,t}^{*i}$. λ_j^i is calculated by $2\pi j/m_i$ and the smoothing parameters are denoted as γ_1^i and γ_2^i.

5.2 The Prophet Package by Facebook

Prophet forecasting models consist of three components, namely trend, seasonal and holidays similar to the model developed by [18]. The formulation of the proposed model can be given as follows:

$$y(t) = g(t) + s(t) + h(t) + \varepsilon_t \tag{29}$$

where $g(t)$ is the trend component, $s(t)$ is the seasonal component, $h(t)$ is the holiday component, and the ε_t is the white noise error.

The linear component $g(t)$ can be calculated by using the basic logistic growth as follows:

$$y(t) = g(t) + s(t) + h(t) + \varepsilon_t \tag{30}$$

The basic trend with logistic grow model can be formulated as in Eq. 31:

$$g(t) = \frac{C}{1 + e^{-k(t-m)}} \tag{31}$$

where C is the carrying capacity, k and m are the growth rate and offset parameters, respectively.

The seasonal component of the series estimated using Fourier series as follows:

$$s(t) = \sum_{n=1}^{N} \left(a_n \cos \cos \left(\frac{2\pi nt}{P} \right) + b_n \sin \sin \left(\frac{2\pi nt}{P} \right) \right) \tag{32}$$

where a_n and b_n are the Fourier parameters with N components. Lastly, the holiday components are added as a regressor to the overall model.

6 Intermittent Time Series

Time series models are essential in predicting demand for tactical, operational, and strategic planning and resource allocation and to provide policymakers with detailed information in various settings; however, traditional time series forecasting techniques may not perform well when a time series includes zeros, taking nonzero values at irregular intervals in time. This problem is mostly encountered when a product moves slowly or its demand is intermittent. An intermittent time series is common in a wide range of industries such as automotive, aerospace, high technology, military, special chemicals, and telecommunication [19], and it violates the main assumptions (e.g., normality and homoscedasticity of error terms, stationarity of data) of traditional time series models [20].

Claiming that SES could fail in forecasting sporadic demand, an ad hoc method (CRSTN) that generates forecasts for two cases separately is proposed: zero demand, nonzero demand [21]. In the latter case, both the demand size and the time between two consecutive demand occurrences are estimated. CRSTN was slightly modified by [22] to generate approximately unbiased forecasts that improve as the mean time between demand occurrences increases. CRSTN was further modified by [23] considering the probability of demand occurrence rather than the expected inter-demand arrivals. This theoretically unbiased method (TSB) allows for the revision of demand size at each time point, regardless of whether a nonzero demand has occurred. A summary of TSB is provided below:

Let X_t and Y_t denote actual demand and demand size for an item in period t, respectively. Suppose that the estimates of these two quantities are X_t^* and Y_t^*. There are basically two cases: positive demand ($X_t > 0$) or no demand ($X_t = 0$) at time t. Let p_t indicate the estimate of the probability of positive demand at the end of period t: $p_t = P(X_t > 0)$. Then, $X_t^* = p_t Y_t^*$ and the values of p_t and Y_t^* are determined depending on whether a demand has occurred. For $X_t > 0$, $p_t = (1-\beta)p_{t-1}$ and $Y_t^* = Y_{t-1}^*$; for $X_t = 0$, $p_t = \beta + (1-\beta)p_{t-1}$ and $Y_t^* = \alpha Y_t^* + (1-\alpha)Y_{t-1}^*$, where α and β are smoothing constants that take on values within the interval [0, 1].

References

1. Tsay RS (2013) Multivariate time series analysis: with R and financial applications. Wiley
2. Marcilio I, Hajat S, Gouveia N (2013) Forecasting daily emergency department visits using calendar variables and ambient temperature readings. Acad Emerg Med 20(8):769–777
3. Blázquez-García A, Conde A, Mori U, Lozano JA (2021) A review on outlier/anomaly detection in time series data. ACM Comput Surv (CSUR) 54(3):1–33
4. Chen C, Liu LM (1993) Forecasting time series with outliers. J Forecast 12(1):13–35
5. Battaglia F, Orfei L (2005) Outlier detection and estimation in nonlinear time series. J Time Ser Anal 26(1):107–121
6. Li S-H, Chan W-S (2005) Outlier analysis and mortality forecasting: the United Kingdom and Scandinavian countries. Scand Actuar J 2005(3):187–211
7. Wang H, Bah MJ, Hammad M (2019) Progress in outlier detection techniques: a survey. IEEE Access 7:107964–108000

8. Robert C, William C, Irma T (1990) STL: a seasonal-trend decomposition procedure based on loess. J Official Stat 6(1):3–73
9. Kim Y, Kim S (2021) Forecasting charging demand of electric vehicles using time-series models. Energies 14(5):1487
10. Beyaztas U, Shang HL, Yaseen ZM (2021) A functional autoregressive model based on exogenous hydrometeorological variables for river flow prediction. J Hydrol 598:126380
11. Makridakis S, Spiliotis E, Assimakopoulos V (2018) Statistical and machine learning forecasting methods: concerns and ways forward. PloS one 13(3):e0194889
12. Chatfield C, Xing H (2019) The analysis of time series: an introduction with R. Chapman and Hall, CRC
13. Cruz A, Muñoz A, Zamora JL, Espínola R (2011) The effect of wind generation and weekday on Spanish electricity spot price forecasting. Electr Power Syst Res 81(10):1924–1935
14. Hyndman R, Koehler AB, Ord JK, Snyder RD (2008) Forecasting with exponential smoothing: the state space approach, Springer Science & Business Media
15. De Livera AM, Hyndman RJ, Snyder RD (2011) Forecasting time series with complex seasonal patterns using exponential smoothing. J Am Stat Assoc 106(496):1513–1527
16. Taylor SJ, Letham B (2018) Forecasting at scale. Am Stat 72(1):37–45
17. Taylor JW (2003) Short-term electricity demand forecasting using double seasonal exponential smoothing. J Oper Res Soc 54(8):799–805
18. Harvey AC, Peters S (1990) Estimation procedures for structural time series models. J Forecast 9(2):89–108
19. Hasni M, Babai MZ, Aguir MS, Jemai Z (2019) An investigation on bootstrapping forecasting methods for intermittent demands. Int J Prod Econ 209:20–29
20. Salinas D, Flunkert V, Gasthaus J, Januschowski T (2020) DeepAR: probabilistic forecasting with autoregressive recurrent networks. Int J Forecast 36(3):1181–1191
21. Croston JD (1972) Forecasting and stock control for intermittent demands. J Oper Res Soc 23(3):289–303
22. Syntetos AA, Boylan JE (2005) The accuracy of intermittent demand estimates. Int J Forecast 21(2):303–314
23. Teunter RH, Duncan L (2009) Forecasting intermittent demand: a comparative study. J Oper Res Soc 60(3):321–329

Erkan Isikli is Assistant Professor in Industrial Engineering at Istanbul Technical University (ITU). He received his Ph.D. in Industrial and Systems Engineering from Wayne State University, USA in 2012. He also holds a master's degree in Business Administration and a bachelor's degree in Mathematics Engineering, both awarded from ITU. His research interests mainly include applied statistics, time series forecasting, and uncertainty quantification. Dr. Isikli has published several papers in refereed academic journals including Journal of Cleaner Production, Energy Sources, Part B: Economics, Planning, and Policy, and Journal of Public Transportation.

Leyla Temizer received undergraduate degree in Statistics from Eskisehir Osman Gazi University, and her MSc degree in Industrial Engineering from Fatih University, Istanbul. She obtained her Ph.D. degree in Industrial Engineering from Istanbul University. Her research interests are applied statistics, forecasting, structural equation modeling and multivariate statistics.

Abdullah Emin Kazdaloglu is a research assistant and a Ph.D. student in indus-trial engineering department at Istanbul Technical University (ITU). He received his B.Sc. degree in management engineering from ITU in January of 2018 and also received his M.Sc. degree in industrial engineering from ITU in August of 2021. He has a passion for data science, machine learning and artificial intelligence and his current research interests are statistics, data science and business analytics methods. He is eager to develop his professional skills and curious to discover and implement new manners, technologies and tools especially in field of machine learning, statistic and computer science ecosystems, data analytics and business knowledge.

Emre Ari is a Research Assistant and a Ph.D. student in the Department of Industrial Engineering at Istanbul Technical University, Turkey. In 2016, he received his M.Sc. Degree in Statistics from Queen Mary University of London in UK and he received his B.Sc. degree in Mathematics from Kahramanmaraş Sütçü İmam University in 2009. His current research interests lie in the area Machine Learning, Deep Learning, Reinforcement Learning, Data Analysis, Financial Approaches and Statistics. He plans to continue his academic career by trying to finding new approaches in his interest areas.

Neural Networks and Deep Learning

Zeynep Burcu Kizilkan, Mahmut Sami Sivri, Ibrahim Yazici, and Omer Faruk Beyca

1 Introduction

Artificial neural network is a well-known machine learning technique inspired by biological neural network structures. It mimics the human brain's working mechanism by artificially forming a neural network. In this book, artificial neural networks are referred to as neural networks. The principal idea of a neural network is to show transformation between input and output as connections between neurons in a sequence (arrangement) of layers [1]. Neural networks are mostly used for prediction, decision-making, pattern recognition, and novelty detection [2]. The first is that without domain expertise, neural networks may assist in estimating function structures and parameters [2].

Neural network is not a new technique. It is a well-searched and developed technique since 1950s. Computational neural networks were founded by McCulloch and Walter Pitts [3]. Both the biological processes in the brain and the application of neural networks in artificial intelligence were the focus of these models [3]. Frank Rosenblatt introduced the term perceptron in 1958 [4]. Perceptrons were the first neural network to be able to learn the weights [3]. In 1980s, developments in neural networks accelerated again [3].

In a biological neuron, there are some parts like axon, dendrites, and synapse [5]. A basic neural network structure consists of input neurons (sometimes called input nodes), hidden layers, and output neurons, whereas some also might include bias values and some activation (transformational) functions to help with the computation of output values using some concepts called gradient descent and backpropagation. Neurons are linked to each other and have individually assigned (multiplicative) weights. With the help of an activation function, input values are transformed into

Z. B. Kizilkan · M. S. Sivri · I. Yazici · O. F. Beyca (✉)
Department of Industrial Engineering, Faculty of Management, Istanbul Technical University, 34367 Macka, Istanbul, Turkey
e-mail: beyca@itu.edu.tr

A. Ustundag et al. (eds.), *Business Analytics for Professionals*, Springer Series in Advanced Manufacturing, https://doi.org/10.1007/978-3-030-93823-9_5

output values in hidden layers. There are many types of functions both for linear and nonlinear modeling since machine learning algorithms are not one size fits all. The purpose of this calculation is to generate an output value that is acceptable in terms of some metrics and requirements. Then, some metrics like accuracy and R^2 are used to measure the model's performance. For example, a higher rate of accuracy is desired, but it is not a requirement since the selection of a model considers more factors.

Dataset is very important for the performance of the model. The best model can be found by changing the number of neurons, the layers, and the type of activation function. Activation functions are used to generate output from original input values. For different purposes, there are various types of neural network structures. Neural networks can generate both linear and nonlinear models. In this chapter, feedforward neural networks, convolutional neural networks, recurrent neural networks, and sequential modeling will be explained in detail.

2 Feedforward Neural Networks

Feedforward neural networks are used to obtain an output value. Feedforward means making a forward pass from input neurons through hidden layers and finally generating an output using the pre-established weight and bias values. Input is propagated through all the hidden layers to reach the output layer. At the output layer, the output computed by the neural network is compared to the actual output. And the difference among them is calculated. This difference is called the cost, and this cost is propagated back into the network so that all the parameters (weight, bias) inside the layers are tuned so that it produces an output that is closer to the actual output. The entire process is called training, and it is repeated until the cost is very small; then, it can be said that the neural network has learned the pattern.

Backpropagation is predominantly used to train feedforward neural networks. A backpropagation technique is employed to determine the gradients [1]. Gradient descent is an optimizer for neural networks that is frequently used. The network parameters may thus be changed using these gradients. Biases and weights of the network which are called parameters are updated by a corresponding loss function derivative. Updating the parameters to a new value in the gradient's direction reduces the loss [1]. Nevertheless, it is important to be aware of the occurrence of local optima problems. As gradient descent leads to the steepest slope, this method attempts to locate local solutions rather than global ones even if gradient descent aimed to locate the global optimum originally [6].

Different structures of neural networks can be built depending on the dataset, data type and output type. Single-layer neural network consists of input neurons directly connected to an output neuron. This type of neural network is also called perceptron, and it uses a linear function to calculate an output. Multi-layer neural network consists of hidden layers between the input and output layers. The best result

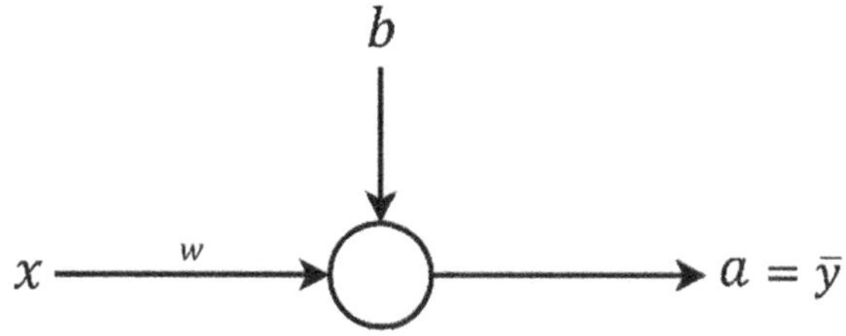

Fig. 1 Single input neuron and single-layered neural network

can be obtained by comparing multiple neural network models. Here are some neural network examples:

2.1 *Single Neuron Single Layer*

In a neural network formed with a single input neuron and a single layer as shown in Fig. 1, there is an input x and output $\overline{y}$. Weight associated with the input is w and bias is b. The output calculated by the input, weight, and bias values is called z, and it is defined as $z = w \times x + b$. Then, an activation function σ is applied to z resulting with a. For a single-layered neural network, a also becomes the final output $\overline{y}$. This single neuron consists of two calculations. The first one is the above equation, and the second is to compute the activation function.

Notations

x input neuron
$\overline{y}$ output neuron
w weight associated with the first layer
b bias associated with the first layer
z output of the first layer
a activation function applied output of the first layer
σ an activation function.

Equations

$$z = \boldsymbol{w} \times \boldsymbol{x} + \boldsymbol{b} \tag{1}$$

2.2 *Multi-layered Neural Networks*

Multiple inputs and multiple neurons in the hidden layer. Each input neuron is connected to all of the neurons in the first layer. It is called a fully connected layer.

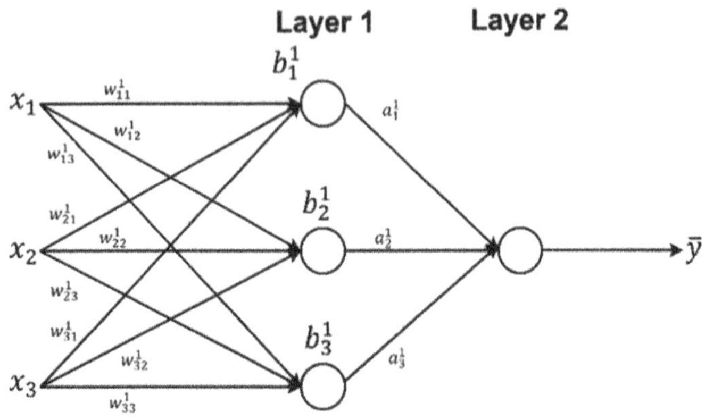

Fig. 2 Multi-layered neural network with multiple neurons

So, there is nxm number of weights formed by combinations among input neurons and neurons in the first layer.

In Fig. 1, for a neural network consisting of three input neurons and three neurons in the first layer, weight associated with x_1 and n_1 is called w_{11}^1 where n_1 is the first neuron in the first layer. Bias of n_1, n_2 and n_3 are b_1^1, b_2^1 and b_3^1 accordingly. The conventional approach is to find z_k^j and a_k^j values separately with the following equations (Fig. 2):

Layer 1

$$z_1^1 = w_{11}^1 \times x_1 + w_{21}^1 \times x_2 + w_{31}^1 \times x_3 + b_1^1 \tag{2}$$

$$z_2^1 = w_{12}^1 \times x_1 + w_{22}^1 \times x_2 + w_{32}^1 \times x_3 + b_2^1 \tag{3}$$

$$z_3^1 = w_{13}^1 \times x_1 + w_{23}^1 \times x_2 + w_{33}^1 \times x_3 + b_3^1 \tag{4}$$

$$a_1^1 = \sigma\left(z_1^1\right) \tag{5}$$

$$a_2^1 = \sigma\left(z_2^1\right) \tag{6}$$

$$a_3^1 = \sigma\left(z_3^1\right) \tag{7}$$

Layer 2

$$z_1^2 = w_{12}^2 \times a_1^1 + w_{22}^1 \times a_2^1 + w_{32}^1 \times a_3^1 + b_1^2 \tag{8}$$

$$a_1^2 = \sigma\left(z_1^2\right) = \overline{y} \tag{9}$$

x_i ith input neuron
$\overline{y}$ output neuron
n_k kth neuron in the first layer
w_{ik}^j ith input neuron' s weight associated with kth neuron in the jth layer
b_k^j bias associated with the kth neuron in jth layer
z_k^j output of the kth neuron in jth layer
a_k^j activation function applied output of the kth neuron in jth layer
σ any activation function
i $= 1, 2, 3$
j $= 1, 2$
k $= 1, 2, 3.$

2.3 Dot Product Solution

If the number of inputs neurons and neurons in the layer are equal, instead of calculating z_k^j values separately if the conditions are satisfied, dot product approach can be applied to the neural network to compute the a_k^j values together.z_k^j matrices seen below can be combined and written as Z^1, which makes the computations faster and more efficient.

$$Z_1^1 = \left[w_{11}^1 \; w_{21}^1 \; w_{31}^1\right] \cdot \begin{bmatrix} x_1 \\ x_2 \\ x_3 \end{bmatrix} + b^1 \tag{10}$$

$$Z_2^1 = \left[w_{12}^1 \; w_{22}^1 \; w_{32}^1\right] \cdot \begin{bmatrix} x_1 \\ x_2 \\ x_3 \end{bmatrix} + b^1 \tag{11}$$

$$Z_3^1 = \left[w_{13}^1 \; w_{23}^1 \; w_{33}^1\right] \cdot \begin{bmatrix} x_1 \\ x_2 \\ x_3 \end{bmatrix} + b^1 \tag{12}$$

$$Z^1 = \begin{bmatrix} w_{11}^1 \; w_{21}^1 \; w_{31}^1 \\ w_{12}^1 \; w_{22}^1 \; w_{32}^1 \\ w_{13}^1 \; w_{23}^1 \; w_{33}^1 \end{bmatrix} \cdot \begin{bmatrix} x_1 \\ x_2 \\ x_3 \end{bmatrix} \tag{13}$$

$$Z^1 = \begin{bmatrix} w_{11}^1 \cdot x_1 \; w_{21}^1 \cdot x_2 \; w_{31}^1 \cdot x_3 \\ w_{12}^1 \cdot x_1 \; w_{22}^1 \cdot x_2 \; w_{32}^1 \cdot x_3 \\ w_{13}^1 \cdot x_1 \; w_{23}^1 \cdot x_2 \; w_{33}^1 \cdot x_3 \end{bmatrix} + \begin{bmatrix} b_1^1 \\ b_2^1 \\ b_3^1 \end{bmatrix} \tag{14}$$

$$A^1 = \sigma\left(Z^1\right) = \begin{bmatrix} a_1^1 \\ a_2^1 \\ a_3^1 \end{bmatrix} \tag{15}$$

Generalized Dot Product Solution for Feedforward

If $i = k$;

$$Z^j = W^j \cdot A^{j-1} + b^j \tag{16}$$

$$A^j = \sigma\left(Z^j\right) \tag{17}$$

$$A^J = \overline{y} \tag{18}$$

2.4 *Loss and Cost Functions*

After running, feedforward $\overline{y}$ is achieved. The next goal is to find the error/cost of predicted output compared with the actual output. Loss function computes the error in the predicted sample $\overline{y}_t$ to the actual value of y_t. y_t is a constant target value. $\overline{y}_t$ is the predicted value computed by an activation function. The weight and the bias values change depending on the neural network. Cost function can be defined as the average of the loss value computed over the number of i samples. The goal is to minimize the cost function. Since y_t is constant, change in cost function highly depends on $\overline{y}_t$ which consists of weight w and bias b. The aim is to find w and b values that conclude a minimum cost or 0. Therefore, the gradient descent method is used.

$C = \frac{1}{i}\sum_{t=1}^{i} L\left(y_t, \overline{y}_t\right)$	(Cost function)
i	number of samples
y_t	actual value
$\overline{y}_t = w \times a + b$	(Predicted value)
$C = f(w, b)$	Cost C is a function of w and b.

2.5 *Gradient Descent*

Gradient descent is the process of maximization or minimization of functions [7]. In neural networks, gradient descent helps with finding the optimal w and b values where the cost function gives the lowest value. In gradient descent, learning equations are used to update the weight and bias values. A derivative of a function on a particular point is equal to the slope of that function for that point [8]. Partial derivatives of the biases and weights concerning the cost function corresponds to the amount of change in weight and bias values. Gradient descent aims to find the lowest value of the cost function, using the amount of change of these parameters to update the weight and bias values in the next iteration using backpropagation. Essentially, it is desired to reach a cost function value of zero, but for some cases, this may not be achievable.

Normally, cost function C mainly depends on weight and biases ($C = f(w, b)$). But to demonstrate it in a two-dimensional graph, it is assumed that cost function is only dependent on weight ($C = f(w)$). The goal is to reach the lowest value of the function which is called global minimum. By updating the weight value after each iteration w gets closer to the global minimum. The derivative of a function at a certain point is equivalent to the slope of the original function [8]. For a value on the graph, taking the partial derivative of the weight with respect to the cost function $\left(\frac{\partial c}{\partial w}\right)$ gives the slope value of the point. A triangle is formed where the line of slope for that point forms the hypotenuse. As in Fig. 3, the steep edge of the triangle corresponds to ∂c while the horizontal edge corresponds to ∂w. The slope can be negative or positive. Aside from the sign of the slope, w becomes closer to the global minimum at each iteration with the use of the learning equations. After each iteration, slope of the updated point is less steep than the previous point since ∂w grows more than ∂c.

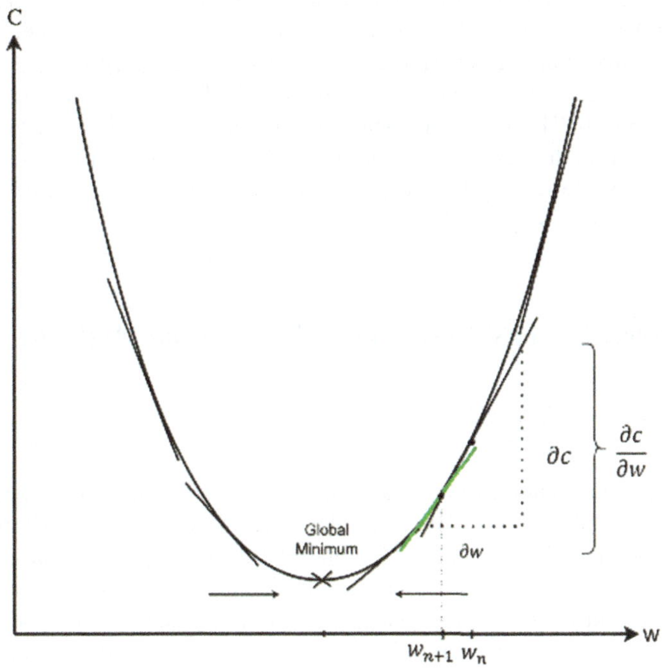

Fig. 3 Demonstration of gradient descent using cost function and weight values

Therefore, the updated amount on the learning equations becomes smaller than the previous iteration. After several iterations, the desired w value that gives the lowest cost value is reached. Then, this value is used in backpropagation to update the neural network.

Learning Equations

$$w_{n+1} = w_n - \alpha \frac{\partial c}{\partial w_n} \tag{19}$$

$$b_{n+1} = b_n - \alpha \frac{\partial c}{\partial b_n} \tag{20}$$

$\frac{\partial c}{\partial w}$ partial derivative of weight with respect to cost
$\frac{\partial c}{\partial b}$ partial derivative of bias with respect to cost
n iteration
α learning rate.

2.6 *Backpropagation*

Backpropagation is a supervised learning algorithm that is used for multi-layered feedforward neural networks [9]. Using the gradient descent approach, the backpropagation algorithm seeks the minimum of the error/cost function in weight space [10]. The error/cost value is propagated through the layers allowing their weights to be adjusted [9]. The same approach is also used for bias values. The chain rule is implemented in the backpropagation stage [10]. In this manner, any sequence of function compositions may be assessed, and its derivative can be retrieved during the backpropagation stage [10].

3 Recurrent Neural Networks and Sequential Modeling

3.1 *Recurrent Neural Networks*

In a feedforward neural network, the input data is passed through the network to produce an output value. Therefore, the incoming data is just forwarded. The computed output value is compared to the real data to determine the error. The weight values on the network are changed depending on the error, and in this way, a model that can output the most accurate result is created. In this type of neural networks, there is no need to have any connection with previous or next information. So, there is no concept of time or order; the only input it is interested in is the current instance at the time.

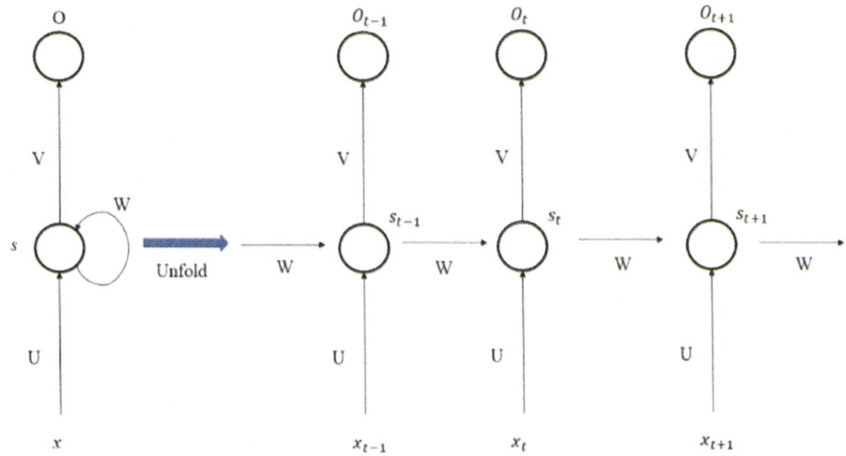

Fig. 4 Representation of an RNN

In recurrent neural networks (RNN), the result is derived not only from the current input but also from other inputs. In RNN, besides the input data at time t, the hidden layer results from time $t-1$ are the input of the hidden layer at time t. The decision made for the input at time $t-1$ also affects the decision to be made at time t. In other words, in these networks, inputs produce output by combining current and previous information. Recurrent networks are used to understand the structure of data coming in a sequence such as text, speech, or statistical data depending on time [11].

A recurring network consists of several copies of the same network, each sending a message to the next network. Figure 4 represents an unfolded RNN.

x_t input of time t
s_t hidden state of time. It is called memory state and is calculated with the previous hidden state and the input at the current step. It is formulated as $s_t = f(Ux_t + Ws_{t-1})$
f activation function such as sigmoid, tan$h$ which provides nonlinearity
U weight of the sequence element
W weight of the memory state
O_t output of time t. It is calculated as $O_t = VS_t$.

RNNs, in fact, conduct the same computation on various inputs of the same sequence at each instance. Since the same parameters are used in each step in the RNN network, the number of parameters that the network needs to learn during the training phase is greatly reduced. Therefore, it also reduces training times.

We have said that recurrent neural networks are suitable for working on sequential input data. Sequential data can be used as inputs, outputs or both in recurrent neural networks. Below are some examples of the different types of recurrent neural networks and the problems in which these types are used (Fig. 5).

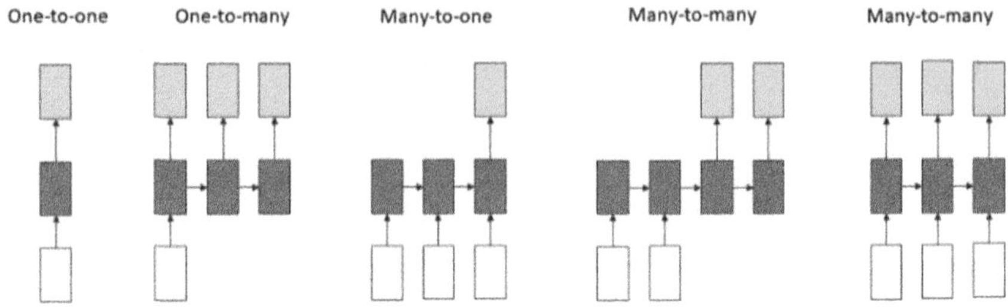

Fig. 5 Different types of RNN

One-to-one is the traditional neural networks and image classification is the most common example for these types of networks. The generation of music is a simple example of how one-to-many RNN may be put to use. Sentiment analysis in which a given text is evaluated as conveying either positive or negative sentiment can be a model for many-to-one architecture. Many-to-many architectures such as machine translation models takes a phrase in a language as input and outputs a sentence in another language. In video classification problems, every frame of the video is labeled. These types of problems can be also modeled like many-to-many RNN. This architecture is also called bidirectional many-to-many RNN.

Recurrent networks are used to make accurate predictions on sequential inputs. These procedures are carried out using the backprop and gradient descent of the error. Backprop is performed by calculating gradient of the error function with respect to weights starting from last layer and going backward to first layer. The learning coefficient is arranged using gradient descent, and the weights are modified to minimize error using this derivative.

Gradient is a value that allows us to change all of the weights. The effect of the error, on the other hand, diminishes significantly in long networks connected to one another, and the gradient may begin to vanish. As a result, it may be impossible to find the correct result. Because all layers and time-dependent steps are multiplied together, their derivatives have the potential to vanish or fly high. The LSTM method is one of the structures designed to solve this problem.

Advantage of RNNs can listed as follows:

- It relates to the previous example. In this way, the entries are progressed without forgetting.
- It has wide usage areas such as text generation, machine translation, time series forecasting, and sentiment analysis.

Disadvantages of RNNs can be listed as follows:

- Gradient exploding and vanishing problems.
- It has difficulty processing long inputs.

3.2 Long Short-Term Memory

As mentioned above, recurrent neural networks have problems associated with vanishing and exploding gradients. Long short-term memory (LSTM) architecture, a type of recurrent neural network, was created by Hochreiter and Schmidhuber in 1997 [12]. This type of neural networks has recently regained popularity in the context of deep learning as it solves the problem of vanishing gradients. LSTM-based networks are ideal for time series estimation and sequential classification problems. It is a structure that receives information outside of the normal flow. This information can be stored, written to the cell and read.

Gates in the LSTM structure determine what will be remembered and what will be forgotten. That is, if the incoming input is unimportant, it is forgotten, if it is important, it is transferred to the next layer. It does this with the help of forget and cell gate. LSTM network has four identical layers. Figure 6 shows the block diagram of a LSTM layer.

Forget Gate

It decides what information is kept or forgotten. Current information and information from the previous hidden layer pass through the sigmoid function. The closer it is to 0, the more it will be forgotten, the closer it is to 1, the more it will be retained. Value of the forget gate is calculated as follows:

$$f_t = \sigma\left(W_f.\left[h_{t-1}, x_t\right] + b_f\right) \tag{21}$$

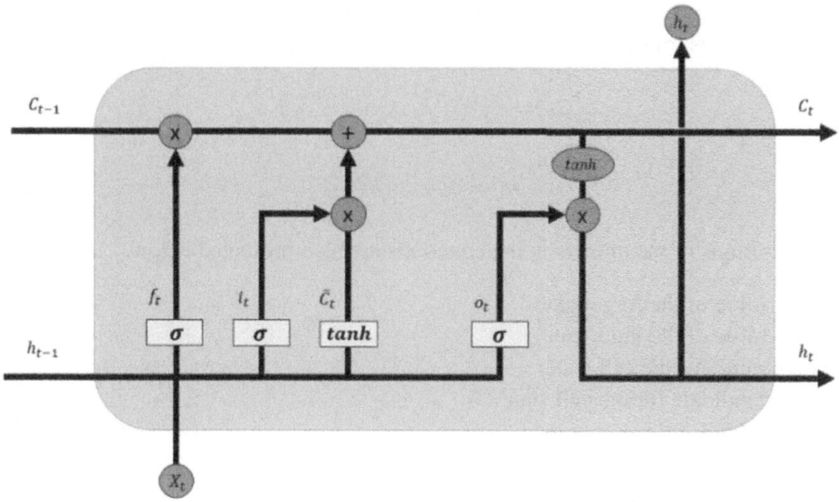

Fig. 6 Block diagram of an LSTM layer

Input Gate

It is used to update the cell state. First, the sigmoid function is applied, and it is decided which information to keep. Then, in order to organize the network, it is reduced to $-1, 1$ with the help of the tan$h$ function and the two results are multiplied. Value of the input gate is calculated as follows:

$$i_t = \sigma\left(W_i.\left[h_{t-1}, x_t\right] + b_i\right) \tag{22}$$

Cell State

It takes the data that needs to be moved and moves it to the end of the cell and from there to the other cells. In other words, we provide the data flow on the network with the help of cell state. First, the result from forget gate is multiplied by the result of the previous layer. It is then summed with the value from the input gate.

$$\tilde{C}_t = \tan h\left(W_c.\left[h_{t-1}, x_t\right] + b_c\right) \tag{23}$$

$$C_t = f_t * C_{t-1} + i_t * \tilde{C}_t \tag{24}$$

Output Gate

It decides the value to be used for the prediction to be sent to the next layer. First, the previous value and the current input pass through the sigmoid function. After the value from the cell state passes through the tan$h$ function, the two values are multiplied and go to the next layer as the "previous value." Value of the output gate is calculated as follows.

$$o_t = \sigma\left(W_o.\left[h_{t-1}, x_t\right] + b_o\right) \tag{25}$$

$$h_t = o_t * \tan h(C_t) \tag{26}$$

Definitions of parameters in the equations are also presented below.

f_t value of the forget gate
i_t value of the input gate
$\tilde{C}_t$ value for the cell state
C_t candidate for the cell state
o_t value for the output gate
x_t input at current timestamp
h_t output of the current block
h_{t-1} output of the previous block
W weight for the neurons
b bias for the corresponding layer.

3.3 Example: Power Consumption Forecast

For the measurements of electrical energy consumption in a household with a sampling rate of one minute over a period of about 4 years, the consumption forecast for the next time period was made using the LSTM method. The dataset contains different electrical quantities and some sub-measurements. Table 1 shows the dataset descriptions [13].

We made one-minute prediction with the LSTM method, using the 2-h information before the predicted time. Below we presented the graph of error reduction in the train and test sets of the LSTM model (Fig. 7).

Finally, Table 2 shows the results of both LSTM and MLP methods in terms of mean absolute error (MAE).

Table 1 Dataset description

Variable name	Variable description
Global_active_power	The total active power consumed by the household (kilowatts)
Global_reactive_power	The total reactive power consumed by the household (kilowatts)
Voltage	Minute-averaged voltage (volts)
Global_intensity	Household global minute-averaged current intensity (amperes)
Sub_metering_1	Active energy for kitchen (watt-hours of active energy)
Sub_metering_2	Active energy for laundry (watt-hours of active energy)
Sub_metering_3	Active energy for climate control systems (watt-hours of active energy)
Sub_metering_4	Remaining watt-hours

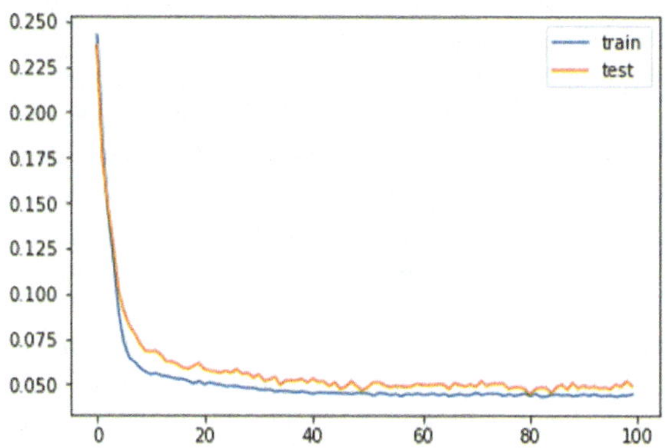

Fig. 7 Mean absolute errors of train and test sets for LSTM iterations

Table 2 The results of LSTM and MLP

Method	MAE
LSTM	0.385
MLP	0.538

4 Convolutional Neural Networks

In machine learning context, neural networks are remarkably one of the most effective tools deployed for many problems in the literature. Neural networks started as simple perceptron; later on, its advanced architecture evolved into other forms such as feedforward networks, recurrent neural networks (RNN), and convolutional neural networks (CNN). Despite RNN and CNN date back to 1990s, their uses were rare due to some reasons. Requirement of big datasets and benchmarks, lack of computational hardware were the main obstacles hindering large-scale deployment of these types of neural networks. However, with the recent advances in algorithmic advances, availability of big datasets due to big data regime and Internet surge, developing high-performance graphics chips such as graphical processing units led to spread use of these networks. These three forces can be seen as driving forces for emerging deep learning advancements [14].

Recurrent neural networks are used for sequence modeling such as speech recognition, text classification, audio processing, and time series forecasting tasks and the most prevalent forms of recurrent neural networks are long short-term memory and gated recurrent units. On the other hand, convolutional neural networks (CNN) are particularly used for image recognition, object detection tasks. Yet CNN has been used for sequence modeling domain, and they have produced promising results for problems in this domain as well.

In this part (or chapter), CNN layer and its main properties will be given in detail. Before getting into CNN, historical background and biological inspiration of the networks may provide some intuition to better conceptualize them. Experiments conducted by Hubel and Wiesel [15] through examining a cat's visual cortex paved the way for convolutional neural network idea [8]. It was found that small regions of the visual cortex are sensitive to specific regions in a visual field by means of special cells on the relevant regions in the cortex. When these small cell regions of the visual cortex are activated, their relevant parts in the visual cortex will also be activated. In addition, these cells are also sensitive to shape and orientation of objects in the visual field which means that they detect edges, corners in the visual field by activating some neuronal cells specific to the shape or orientation of the visual field. A horizontal edge triggers the neuronal cells, relevant to horizontal edge, to be excited. This property is also the same for vertical edge and their specific neuronal cells in the visual cortex. On the other hand, by examining the cells formation through using a layered architecture, this finding led to the idea which different levels of abstractions constructed by portions of a visual can be obtained through these different layers. Accordingly, this idea is similar to hierarchical feature extraction in machine learning context in the way that while earlier layers of CNN detect primitive shapes in a visual, last layers, on the other hand, detect more complex shapes in the visual [16].

Fundamentally convolutional neural networks consist of three main layers:

- Convolution Layers
- Pooling Layers
- Dense Layers.

4.1 Convolution Layer

CNN were introduced by LeCun et al. [17] for an image recognition task. CNN is a special type of neural network used for data having grid-like topology. CNN makes use of sparse connections and parameter sharing properties effectively, and, in turn, produce promising results with less memory requirement that will be mentioned in the following sections. Rather than connecting all states in a specific layer indiscriminately to the states in the previous layer to get a feature map value for a visual, CNN connects a feature value in a particular layer to a local spatial region in the previous layer through parameter sharing property throughout the whole visual footprint. This mechanism construction is in line with the insights obtained by Hubel and Wiesel's cat visual cortex experiments [16].

CNN work with grid-like topologies as mentioned earlier, and these topologies may be in the form of time series data as 1D grid, in the form of an image data as 2D grid of pixels, or in the form of a video data as 3D grid of pixels. In mathematical sense, CNN employs a mathematical operation called convolution; hence, they use convolution over the topologies to get feature maps of given input. Using convolution operation instead of general matrix multiplication in one of their layers makes CNN distinct from conventional neural network types [8].

The convolution operation is just a linear operation over two functions of a real-valued argument. As an example of convolution operation, tracking a car's position with GPS may be considered. A signal for spotting location of the car is received as a single output that is $x(t)$ at time t. $x(t)$ and t are both real-valued arguments that different values can be obtained at any instant time t. On the other hand, the received signals might contain some level of noise. Hence, to get more accurate measurements for the location of the car, several measurements obtained can be averaged. In this sense, more weight must be given to more recent measurements because they are more relevant to desired output. Giving more weights to more recent measurements is done by introducing a weighting function $w(a)$. In the weighting operation, a corresponds to age of a measurement. Then, less noisy measurements of the location of the car can be obtained by taking weighted average of measurements at every time point as obtained function f represents smoothed estimate of several measurements. The formula to obtain f is given in Eq. 27:

$$f(t) = \int x(a)w(t-a)d(a) \tag{27}$$

The formula given in Eq. 27 corresponds to convolution operation. In the compact form, convolution between two functions is typically represented by an asterisk as per given in Eq. 28:

$$f(t) = (x * w)(t) \tag{28}$$

In these formulas, w has to denote a valid probability density function. In addition, w has to be 0 for all negative arguments. In the sense of the problem framed for spotting the location of the car, mentioned limitations are specific to the problem. In a broader sense, convolution operation may be applied to any functions that the integral given in Eq. 27 is defined. The operation may be used beyond the aim of taking weighted averages as well [8].

In the context of convolutional networks, the first argument of the car example, that is the function x, is called as the input, and the second argument of the car example, that is the function w, is called the kernel. The resultant output of this convolution operation is oftentimes called as the feature map [8].

In the example of spotting the location of the car, signal receiving may be in discrete time rather than continuous one; hence, convolution operation can be represented by using summation operation for this discrete time event rather than by using integral operation. In this sense, t can take on only integer values, thereby forming the formula as per given in Eq. 29:

$$f(t) = (x * w)(t) = \sum_{a=-\infty}^{\infty} x(a)w(t-a) \tag{29}$$

The input of convolution operation is a multidimensional array of data used, and the kernel is a multidimensional array of parameters obtained by the learning algorithm used. The multidimensional arrays are simply dubbed as tensors.

In more general cases, convolution operations are performed over more than one axis at a time. The most prevalent example of this utilization is using two-dimensional image I as an input so that two-dimensional kernel K utilization will need as per given Eq. 30:

$$S(i,j) = (I * K)(i,j) = \sum_p \sum_q I(p,q)K(i-p, j-q) \tag{30}$$

Convolution operation also holds the commutative property; hence, it can be written in the form as in Eq. 31:

$$S(i,j) = (K * I)(i,j) = \sum_p \sum_q I(i-p, j-q)K(p,q) \tag{31}$$

The formula in Eq. 31 is more straightforwardly applied in a machine learning library due to less variation in the range of valid p and q values [8].

The commutative property emerges as the kernel relative to the input is flipped since the index into the input increases along as a result of m's increase. On the other hand, the index into this kernel decreases. Hence, flipping the kernel enables to get the commutative property. Yet, in machine learning context, holding this property is not significant. Due to this reason, cross-correlation operation, which is the same with convolution but does not perform flipping the kernel, is usually used in neural network library implementations. This operation's formula is given in Eq. 32:

$$S(i, j) = (K * I)(i, j) = \sum_p \sum_q I(i + p, j + q)K(p, q) \tag{32}$$

Convolution and cross-correlation terms are interchangeably used in the context of machine learning, because convolution with flipping is not relevant in this context. Hence, many machine learning libraries employ cross-correlation in their implementations without destroying the essence of the convolution operation. In addition, convolution operation is used in hybrid form with other functions simultaneously, and combination of these two functions has no commutative property without considering whether the kernel is flipped or not [8].

An illustrative example of convolution without kernel flipping applied to 2D data can be seen in Fig. 8.

In Fig. 8, the kernel is slided over the entire image, resultant outputs of convolution operations are given. In this figure, the outputs of the kernel are restricted to the only position in which the kernel can cover the whole image. In this case, each slide operation is performed by one-shift stride [8].

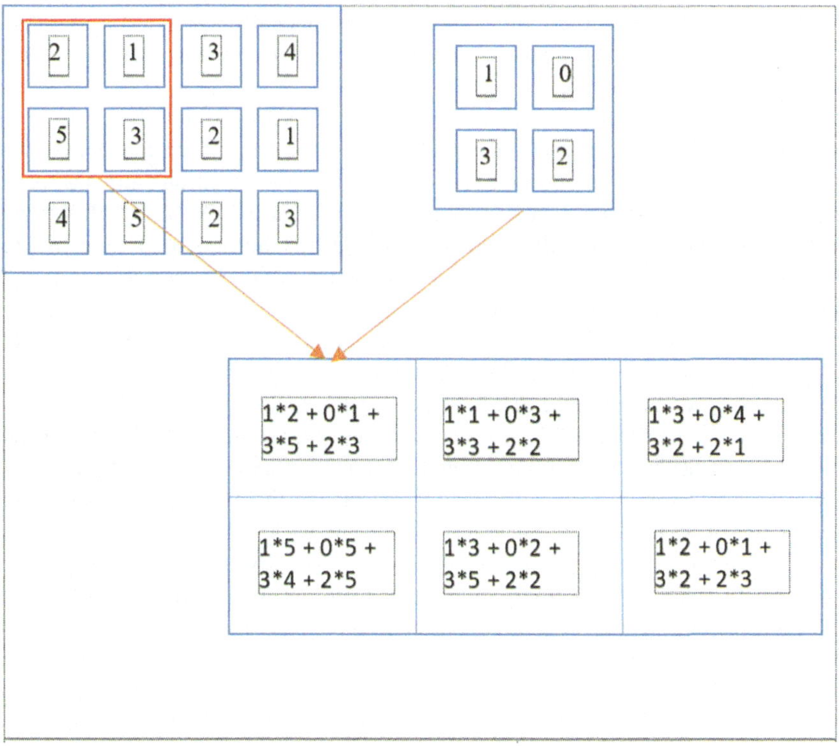

Fig. 8 An illustrative example of the convolution

Structural Properties of CNNs

In the former sections, CNN reaps the benefits of parameter sharing, sparse interactions and equivariant representations along with convolution operations that contribute to high performance of CNN in many tasks. In traditional neural networks, dense connections which each input entry is linked to each output entry by means of matrix multiplication constitute layers of the networks. Parameters of this layer are stored in the weight matrix as the result of matrix multiplication. However, CNN makes use of sparse connections between input and output entries through using kernels smaller than inputs. In other words, by using small kernels having tens or hundreds of pixels, meaningful features such as edges and corners can be detected from an image having thousands or millions of pixels. This feature contributes to memory requirements of the model, and statistical efficiency of the model increases as well. Along with the mentioned advantages, less computation is needed as a result of this feature. For example, matrix multiplication operation for m inputs and n outputs require mxn parameters; hence, time complexity of the operation per example is $O(mxn)$. On the other hand, if number of connections of each output are restricted to k, then time complexity of the operation decreases to $O(kxn)$. Hence, k very much smaller than m helps decrease time complexity dramatically.

Graphical illustrations of sparse connectivity, both from below and above view, are exhibited in Figs. 9 and 10.

In Fig. 9, at the bottom of the figure, dense connection of one entry to its successor layer is seen with boldly depicted neurons [8]. All other neurons are connected to the succeeding layer with full connection as well. On the other hand, an entry has sparse connection to neurons in its succeeding layer is seen at the top in Fig. 9. While a matrix multiplication in which the relevant entry affects the all outputs forms weight matrix for the dense layer, a convolution with kernel size of 3 is performed through sparse connectivity for convolution layer [8].

In Fig. 10, outputs of both at the above and below layers affected by their predecessor layers are represented to show the difference between sparse connection and dense connection in the sense that how the outputs are affected by each connection type relevant to its preceding layers. Output in the layer at the bottom, s_3, is affected by all of its predecessor inputs through dense connections (via matrix multiplications) while its counterpart at the above one is only affected by three of its predecessors with kernel size of 3 via using sparse connections. The units that one output is affected by is referred to as receptive field of the output. Receptive fields of a convolution layer are broader than a densely connected layer, and this feature contributes to performance of CNN to extract more meaningful features from given input. Another remarkable feature of a convolution layer is the parameter sharing property. With this property, instead of performing a separate set of parameters learning for every location for an image, only one set of parameters for every location is learned by convolution operation. Figure 11 shows an example of parameter sharing property in a convolution layer in comparison with a densely connected layer [8].

Connections between input entries and output entries are indicated with black arrow(s) in Fig. 11. These connections have particular parameters in each model.

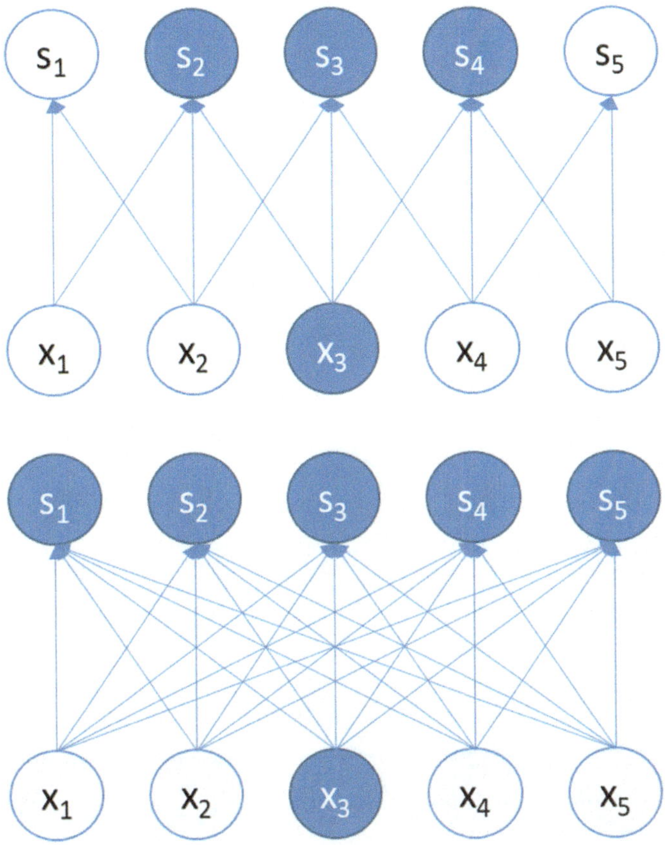

Fig. 9 An example of sparse connections viewed from below

At the top layer, the central element of a specific width kernel, that is the kernel of width 3 here, is the input connected by a black arrow to successive layer. This individual parameter is shared and used at the whole locations; hence, parameter sharing property for the model is provided. On the other hand, at the bottom one, the central element of a weight matrix is used once since this layer has no parameter sharing property [8]. Parameter sharing property enables the model to detect a particular shape of any part of an image in the other locations if the particular shape is available in any other parts of the image regardless of its spatial location [16].

The last property to mention about convolution operation is equivariance to translation. This property implies that if the pixels of an input image are shifted by one unit in any direction, then resultant output of applied convolution will be shifted with this shifted input values due to parameter sharing property [16]. When equivariance to translation is considered in the context of one-dimensional time series data, if one-step-shift is applied to an input of time series, then resultant output will appear

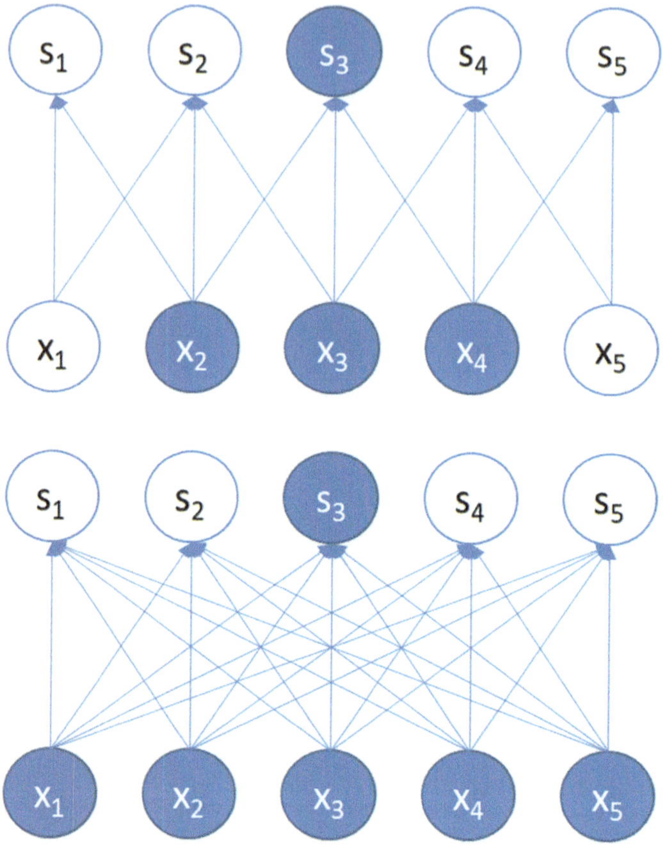

Fig. 10 An example of sparse connections viewed from above

as one-step shifted output in accordance with one-step-shift in the input. When two-dimensional image is considered for the same property, convolution operation will produce a 2-D feature map of the image. If the object in the input is moved, the corresponding representation of the input will be shifted within the same amount [8].

4.2 *Pooling Layer*

Pooling operation can be seen as one of the fundamental layers in convolutional neural networks. Typical CNN are made up of three stages. In the first stage, convolutional layer(s) produces some feature maps (outputs) by producing some set of linear activations. This is followed by transforming the set of linear activations into nonlinear activations, and this transformation stage is referred to detector stage. In

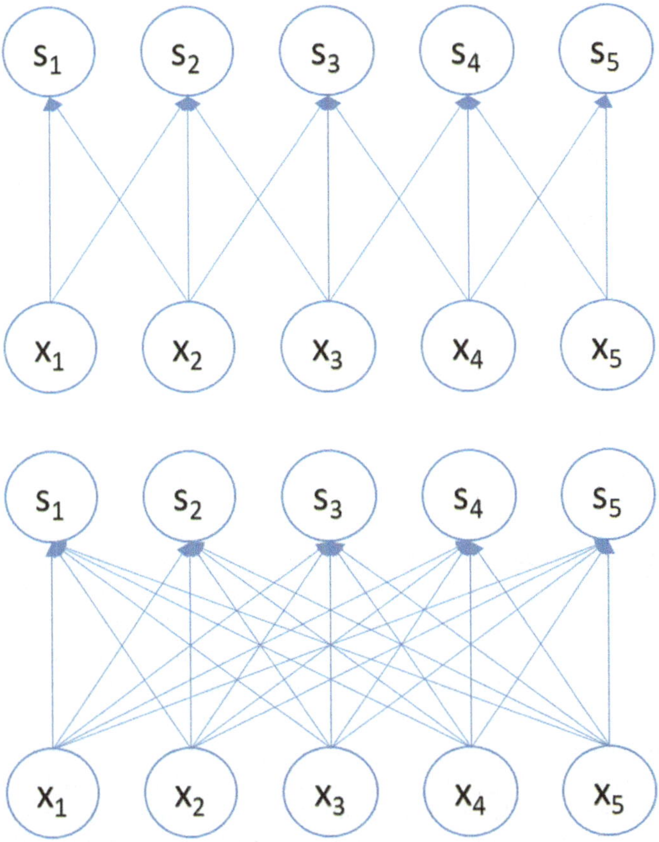

Fig. 11 Parameter sharing property visualization

the last stage, a pooling function is used to produce a new kind of output by reducing the spatial location of given activation maps [8, 16].

Simply, new output, which is a summary statistics of nearby outputs, is produced by applying pooling function for an output [8]. The outputs are generated by feature maps, and pooling is applied to each feature map (each new output) independently. As a result of the pooling function, the number of the feature maps will remain the same before pooling operation; however, the spatial size of the feature maps will reduce. There are prevalently two types of pooling operations; max-pooling, and average pooling. A typical example of max-pooling operation is depicted in Fig. 12.

Max-pooling operation results in an output that is maximum among given input within a rectangular neighborhood. On the other hand, average pooling operation results in an output that is the average of given input within a rectangular neighborhood. In max-pooling, the maximum element within pooling kernel in each feature map is activated, and an output set of these pooling operations then forms new outputs. In Fig. 12, different resultant feature maps due to different strides by the

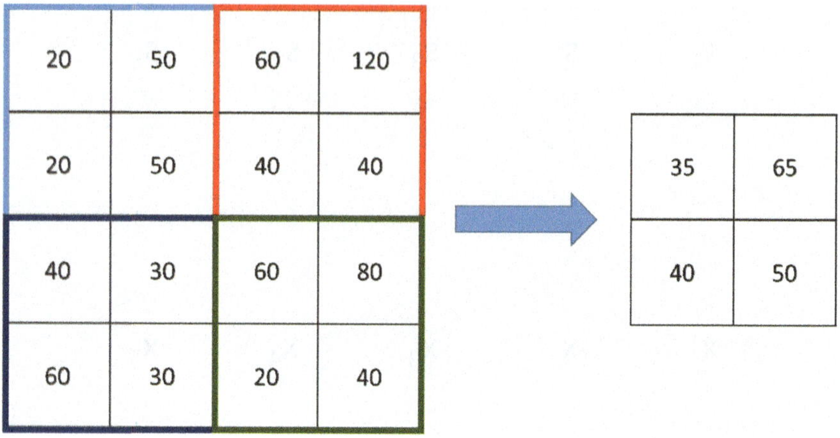

Fig. 12 An example of max-pooling with 3 × 3 kernel

same-width pooling can be seen. The spatial dimension decreased when the pooling produced new feature maps from the input in Fig. 12. Meanwhile, the spatial dimension of resultant outputs for stride 2 was smaller than its counterpart for stride 1. In the outputs for stride 2, there is no overlapping for pooling kernels while the other one has some overlapping locations. Both of them can be used in CNN applications; however, using overlapping locations with stride 1 can reduce the risk of overfitting of this approach [16].

In the average pooling, the average of elements within the pooling kernel in each feature map is produced, and an output set of these pooling operations forms new outputs. The latter one performs some kind of smoothing function due to its averaging operation. In many deep learning applications, max-pooling is mostly preferred pooling operation over average pooling.

Pooling operations have some features, and one of them is translation invariance. Translation invariance means that a significant change in the resultant feature map is not seen when an image is shifted. This idea stems from that similar images oftentimes have some characteristic shapes within them, but these shapes can be in different locations within each image. Hence, translation invariance feature enables to classify these similar images in the same manner. For instance, translation invariance helps classify a dog regardless of its location within the image. Another feature of pooling is the size of receptive field of a layer is enlarged while the spatial size of the layer is reduced. To extract more high-level features from complex features, this receptive field enlargement enables capturing larger portions of the image. In addition, max-pooling provides nonlinearity characteristics to feature maps as well, and it has been used in many CNN applications [16].

4.3 Dense Layer

In a typical CNN architecture, the last layer usually contains one or more densely connected layers. The dense layers connect each feature in the final spatial layer to hidden units in the first dense layer.

Extracted features from a given input are finally processed in these fully connected layers whether they perform classification or regression tasks. Logistic, softmax, linear, or sigmoid activation functions may be used at the final layer depending on the task (i.e., multi-class, binary classification, or regression). These layers perform like feedforward neural networks as they have dense connections. In CNN, fully connected layers account for most of the parameters of the network. For example, between two fully connected layers, if two layers have 2048 hidden units, then the corresponding connections between these layers will have more than 4 million weights [16].

Some distinguished papers use CNN for image recognition with one fully connected layer at the end of their architecture. For example, remarkable architectures such as EfficientNet-B0 [18], ResNet [19], Inception-v4, Inception-ResNet-v1, and Inception-ResNet-v2 [20] used single dense layer in their models. On the other hand, other prominent CNN architectures for image recognition such as MobileNet [21], DenseNet [22], Inception-v2 [23], VGGNet [24], and GoogLeNet [25] used more than one dense layers in their models. Because there is no rule of thumb in spotting how many dense layer(s) will be used in any architecture, inclusion of dense layer to model can depend on the application domain since the number of fully connected layers may be sensitive to application area. For instance, there may be differences in the nature of a fully connected layer for a classification task when compared to a segmentation task [16].

References

1. White L, Togneri R, Liu W, Bennamoun M (2019) Neural representations of natural language, Vol 783. Singapore: Springer Singapore
2. Si S (2010) Data mining techniques for the life sciences, vol 1. Humana Press, Totowa, NJ
3. Tappert CC (2019) Who is the father of deep learning? In: Proceedings of the 6th annual conference computer science computational intelligence CSCI, pp 343–348. doi: https://doi.org/10.1109/CSCI49370.2019.00067
4. Rosenblatt F (1958) The perceptron: a probabilistic model for information storage and organization in the brain. Psychol Rev 65(6):386–408. doi: https://doi.org/10.1037/h0042519
5. Aggarwal CC (2018) Neural networks and deep learning. Springer International Publishing, Cham
6. Mirjalili S (2019) Evolutionary algorithms and neural networks, vol 780. Springer International Publishing, Cham
7. Paper D (2018) Gradient Descent. Data science fundamentals for python and MongoDB, Berkeley. Apress, CA, pp 97–128
8. Goodfellow I, Bengio Y, Courville A (2016) Deep learning. The MIT Press
9. Pinjare SL, Kumar M (2012) Implementation of neural network back propagation training algorithm on FPGA. Int J Comput Appl 52(6):1–7. doi: https://doi.org/10.5120/8203-1599

10. Benvenuto N, Piazza F (1992) On the complex backpropagation algorithm. IEEE Trans Signal Process 40(4):967–969. https://doi.org/10.1109/78.127967
11. Zaccone G, Karim MR, Menshawy A (2017) Deep learning with TensorFlow. Packt Publishing Ltd.
12. Hochreiter S, Schmidhuber J (1997) Long short-term memory. Neural Comput 9(8):1735–1780
13. Hebrail AB, Georges, "No Title."
14. Chollet F (2017) Deep learning with python. Simon and Schuster
15. Hubel DH, Wiesel TN (1959) Receptive fields of single neurones in the cat's striate cortex. J Physiol 148(3):574–591
16. Charu CA (2018) Neural networks and deep learning: a textbook. Spinger
17. LeCun Y, Bottou L, Bengio Y, Haffner P (1998) Gradient-based learning applied to document recognition. Proc IEEE 86(11):2278–2324
18. Tan M, Le Q (2019) Efficientnet: Rethinking model scaling for convolutional neural networks. In: International conference on machine learning, pp 6105–6114
19. He K, Zhang X, Ren S, Sun J (2016) Deep residual learning for image recognition. In: Proceedings of the IEEE conference on computer vision and pattern recognition, pp 770–778
20. Szegedy C, Ioffe S, Vanhoucke V, Alemi AA (2017) Inception-v4, inception-resnet and the impact of residual connections on learning. In: Thirty-first AAAI conference on artificial intelligence
21. Howard AG, Zhu M, Chen B, Kalenichenko D, Wang W, Weyand T et al (2017) Mobilenets: efficient convolutional neural networks for mobile vision applications. arXiv Prepr. arXiv1704.04861
22. Huang G, Liu Z, Van Der Maaten L, Weinberger KQ (2017) Densely connected convolutional networks. In: Proceedings of the IEEE conference on computer vision and pattern recognition, pp 4700–4708
23. Szegedy C, Vanhoucke V, Ioffe S, Shlens J, Wojna Z (2016) Rethinking the inception architecture for computer vision. In: Proceedings of the IEEE conference on computer vision and pattern recognition, pp 2818–2826
24. Simonyan K, Zisserman A (2014) Very deep convolutional networks for large-scale image recognition. arXiv Prepr. arXiv1409.1556
25. Szegedy C, Liu W, Jia Y, Sermanet P, Reed S, Anguelov D et al (2015) Going deeper with convolutions. In: Proceedings of the IEEE conference on computer vision and pattern recognition, pp 1–9

Zeynep Burcu Kizilkan is a Research Assistant and a master's student in the Department of Industrial Engineering at Istanbul Technical University. She holds Bachelor of Science (B.Sc.) degrees in both Industrial Engineer-ing and Software Engineering from Bahcesehir University. Both of her un-dergraduate thesis focused on Neural Networks. Her research interests include data science and machine learning as well as decision making. She is looking forward to completing her master's degree and continue her ac-ademic career.

Mahmut Sami Sivri is currently a lecturer at Industrial Engineering Depart-ment in Istanbul Technical University. He is also working on some R&D projects as Director of Data Analytics at ITU's Technopark. He received the BS degree in Computer Engineering and the MSc degree in Engineer-ing Management from Istanbul Technical University. He worked in various companies and positions in the IT industry since 2008. His current re-search interests include machine learning, sentiment analysis, deep learn-ing, big data and its applications, Industry 4.0, financial technologies, da-ta analytics, supply chain and logistics optimization.

Ibrahim Yazici received B.Sc. degree in industrial engineering from Kocaeli University, M.Sc. in industrial engineering from İstanbul Technical University. Currently, he is doing Ph.D. in industrial engineering at Istanbul Technical University. His research interests span deep learning, deep rein-forcement learning and machine learning.

Omer Faruk Beyca received his B.S. degree in industrial engineering from Fatih University, Istanbul, Turkey, in 2007, and the Ph.D. degree from the School of Industrial Engineering and Management, Oklahoma State Uni-versity, Stillwater, OK, USA, in 2013. He is with the Department of Indus-trial Engineering, Istanbul Technical University, as an Assistant Professor. His research has been published in international peer-reviewed journals, conference proceedings, and book chapters.

Feature Engineering

Alp Ustundag, Mahmut Sami Sivri, and Kenan Menguc

1 Introduction

As the amount of data generated and collected grows, analyzing and modeling so many input variables get more difficult. So, it is important to reduce model complexity and establish simple, accurate and robust models. Feature engineering is the process of using domain knowledge to extract input variables from raw data, prioritize them and select the best ones so that machine learning algorithms work well and model performance is improved.

Feature engineering has been defined in different ways in the literature. In this chapter, all processes related to the variables of a model are defined as feature engineering which consists of three main components (Fig. 1):

- Data Preprocessing is executed to clean the data by handling missing and noisy values and is also used to transform data into understandable format for computing via methods such as aggregation, normalization, discretization, vectorization.
- Feature Extraction is used to generate new candidates as well as to transform existing ones to make them more suitable for modeling.
- Feature Selection is used to discover subsets from the feature pool to increase prediction performance while lowering the size and complexity of the model.

Data preprocessing, which is the first component of feature engineering, is the preparatory phase that organizes, sorts and merges all accessible data. Data cleaning, a critical part of preprocessing, refers to techniques for locating, deleting and restoring corrupted or missing data. Raw data may contain erroneous or missing numbers, as well as redundant information. The most often encountered issues with raw data fall into three categories: missing, noisy and inconsistent. In this phase, numerous

A. Ustundag (✉) · M. S. Sivri · K. Menguc
Department of Industrial Engineering, Faculty of Management, Istanbul Technical University, 34367 Macka, Istanbul, Turkey
e-mail: ustundaga@itu.edu.tr

A. Ustundag et al. (eds.), *Business Analytics for Professionals*, Springer Series in Advanced Manufacturing, https://doi.org/10.1007/978-3-030-93823-9_6

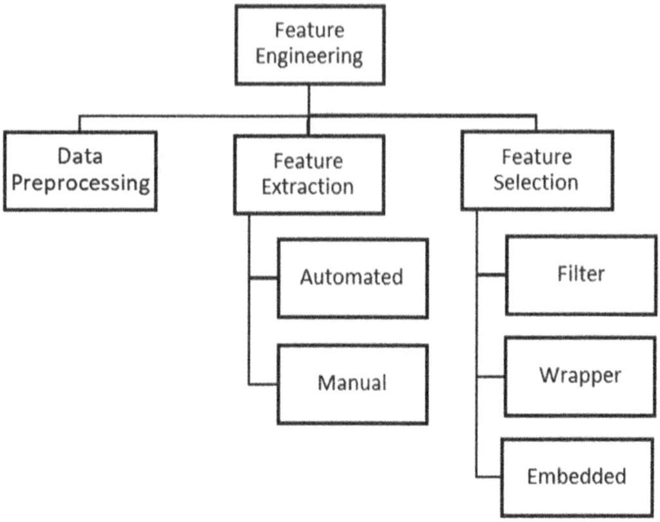

Fig. 1 Components of feature engineering

approaches, including imputation, binning, log transform, feature splitting, outlier handling and scaling, can be employed.

As the second component of feature engineering, feature extraction is executed manually or automatically. Manual feature extraction entails defining and describing the relevant features regarding the model and applying the proper technique to extract them. In many circumstances, the high level of domain knowledge is very important for successful results. On the other hand, automated feature extraction employs ML algorithms like deep networks to extract features automatically from unstructured data such as text or images without the need for human interaction.

For example, let us assume that a model is developed for daily stock direction forecasting for financial markets. The feature pool consists of historical price, transaction data, company information, economic indicators and expert opinions. When new features are derived from existing price and transaction data for the model, it can be considered as a manual feature extraction approach. However, the question of which variables from the large feature pool will be used in the model can be solved with feature selection techniques.

Feature selection, the third component, lowers the dimensionality of data by constructing a model using just a subset of the input features and thus it provides enhanced prediction performance. Three types of feature selection algorithms exist. The filter type feature selection algorithm quantifies the value of features depending on their properties, such as feature variance and output relevance. Information Gain, Chi-square Test, Correlation Coefficient, Variance Threshold and Mutual Info are the most known methods for filter type which are fast and scalable. The wrapper type feature selection algorithm begins with a subset of variables and then adds or eliminates variables based on a selection criterion. Recursive Feature Elimination,

Forward Selection, Backward Elimination and SHAP are most used techniques for wrapper type. The embedded type feature selection methods such as Random Forest and Lasso determine the significance of inputs as part of the model learning process.

The dimensionality reduction is an important aspect of feature engineering which is used to describe lowering the dimension of input features. Both methods, feature selection and extraction can be used for dimensionality reduction. The primary difference between feature selection and extraction is that feature selection keeps a subset of the original variables while feature extraction creates totally new ones. However, feature extraction entails transformation of the features, which is generally irreversible due to information loss during the dimensionality reduction process.

In summary, there are several benefits of feature engineering:

- Reducing model complexity
- Increasing model accuracy
- Reducing overfitting risk
- Reducing information redundancy
- Easier model training and faster convergence
- Increasing computational performance and reducing the cost
- Improving model interpretability and explainability
- Increasing data visualization capacity.

In this section, important feature engineering techniques for numerical data are briefly explained and several examples are given.

2 Feature Extraction and Dimensionality Reduction

Feature extraction is used for dimensionality reduction by creating a new, smaller set of features that stills captures most of the useful information. In this section, three main techniques, Principal Component Analysis (PCA), Linear Discriminant Analysis (LDA) and Autoencoders (AE), will be explained.

PCA is an unsupervised technique that generates linear combinations of the original features. The objective is to extract critical information from the data and express it as a collection of summary features called principal components. The new features are orthogonal and so they are uncorrelated. Additionally, they are ranked by their explained variance. PCA calculates the eigenvectors of a covariance matrix with the greatest eigenvalues and then uses them to project the data onto a new subspace with equal or less dimensions. It can also be expressed that PCA transforms a matrix containing n features into a new dataset containing less than n features.

LDA is another linear transformation technique as PCA used for dimension reduction. However, LDA is a supervised method and uses class label information. As PCA represent the direction of maximum variance in data, LDA describes the direction that maximizes class separability (Fig. 2).

Autoencoder is a member of ML algorithms which is also used as a dimensionality reduction technique. The primary distinction between Autoencoders and other

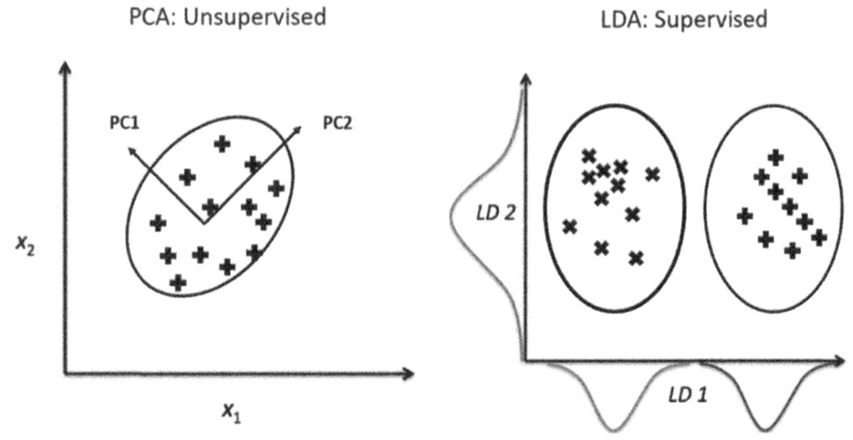

Fig. 2 PCA versus LDA

dimensionality reduction approaches is that Autoencoders project data from a high dimension to a lower dimension via non-linear transformations. An Autoencoder's fundamental architecture may be split down into two distinct components. The first component, Encoder, takes the incoming data and compresses it to eliminate any noise or useless information. The Encoder stage's output is called latent space. The second component, Decoder, takes as input the encoded latent space and tries to reproduce the original Autoencoder input (Fig. 3).

The above explained three feature extraction techniques are applied on the German Credit Dataset, which contains 1000 loan applications with 20 categorical/symbolic attributes prepared by Hofmann [1]. Each entry in the dataset represents a person who takes a credit by a bank. There is a classification whether an applicant has a Good or Bad credit risk. To determine the credit risk, 20 features are given such as age, sex, occupation, account balance. In this case, the objective is to try to predict if there is a credit risk or not by looking at the given features. The input data is

Fig. 3 Autoencoder basic architecture

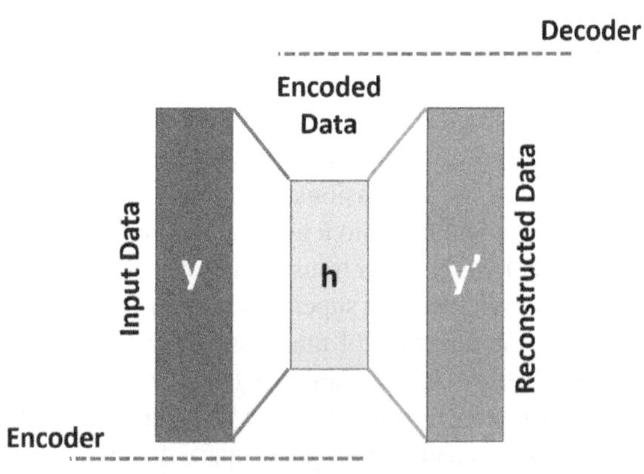

Table 1 Comparison of results

Methods	Accuracy (%)	$F1$ score (creditability: 0) (%)	$F1$ score (creditability: 1) (%)
No dimension reduction	75	50	84
PCA ($n = 8$)	77	56	85
LDA	68	51	77
AE	63	33	75

divided into train (70%) and test sets (30%). Random Forest classifier is used to predict the output on the test data using dimension reduction techniques. As shown below, training a Random Forest classifier using PCA ($n = 8$), led to 77% accuracy and produced better results comparing to other methods (Table 1).

3 Feature Selection

In data analytics, feature selection is critical to reducing the complexity created by irrelevant factors in the data and to avoiding scenarios like multicollinearity, which can be caused by related variables. Feature selection approaches are commonly divided into three categories in the literature, as explained below. Table 2 also outlines the benefits and drawbacks of the techniques [2].

In this section, main feature selection methods are explained in detail, and tested these methods in a credit risk scoring dataset. The explanations of the variables in the dataset are given in Tables 3 and 4.

Table 2 Comparison of feature selection methods

	Filter-based	Embedded	Wrapper
Computation	Low	Mid	High
Running time	Fastest	Faster	Slower
Overfitting	Lower risk	Higher risk	Lower risk
Feature dependency	Ignores	Considers	Considers
Prediction method	No interaction	Interacts	Interacts
Generalization	Good	Better	Best

Table 3 Features of credit risk dataset

Feature	Type
Purpose	Categorical
Loan duration	Continuous (months)
Account balance	Continuous
Pension funds	Continuous
Account length	Continuous (years)
Sex	Categorical
Marriage status	Categorical
Age	Continuous (years)
House	Categorical
Job	Categorical
Employed duration	Continuous (months)

Table 4 Encoding of categorical variables

#	Purpose	Sex	Marriage	House	Job	Risk
0		Female				High
1	Business	Male	Single	Own	Blue collar	Low
2	Combine debts		Divorced	Rent	White collar	
3	Renovation		Married	Other	Other	
4	Marriage					
5	Medical					
6	Mortgage					
7	Vehicle					
8	Vacation					
9	University					
10	Unexpected					

3.1 Filter-Based Methods

In filter-based methods, the relevance score of the features is evaluated by calculating a ranking criterion. Low-scoring features that fall below a certain threshold or exceed a certain number of features are removed.

Pearson Correlation: Pearson correlation coefficient, which is a filter-based feature selection method, is determined by calculating the covariance matrix between the variables and finding a correlation coefficient between -1 and $+1$ for each variable. Variables are ordered according to the absolute value of the coefficient between the variable and target variable [3]. Also, the interpretation of the correlation coefficients is given below.

- 0 shows no correlation
- A value closer to 1 shows stronger positive correlation
- A value closer to − 1 shows stronger negative correlation

Pearson correlation coefficient formula calculated as follows:

$$P_{ij} = \frac{C_{ij}}{\sqrt{C_{ii} * C_{ij}}} \tag{6.1}$$

P_{ij} Pearson correlation coefficient.
C_{ij} Covariance value between variables i and j.

Chi-square Test: In mathematical statistics, the Chi-square test is used to determine the independence of two variables. In a feature selection process, the existence and degree of a relevance between the variables and the target variable are detected. Next, variables eliminated after computing the chi-square statistics between each variable in the data set and the target variable.

Mutual Information: The quantity of knowledge that one variable knows about another can be determined by mutual information [4]. If the value found is 0, this indicates that these variables are unrelated to one another. It is a valuable feature selection approach since it allows you to assess the importance of a variable. The following is a formula for the mutual information:

$$I(x, y) = \sum_{i=1}^{n}\sum_{j=1}^{n} p(x_i, y_i)\log\left(\frac{p(x_i, y_j)}{p(x_i), p(y_j)}\right) \tag{6.2}$$

$I(x, y)$ Mutual information between x and y.
$p(x_i, y_j)$ is the joint mass probability between x_i and y_j.

3.2 *Embedded Methods*

Embedded approaches use the structure of prediction algorithms to calculate feature importance.

Least Absolute Shrinkage and Selection Operator (LASSO): LASSO entails penalizing the regression coefficients' absolute values [5]. This penalty is known as L1 norm or L1 penalty. While there are some similarities between LASSO and Ridge regression, the primary distinction is that LASSO decreases the coefficient of the less important feature to zero and eliminates some features completely. As a result, it is projected to perform better when dealing with a high number of variables. In classification problems, logistic regression with L1 penalty can be used.

$$\sum_{i=1}^{M}(y_i - \hat{y}_i)^2 = \sum_{i=1}^{M}\left(y_i - \sum_{j=0}^{p}(\beta_j * x_{ij})^2\right)^2 + \lambda\sum_{j=0}^{p}|\beta_j| \tag{6.3}$$

Ridge Regression: The square of the coefficients is penalized in ridge regression [6]. This penalty is known as L2 norm or L2 penalty. The distinction between the LASSO and ridge is illustrated in equations.

$$\sum_{i=1}^{M}(y_i - \hat{y}_i)^2 = \sum_{i=1}^{M}\left(y_i - \sum_{j=0}^{p}(\beta_j * x_{ij})^2\right)^2 + \lambda\sum_{j=0}^{p}\beta_j^2 \tag{6.4}$$

Decision Tree Feature Importance: The decision tree is a structure that has collections of internal nodes and leaves. Choosing a feature in the internal node tells the decision tree how to split itself that is, how to divide it into two independent sets. Relative importance of features can be calculated by using Gini index or information gain for classification problems, variance reduction for regression. Thus, it is calculated that how much the variable reduces the variance or impurity in the data. In the ensemble methods, the average over all trees is used to determine the importance of a feature.

3.3 Wrapper Methods

Wrapper approaches, unlike filter-based methods, require a prediction algorithm. The relevance of a feature is determined by its prediction performance.

Recursive Feature Elimination: The recursive feature elimination is an iterative approach. After the predictive performance calculated using all variables, the weakest performance feature is eliminated at each iteration. The method relies on re-running the model after each step. Therefore, the running time may take longer.

SHAP Values: SHAP values is an alternative to the importance of permutation features. The relevance of the permutation feature is determined by the decline in model performance. The SHAP score is determined by the magnitude of feature attributions. The Shapley value is the average of all marginal contributions to all potential coalitions divided by the number of coalitions [7]. The amount of time it takes to compute each feature increases exponentially with the number of features.

In the example, all explained methods and all possible number of features was tested. Table 5 shows the best results of each method.

Table 6 shows the combination of variables for the method that gave the best results in the tests.

Trying all combinations of number of features where the dataset is very large and the number of variables is too big will cause too much processing cost. Therefore, instead of the above approach, in the next section, the question of how to reach the optimum number of features with less processing time will be answered.

Table 5 Comparison of feature selection methods

Method	No. of features	Accuracy (%)
No selection	11	91.77
Correlation	6	96.47
Chi-square	6	98.82
Mutual info	6	97.65
Lasso	7	97.65
Ridge	7	97.65
Tree	10	96.47
Recursive	10	96.47
Shap	9	96.47

Table 6 Feature combination giving the best result

Features
Purpose
Loan duration
Account balance
Pension funds
Marriage status
Employed duration

4 Global Search Methods

The ideal number of features in a prediction model is determined using some optimization algorithms in this section. Heuristic algorithms will be used for optimization. These algorithms are Genetic algorithm, Randomized search algorithm and Hill Climbing algorithm.

To demonstrate global search methods, credit loan data will be used. The bank wants to determine the credit status of the customer who applies for a loan. There are two classes in dataset according to whether they are suitable for credit or not. What is the probability of each credit class according to bank customer's predictor data? The dataset created by the bank is used as a train set and a test set for its previous customers. Cross validation method is used to optimize the feature selection.

Cross Validation: The K-Fold CV method first randomly divides the data sets into K groups. The selected group is used as the validation set and the other part, the $K-1$ group, is used as a train set. An average performance score is generated by creating K data sets as train and validation. Figure 4 illustrates this usage schematically.

The K-Fold CV approach takes us out of overate status for machine learning and deep learning applications. The K value was determined as 5 because the data set was not large in the studies. 20% of the data has been reserved as test data, then the remainder has been used as train data.

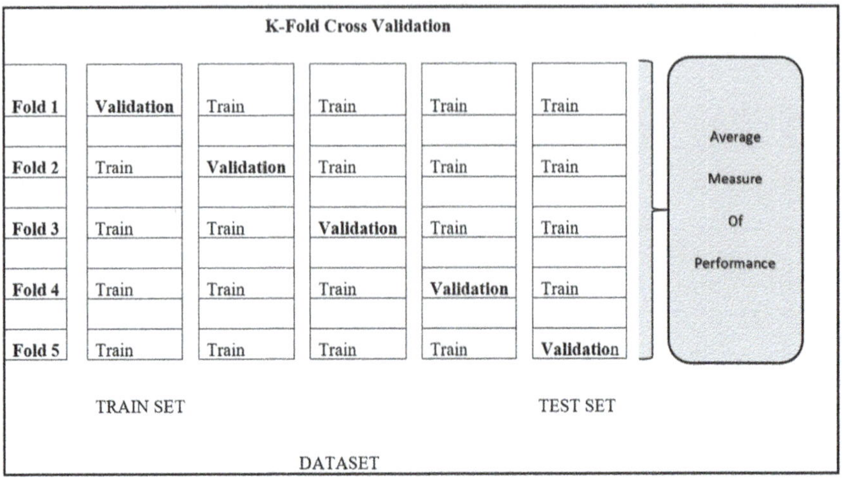

Fig. 4 *K*-fold cross validation

The accuracy score was concluded 0.75 and ROC score was concluded 0.69 by using only Naive Bayes for the data set with 11 features. Accuracy score is a value above 0.7. However, the ROC score was found to be 0.69. Our dataset consists of 2 different classes that indicate whether credit will be given or not. A score of 0.5 can be taken if a random decision is made when there are no features. As a result, if the model finds higher ROC values, it can easily decide if the credit will be given (Fig. 5).

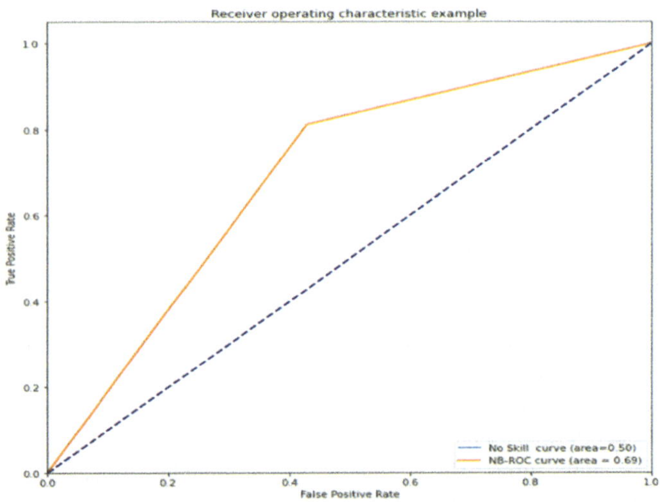

Fig. 5 ROC score of pure Native Bayes model

4.1 Genetic Algorithm (GA)

Biological science has shown that those who get adopted well to nature can survive. Genetic algorithm (GA) is an optimization tool based on gene differentiation and evolution of living things [8, 9]. GA can find optimal or near-optimal solutions of complex problems [10] (Fig. 6).

There are *n* genes in a chromosome with *n* features according to genetic algorithm. Whether or not each feature is in the model is a binary situation. Therefore, gene pairs of size *N* consisting of 0 and 1 try to find the optimal number of inputs with matches. Mating means exchanging some genes from 2 parents. Genes are diversified by the crossover process and the rate of this crossover is determined as the crossover parameter in the algorithm. This parameter determines the percentage of property change for the new solution proposal. Different solutions can be obtained by using random gene changes outside the parent's genes which are provided by the mutation rate parameter. The population parameter determines the amount of different individuals who will search solution. The Iteration parameter provides the number of the search repeated (Table 7).

There are 11 different features in our data set, so our gene length is 11 units. The algorithm parameters have been determined as 0.5 crossover value, 1000 generation number and 0.05 mutation rate. The model has been run by assigning the CV value to 5 for two different performances. (Accuracy score, ROC score) criteria (Table 8).

Number of valid variables is 4 and accuracy score of model is 0.86. This score is higher than pure Native Bayes model score (0.75). The credit status has been determined according to 11 features in the pure Native Bayes model. Genetic algorithm is used as selector and Naive Bayes is used as estimator in the new model. This hybrid model found a higher accuracy score according to fewer features. Reducing the number of features reduces the cost of using the model (Table 9).

Number of valid variables is 4 and ROC score of model is 0.90. This score is quite high compared to the pure Naive Bayes model. The reduction in the number of features will greatly reduce the cost of using the model. It would be very economical in terms of time for the bank to collect 4 data instead of 11 data (Fig. 7).

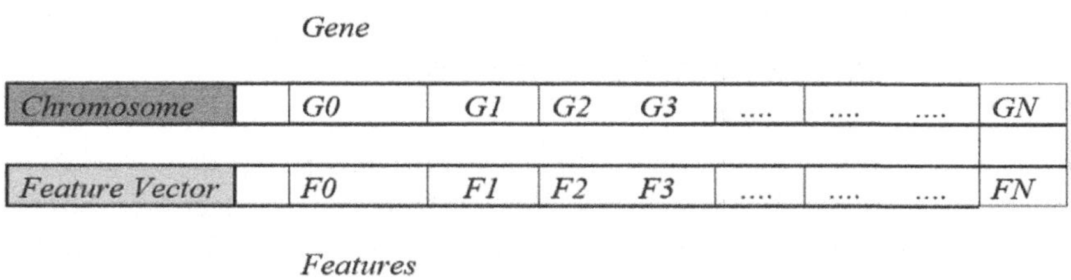

Fig. 6 Gene schema of genetic algorithm

Table 7 A resampling scheme for GA feature selection

```
Create an initial random feature subset;
for number of GA iterations do
    Create internal resample based on training set;
    Perturb the current subset;
    Fit model (Naive Bayes) on internal analysis set;
    Predict internal assessment set and estimate performance;
    if performance is better than the previous subset then
        Accept new subset;
    else
        Calculate acceptance probability;
        if random uniform variable > probability then
            Reject new subset;
        else
            Accept new subset;
        end
    end
end
Determine optimal feature set;
Fit final model (Naive Bayes) on best features using the training set;
```

Table 8 Results of genetic algorithm and Naive Bayes model (accuracy)

3	7	9	11
Account balance	Marriage status	House	Employed duration

Table 9 Results of genetic algorithm and Native Bayes model (ROC)

3	10
Account balance	Job

4.2 Randomized Search Algorithm (RSA)

Random search gives successful results in hyper parameters, whether to determine optimal parameters or feature reduction. Random search initially generates and evaluates random samples of the objective function. It is particularly effective for complex problems because it assumes nothing about the structure of the objective function. Random search is also embedded in many algorithms. It is a method in which random combinations of algorithms are chosen and used to train a model. The algorithm finds the best random combinations of hyper parameters.

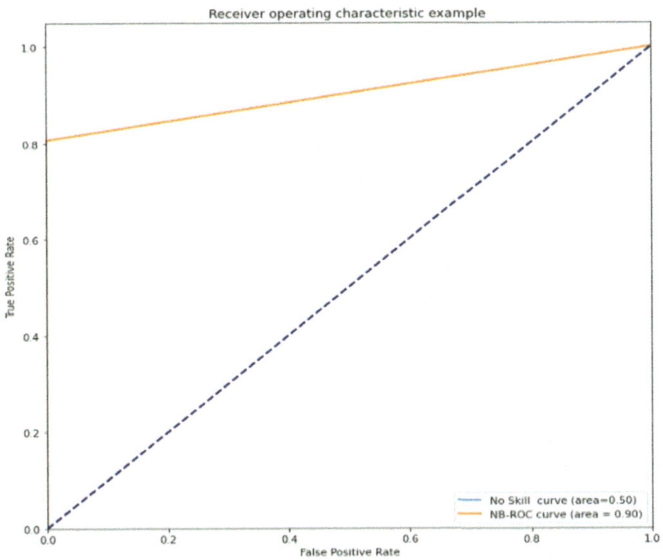

Fig. 7 ROC score of GA and Native Bayes model

Table 10 Results of RSA and Naïve Bayes model (accuracy)

3	7	11
Account balance	Marriage status	Employed duration

Random search algorithm is used as selector and Naive Bayes is used as estimator in the new model. The data set is divided into 20% test data and 80% train data.

The accuracy and ROC scores for the model where the alpha value is 1 and the CV value is 5 have been calculated. Accuracy score is 0.85 and the preferred features of the model are as follows (Table 10).

ROC score is 0.71 and the preferred features of the model are as follows (Fig. 8; Table 11).

4.3 Hill Climbing Algorithm (HCA)

This algorithm is used for both maximization and minimization problems. A simple hill climbing algorithm determines the direction according to their location by checking neighbor's value. If our problem is a maximization problem, hill climbing algorithm walks toward the higher value of neighbors.

Simple Hill Climbing: The algorithm examines the neighboring nodes one by one and selects the first neighbor node as the next node which optimizes the current cost.

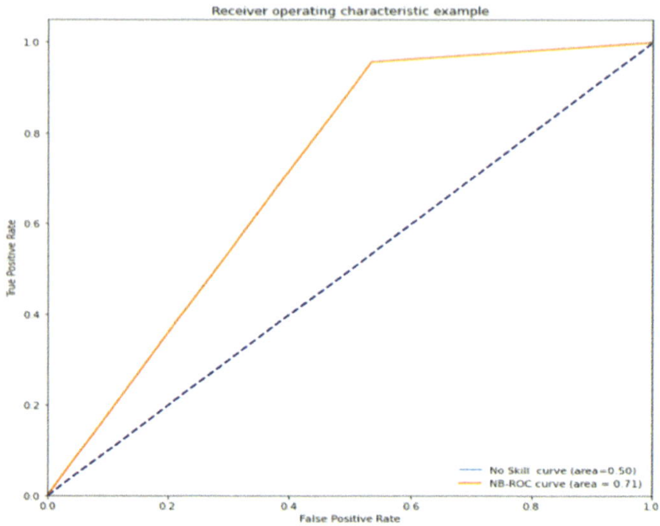

Fig. 8 ROC score of RSA and Native Bayes model

Table 11 Results of RSA and Naïve Bayes model (ROC)

2	3	6	9	10
Loan duration	Account balance	Sex	House	Job

Steepest-Ascent Hill Climbing: The algorithm first examines all neighboring nodes. The algorithm steps to the closest better solution. The small step ensures that the algorithm does not miss some solutions. Therefore, the algorithm is more likely to find the maximum or minimum points with the smallest step than with simple hill climbing.

Stochastic Hill Climbing: The algorithm does not examine all neighboring nodes before deciding which node to choose. The algorithm just randomly picks a neighboring node. The algorithm decides whether to move to the new node or examine another node based on the amount of improvement in solution. The algorithm performs random steps by keeping a solution in memory. If the new solution is equal to or better than the current solution, it remembers that solution. Although the algorithm is a fast algorithm, it can be stuck at local minimum and maximum points. Step length is an important parameter for this algorithm (Fig. 9).

The algorithm was run twice for both accuracy and roc score for credit selection. Stochastic hill climbing was used as the selector, while Naive Bayes was used as the estimator. 1000 iterations and the largest step size to be taken is set to 0.02. The following solution has been found under these conditions.

Accuracy score is 0.81 and the preferred features of the model are as follows (Table 12).

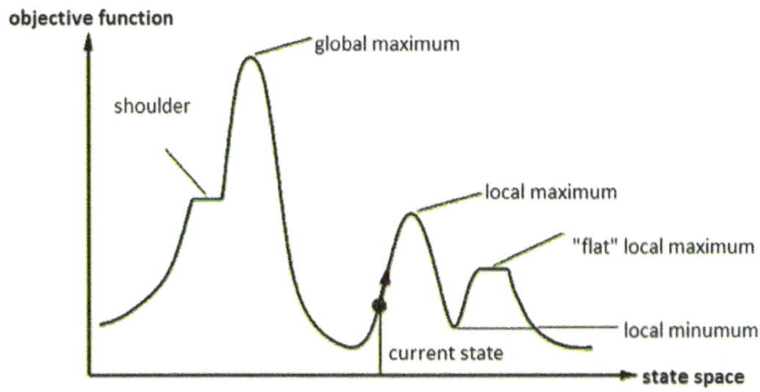

Fig. 9 Critical points of HCA

Table 12 Results of HCA and Naïve Bayes model (accuracy)

1	3	6	9	11
Purpose	Account balance	Sex	House	Employed duration

ROC score is 0.76 and the preferred features of the model are as follows (Fig. 10; Table 13).

The ROC score of the genetic algorithm was found to be higher among the 3 different approaches. Therefore, this value distinguishes the situation if giving credit is much better in making a binary decision.

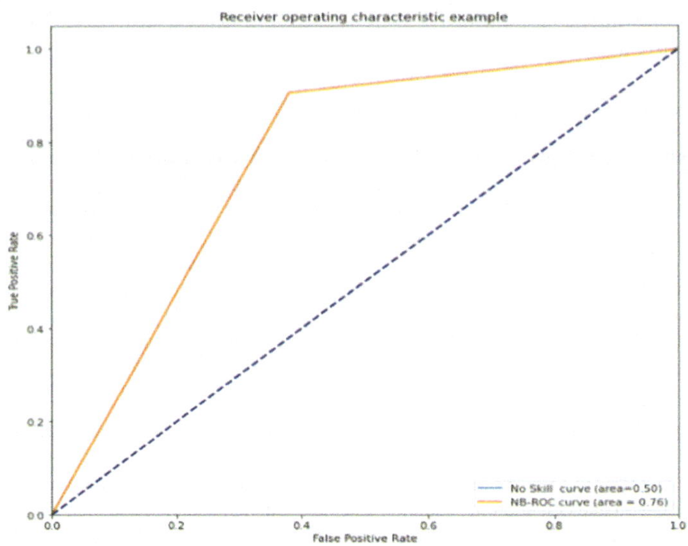

Fig. 10 ROC score of HCA and Native Bayes model

Table 13 Results of HCA and Naïve Bayes model (ROC)

3	6	9	11
Account balance	Sex	House	Employed duration

References

1. UCI (2021) Hofmann H. https://archive.ics.uci.edu/ml/datasets/statlog+(german+credit+data). Accessed 14 Oct 2021
2. Biswas S, Bordoloi M, Purkayastha B (2017) Review on feature selection and classification using neuro-fuzzy approaches. Int J Appl Evol Comput 7:28–44. https://doi.org/10.4018/ijaec.2016100102
3. Zhang L, Wang X, Qu L (2008) Feature reduction based on analysis of covariance matrix. Proc Int Symp Comput Sci Comput Technol ISCSCT 1:59–62.https://doi.org/10.1109/ISCSCT.2008.17
4. Battiti R (1994) Using mutual information for selecting features in supervised neural net learning. IEEE Trans Neural Netw 5:537–550. https://doi.org/10.1109/72.298224
5. Tibshirani R (1996) Regression shrinkage and selection via the lasso. J R Stat Soc Ser B (Methodol) 58:267–288. http://www.jstor.org/stable/2346178
6. Ng AY (2004) Feature selection, l1 vs. l2 regularization, and rotational invariance. In Proceedings of the twenty-first international conference on machine learning ICML'04. Association for Computing Machinery, New York, NY, USA, p 78. https://doi.org/10.1145/1015330.1015435
7. Lundberg SM, Lee SI (2017) A unified approach to interpreting model predictions. In Guyon I, Luxburg UV, Bengio S, Wallach H, Fergus R, Vishwanathan S, Garnett R (eds) Advances in neural information processing systems, vol 30. Curran Associates, Inc., pp 4765–4774. http://papers.nips.cc/paper/7062-a-unified-approach-to-interpreting-model-predictions.pd
8. Mitchell M (1998) An introduction to genetic algorithms
9. MIT Press. Haupt SE, Haupt RL (1998) Optimizing complex systems. In: 1998 IEEE aerospace conference proceedings (Cat. No. 98TH8339), vol 4, pp 241–247
10. Mandal D, Bolander ME, Mukhopadhyay D, Sarkar G, Mukherjee P (2006) The use of microorganisms for the formation of metal nanoparticles and their application. Appl Microbiol Biotechnol 69(5):485–492

Alp Ustundag is the Head of Industrial Engineering Department of Istanbul Technical University (ITU) and the coordinator of M.Sc. in Big Data and Business Analytics Program. He is also the CEO of Navimod Business Analytics Solutions located in ITU Technopark (http://navimod.com/). He has worked in IT and fi-nance industry from 2000 to 2004. He continued his research studies at the Uni-versity of Dortmund between 2007-2008 and completed his doctorate at ITU in 2008. He has conducted a lot of research and consulting projects in the finance, retail, manufacturing, energy, and logistics sectors. His current research interests include artificial intelligence, data science, machine learning, financial and supply chain analytics. He has published many papers in international journals and pre-sented various studies at national and international conferences.

Mahmut Sami Sivri is currently a lecturer at Industrial Engineering Department in Istanbul Technical University. He is also working on some R&D projects as Direc-tor of Data Analytics at ITU's Technopark. He received the B.Sc. degree in Computer Engineering and the M.Sc. degree in Engineering Management from Istanbul Technical University. He worked in various companies and positions in the IT industry since 2008. His current research interests include machine learning, sentiment analysis, deep learning, big data and its applications, Industry 4.0, financial technologies, data analytics, supply chain and logistics optimization.

Kenan Menguc completed his undergraduate education in 2016 at Doğuş University, Department of Industrial Engineering and his M.Sc. in 2018 at Istanbul University, Department of Industrial Engineering. He worked as a lecturer in the Department of Logistics at Beykent University between 2018 and 2020. While he currently conducts his Ph.D. study at Yıldız Technical University, he works as a research assistant at Istanbul Technical University.

Text Analytics

Mahmut Sami Sivri and Buse Sibel Korkmaz

1 Introduction

In the age of big data, organizations and businesses have had to manage and make sense of data generated by a wide variety of systems, processes and transactions. The data contained in traditional relational databases is rather small compared to various sensor or social media data. Therefore, one of the most important factors triggering this era is the text data and its storage and analysis. For example, today there is a huge amount of data in the form of news, blogs, tweets on social media, status messages, hashtags, articles, wikis and much more. Even e-commerce stores and many industries generate a lot of textual data, from new product information to customer reviews and feedback.

2 Text Processing

Since unstructured text data can contain many special characters (especially tweets, user comments), the data may need to be preprocessed. The data is mostly dirty and noisy. It is needed to convert the raw text data into an understandable format for further analysis. In the below section, some text processing approaches are listed. In this section, these steps are applied on the sentence written as "the paragrahps, need to be preprocessed.!".

M. S. Sivri (✉)
Department of Industrial Engineering, Faculty of Management, Istanbul Technical University, 34367 Macka, Istanbul, Turkey
e-mail: sivri@itu.edu.tr

B. S. Korkmaz
Department of Informatics, Faculty of Engineering and Architecture, Technical University of Munich, 85748 Garching, Germany

A. Ustundag et al. (eds.), *Business Analytics for Professionals*, Springer Series in Advanced Manufacturing, https://doi.org/10.1007/978-3-030-93823-9_7

2.1 Data Cleansing

Punctuation marks are cleaned during data cleansing as they do not contain any extra value or information for analysis. Additionally, by reducing the quantity of the data, deleting these flags will aid in increasing computing performance. When the punctuation marks are removed in the example sentence, the sentence as "the paragrahps need to be preprocessed" is obtained.

2.2 Stop Words Removal

In text analysis, common words provide little contextual information compared to other keywords. Therefore, these are removed from the data although common terms such as "the", "a" serve a grammatical purpose. In this way, it is ensured that the algorithms focus on important keywords instead of them in the analysis. The example turned to "paragrahps need preprocessed", after applying this step.

2.3 Spelling Correction

Words written incorrectly or incompletely in the text can be corrected with the help of libraries. Most of the text data, especially in comments, blogs and social media are data sources with short words and a lot of typos. Data from these data sources can be corrected by subjecting them to spelling correction. Apparently, the word "paragraph" was misspelled in the sentence. Finally, correct spelling as "paragraphs need preprocessed" is obtained.

2.4 Lemmatization

Lemmatization is the process of extracting the root word by considering the vocabulary. For example, the word "better" is analysed for "good". By applying this operation, plural suffix is disappeared. The sentence became "paragraph need preprocessed".

2.5 Stemming

These operations are applied to find the root word. For example, when words such as "fishing" and "fishes" in English are subjected to this process, it provides "fish" as a root word [1]. With the stemming process, the attachments added to the root are cleaned. The first form of the verb in the sentence is obtained, and it resulted as "paragraph need preprocess".

2.6 Tokenization

This process, which is one of the starting points of natural language processing, means dividing the text into minimum meaningful units. Separation into strings can be done on the basis of sentences, phrases or words. Although words in English and similar languages can be distinguished by spaces, in some languages, different methods may be required for words [2]. After tokenizing the final phrase, root words as ["paragraph", "need", "preprocess"] are obtained. Text processing steps and their results are given in Table 1.

3 Text Representation

Computers are very well programmed to work with numbers. In the case of natural language processing, almost all sources include unstructured data such as text or speech. In order to decode the patterns within the data and do calculations belonging to machine learning algorithms, natural language source data should be converted to numerical form before starting to process it. This is called feature extraction or feature encoding. There are several text representation techniques with a trade of easiness and accuracy. They will be explained here with a top-down approach.

Text representation methods are mainly twofold: Discrete text representation and distributed/continuous text representation.

Table 1 Text processing steps and their results

Step	Resulting sentence
	"the paragrahps, need to be preprocessed.!"
Data cleansing	"the paragrahps need to be preprocessed"
Stop words removal	"paragrahps need preprocessed"
Spelling correction	"paragraphs need preprocessed"
Lemmatization	"paragraph need preprocessed"
Stemming	"paragraph need preprocess"
Tokenization	["paragraph", "need", "preprocess"]

3.1 Discrete Text Representation

In this type of representation, words are interpreted as numbers regarding their relevant indexes in a dictionary from a larger body of text or corpora.

Techniques that are used widely under this category are:

- One-hot encoding
- Bag-of-words
- Count vectorizer
- Term frequency—Inverse document frequency (TF-IDF).

One-hot encoding

As a first step, it orders all of the words in the vocabulary as vector indexes. Then it creates a vector for each word in the text with a length of the number of unique words in the text, which is a vocabulary size. Lastly, it puts 1 or 0 according to whether the word is a corresponding word of the relevant index.

For demonstration purposes, look at the following sentence: "I love machine learning". The vocabulary would be formed by "I", "love", "machine" and "learning" words. Each word in the sentence would be interpreted as in Table 2.

The complete sentence is interpreted as a one-hot encoding vector in Table 3.

This method is not the most convenient technique for large text bodies since word vector representation enlarges with the growth of vocabulary size which depends on the number of unique words in the text and gets computationally expensive. Furthermore, it examines each word in a standalone manner, so it does not capture a relationship between the words and the context of the whole text body. Despite the disadvantages, this method is the easiest method to understand and implement in addition to being a good start point to understand text representation.

Bag-of-words

The BoW is a simple and popular method of feature encoding. It is a text representation technique that describes the occurrence of words which is called word frequency

Table 2 One-hot encoding per words

Words	One-hot encoding
I	[1, 0, 0, 0]
Love	[0, 1, 0, 0]
Machine	[0, 0, 1, 0]
Learning	[0, 0, 0, 1]

Table 3 One-hot encoding for the complete sentence

Sentence	One-hot encoding
I love machine learning	([1, 0, 0, 0], [0, 1, 0, 0], [0, 0, 1, 0], [0, 0, 0, 1])

Table 4 Dictionary of the given poem

Words	
It	Age
Was	Wisdom
The	Foolishness
Best	
Of	
Times	
Worst	

within a document or large text body [3]. It is called a "bag" of words because the method basically puts all words in a bag then counts the presence of each unique word in the bag.

To build concrete knowledge, a step-by-step example is given as follows. The following snippet is taken from "A Tale of Two Cities" by Charles Dickens for the data gathering step.

It was the best of times.
It was the worst of times.
It was the age of wisdom.
It was the age of foolishness.

If each line is treated as a different document, the combination of four lines form the entire corpus. To further and model a BoW representation, the dictionary would be created as in next by ignoring punctuation and case (Table 4).

The scoring of the document and creating the document vector is the third and final step for BoW text representation. Although there are multiple simple methods to score the document, scoring regarding the word frequency would be the most insightful technique. This ability filter out uncommon and irrelevant words easily by helping model to train and converge faster. However, still, it will suffer from not capturing the context because it discards the order of the words. For example, if a verb has a "not" negative connotation before itself, this would be a meaning changer case in the context. Also, articles—a, an, the—and prepositions—of, on, etc.—are the most frequent words in texts and dominates BoW, but they do not consist of any "informational content".

The frequency of the words is calculated by the presence of each word out of all the words in a corpus (Table 5).

The most popular implementation of text representation based on frequencies is the word clouds. They include unique words in the source by different colours according to being a single word or *n*-grams and sizes regarding the frequency of the corresponding word [4]. Example word clouds for positive and negative connotated Turkish financial news could be seen in (Figs. 1 and 2).

Table 5 Frequencies per words

Words	Frequencies
It	4/24
Was	4/24
The	4/24
Best	1/24
Of	1/24
Times	1/24
Worst	1/24
Age	1/24
Wisdom	1/24
Foolishness	1/24

Fig. 1 Negative financial news word cloud

Count vectorizer

This approach is based on the classical BoW approach. Additionally, Python implementation with the sklearn package includes *n*-grams in the dictionary. For example, if one specifies the *n*-gram parameter as three in the CountVectorizer module in Python, the text representation would also include "it was the" 3-g that has 4/24 word frequency. Moreover, one may limit the maximum feature number that models the text with an aim of not including non-impactful words.

Fig. 2 Positive financial news word cloud

TF-IDF

Term frequency-inverse document frequency is an advanced variant of the BoW technique. With the ability of suppressing the high-frequency words and ignore the low-frequency words, the given approach is a good representation method for natural language processing. TF-IDF works as a combination of two frequency metrics:

TF frequency of the word in the current document (a line in the previous example)

IDF frequency of rareness of the corresponding word over entire corpus (four lines in the previous example)

The intuition behind the TF-IDF is that a rare word in a document has a more impact on the meaning of the text. However, it still does not capture positional information that leads to not capture the context. It assumes all words are independent of each other except *n*-grams and does not benefit from co-occurrence statistics between words. Also, it is a method that highly depends on the source data. A representation generated through football data cannot be used for volleyball. Hence, TF-IDF requires high-quality source data. Besides of given drawbacks, this method is simple, easy to understand and implement. Furthermore, it was very commonly used in the search engines such as Google and Yahoo until the recent developments in the deep learning field.

3.2 Distributed Text Representation

Unlike discrete text representation, distributed text representation does not ignore the meaning dependency of words on each other hence does not treat them as independent or mutually exclusive. So, it interprets a word as distributed across the vector. The major advantage of distributed text representation is the ability of capturing context and meaning.

The mostly utilized distributed text representation techniques are:

- Co-occurrence matrix
- Word2Vec
- GloVe.

Co-occurrence matrix

This matrix representation considers the co-occurrence of entities including single words, bi-gram or phrase near to each other. When individual words are used to create the co-occurrence matrix, it assists to carve-out the association between different words in a source data. In implementation, it works as matrix multiplication of count vectorizer and transpose of it.

For example, following five sentences could be considered as documents and combination them would form the corpus (Table 6).

After creating the count vectorizer for given corpus and multiplying it with its transpose, insightful information could be obtained by examining with adjective and products associations. Also, since the given source data includes "recommend" itself as a feature, it is possible to analyse co-occurrence of "recommend" and product names. The resulting matrix size would be *n* x *n* where *n* is the feature size which is a number of unique words in the corpus. In Table 7 product names, "recommend" and feature associations are given after obtaining the co-occurrence matrix. Complete matrix and implementation could be found on Jupyter Notebook.

According to filtered co-occurence matrix of the given corpus, product_x has more positive associations than product_y which may direct customer decisions to buy product_x instead of product_y.

Despite of its simplicity, ability of capture the context and being a global representation by using entire corpus, co-occurrence matrix is not a scalable method for larger vocabularies and computationally inefficient since it creates a sparse matrix

Table 6 Example sentences

Sentences
"product_x is awesome"
"product_x is better than product_y"
"product_x is disappointing"
"product_y beats product_x by miles"
"ill definitely recommend product_x over others"

Table 7 Co-occurrence matrix

Words	product_x	product_y	Recommend
Awesome	1	0	0
Beats	1	1	0
Better	1	1	0
Definitely	1	0	1
Disappointing	1	0	0
Ill	1	0	1
Miles	1	1	0
product_x	5	2	1
product_y	2	2	0
Recommend	1	0	1

(i.e. a matrix includes zeros). Furthermore, some associations are not understandable from co-occurrence matrix. For example, the association between "beats" and "product_x" is not clear in the given matrix.

Word2Vec

It is the most popular text representation algorithm in natural language processing. The most distinct feature of Word2Vec comparing to previous methods is that Word2Vec is a prediction-based method instead of count-based techniques such as a co-occurrence matrix. Each word is interpreted with a fixed-length vector while capturing underlying semantic and syntactic relations with other words [5]. Word2Vec's shallow one layer network mechanism operates similarly to an autoencoder that is a machine learning algorithm used for dimensionality reduction mostly.

Word2Vec builds the vector representation through two opposite approaches: CBOW and skip-gram [6]. The first one predicts a word by the context of surrounding *n*-words (*n* is adjustable). The underlying algorithm could be put in a summary such as filling a word blank as the word will be more suitable regarding the surrounding context. Word2Vec works more efficiently with smaller datasets and faster to train compared to skip-gram. The latter one predicts surrounding words by a target word. This method outperforms CBoW generally but tends to have larger training time.

Word2Vec is able to capture syntactic relationships such as present or past tense as well as semantic relationships such as country-capital or object-aliveness. It is capable of extracting the similarity between multiple words with the help of simple vector operations. There are two methods to gauge similarity between the vector representation of selected words. If the vectors are normalized, then their dot products provide the similarity between them. Another technique is that if vectors are not normalized, the cosine similarity of two vectors gives the similarity between two words.

In addition to the advantage of interpreting the syntactic and semantic relationships between words, Word2Vec is also an unsupervised machine learning algorithm. So,

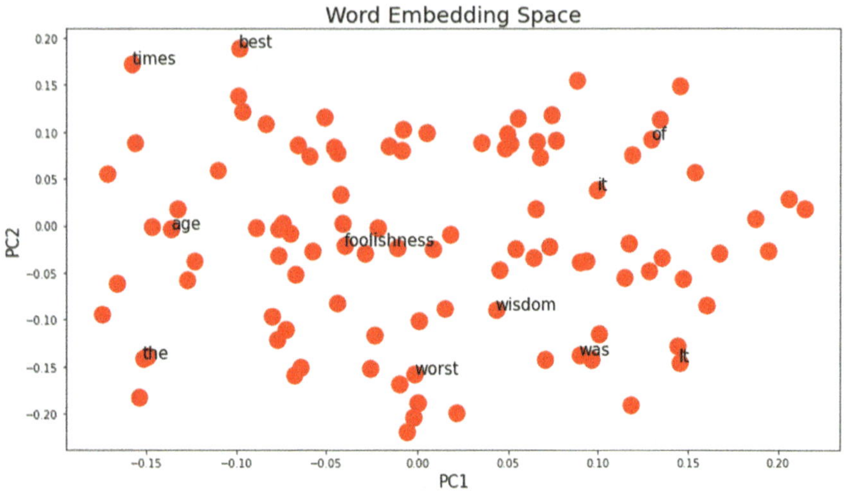

Fig. 3 Word2Vec embedding space example of the poem

it does not require human effort to tag the source data. However, it does not perform well when handling out-of-vocabulary words. It assigns random vector representation to out-of-vocabulary words that may cause sub-optimality. Moreover, the vector representations depend on local information such as the context of neighbour words that is another cause to obtain a sub-optimal representation. Word2Vec is a method that suits more to larger corpus due to better convergence of the algorithm, especially if skip-gram is chosen.

The following example is the two-dimensional Word2Vec representation of previously given poem of Charles Dickens. In Fig. 3, the oppositeness of "best" and "worst" keywords is seen as in being below and above in the space.

Google as a developer of the Word2Vec algorithm provides 300-dimensional pre-trained embeddings on more than a billion Google News. The implementation of Word2Vec embeddings with pre-trained vector representation can be found gensim library in Python. An example is given on Jupyter Notebook.

GloVe

GloVe holds an abbreviation for Global Vectors that is a technique developed by Stanford professors to overcome the local information dependency of the Word2Vec algorithm. To accomplish the aim of benefiting from global knowledge, GloVe utilizes the best count-based technique which is the co-occurrence matrix. In addition to the co-occurrence matrix, it leverages the advantages of Word2Vec. Hence, GloVe is referred as a hybrid technique for continuous word representation. The major difference of GloVe is the derivation of the objective function that will not be explained here but can be found on the original paper [7].

An example result with the same poem can be seen in Fig. 4. This embedding space was obtained through 25-dimensional pre-trained embeddings by gensim

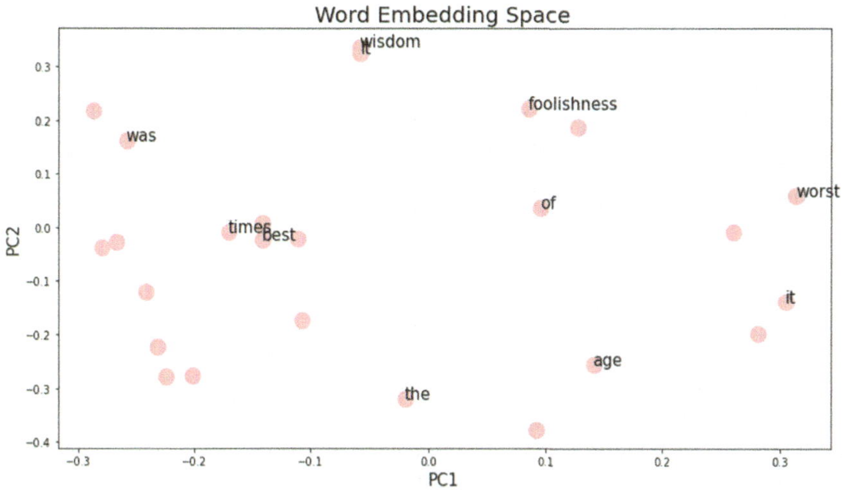

Fig. 4 GloVe embedding space example of the poem

library. Since the behind algorithm and input source is different from Word2Vec, the embedding space also differs.

GloVe tends to outperform Word2Vec in analogy tasks. Furthermore, it considers word pair to word pair relationship while constructing the vectors. Hence, GloVe tends to add more meaning to the vectors when compared to vectors constructed from word to word relationships. Also, GloVe has shorter training time comparing to Word2Vec, while it costs computationally more because of the utilization of co-occurrence matrix and global information.

4 Text Classification

Text classification is one of the commonly used NLP techniques in varying business problems such as detection of spam emails, automated tagging of customer inquiries and topic classification of news articles. Some text classification areas have their own research fields. For example, sentiment analysis, which is another widely known application of NLP, also falls under text classification in the categorization of NLP tasks. Moreover, classifying news articles into pre-defined topics is an application of topic modelling techniques in natural language processing.

A common nature of these given examples is all of them can be analysed with supervised machine learning algorithms. Hence, text classification can be considered under supervised machine learning tasks. It includes a labelled dataset which could be in different forms such as spam or not spam label for spam classification or topic classes for topic modelling. A complete text classification pipeline consists of three main steps:

- Dataset preparation
- Feature engineering
- Model training
- Improve performance of trained model (optional but highly recommended for all kind of machine learning applications).

All steps could be explained with a comprehensive spam classification example.

4.1 Dataset Preparation

For illustration purposes, the UCI Machine Learning SMS Spam Collection dataset has been used. This corpus was formed initially by a collection of 425 SMS, which includes only spam messages, 3375 legitimate (ham) SMS, which were gathered by the Department of Computer Science at the National University of Singapore dataset, and other 1452 SMS, which includes 450 spam messages [8].

Since the dataset already has labels, it does not require labelling. Table 8 shows some examples of SMS messages with their labels.

Advanced text representation techniques will be applied to messages in the feature engineering part. In addition to this, there is a need to change labels from text form to numerical. The label "Spam" could be replaced with 1 and "ham" with 0.

4.2 Feature Engineering

Since machine learning classifiers work with numbers, given messages will be converted to descriptive numerical features that allow advanced models to capture patterns within the text. Before continuing to preprocess the text and make it understandable to a machine, world clouds could be created as in Figs. 1 and 2 to have better insight into classes.

As in Fig. 5, the words "free", "call", "text" and "now" are mostly used in spam messages as they urge actions without further thoughts, whereas there is not a dominant word in non-spam context (Fig. 6).

Table 8 Example messages from spam dataset

SMS	Label
WINNER!! As a valued network customer you have been selected to receive a £900 prize reward! To claim call 09061701461. Claim code KL341. Valid 12 h only	Spam
Had your mobile 11 months or more? U R entitled to Update to the latest colour mobiles with camera for free! Call the mobile update co FREE on 08002986030	Spam
I HAVE A DATE ON SUNDAY WITH WILL!!	Ham

Fig. 5 Word cloud of spam messages

The feature representation methods BoW and TF-IDF have been explained previously under the "text representation" title. They are applied to the messages after stemming, lemmatization and removing stop words. Example preprocessed messages are shown in Table 9.

First, the count vectorizer method and chi-square statistical feature selection methods are applied on the given dataset with a $p = 0.99$ significance threshold. Unigrams, bigrams and trigrams that confirm the p-score threshold are chosen to continue machine learning classifiers. Later, TF-IDF, which is a more advanced method compared to Count Vectorizer, is applied. Selected features by count vectorizer and TF-IDF are given in Tables 10 and 11, comparatively.

4.3 Model Training

After choosing features with feature engineering techniques, a machine learning classifier could be trained with selected features. At this point, it is also possible to benefit from more simple methods such as taking an average of p-values of a sentence or just counting significant words in a sentence and classifying regarding the majority

Fig. 6 Word cloud of non-spam messages

Table 9 Example preprocessed messages from spam dataset

SMS	Label
Winner valu network custom select receiv 900 prize reward claim call 09061701461 claim code kl341 valid 12 h	Spam
Mobile 11 month u r entitl updat latest colour mobil camera free call mobil updat co free 08002986030	Spam
Date sunday	Ham

Table 10 Selected features through count vectorizer + chi-square ($p = 0.99$)

Feature	*p*-score	Label
Date	1.00	0
Holiday	1.00	0
Guarantee call	1.00	0
Deliver	1.00	0
Award	1.00	1
Camera	1.00	1
Camera phone	1.00	1
Cash	1.00	1
Chance win	1.00	1

Table 11 Selected features through TF-IDF + chi-square ($p = 0.99$)

Feature	p-score	Sentiment score
Send stop	0.99	0
UK	0.99	0
Club	0.99	0
Receive	0.99	0
Claim	1.00	1
Urgent	1.00	1
Prize	1.00	1

of features belonging to a label. Moreover, advanced deep learning methods could be utilized to classify the text.

Here, only machine learning classifiers implementation has been explained. However, other techniques have been described under the "sentiment analysis" title.

Several machine learning classifiers such as k-neighbours, decision tree, random forest, multi-layer perceptron, AdaBoost, and Naive Bayes algorithms have been built. All models are trained, and the result of the best-performed algorithm, which is the AdaBoost classifier, is given in Figs. 7 and 8. Since the features of spam messages are distinctive, as seen in Tables 12 and 13, the expected accuracy of explained task is high. As a parallel to intuition, 96% accuracy has been reached with the AdaBoost algorithm.

To understand the feature selection significance, the Lime algorithm, which is an explainable artificial intelligence tool, is given as an example in Fig. 9. The example sentence is selected as "Sorry, I'll call later". Regarding the Lime algorithm, the words "I", "ll", and "Sorry" do not have any effect on spam classification. On the other hand, the "later" has a very non-spam context, and the "call" is close to the spam context. Since later is dominant than call, the overall sentence label is "ham".

```
---------------Prediction with AdaBoost---------------
Accuracy: 0.96
Auc: 0.9
Detail:
              precision    recall  f1-score   support

           0       0.97      0.99      0.98      1453
           1       0.91      0.81      0.86       219

    accuracy                           0.96      1672
   macro avg       0.94      0.90      0.92      1672
weighted avg       0.96      0.96      0.96      1672
```

Fig. 7 Accuracy, precision, recall and $F1$-score metrics of spam classification

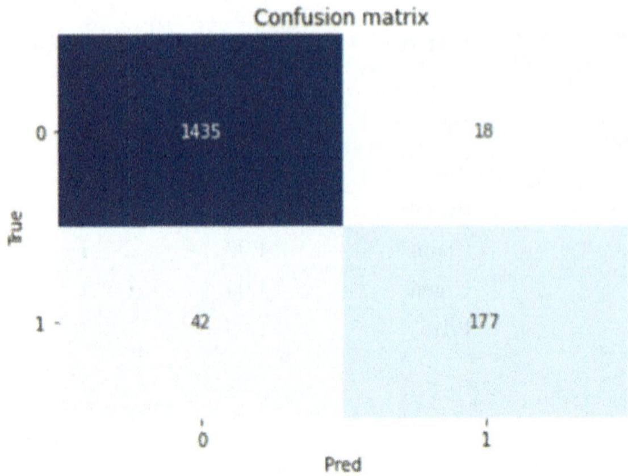

Fig. 8 Confusion matrix of AdaBoost classifier

Table 12 Model performance of spam classification

Label	Precision	Recall	$F1$-score
0	0.97	0.99	0.98
1	0.91	0.81	0.86

Table 13 Confusion matrix of spam classification

True/pred.	0	1
0	1435	18
1	42	177

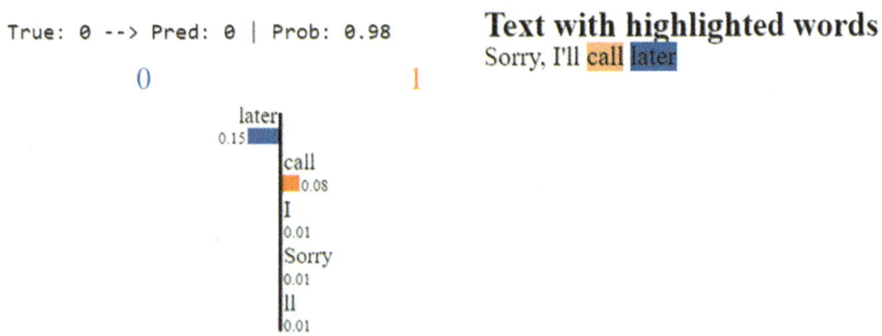

Fig. 9 Lime representation of example sentence

5 Topic Models

Topic modelling, similar to clustering, is an unsupervised learning method that helps to find the topics in the text when the searched groups are not known. In this method, the main goal is to utilize mathematical and statistical methods to identify hidden and latent semantic patterns in a corpus of data. In addition, it may assist with the following tasks:

- Identifying hidden patterns within the documents.
- Organizing the papers according to the topics that have been identified.
- Organizing, summarizing and searching for materials is accomplished via categorization.

In this section, three topic models are presented. These three models are tested on customer complaints data published by the US government [9]. The data is a collection of complaints about consumer financial products and services. It consists of a sample of 10,000 complaints.

Before applying topic models, some preprocessing steps explained below should be followed.

1. Punctuation removal
2. Tokenization
3. Stop words removal
4. Lemmatization
5. Computing term frequencies or TF-IDF
6. Topic model.

5.1 Latent Dirichlet Allocation (LDA)

Using Dirichlet distributions, it creates a topic per document model and a words per topic model, which are both used in the analysis. The model uses the topic for each word, the distribution over topics for each document and the distribution of words per topic [10]. The top terms of the five topics found with the LDA model are given in Table 14.

Table 14 Topics and top terms of the topics obtained from LDA model

Topic	Top terms
1	Loan, payment, mortgage, bank, account
2	Call, debt, phone, number, company
3	Request, file, complaint, information, account
4	Credit, report, debt, credit report, account
5	Card, credit card, account, credit, charge

Table 15 Topics and top terms of the topics obtained from LSI model

Topic	Top terms
1	Account, credit, payment, loan, report
2	Credit, report, loan, credit report, payment
3	Account, debt, loan, card, bank
4	Debt, loan, credit, report, credit report
5	Account, card, credit card, loan, credit

5.2 Latent Semantic Indexing (LSI)

This method accepts as input a collection of documents. The document co-occurrence matrix is used to generate a word-document matrix. It makes advantage of the TF-IDF conversion to eliminate superfluous high-frequency terms from the word-document matrix. Then, each document's weight is normalized to its unit length (normalization process). Finally, the single value decomposition (SVD) technique is used to reduce the file size. This method selects the biggest single values [10]. Result of the model on consumer complaint data is presented in Table 15

5.3 Non-negative Matrix Factorization (NMF)

NMF factorizes higher dimensional matrices into lower dimensional factors. Coefficients of low-dimensional matrices are not negative. Let us take a matrix where there is a matrix of articles by words. When this matrix is decomposed, articles by topics and topics by words are obtained [10]. Top terms found with NMF model are listed in Table 16.

Table 16 Topics and top terms of the topics obtained from NMF model

Topic	Top terms
1	Loan, payment, mortgage, would, told
2	Credit, report, credit report, reporting, info…
3	Account, bank, information, check, money
4	Debt, collection, company, call, agency
5	Card, credit card, credit, charge, fee

6 Sentiment Analysis (Opinion Mining)

Sentiment analysis is the mining of a text to extract underlying emotions of subjective information. It helps businesses to understand customer reactions analytically and shape their products around customer behaviour. Also, firms could assess their market position against competitors through customer sentiments. The source material could be any type of text such as plain text, emojis or symbols which are the equivalent of the emojis. This type of source information could be gathered easily from social media, news, blogs and customer calls.

Despite of having simply accessible sources, sentiment analysis requires deep understanding of natural language processing. The research area was not so developed until the recent advancements in deep learning. Previous methods were based on mostly counting techniques. Since sentiment analysis a sub-problem of text classification field, currently popular techniques in sentiment analysis benefit from developments in deep learning techniques which are used widely in text classification. In this chapter, several sentiment analysis approaches will be shared in a wide range from counting-based simple techniques to advanced transformers architecture.

All sentiment analysis methods could be grouped under three categories:

- Rule-based methods
- Statistical (machine learning including) methods
- Deep learning-based methods.

6.1 Rule-Based Sentiment Analysis

Rule-based sentiment analysis is found on pre-defined classification rules. These rules could be defined on a corpus with a context or a dictionary. The former is called corpus-based sentiment analysis that matches sentiments such as negative, neutral and positive with individual words regarding their meaning in the given context. For example, the "increase" word in a sales context has a positive connotation, whereas a negative in a cost-related corpus. This kind of context-dependent sentiment change is addressed by also domain-specific dictionaries. Moreover, these dictionaries are introduced to embed domain expert knowledge in sentiment analysis to overcome situations such as differing connotations of a specific term.

The first general use dictionary has been proposed by Harvard as Harvard General Inquirer (Harvard GI) [11]. Since Harvard GI is not capable to handle context-dependency, domain-specific dictionaries followed it. An example domain-specific dictionary was formed by Loughran and McDonald as Financial Sentiment Dictionary [12]. These dictionaries empowered the dictionary-based sentiment analysis against corpus-based since its value proposition was built on being context-aware.

Before using a dictionary or corpus-based lexicon, the source data should be preprocessed to ensure the result will not contain any noise. This preprocessing includes tokenization of the text. Tokenization means splitting each document into

Fig. 10 Rule-based sentiment analysis steps

sentences and further splitting it into individual words. After obtaining "tokens", stop words would be detected through a stop word dictionary. The famous stop word dictionary is hosted by NLTK in Python [13]. Next to the removal of stop words, punctuation removal is to be considered. Punctuation has significance if it enriches the meaning such as "!" or "?". Other punctuations such as "." generally do not have an effect on meaning so must be removed before start running the preprocessed text with sentiment lexicon to decide its polarity. All steps are summarized in Fig. 10.

For demonstration, consider this problem instance: "Pillows nice, great welcoming and service." from hotel reviews in Europe dataset in Kaggle [14]. First, it is needed to be tokenized as in Table 17. In this step, plural words will be converted to a singular form in addition to transforming uppercase letters to lowercase letters. Even though having lowercase words is not mandatory, it generally improves the performance. Likewise, most sentiment dictionaries do not include both uppercase and lowercase versions of an expression at the same time. Also, verb conjugations are removed as a convention because they do not change the original connotation of the original form of the word.

As a second step, simple feature engineering will be applied to remove words, conjunctions and punctuations with no significance for sentiment analysis (Table 18).

Finally, preprocessed tokens in Table 19 will run with lexicon. This lexicon provides a number for each token and the model provides a compound score as a sentiment index of the given text. This sentiment index could change from dictionary to dictionary or corpus to corpus. In here, a simple count-based example is given in Table 20. The sentiment score of given text would be 0.6 through counting.

Table 17 Tokenization of "pillows nice, great welcoming and service"

Token
Pillows
Nice
,
Great
And
Welcoming
Service
.

Table 18 Preprocessing of "pillows nice, great welcoming and service"

Token	Preprocessing
Pillows	Plural to singular
Nice	
,	Punctuation
Great	
And	Conjunction
Welcome	Verb conjugation
Service	
.	Punctuation

Table 19 Removal of punctuations and conjunctions in "pillows nice, great welcoming and service"

Preprocessed token
Pillow
Nice
Great
Welcome
Service

Table 20 Sentiment match of tokens

Preprocessed token	Sentiment
Pillow	Neutral
Nice	Positive
Great	Positive
Welcome	Positive
Service	Neutral

Python natural language libraries provide their own sentiment analysers such as Vader of NLTK [15]. A more sophisticated result for the given example would be in Table 21 with using Vader.

Rule-based sentiment analysis is simple yet somehow powerful method to analyse social media data. It would not work on more sophisticated data such as financial news. It is used mainly as a tool for the unsupervised approach. Hence, there is no need to prior training. This property brings faster execution, with hash-table data structure especially. Furthermore, unlike the machine learning-based techniques that need large dataset to get itself trained and achieve meaningful accuracies, a rule-based

Table 21 Vader result by NLTK

Example review	Negative	Neutral	Positive	Compound
Pillows nice, great welcoming and service	0.000	0.345	0.655	0.6908

method could work in any size data from small to big data effectively. Addition to that, it offers lesser risk since already tested and widely used.

Despite being simple and less risky, dictionaries or lexicons are specific to the domain. Best performance of Vader on social media problems depict an example of that. Many lexicons need to be tried and customization in most cases is not possible. Moreover, if a new lexicon would be created, an expert would be required to state the rules in the first place. This limitation would apply when also updating existing rules or adding new ones. Another important point to state is that rule-based approaches have no learning capability.

6.2 *Statistical Sentiment Analysis*

Statistical sentiment analysis is formed by two steps: First, feature engineering is to be applied by statistical methods such as TF-IDF and count vectorizer, which are explained techniques under text representation title in this book. Second, a machine learning classifier would be used to classify sentiment labels of given features which are the outputs of the feature engineering part. Since the feature engineering part utilizes statistical methods such as *p*-score to identify the most relevant features, these methods are named as statistical sentiment analysis in this book.

For a demonstration of statistical sentiment analysis implementation, a metal news dataset has been used. Example news are given in Table 22.

As in working with rule-based techniques, text preprocessing increases the accuracy of statistical sentiment analysis too. Hence, text preprocessing methods such as stop words removal, tokenization, lemmatization and stemming would be applied to source information before feature engineering. Preprocessed news example is given in Table 23.

Then, a count vectorizer or TF-IDF would be used to vectorize text data to numerical form. Here, it is possible to state *n*-grams (phrases include *n*-words), and it is given to the vectorizer one to three *n*-grams in the example sentiment analysis. Since not all features have significance on the result, some of them would be excluded by applying a feature selection technique. In this example, the chi-square statistic has been utilized. A feature would be included or not regarding its associated *p*-value,

Table 22 Example news headlines in metal news dataset

News	Sentiment score
Kazakhstan's Jan–Aug copper output up; zinc, steel production falls	– 1
Base metals fall as weak China data weighs on demand outlook	– 1
Copper eases on potential delay in U.S.-China trade deal	– 1
Commerzbank sees aluminium prices averaging $1675/T in 2020, $1750/T in 2021	1
Shanghai aluminium jumps to 2-1/2-month high on supply worries	1

Table 23 Preprocessed example news headlines in metal news dataset

News	Sentiment score
Kazakhstan Jan–Aug copper output zinc steel product fall	− 1
Bas metal fall weak china data weigh demand outlook	− 1
Metalscopp eas potenti delay uschina trade deal	− 1
Commerzbank see aluminium pric average 1675 T 2020 1750 T 2021	1
Shanghai aluminium jump 212 month high suppli worri	1

which should be higher than chosen 0.05 p-value limit. If an expression has a p-score associated with a class (negative, neutral or positive) upper than 0.05, it will not be included in the features. In Jupyter Notebook, p-score is interpreted as 1 − (p-score) to emphasize the correlation between expressions and their significance such that a higher p-score means higher association. Example selected features through count vectorizer and TF-IDF and chi-square feature selection are given in Tables 24 and 25, respectively.

After choosing features as above, they would be given to a machine learning classifier that learns sentiments of features and decides a sentence's sentiment score

Table 24 Selected features through count vectorizer + chi-square ($p = 0.95$)

Feature	p-score	Sentiment score
Metalscopp fall	1.00	− 1
News metalscopp fall	1.00	− 1
Fall	1.00	− 1
Tension	1.00	− 1
Hope	0.99	− 1
Metalslondon copper edge	0.96	1
Deal hope	0.96	1
Trade deal hope	0.96	1
Retreat	0.96	1

Table 25 Selected features through TF-IDF + chi-square ($p = 0.95$)

Feature	p-score	Sentiment score
Metalscopp fall	0.98	− 1
News metalscopp fall	0.98	− 1
Fall	0.96	− 1
Commod	1.00	0
Disrupt fear	0.99	0
Begin	0.99	0
Hope	0.95	1

Table 26 Model performance of metal news classification

Label	Precision	Recall	$F1$-score
− 1	0.80	0.56	0.66
0	1.00	0.06	0.11
1	0.67	0.93	0.78

Table 27 Confusion matrix of metal news classification

True/pred.	− 1	0	− 1
− 1	35	0	27
0	3	1	13
1	6	0	81

regarding the features it includes. In this example, several machine learning classifiers are compared, and the best performer is given here. Any machine learning classifiers such as a k-neighbours, decision tree, random forest, multi-layer perceptron, AdaBoost and Naive Bayes could be utilized in this part. Also, deep neural networks may be built for the classification task but not recommended due to not so sophisticated nature of the problem. The outcome of the Naive Bayes algorithm for the financial sentiment analysis can be seen below with accuracy 71% (Tables 26 and 27).

The statistical sentiment analysis reached good performance even though having an imbalanced dataset that skewed for negative news. To demonstrate the decision mechanism of the classifier, a news instance ("METALS-Aluminium bounces on short-covering after Rusal news") and the trained classifier are given to the Lime representation tool [16]. As in Fig. 11, the classifier decides the polarity of complete sentences by the associated sentiment scores of individual words. In the given

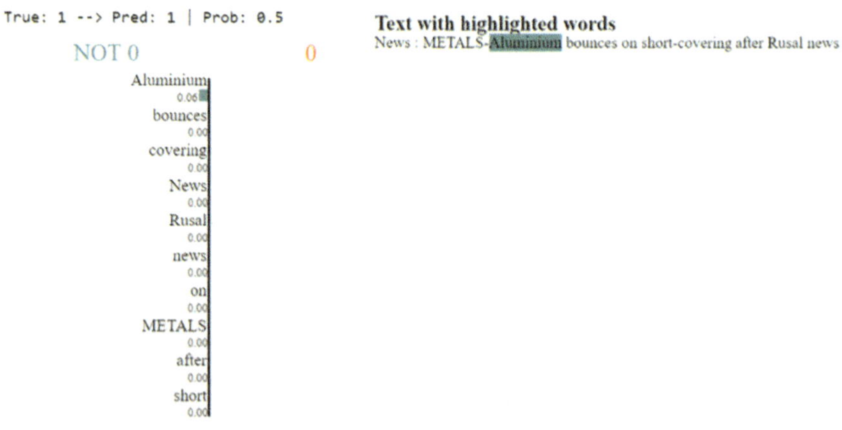

Fig. 11 Lime representation of the classifier's decision mechanism

example, "aluminium" has a positive polarity, and other features are neutral, so the predicted sentiment would be positive which is correct.

The performance of statistical sentiment analysis methods is determined by feature extraction and feature selection parts. It requires fine-tuning to obtain good results. Moreover, model behaviour with new features is not predictable. Also, it generalizes the sentiment score of an expression that depends on the context of the corresponding sentence or corpus.

Yet, they have simple development and better performance against rule-based approaches. They have widely used methods until very recent developments in transformers architecture and attention mechanism [17]. Statistical sentiment analysis could be a very suitable option to work on small datasets.

6.3 Deep Learning-Based Sentiment Analysis

Previously mentioned methods of sentiment analysis operate without having a memory which means that every single instance is processed independently with no relation to others. This is the opposite way of how the human brain works when reading a text, since memorizing previous content helps to understand the next ones and easier capture the context. Recurrent neural networks (RNNs) adopt the same principle. RNNs iterate along with the sequence elements by processing and memorizing relative information to previously processed ones until that point. It enables to achieve more accurate results. The memorization mechanism has an important effect on the performance of the model.

In recent years, deep learning-based NLP solutions had a breakthrough with the introduction of transformers architecture. Attention mechanism in transformers replaces the forget and update gates in RNNs while proposing a more sophisticated approach that emphasizes individual sequence items which have greater importance on the classification of the text or capture the context. Even though these models rely on revolutionary logic, they still cannot process raw text. At this point, they benefit from embeddings with different options such as lowercase or uppercase. These deep learning architectures with provided language models are called pre-trained deep learning models.

Pre-trained deep learning models offer great flexibility with no need to create embeddings or train the overall architecture that consists of several layers to extract descriptive features of the source information and a classification layer at the end for classifying the extracted features. Google search engine even uses one of the pre-trained deep learning models as of 2021 [18]. Another one also belongs to Facebook [19]. The reason behind the tech giants benefiting from this model is their current state-of-the-art positions.

These state-of-the-art approaches also have been applied for sentiment analysis tasks in the published literature.

Here, the previous metal news dataset will be used to explain how to take the privilege of using the mentioned models. As mentioned before, there are different

Table 28 Pre-trained language model selection for each DL model

DL model	Language model
BERT	Bert-base-uncased
Flair	Distilbert-base-uncased
XLNET	XLNet-large-cased
ULMFit	Trains its own language model

Table 29 Results of DL-based sentiment analysis techniques

DL model	Accuracy for negative news	Accuracy for neutral news	Accuracy for positive news	Overall accuracy
BERT	0.71	0.0	0.84	0.73
XLNet	0.71	0.11	0.81	0.72
Flair	0.64	0.06	0.79	0.72
ULMFit	0.58	0.10	0.76	0.64

language models based on trained embeddings. Since each pre-trained deep learning (DL) model has different characteristics, used language models are also not the same. For the metal news dataset example, utilized language models are given in Table 28.

Hyperparameter tuning is applied above DL models and classification layers trained with source information along ten epochs. The results are given in Table 29.

7 Advanced Topics in Natural Language Processing (NLP)

7.1 Similarity

- Cosine Similarity: It provides us with the measure of the cosine of the angle that exists between two vectors.
- Jaccard Index: It is determined as the intersection of words divided by the union of the words.
- Levenshtein distance: It is found as the fewest possible insertions, deletions and substitutions needed to convert first text to second text.
- Hamming distance: It is defined as the number of positions that have different characters in both strings with same length

The results of two sample words ("test", "text") with different similarity distance are given in Table 30.

Table 30 Results of similarity methods

Method	Similarity
Cosine	0.75
Jaccard	0.6
Levenshtein	1
Hamming	1

7.2 *POS Tagging*

Part-of-speech (POS) tagging is an important part of natural language processing, which includes tagging words with clauses such as nouns, verbs, adjectives and the like. This method forms the basis of many advanced natural language processing applications, especially sentiment analysis. For example, when this process is performed, tagging operations can be done as the word "run" is a verb. NLTK library contains built-in tags for this operation. It is going to be used to tag a sample sentence "This book has written by academic members from Istanbul Technical University in Turkey". Results of POS tagging operation are presented in Table 31.

7.3 *Named Entity Recognition (NER)*

It can be described as the process of finding out which entity a word in the text refers to. There are many libraries for this operation such as NLTK chunker, StanfordNER, SpaCy, OpenNLP and NeuroNER. In addition, this process can be performed with many web services such as WatsonNLU, AlchemyAPI, NERD, Google Cloud NLP API [1]. When SpaCy library is used on the previous sample sentence, two tags which are presented in Table 32 are obtained.

Table 31 Results of POS tagging

Word	Tag	Tag definition
This	DT	Determiner
Book	NN	Noun, common, singular or mass
Written	VBN	Verb, past participle
Academic	JJ	Adjective or numeral, ordinal
Members	NNS	Noun, common, plural
Istanbul	NNP	Noun, proper, singular
Technical	NNP	Noun, proper, singular
University	NNP	Noun, proper, singular
Turkey	NNP	Noun, proper, singular

Table 32 Results of NER

Phrase	Tag	Tag definition
Istanbul Technical University	ORG	Companies, agencies, institutions, etc.
Turkey	GPE	Countries, cities, states

7.4 *Clustering*

In the natural language processing, cluster analysis is used for document clustering problems. In these type of problems, unsupervised machine learning concepts and techniques are applied on text. For the purpose of categorizing the texts into different clusters, unsupervised machine learning techniques are employed. They include characteristics that make documents contained within one cluster more comparable and connected to each other than papers contained within other clusters, as compared to documents contained inside other clusters.

In this section, same consumer complaints dataset with section of topic models are used. Clustering application includes some preprocessing steps explained below.

- Punctuation removal
- Tokenization
- Stop words removal
- Lemmatization
- Computing term frequencies or TF-IDF
- K-means clustering.

After completing above steps, clusters and key features of these clusters that are presented in Table 33 are obtained.

7.5 *Disambiguating*

Words can have different meanings in different contexts. This situation presents us with the problem of disambiguation. The Lesk algorithm is one of the most effective

Table 33 Obtained clusters

Cluster	Documents	Key features
1	654	Debt, collection, company, credit, account
2	368	Account, bank, call, nt, payment
3	270	Loan, mortgage, payment, home, month
4	220	Credit, report, credit report, account, information
5	176	Card, credit card, credit, account, charge

Table 34 Different meanings of a word

Sentence	Word	Meaning
The fruits of the cherry tree have ripened	Fruit	The ripened reproductive body of a seed plant
Finally graduating, he reaped the fruits of his work	Fruit	The consequence of some effort or action

algorithms for disambiguating word senses. How the same words have different meanings on two sample sentences using the NLTK Library is analysed (Table 34).

7.6 Language Identification and Translation

In the globalizing world, companies can acquire customers from many parts of the world, and at the same time, they need to follow the news from many locations. At this point, there may be a need to analyse text from many different languages. The importance of language identification and language translation in text analysis comes from this situation. The textblob library can be used as limited version to do these types of operations.

References

1. Kulkarni A, Shivananda A (2017) Natural language processing recipes. Springer, New York, NY, pp 38–151
2. Sun S, Luo C, Chen J (2017) A review of natural language processing techniques for opinion mining systems 36:10–25
3. Zhang Y, Jin R, Zhou ZH (2010) Understanding bag-of-words model: a statistical framework. Int J Mach Learn Cybern 1(1–4):43–52
4. Oesper L, Merico D, Isserlin R, Bader GD (2011) WordCloud: a cytoscape plugin to create a visual semantic summary of networks. Source Code Biol Med 6(1):7
5. Mikolov T, Chen K, Corrado G, Dean J (2013) Efficient estimation of word representations in vector space. In: ICLR workshop papers
6. Mikolov T, Sutskever I, Chen K, Corrado G, Dean J (2013) Distributed representations of words and phrases and their compositionality. In: NIPS, pp 3111–3119
7. Pennington J, Socher R, Manning CD (2014) GloVe: global vectors for word representation. In: EMNLP
8. Almeida AT, Hidalgo JMG, Yamakami A (2011) Contributions to the study of SMS spam filtering: new collection and results. In: Proceedings of the 2011 ACM symposium on document engineering (DOCENG'11), Mountain View, CA, USA
9. Data.Gov. (2021) Consumer complaint database. https://catalog.data.gov/dataset/consumer-complaint-database
10. Sarkar D (2019) Text analytics with python. APress, Karnataka, India
11. Stone PJ, Hunt EB (19623) A computer approach to content analysis: studies using the general inquirer system. In: Proceedings of the May 21–23, 1963, spring joint computer conference [AFIPS'63 (Spring)]. Association for Computing Machinery, New York, NY, USA, pp 241–256

12. Loughran T, McDonald B (2015) The use of word lists in textual analysis. J Behav Financ 16(1):1–11
13. Bird S, Klein E, Loper E (2009) Natural language processing with python: analyzing text with the natural language toolkit. O'Reilly Media, Inc.
14. https://www.kaggle.com/jiashenliu/515k-hotel-reviews-data-in-europe. Accessed 20 Jun 2021
15. Hutto CJ, Gilbert E (2014) VADER: a parsimonious rule-based model for sentiment analysis of social media text [online]. Available at http://comp.social.gatech.edu/papers/icwsm14.vader.hutto.pdf. Accessed 20 Jun 2021
16. Ribeiro MT, Singh S, Guestrin C (2016) Why should I trust you? In: Proceedings of the 22nd ACM SIGKDD international conference on knowledge discovery and data mining—KDD'16 [online]. Available at https://arxiv.org/pdf/1602.04938.pdf. Accessed 20 Jun 2021
17. Vaswani A, Brain G, Shazeer N, Parmar N, Uszkoreit J, Jones L, Gomez A, Kaiser Ł, Polosukhin I (2017) Attention Is all you need. In: Advances in neural information processing systems [online]. Available at https://papers.nips.cc/paper/2017/file/3f5ee243547dee91fbd053c1c4a845aa-Paper.pdf. Accessed 20 Jun 2021
18. Devlin J, Chang MW, Lee K, Google K, Language A (2019) BERT: pre-training of deep bidirectional transformers for language understanding. In: Proceedings of NAACL-HLT 2019 [online], pp 4171–4186. Available at https://aclanthology.org/N19-1423.pdf. Accessed 8 Aug 2021
19. Akbik A, Bergmann T, Blythe D, Rasul K, Schweter S, Ai R, Berlin R, Vollgraf R (2019) FLAIR: An easy-to-use framework for state-of-the-art NLP. In: Proceedings of NAACL-HLT 2019 [online], pp 54–59. Available at https://aclanthology.org/N19-4010.pdf. Accessed 8 Aug 2021

Mahmut Sami Sivri is currently a lecturer at Industrial Engineering Department in Istanbul Technical University. He is also working on some R&D projects as Director of Data Analytics at ITU's Technopark. He received the B.Sc. degree in Computer Engineering and the M.Sc. degree in Engineering Management from Istanbul Technical University. He worked in various companies and positions in the IT industry since 2008. His current research interests include machine learning, sentiment analysis, deep learning, big data and its applications, Industry 4.0, financial technologies, data analytics, supply chain and logistics optimization.

Buse Sibel Korkmaz is a PhD student at Imperial College London and a machine learning engineer at Comcast. She is studying autonomous systems and artificial intelligence in her PhD thesis for automated synthesised supervisory controllers for industrial plants. She completed her undergraduate education in Istanbul Technical University with a double degree in B.Sc. in Computer Engineering and Industrial Engineering. She studied natural language processing to understand the effects of news on financial instrument prices under su-pervision of Prof. Alp Ustundag and published conference papers on natural lan-guage processing and gaussian process modelling.

Image Analysis

Nurullah Calik and Behcet Ugur Toreyin

1 Introduction to Image Processing

One of the biggest impacts on the advancement of information technologies is the easy acquisition and processing of information. Many phenomena occurring around us leave traces of information that reveals itself. By following and analyzing these traces, many problems [1–4] can be solved with the help of computers. Computers, representing facts and happenings in numbers, can perform intelligent actions thanks to advanced algorithms. Developing algorithms suitable for the characteristics of phenomena is of utmost importance. Images acquired by various sensing modalities provide a plethora of information witnessing nature. Meaningful knowledge from images is extracted by (digital) image processing methods.

The main purpose of image processing is to ensure that the functions performed in the human visual cortex are performed by machines. The first step toward this goal is the mathematical nature of the image. The digital image is composed of numbers obtained by the cameras converted into light intensity in pixels. Different values of these numbers correspond to different color tones. Monochrome (gray-level) color means an image with gray tones. Eight-bit-quantized images contain pixel values within a dynamic range of [0, 255] representing gray levels. As seen in Fig. 1, each value corresponds to a different shade of gray on the screen. The computer actually "sees" an image as a bunch of numbers. This stack of numbers is called a matrix.

Screens don't just produce gray-level images. Color images are formed by mixing the basic colors, namely, Red, Green, and Blue (RGB) in certain proportions. A pixel on the screen actually contains RGB sub-pixels. When these sub-pixels are lit at different intensities, other intermediate colors are formed. Gray-level images are

N. Calik
Department of Biomedical Engineering, Istanbul Medeniyet University, Istanbul, Turkey

B. U. Toreyin (✉)
Informatics Institute, Istanbul Technical University, Istanbul, Turkey
e-mail: toreyin@itu.edu.tr

A. Ustundag et al. (eds.), *Business Analytics for Professionals*, Springer Series in Advanced Manufacturing, https://doi.org/10.1007/978-3-030-93823-9_8

Fig. 1 Gray-level image example

represented by only one matrix, while RGB images are described by three matrices arranged in a channel (cf. Fig. 2).

Histogram: The histogram is the most basic physical information that can be obtained from an image. Counting the brightness (intensity) of pixels yields critical information about the image. The histogram of the image given in Fig. 3a is given in Fig. 3b. In the graph in Fig. 3b, 0 corresponds to black pixels and 255 white pixels, while values in-between correspond to different gray levels. As can be seen from the histogram graph, the amount of pixels after 250 is higher than the others. This is due to the whiteness of the leaves of daisies.

Contrast is another instrumental concept. Inspecting Fig. 4, it can be seen that it is hazier compared to the image in Fig. 3. Looking at the histogram of this image, it is seen that a numerical equivalent of this haze can also be obtained. When Fig. 4b is compared with Fig. 3b, it is interesting that there is a fundamental difference between the two, such as width and narrowness. The mathematical equivalent of this is the standard deviation of the image. Thus, if an image has a high standard deviation, it

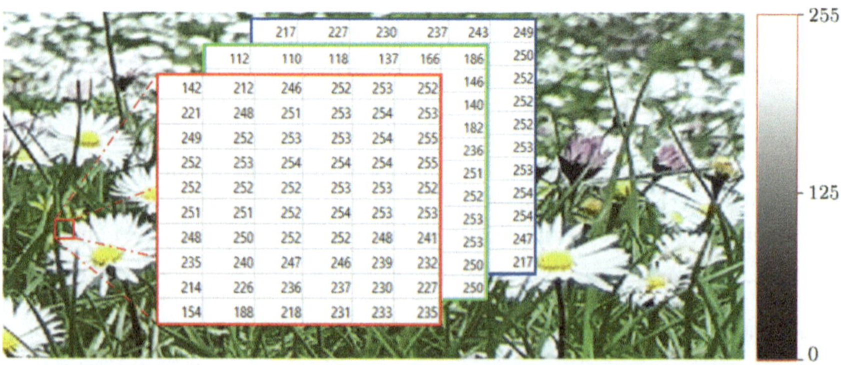

Fig. 2 Color images are composed of three matrices corresponding to color channels. Values in each channel matrix define relevant color intensity

Fig. 3 Histogram of a gray-level image. (a) Input image (b) Histogram plot

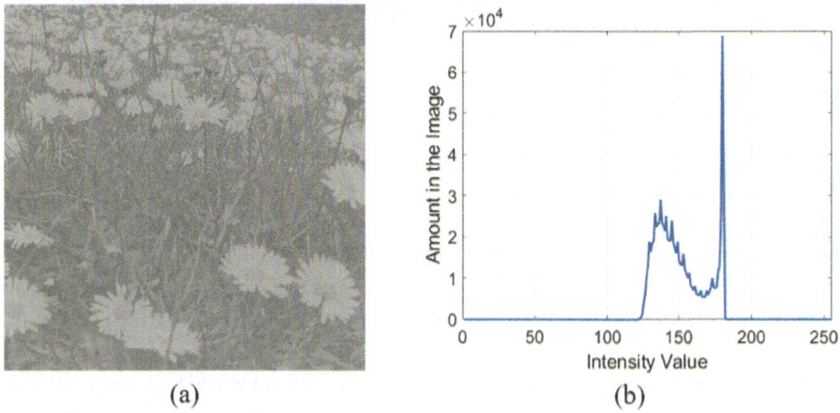

Fig. 4 If the input image (a) has low contrast, the standard deviation of its histogram (b) is narrow

has high contrast; otherwise, it can be said to have low contrast. Images with high contrast are clearer and details can be observed easily. A computer can now evaluate the contrast difference that can be made with the visual cortex over the standard deviation value.

Point Operations: As the name suggests, point operations deal with operations applied on pixel intensity values. As such, relations with neighboring are not considered (cf. Fig. 5). An example of this is given in Fig. 5. Figure 5a has been reprinted for comparison purposes as the original image. In Fig. 5b, the function in Eq. (1) is applied.

$$T_1(x) = \begin{cases} x, & 100 \leq x \leq 220 \\ 0, & \text{otherwise} \end{cases} \tag{1}$$

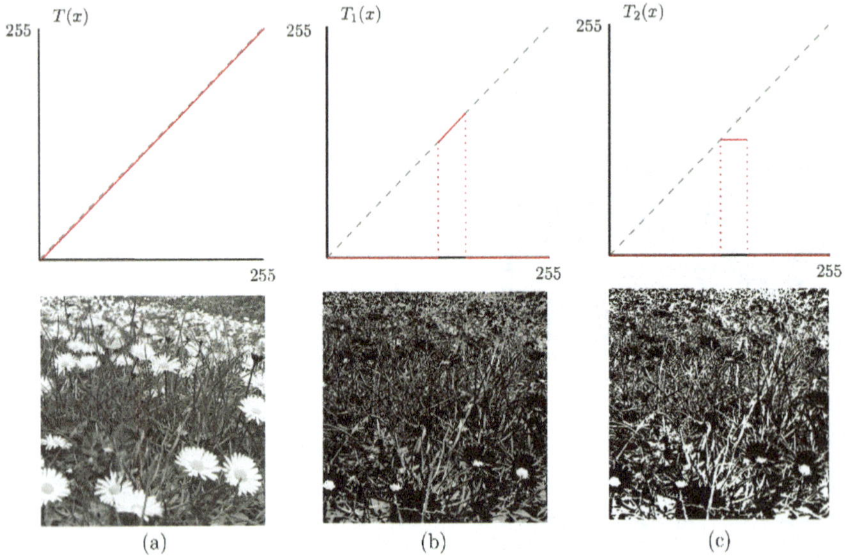

Fig. 5 Some point operations and their effects on image. Since the transformation in (a) is linear, the intensities in the image are preserved. In (b), only the range of certain pixel values is preserved while the others are set to zero. In the transformation in (c), the pixel values in the selected region are assigned to a single gray level, while the others are zero

The transfer function in (1) keeps the pixel values between [100, 220] and makes background the other values. An example of this is given in the image of Fig. 5b. Herein, it is seen that values close to white are preserved, especially for grasses and daisies, below 220.

In Fig. 5c, the pixels' own values are not preserved, but instead they are assigned a value of 1 by using transformation Eq. (2). In this way, a binary image was obtained. Binary images are an important area of image processing. Special mention is made of this in the following sections.

$$T_2(x) = \begin{cases} 1, & 100 \leq x \leq 220 \\ 0, & \text{otherwise} \end{cases} \tag{2}$$

Many improvements can be made owing to point operations [5, 6]. However, operations with pixel values alone are not sufficient for pattern recognition. In order to reveal the pattern, it is necessary to compare the pixel with neighboring pixels. With the introduction of coordinate information, advanced features can be extracted from the image. As a result of these processes, computer vision applications become feasible. These processes are covered under image filtering.

2 Image Filtering

2.1 *Spatial Filtering*

Point operations are only concerned with the brightness value of a pixel in the input image, as seen in Fig. 6a. This transformation can be represented as $T(I[x_0, y_0])$. Notice that $T(\cdot)$ does not deal with coordinate information. In spatial filtering, the intensity values of neighboring pixels are also included in the calculation of the output. This situation is illustrated in Fig. 6b.

With what transformation will the values of neighboring pixels be taken into account and form the output? At this point, the most mathematical interpretation of which are familiar with is the weighted summation. Of course, an output could be obtained by making exponential calculations of these points and adding them with the logarithm of the center pixel. However, the interpretability and predictability of the process are important. At this point, since the weighted sum operation can take a matrix form, linear algebra's vast toolbox can be used to interpret operations.

A weight sum operation requires two inputs: (1) Input image (2) kernel matrix. These are given in Fig. 7a, b, respectively. The values in the kernel matrix represent the contribution of the neighbors around the center pixel and the center to the output

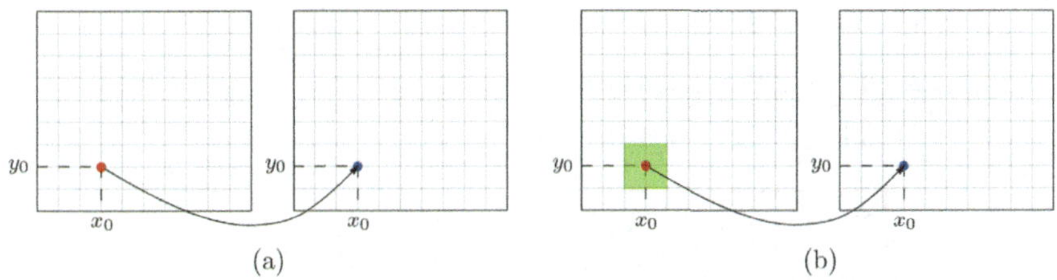

Fig. 6 Comparison of point and pixel neighborhood-based operation. While point operations produce results by only processing the value in the relevant coordinate (a), neighborhood-based operators generate output by including the neighborhoods around the center pixel (b)

1	8	11	5	12	62	3
15	36	11	22	9	7	3
2	1	1	6	13	41	73
21	11	22	62	8	15	1

(a)

-1	0	0
-1	0	0
0	1	1

(b)

Fig. 7 The input image (a) and kernel matrix (b) required for the convolution operation

value. In the example given in Fig. 7, while there is no contribution from the center, only negative and positive contributions come from certain neighbors.

The kernel produces the output by multiplying and summing the pixels in the overlapping area in the input image. After the process is finished, it shifts to the right and continues the same step. The output image is obtained after all the pixels are navigated with the sliding operation. This custom-defined operation is called convolution. An example for three center pixels is shown in Fig. 8.

When the kernel is matched with the region with 36 central pixels, a value of -14 is produced for the output. The kernel then shifts one right and outputs -37 for the 11-value center pixel. This process continues throughout the entire image. Here are some issues that need attention. Although pixel values have been used as positive until now, negative numbers can come from the differences between them. While the image resulting from the convolution process is printed on the screen, these values are drawn to the [0, 255] boundaries with certain operations. Convolution is one of the most important operators for extracting information from images. The aspect of this process that is open to interpretation and prone to designing new processes due to the frequency domain relationship that we will see in the following sections offers us a strong analysis and synthesis framework.

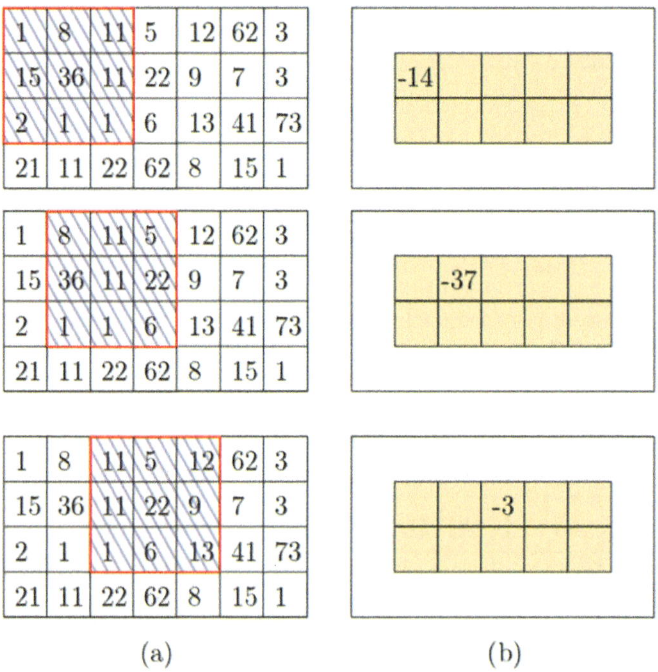

Fig. 8 Example of convolution process for three points. In (a), the intersecting regions of the kernel and the input image are shown. In (b), the values calculated in the output matrix are given

(a) (b) (c)

Fig. 9 Edge detection example produced by convolution. After filtering the input image (a) with h_x and h_y, images containing the edge information in (b) and (c), respectively, are formed

One of the most basic applications that can be done with the convolution operation is edge detection. For this, the kernels in Eq. (3) are applied to the image separately.

$$h_x = [-1, 0, 1] \quad h_y = h_x^T \tag{3}$$

Here, h_x and h_y are two separate kernels that are transpose of each other. If the operation on h_x is interpreted, the kernel looks at the difference between the neighbors in the x-direction while multiplying the central pixel with a coefficient of 0. If this difference is close to each other, the difference will be low. That is, similar pixels will be drawn into the dark tone region. If the difference between neighbors is high, these areas will become prominent in the white tone region and edge regions where sudden transitions occur. So, this kernel is *a filter for edge detection.*

An example for edge detection is given in Fig. 9. When the h_x and h_y filters are applied to the input image in Fig. 9a, respectively, the outputs in Fig. 9b, c are revealed. Notice that, while vertical edge components appeared in Fig. 9b, horizontal components became evident in Fig. 9c. Neighboring pixel values close to each other are eliminated by filtering. When visualizing Fig. 9b, c, it needs to be specified that the abs(·) function is used. Otherwise, negative boundaries do not appear.

The coefficients considered so far were fairly simple and easy to interpret in the spatial domain. So, how will the resulting effects be interpreted when different coefficients are given? Or how to obtain the filter coefficients that will reveal a desired effect? Spatial domain analysis does not offer this opportunity. A different analysis framework is needed to interpret the 2D convolution operation. At this point, Frequency Analysis, which can be done with Fourier Transform, reveals the working principles of filters. Not only that, it also opens the door to the emergence of new filter types.

2.2 Frequency Analysis

The 2D Fourier Transform provides a powerful framework for processing and interpreting images in a domain other than the spatial domain. Concepts and problems that cannot be easily understood in spatial terms can be solved by frequency analysis techniques. In addition, the working mechanism of spatial filters made by convolution is also revealed.

Frequency Analysis of Images: Fourier transform (FT) of images is directly related to 2D convolution. The operation of the convolution mask over the pixels becomes more understandable with frequency analysis. In addition, thanks to this technique, new filters can be designed and many groundbreaking applications can be paved. Also, it should be noted that the 2D convolution process is the basis of convolutional neural networks.

Understanding the 2D sinusoidal signals first has a critical role before the Fourier analysis of images. Although it is not different from 1D ones as a function, it has a directional structure since it contains arguments over two dimensions. In the spatial domain, sinusoidals have two different angular frequencies in both the x- and y-directions. As an example, let's generate a signal whose periods in the x- and y-directions are $N_{x0} = 16$ and$N_{y0} = 32$, respectively. For this, it is needed to determine the angular frequencies of the signal.$w_{x0} = 2\pi \times \frac{1}{16} = \frac{\pi}{8}$, while $w_{y0} = 2\pi \times \frac{1}{32} = \frac{\pi}{16}$. From here, the 2D signal can be written as $I_1 = \sin\left(w_{x0} \times n_x + w_{y0} \times n_y\right)$. In addition, let a signal be defined as $I_2 = \sin\left(\frac{2\pi}{32} \times n_x + \frac{2\pi}{128} \times n_y\right)$ over $N_{x0} = 32$ and $N_{y0} = 128$. The visualization of these sinusoidals is given in Fig. 10. The image size consists of $N_x = 128$ and $N_y = 128$ pixels.

When Fig. 10a is examined, it is seen that there is a total of 8 cycles in the x-direction. This is because the ratio of the N_x pixel length of the image to the period

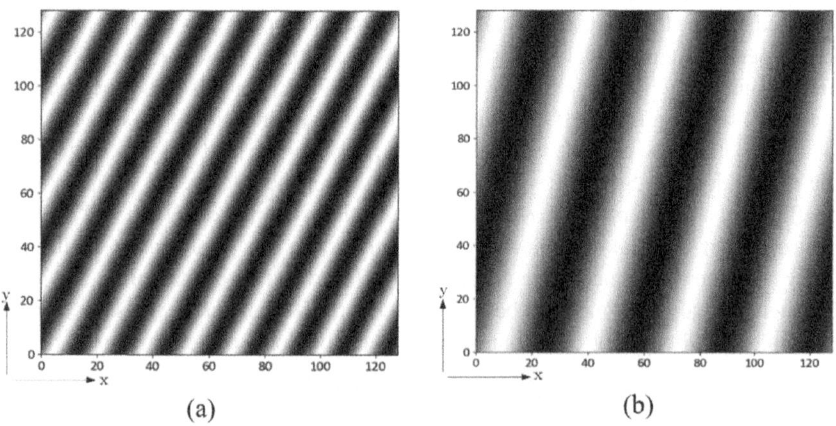

Fig. 10 Depicting 2D sinusoids by using constant periods (a) $N_{x0} = 16$, $N_{y0} = 32$ and b) $N_{x0} = 32$, $N_{y0} = 128$. Directional behavior of sinusoidal signals under different angular frequencies. The range of values in white and black is $[-1, 1]$

of the sinusoidal in the x-direction $\left(\frac{N_x}{N_{x0}} = \frac{128}{16}\right)$. When the y-direction is investigated in Fig. 10b, it is seen that there is only 1 cycle. Because it is defined as $N_{y0} = 128$ for the I_2. Since directional sinusoidals contain two different independent frequencies, the Fourier Transforms must also be over these two frequency arguments (ω_x and ω_y). Figure 10 is actually a three-dimensional shape. The amplitude information of the sinusoidal signal is kept in the z-axis direction. In pictorial representation, white and black colors correspond to the range $[-1, 1]$.

It is obtained from the equation in DTFT (4) for images. This transformation is a transformation from the x or y spatial coordinate space to the frequency space ω_x and ω_y.

$$F_I\left(e^{j\omega_x}, e^{j\omega_y}\right) = \sum_{y=-\infty}^{\infty} \sum_{x=-\infty}^{\infty} I\left[n_x, n_y\right] e^{-j(\omega_x \times n_x + \omega_y \times n_y)} \quad (4)$$

What will be explained hereafter is about what to consider in the implementation of the FFT algorithm and how to interpret the magnitude spectrum instead of using the equations analytically.

For the analysis of a oriented sinusoidal, the $I_3 = \cos\left(\frac{2\pi}{N_{x0}} \times n_x + \frac{2\pi}{N_{y0}} \times n_y\right)$ signal taken as N_{x0} and N_{y0} periods 3 and 32 are generated. Thus, $f_{x0} = \frac{1}{3}$ and $f_{y0} = \frac{1}{32}$ here, it is expected that the frequency in the x-direction will be higher. It should be noted that the *highest relative frequency for discrete-time signals can be 0.5 (angular frequency π.).* At frequencies above 0.5, aliasing occurs. In Fig. 11, the signals and its' FT are given. Figure 11a It is seen that the I_3 oscillates more frequently in the x-direction. Figure 11b shows FT for the f_x and f_y directions. 2D sinusoidal signals have Dirac as well as 1D ones [5, 6]. Therefore, a shape is only seen as a single point because of aerial viewpoint. The reason why there are two points is that the

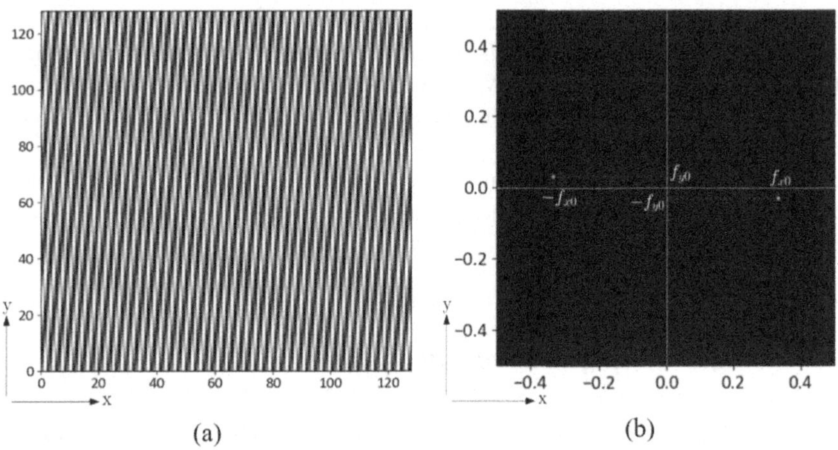

Fig. 11 Spatial representation of I_3 signal (**a**) and its magnitude spectrum (**b**)

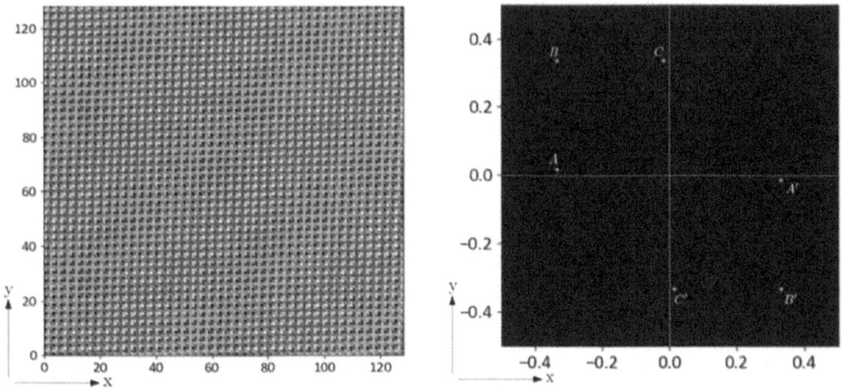

Fig. 12 Spatial drawing (**a**) and its amplitude spectrum (**b**) of the I_4 signal generated with three different oriented frequencies. The I_4 is generated over the periods $N_{x0} = [3, 3, 64]$ and the corresponding $N_{y0} = [64, 3, 3]$ periods. A and A' in (**b**), [3, 64]; B and B', [3, 3]; C and C' are the frequency points found for the [64, 3] periods

signals are considered in the spatial domain consist of real numbers. Therefore, FT is symmetric with respect to zero. The amplitude spectrum is evaluated after taking the FT of the signal to obtain Fig. 11b. Thus, there is no phase information in this image. It contains only the magnitude amounts of the components at the relevant frequency. Since there is only one component inside I_3, its information is seen in the spectrum.

Also, signals containing more than one component can also be analyzed with FT. Figure 12 is a good example of this. The sinusoidals generated from the frequency points $N_{x0} = [3, 3, 64]$ and $N_{y0} = [64, 3, 3]$ are added to form Fig. 12a. When looking at this figure, while it is difficult to obtain any information about the components, they can be examined easily thanks to FT (Fig. 12b).

In the magnitude spectrum of I_4, A and A' points correspond to the component formed with the period $N_{x0} = 3$ $N_{y0} = 64$, while C and C' point to the sinusoidal formed with $N_{x0} = 64$ $N_{y0} = 3$. The same interpretation applies to B and B'. In this example, a complex pattern could be revealed using only three different frequency components. When these signals are added with the necessary amplitude and phase components, square wave and other periodic signals can be obtained in a directed manner.

2.3 Image Filtering in Frequency Domain

It was emphasized that different effects occur with different kernels by using the neighborhoods of pixels in spatial domain filtering [7]. We saw that some of these

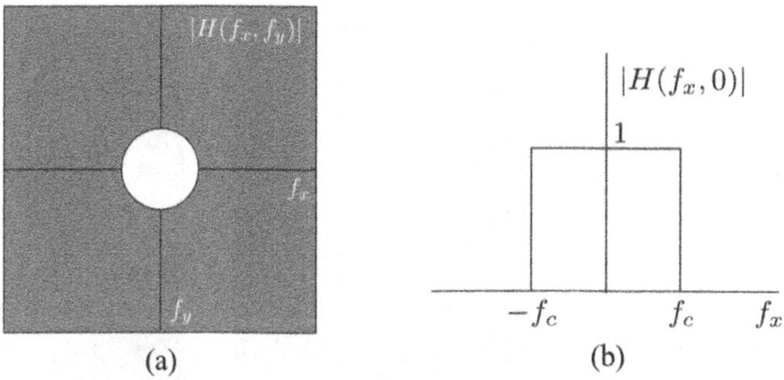

Fig. 13 Frequency response of ideal 2D low-pass filter with f_c cutoff frequency (**a**). Its cross section over f_x is given in (**b**). White and gray colors represent 1 and 0 values

make the picture smoother, while others detect the edges. We explained their kernel work over the spatial domain. Now, we will look at the interpretation of the effect of kernels with convolution in the frequency domain. In this direction, three different filtering types will be examined and examples will be given on some special kernels.

The first filter is the low-pass filter. The frequency characteristic of an ideal low-pass filter (LPF) is given in Fig. 13. While LPF passes frequencies around 0 up to a certain cutoff frequency, it does not pass those above it. The filters obtained with kernels are not ideal low-pass filters. It reveals effects that are close to it. One of them is the moving average (MA) filter. The operation of the MA filter depends on the size of the mask. If a wide mask is used (11×11 etc.), the cutoff frequency is low, so the blur effect will be high. On the other hand, since the 3×3 mask is narrow in spatial, it has a wider bandwidth in frequency.

Another filter is the Gaussian filter. In the spatial domain it is created according to Eq. (5). The σ parameter sets the filter width. If σ is large, kernel size is large and vice versa. Thanks to σ, the filter effect can be adjusted. An example of this is given in Figs. 14 and 15.

$$h_{gs}(x, y) = \frac{1}{\sqrt{2\pi\sigma^2}} \mathrm{e}^{-\frac{x^2+y^2}{2\sigma^2}} \tag{5}$$

In Fig. 14, a Gauss LPF is generated by using $\sigma = 5$. The filter application to image can be done with two ways. (1) the image is convolved by using filter kernel. (2) Firstly, FTs of image and kernel are calculated, then they are multiplied. After this process, inverse FT is taken. In this application, 2nd way is chosen. It is important to determine the filter sizes according to σ value when creating a Gaussian LPF in the spatial plane. In practice, Gauss is assumed that it becomes 0 after 3σ. Therefore, it is sufficient to use a kernel size of $(2 \times 3\sigma) \times (2 \times 3\sigma)$. Getting much size places a burden on unnecessary computational complexity. Multiplying by 2 is due to the fact that the Gaussian signal is symmetrical. When the FT of the input image given in Fig. 14a is multiplied by the frequency response of the filter, it is seen that low frequencies remain around 0. Therefore, the image becomes smoother (Fig. 14c). In

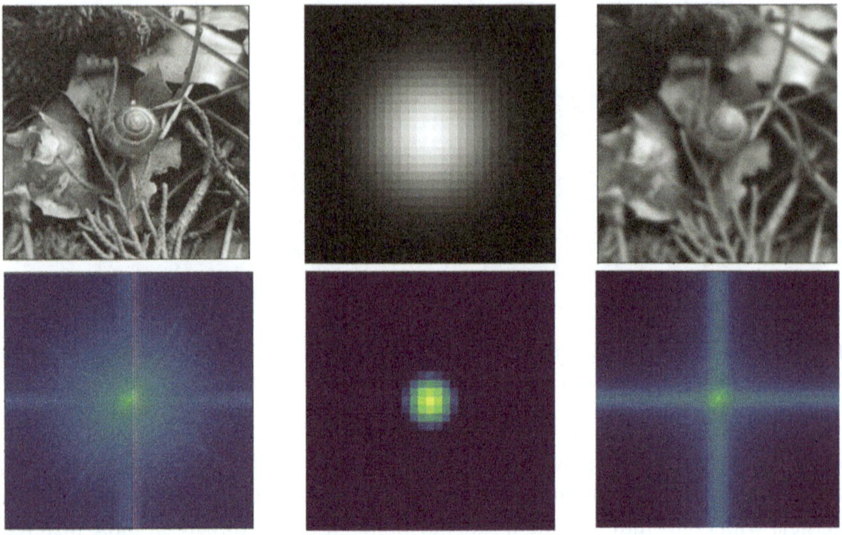

(a) (b) (c)

Fig. 14 Image filtering with Gauss LPF ($\sigma = 5$). When input image (a) is filtered with (b), blur effect is revealed on image (c). Because, Gauss LPF attenuates high-frequency region. Their spectrums are given below images

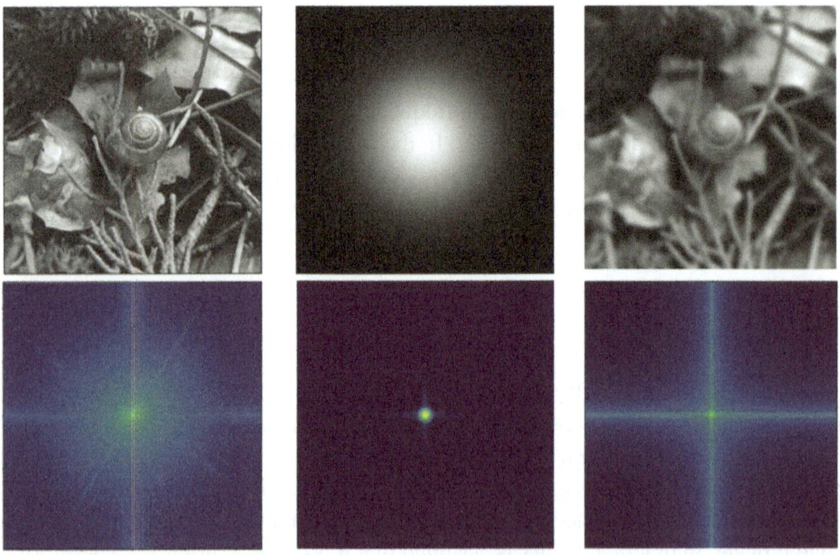

(a) (b) (c)

Fig. 15 Filtering the input image (a) with a Gaussian LPF (b) with a high σ value. Since the frequency response of the filter is narrow, the components of the image are also extinguished (c). Therefore, information about the image cannot be obtained

this example, it is observed that the dominant frequency components of the image are preserved. For this reason, rough information about the input image can still be seen.

Note: Fourier transform of 2D Gauss function is again a Gauss. Hence, it can be said that Gauss is an eigenfunction of FT. If σ is low in the spatial domain, filter bandwidth will be large.

Another example is given in Fig. 15. Herein, $\sigma = 20$ is selected in the spatial domain. Thus, filter size becomes larger (Fig. 15b). Contrast to this, the frequency response of the filter has a narrow bandwidth. It means that this filter passes only very low frequencies. Its effects are seen Fig. 15c. Since the fundamental frequency components belonging to the input image are also lost, nothing about the image is seen (Fig. 15c).

One of the most useful applications of low-pass filters is the noise removal of images. An example of this is shown at Fig. 16. The input image affected by $\mathcal{N}(0, 0.5)$ noise, and is filtered with LPF of $\sigma = 3$. As can be seen in Fig. 16c, the input image, although it is not perfect, can be reconstructed remarkably from the noisy image. If the work done by the filter is interpreted, it is based on removing high-frequency harmonics of noise. This is suitable to filter due to the dominant components of the image being located in the low-frequency region. Of course, noise is also effective at frequencies around 0. These harmonics cannot be filtered because the image also

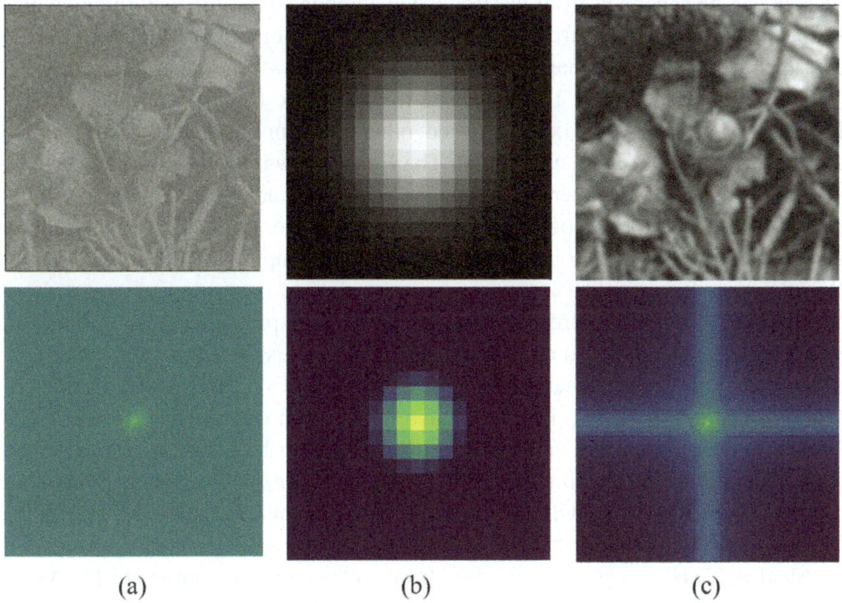

Fig. 16 Noise removal by using LPF. The noise appearing in the input image (a) is dispersed in the frequency domain. If the filter shown in (b) is applied to (a), the noise-reduced output is obtained as in (c)

(a) (b) (c)

Fig. 17 A text is hidden in (**a**) by using sinusoidals with very high amplitude. Magnitude spectrum of the cluttered image is presented in (**b**). However, only sinusoidal harmonics are seen. To obtain image info, $\log_{10}(\cdot)$ function should be used (**c**)

contains information. Therefore, the disturbance effects are still visible in the output image.

Another LPF application is shown in Fig. 17. Figure 17a is created by using a hidden text image and some sinusoidals with $A = 500$ and $N_{x0}, N_{x0} = [4, 8, 18]$ periods. Because of the drastically high amplitude of sinusoidal, text information cannot be seen in Fig. 17b. After taking $\log_{10}(\cdot)$ of spectrum, text components over low frequency appear (Fig. 17c). The frequency region where the text information is located can be separated from the frequency components of the disturbing signals. Using the LPF, only the frequency region can be filtered. Thus, the text can be extracted from cluttered image.

The filter given in Fig. 18b was used to extract the text from Fig. 18a image. LPF has just filtered where txt was placed, other high amplitude level components—except for only one—were eliminated. The sinusoidal with $N_{x0}, N_{y0} = 16$ period is very close to the text frequency region. Therefore, it is able to enter the asymptotic part of the filter. Although the filter suppresses the power of the disturbing signal, it is somewhat visible in the output image because the amplitude of the signal is high (Fig. 18c).

Besides LPFs, another frequently used filter is high-pass filter (HPF), which are useful for finding edges and revealing details. Why HPFs work this way can only be understood by looking at their frequency response. HPFs are one of the most useful filters in image processing. Discriminative information of an image lies on edges. Extracting edge gives critical point of objects that can be recognized or detected. When the kernels of convolutional neural networks are examined, many of them are operated as a HPF. As known, feature maps of these kernel consist of edges of image. In this chapter, learn how a kernel can detect edges using frequency analysis.

An ideal 2D HPF is given in Fig. 19a. It works the opposite version of LPF. Again, there is a cutoff frequency, but HPF passes high-frequency region over the spectrum.

Although the ideal HPF is as shown in Fig. 19, HPF can be produced with many kernels. An example of this is the Gaussian function. While Gaussian 2D function has LPF characteristic, HPF in frequency domain can be obtained with $1 - H_{\text{LPF}}(f_x, f_y)$

Fig. 18 Extracting hidden text from clutter image (a) can be used by a proper LPF. Herein, LPF's cutoff frequency (b) is adjusted to pass only text harmonics (c) that are placed around (0,0)

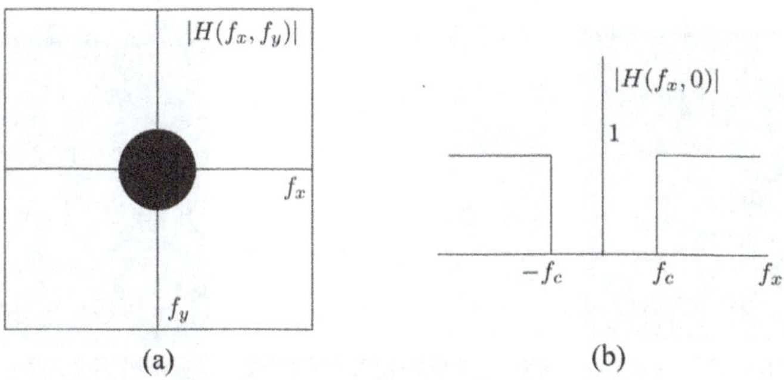

Fig. 19 Frequency response of an ideal HPF (a), and its section in the f_x direction (b). HPF passes values above a certain cutoff frequency. Components in the low-frequency region are eliminated from the image

process. In addition, HPF can be operated by using Prewitt Eq. (6) and Sobel Eq. (7) kernels [8].

$$h_x^P = \begin{bmatrix} -1 & 0 & 1 \\ -1 & 0 & 1 \\ -1 & 0 & 1 \end{bmatrix} \tag{6}$$

$$h_x^S = \begin{bmatrix} -1\,0\,1 \\ -2\,0\,2 \\ -1\,0\,1 \end{bmatrix} \tag{7}$$

These kernels are also called directional filters because they can detect edges in the direction they are selected. Equations (6) and (7) can access only x-direction edges. To obtain y-direction edges, their transposed versions are used. The way they work can be understood by looking at the magnitude spectrum. Let's handle Sobel x-direction filter in Eq. (7). Figure 20 demonstrates the working mechanism of Sobel filter. When an image (Fig. 20a) is filtered by Fig. 20b kernel, output includes x-direction edges (Fig. 20c).

This is because Fig. 20b filter only passes high frequencies in the x-direction. When the frequency characteristic of Fig. 20b is examined, this situation is observed better. Since all the components in the y-direction are damped, the sinusoidals in this direction cannot combine to form edges in the full y-axis. The opposite of this is true. If the kernel's transpose is taken, and applied to the image, this time only the components in the y-direction are passed.

The passing of the edges in all directions is with the Laplacian filter [9]. The filter whose kernel is given in Eq. (8) is the discretized form of the 2nd order derivative operator. By using this filter, the details of the image can be obtained with a single operation.

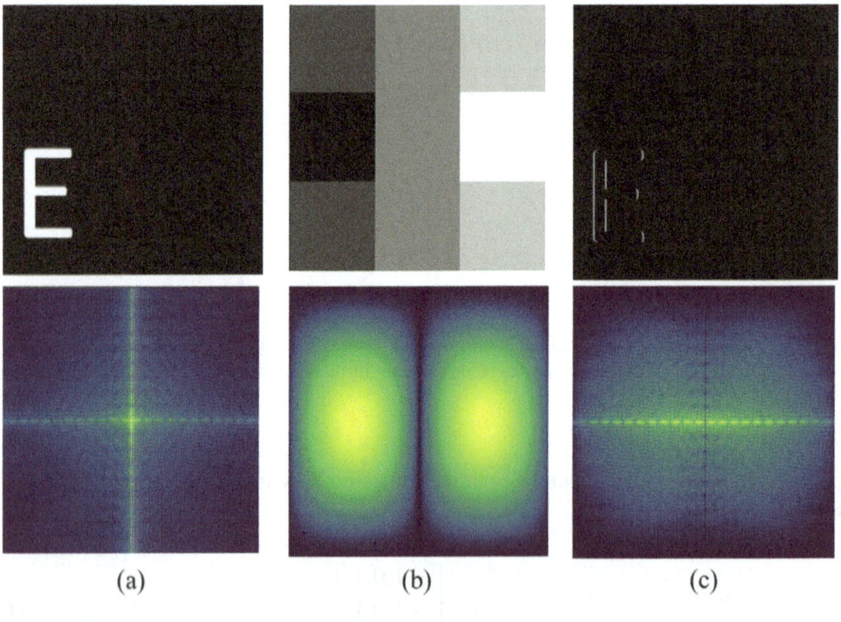

(a) (b) (c)

Fig. 20 Application of Sobel filter over letter "E" (a). By using frequency analysis of filter (b), it can be understood why this filter generates (c)

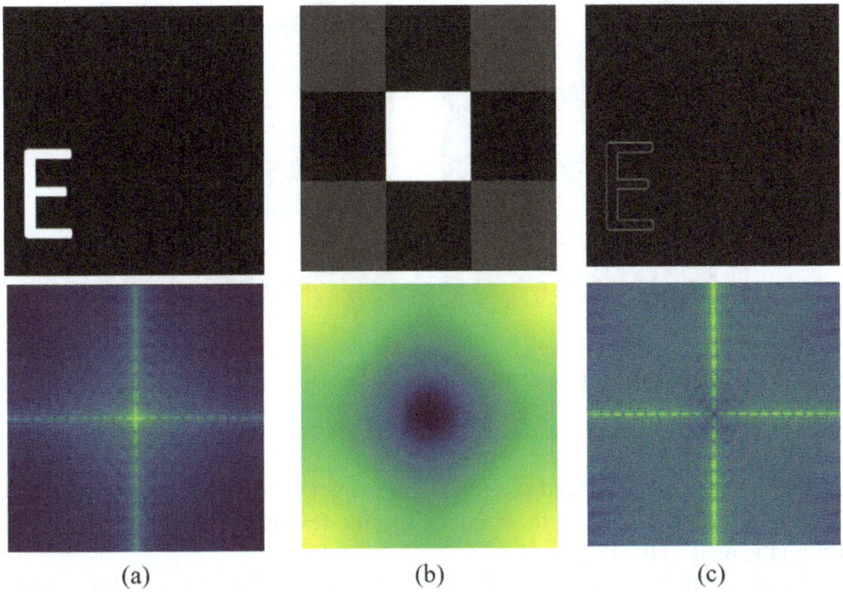

Fig. 21 Finding the edges of the letter "E" (a) using the Laplacian filter. As can be seen in the Laplacian filter (b), it passes high-frequency components in all directions. This causes the edges of objects to be found (c)

$$h^L = \begin{bmatrix} 0 & -1 & 0 \\ -1 & 4 & -1 \\ 0 & -1 & 0 \end{bmatrix} \tag{8}$$

The working of this filter is shown in Fig. 21. When the frequency response of the filter is examined, it is seen that the filter only suppresses around 0. High-frequency regions, on the other hand, are made more dominant. For this reason, values close to each other in the image were filtered, and the borders where instantaneously changes were observed were visible. It is also observed from the frequency response that the filter performs this operation in every direction without choosing a direction.

Figure 22 is a notable example of using high HPFs. Using Eq. (8), the details of the image can be found. If these details are added to the original image, the image is made sharper. In this case, the kernel to be applied to the input is given in Eq. (9). With this kernel, the sharpness of blurry images can be increased.

$$h^{SH} = \begin{bmatrix} 0 & 0 & 0 \\ 0 & 1 & 0 \\ 0 & 0 & 0 \end{bmatrix} + \begin{bmatrix} 0 & -1 & 0 \\ -1 & 4 & -1 \\ 0 & -1 & 0 \end{bmatrix} = \begin{bmatrix} 0 & -1 & 0 \\ -1 & 5 & -1 \\ 0 & -1 & 0 \end{bmatrix} \tag{9}$$

This situation can be seen when the details in Fig. 22 are examined. Leaf details that are more blurred in Fig. 22a are more prominent in Fig. 22b. In addition to LPFs

Fig. 22 The image becomes sharper when the details found with the Laplacian filter are added to the original image (**a**). This can be seen in (**b**)

and HPFs, there are also bandpass filters that extract features from the image in the desired direction. Thanks to these filters, specific feature engineering can be done within the image [10, 11].

The ideal bandpass filter (BPF) is given in Fig. 23. Unlike LPF and HPF, it needs two cutoff frequencies as f_L and f_H. In fact, a BPF can also be thought of as a cascade of LPF and HPF. The frequency region in between f_L and f_H is filtered and only the relevant harmonics are given to output. Figure 23b is obtained when the cross section of the BPF given in Fig. 23a is taken at any angle.

Another type of BPF is Gabor Filters (GF). GFs have a special place among other filters. Using these filters, only the region around a center frequency (f_{x0}, f_{y0}) is filtered. The importance of this is that feature values can be extracted from the image at the desired frequency points.

In this way, discriminative features for image recognition are obtained. A GF consists of two functions. These are the 2D Gaussian function, which allows local filtering, and the modulator signal cosine, which determines the center frequency of the filter. These are combined in Eq. (10).

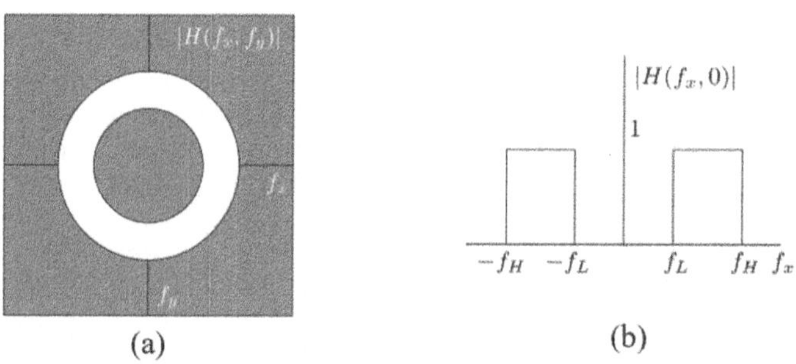

Fig. 23 The bandpass filter (a) needs two cutoff frequencies. It provides selective permeability by filtering the places between this region (b)

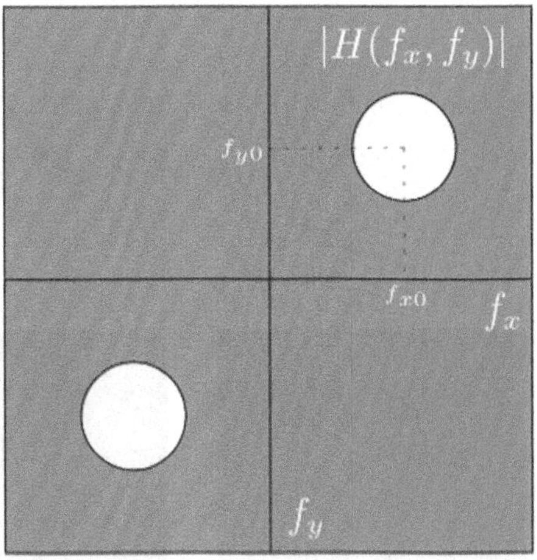

Fig. 24 Gabor Filters are localized on a certain center frequency (f_{x0}, f_{y0}) where they filter an area in the σ bandwidth. The reason why two circles appear in the figure is that the filters created in the spatial domain consist of real numbers

$$h(x, y) = e^{-\frac{x^2+y^2}{2\sigma^2}} \times \cos\left(\omega_{x0} \times x + \omega_{y0} \times y\right) \tag{10}$$

The Gaussian function creates a low-pass filter in the frequency domain with a bandwidth dependent on σ. Beside, cosine shifts this filter with center frequency (0, 0) to the point (f_{x0}, f_{y0}). Let's examine Fig. 24 to see the behavior of the GF.

Figure 25 image was created by concatenating 9 different sinusoids using $N_{x0}, N_{y0} = [4, 16, 64]$ frequencies over $N_x, N_y = 128$. This example is also important in this respect. If these signals were given in the image not as concatenate but as a summation, similar magnitude spectrum would be obtained. It should be noted that FT gives an amount of sinusoidal in an image. It does not concern their position. This topic is dealt with by wavelet transform. The central frequency of harmonics in Fig. 25 is given in (10). GF allows to select which pattern will be filtered.

(a) (b)

Fig. 25 The image formed by concatenating sinusoidals obtained over nine different frequencies (a), and its magnitude spectrum (b)

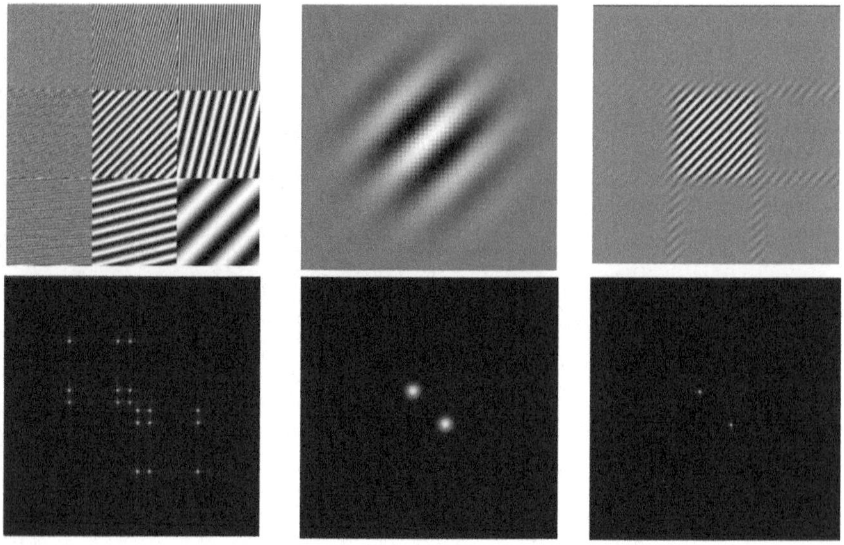

(a) (b) (c)

Fig. 26 Filtering the input image (a) using $\sigma = 10$. Since the bandwidth of the filter is narrow (b), it can only filter the component at the center frequency on which it is located. Therefore, output image contains only relevant frequency pattern (c)

$$\begin{bmatrix} \left(\frac{1}{4}, \frac{1}{4}\right) & \left(\frac{1}{16}, \frac{1}{4}\right) & \left(\frac{1}{64}, \frac{1}{4}\right) \\ \left(\frac{1}{4}, \frac{1}{16}\right) & \left(\frac{1}{16}, \frac{1}{16}\right) & \left(\frac{1}{64}, \frac{1}{16}\right) \\ \left(\frac{1}{4}, \frac{1}{64}\right) & \left(\frac{1}{16}, \frac{1}{64}\right) & \left(\frac{1}{64}, \frac{1}{64}\right) \end{bmatrix} \tag{11}$$

Let's generate a GF by using $\sigma = 10$ and $f_{x0}, f_{y0} = \left(\frac{1}{16}, \frac{1}{16}\right)$ parameters. Unit impulse response and frequency response of this filter are shown in Fig. 26b. Due to $\sigma = 10$, bandwidth of filter is narrow. Therefore, filter passes only the harmonic with $\left(\frac{1}{16}, \frac{1}{16}\right)$ central frequency (Fig. 26c).

In Fig. 27, GF central frequency is same, but now $\sigma = 5$ is preferred. This causes wide bandwidth. Thus, neighbors of $\left(\frac{1}{16}, \frac{1}{16}\right)$ can pass over the filter. Figure 27c also includes neighbor outputs. At this point, it should be noted that the component in the $\left(\frac{1}{16}, \frac{1}{16}\right)$ center frequency is paler than the other neighbors. It is due to the fact that $\left(\frac{1}{16}, \frac{1}{16}\right)$ stays on the diagonal of GF, which has a circular projection. Therefore, $\left(\frac{1}{64}, \frac{1}{64}\right)$ is farther from *t* center $\left(\frac{1}{16}, \frac{1}{16}\right)$. than the others, and is multiplied by a lower coefficient. That phenomena can be seen in Fig. 27c.

What happens by lowering the sigma parameter of the GF is shown in Fig. 28. With the decrease of σ, the bandwidth of the local filters increased, and they overlapped each other. Therefore, the central region is filtered as well as other neighboring frequencies. These components are seen in the output image at Fig. 28c. Another thing from here is that elliptical filters can be created with Gabor atoms. The combination of two overlapped GF yields an approximate elliptical filter. With this understanding,

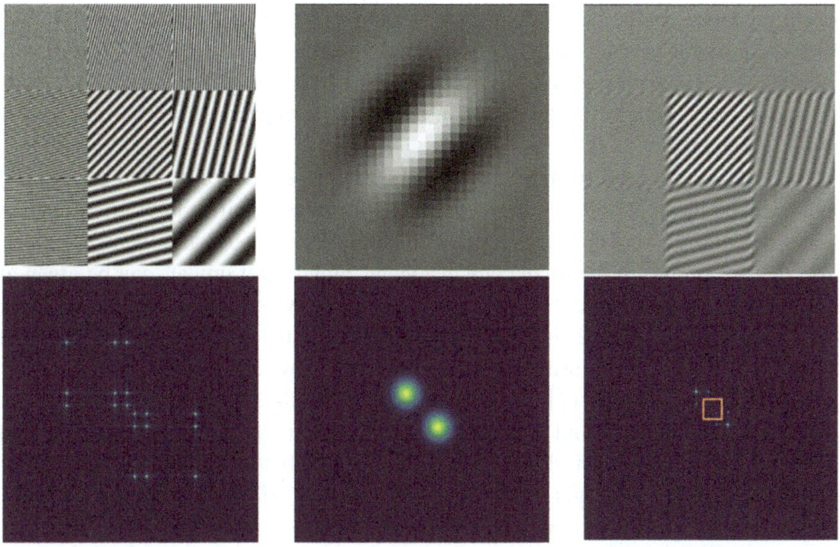

(a) (b) (c)

Fig. 27 Filtering the input image (a) using $\sigma = 5$. If $\sigma = 5$ is selected, the filter bandwidth (b) increases. Thus, neighboring components are now also filtered out. Therefore, they begin to appear on the output image. Since $\left(\frac{1}{16}, \frac{1}{16}\right)$ neighbor is farther from the center than the others, it is more damped by the filter. This is seen in (**c**)

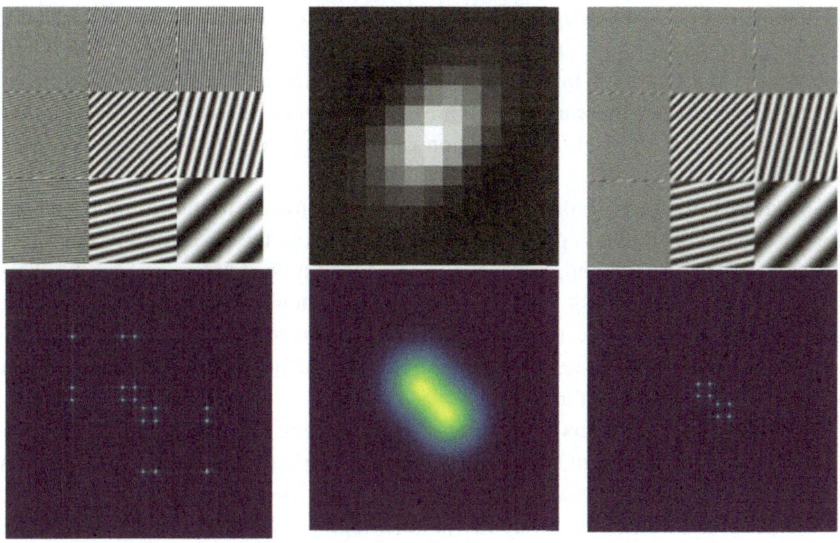

(a) (b) (c)

Fig. 28 Filtering the input image (a) using $\sigma = 2$. When $\sigma = 2$ is selected, the GFs overlap each other (b). Therefore, components along the relevant axis are filtered. The images in (**c**) illustrate this. Thus, elliptical filtering is done around (0, 0)

new filter types can be derived through frequency analysis. It is very difficult, and often not possible to make such interpretations through kernels in the spatial domain. Owing to FT, a new perspective has been obtained in the analysis of images. By using the approaches brought by this domain, the working mechanism of kernels can be easily understood.

GFs are widely used in the literature as a feature extraction tool in the field of pattern recognition [12–15]. It paves the way for many cognitive operations such as object detection and classification by acquiring the oriented components. Besides, GFs are also preferred for image enhancement. Fingerprint images can be given as an example [16]. When these images are handled as raw, they contain many artifacts. GFs are used to eliminate them and obtain clearer images. GFs also have a close bond with physics. They are also known as the narrowest projection filters in GF time–frequency analysis [17]. As has been learned so far, if a filter expands in time, it contracts in frequency and vice versa. For this reason, the product of the width of the filter in time and its width in frequency cannot go below a certain value. This is also evident in Heisenberg's uncertainty principle [18, 19]. In addition to all these, GFs are used in the development of communication technologies.

2.4 Advanced Filtering Techniques

Filtering noise in images is a vital task that needs to be done as a preprocess, especially in applications such as recognition and object detection. Pattern information cannot be preserved due to noise and models cannot produce accurate results, even if they work well. At this point, it may be necessary to use some advanced techniques beyond linear filtering. Since these algorithms are discussed in the field of Image Restoration (enhancement), more useful information can be obtained when research is carried out with the relevant title.

Median Filtering: Median filtering is a very convenient algorithm, especially for hardware-constrained applications. It filters without the need for any arithmetic operation. On the one hand, it cleans the noise, and it does not more blur the sharp lines in the image. A mask of size $N \times N$ shifts through the image, the same as in convolution. Then, by entering the mask, the numbers are sorted in descending order and the middle value is taken. This type of nonlinear filtering provides an effective solution for fast and non-intense noise.

Non-Local (NL) Means Denoising [20]: The aim of the algorithm is to preserve the textural information in the image while filtering out the noise. It does this by finding non-local samples of the pixel region to be filtered with similar statistics. Noise behaves statistically differently in other regions of the image. The NL Means algorithm uses this feature to filter by trying to preserve the texture.

Total Variation Denoising [21]: It does not use linear filters as in the NL Means algorithm. Instead, it defines reconstruction as an optimization problem. The purpose of the TV is to produce an output that minimizes the difference between neighboring samples along the signal. While doing this, it also tries to preserve the patterned structures by using regularization terms.

Wavelet Transform [5]: The filtering of images in the Fourier domain is insufficient in the face of complex problems obtained in real life. One of the reasons for this is the global transformation side of FT. Noise in the image may be non-stationary. For this, algorithms that handle the image locally and apply the necessary operation are needed. The wavelet transform suggests that the image can be expressed in terms of locally defined orthogonal wavelets, not sinusoidals between $(-\infty, \infty)$. Thanks to these well-localized bases, regional operations can be performed more successfully within the image.

3 Image Segmentation

3.1 Binary Image Processing

Morphology is the study of the shape of objects. This concept, which deals with morphological properties, is often used in image processing to examine shapes that appear in binary images. It also applies to gray-level images. Significant features of objects obtained in binary images are extracted with mathematical operators in set theory, on which morphology is based. Owing to these features, pre- or post-processing of the objects can be done, or geometric information of the objects can be obtained. In this section, basic morphological operators such as dilation, erosion, opening and closing will be addressed, and useful examples from practical life will be given for each example [5, 6].

Binary images are obtained by thresholding monochrome and color images. This thresholding can actually be thought of as the point operator seen in Fig. 29. Figure 29a, b shows the transformation function that creates two binary images.

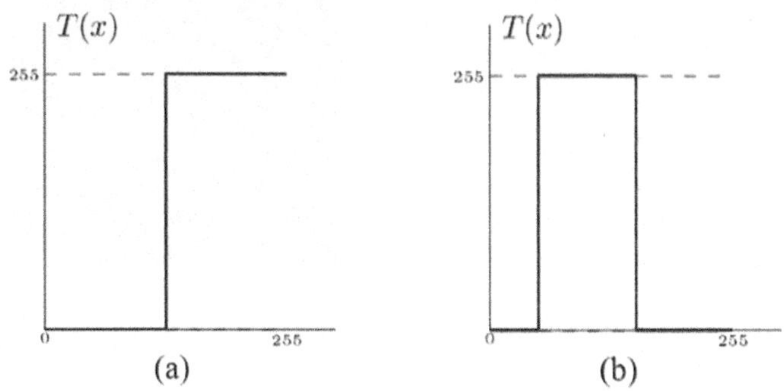

Fig. 29 As seen in (a), thresholding functions can be used for binarization over a certain value, while some functions pass only a pixel range through the threshold (b). These functions can be like those drawn with black and red dashed lines

Figure 29a is the most frequently used function, and if the intensities in the image are above or below a certain value, it writes the value of 255, not 0 to the coordinate where that pixel is located. It can also be used to create binary images in Fig. 29b. This function maps a certain pixel range to a value of 255, sets other intensity values to 0, and vice versa. In this case, actually the transition to binary image can be meant as an intensity-wise segmentation. After the binary images are acquired, the pixel values are assigned as 0 and 1. Thus, while minimizing the memory requirement of the binary image, it is also possible to use useful operators of mathematical set theory.

Figure. 30 can also be given as an example of the binarization process. The morphological features of cells in a medical image contain important information to diagnose the disease. Vital information can be deduced from geometric distribution of nuclei. Figure 30b is obtained by passing the relevant gray image through a threshold value of 125. It should be noted here that since our object of interest is cells, values below 125 are taken as 1 (foreground) and the others as 0 (background). After converting to 1 and 0 values, morphological image processing rules come into play. In the images, operations are performed on concepts such as connection points, neighborhood and structural elements.

In binary images, the concept of neighborhood is used with an approach that represents the connection points of the pixels. While pixels can be defined with 4-connected conditionals such as right, left, up and down, they can also be expressed as 8-connected neighborhoods in which diagonals are also considered. The structural element is the kernel definitions, which will be processed on the images and extract information in the desired form. Figure 31 is given as an example of these structures. Here the marker point represents the center point of SE which determines where the value obtained as a result of the operation performed on the input image of SE will be assigned in the output image.

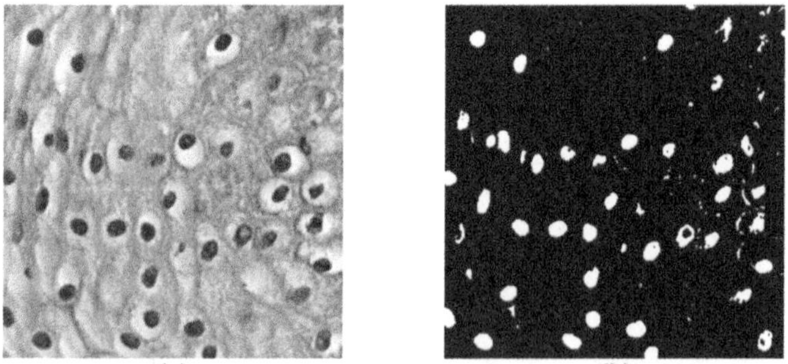

(a) (b)

Fig. 30 A histopathology image (a), and its binary output (b) after thresholding. *Image source* [22]

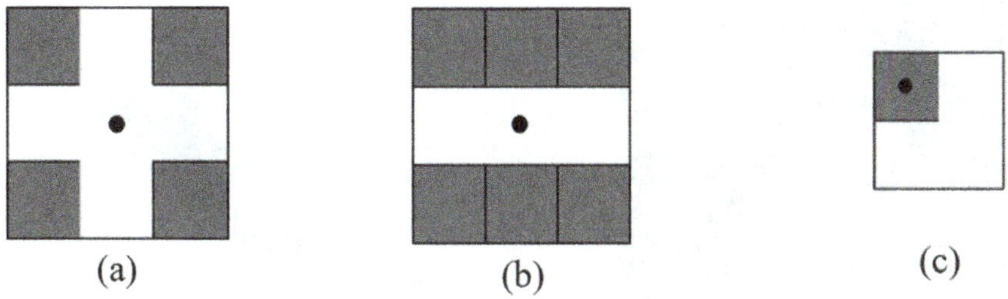

Fig. 31 The structural elements given in (a)(b) and (c) perform different functions in the binary image according to their shapes

3.2 Image Segmentation via Unsupervised Learning

Intensity can be expressed with a value in monochrome (gray level) images. It was enough to select a value as a threshold when creating the binary image. In this way, pixels are clustered according to 2 different classes (foreground, background). In color images, the pixel is defined by a vector such as $x = [x_R, x_G, x_B]^T$. An example of this is given in Fig. 32. In a 3D space, a binary image is created using a plane, not a single value. This plane is defined by the coefficients w.

Now let's examine this fact through a real image. All pixels in the image given in Fig. 33a are shown in Fig. 33b. Since there are no vivid colors in the image, the pixels are concentrated on a diagonal path rather than the corner points of the 3D space. To create a binary image over these pixels, the pixel cloud needs to be separated into two different clusters. This can be done with the k-means [23, 24] algorithm.

Fig. 32 While only 1 value is sufficient for thresholding in gray-level images, this calculation can be done by using different geometries in 3D space. A plane shown in this example can do thresholding

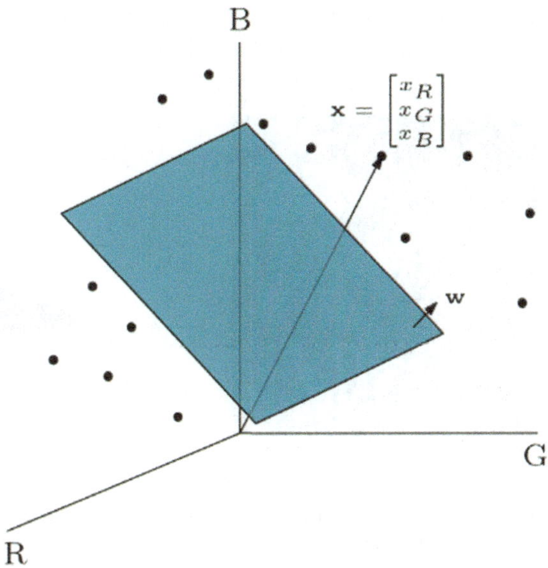

Fig. 33 3-channel color image (a) and its pixel distribution in 3D space (b)

The k-means algorithm tries to divide a cluster into k bunches. This should be noted that the method is unsupervised. k-means assigns points to k classes without knowing their classes. An example is given by choosing $k = 2$ over Fig. 34a. A binary image is formed because the pixels are divided into two different groups. The reason why the pixel values do not appear as white and black in the image in Fig. 34a is the assignment of the center pixel value of the pixel clusters.

The k-means clusters similar pixel groups. The center pixels of these clusters appear as the class center of the group. In this way, similar colors can be segmented. Another effective application of the k-means is color quantization (CQ). The purpose of CQ is to express similar pixels as a single-pixel center, instead of storing the 3D color vector belonging to each coordinate in the image, by encoding the $N \times M$ size image with class IDs.

An example of this is given in Fig. 35. First, Fig. 35a was created by choosing $k = 5$. Instead of their original values, the pixel values of the center to which the

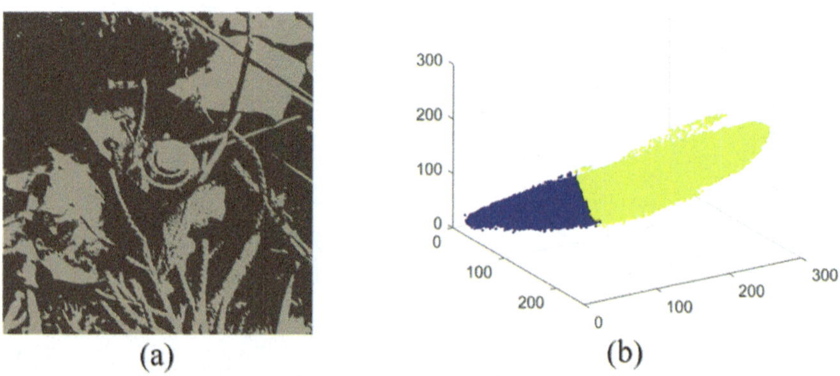

Fig. 34 The image obtained by selecting $k = 2$ (a). Pixel distribution is divided into 2 different clusters (b). Then the center value of clusters is assigned in place of grouped pixels

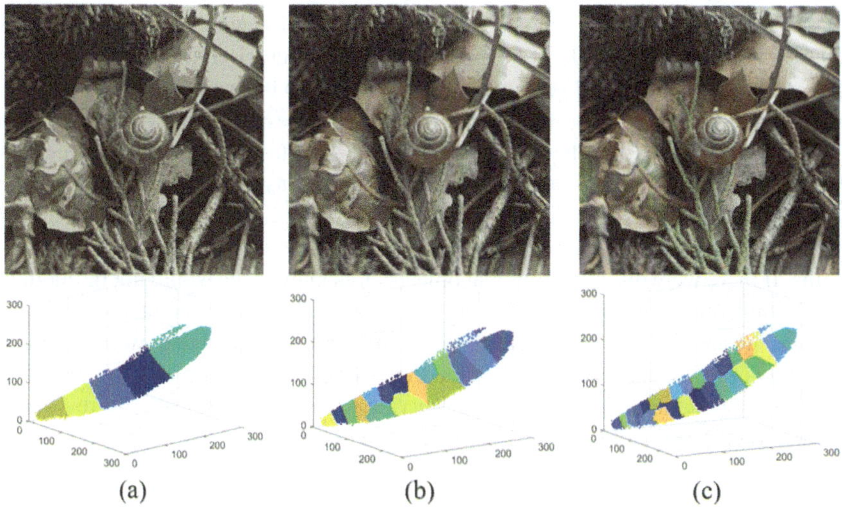

Fig. 35 Example of CQ made using k-means. The k value is chosen as 5, 16 and 32 in (a), (b) and (c), respectively

pixels belong are embedded in the image. This image is quite far from the original. When $k = 16$ is selected, the number of clusters increases (Fig. 35b). The image has become more distinct as the pixel values are closer to the original ones. At $k = 32$, the image is largely the same, except for slight quantization errors. There isn't much of a noticeable difference in Fig. 35c. However, data can be compressed up to 3 times. Especially in systems where stream video is processed, if the transmission speed is low, similar compressions are made by sacrificing the image quality.

K-means is just one of the clustering algorithms. There are more robust algorithms than k-means against cases where the pixel distribution in 3D space is nonlinear. Some of these are algorithms like Gaussian Mixture Model (GMM), Density-Based Spatial Clustering of Applications with Noise (DBSCAN) and Agglomerative Clustering. These methods are examined under the name of unsupervised learning.

4 Feature Extraction

Image segmentation enables objects or regions of interest to be handled individually. The next step that can be done in these areas is the extraction of features. The attribute can be geometrical features such as the area, circumference, kurtosis of the object, or it can be obtained through textural information [6]. The feature extraction step is the gateway for computers to cognitive operations over images. In this way, advanced applications such as classification and detection can be made under image understanding.

After the digital image is created on the computer, the numerical equivalents of the visible objects appear. Every feature that is easily perceived in the brain, such as colors, edges, corners, and geometric features, can also be expressed with numbers. For example, let's try to develop an algorithm that classifies coins. Their radii and therefore their areas will be different. Coin classification can be made using this numerical information. Of course, in order for the coins to be handled one by one through the image, they will need to go through segmentation processes and, if necessary, morphological operations. Only area or radii is enough for classification coins. If there were different elliptical objects, area knowledge would not be enough. In addition, minor and major axes lengths would have to be added. Another example is the classification of fruits on a sliding belt in a fruit crate factory. For now, let's assume that only oranges and apples will be classified. In such a case, color information will also be included in the work. All of these values are brought together to reveal the feature vector in Eq. (12).

$$\boldsymbol{x} = [x_1, x_2, x_3, \dots, x_N]^{\mathrm{T}} \tag{12}$$

Each element in the vector corresponds to attributes obtained from the entire image or locally from a region or an object. In the vector given in (11), a total of N features are stored. What these attributes might be will be discussed next.

4.1 Geometric Features

Boundary: The circumference is one of the first attributes that can be acquired over objects. Objects with different circumference characteristics in the picture of the interest can be easily classified based on this value. Figure 36a, b are given as

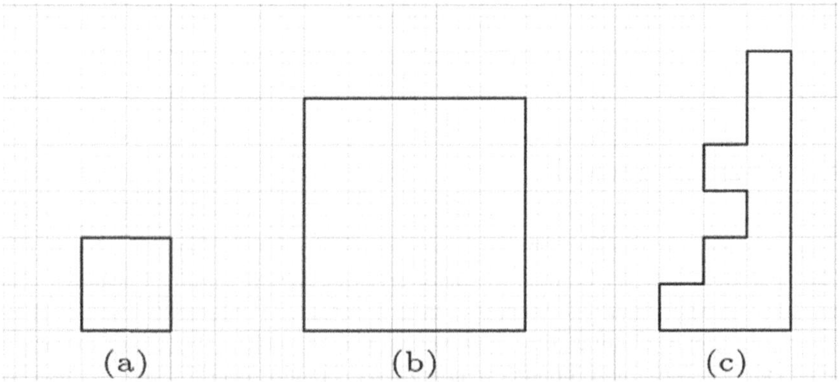

Fig. 36 Using the boundary as a feature. Different shapes can have different circumference sizes (like (a) and (b)). But this is not always the case like in (**b**) and (**c**)

an example. On the other hand, the perimeters of both Fig. 36b, c are 20 units. Circumference knowledge alone is not sufficient to solve complex problems. In addition, an attribute value or vector must be produced that includes local changes on the edges of the object. The chain code algorithm works well for this.

Chain code uses directional line segments to define a boundary. These segments, in connection to each other, express the total circumference, including local variations. Orientations are translated into numerical values according to the number scheme expressed in Fig. 37. The sequence of directional numbers presents the Freeman Chain Code [25]. This code defines the vector form of object boundary information.

Depending on the connection type, coding can be done over different number schemes. Figure 37a is used for 4-connectivity and Fig. 37b is used for 8-connectivity. Angle resolution, on the one hand, presents the border changes detail well, on the other hand, it increases the length of the feature vector.

Let's code Fig. 38a, b using the Fig. 37a number scheme. Of course, where the starting point will be is an important uncertainty. For now, let's select this as the lower left corner. Starting from here, the code of Fig. 38a will be,

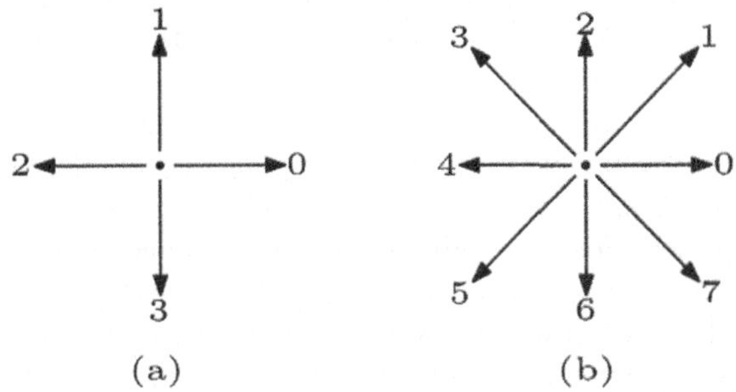

Fig. 37 Freeman chain code vectorizes shapes over the environment in the systematic framework shown in (**a**) and (**b**). In this way, instantaneous changes in the environment are represented in this vector as information

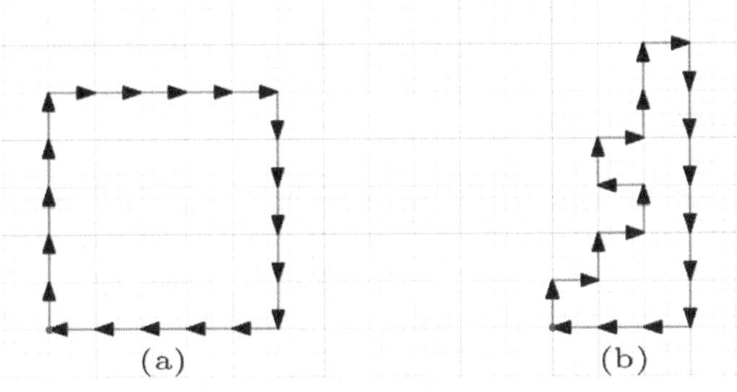

Fig. 38 Although (a) and (b) have the same perimeter, the chain code is different due to instant orientation

$$x_a = [1, 1, 1, 1, 1, 0, 0, 0, 0, 0, 3, 3, 3, 3, 3, 2, 2, 2, 2, 2] \tag{13}$$

Figure 38b, on the other hand, is expressed as,

$$x_b = [1, 0, 1, 0, 1, 2, 1, 0, 1, 1, 0, 3, 3, 3, 3, 3, 3, 2, 2, 2] \tag{14}$$

Although the two boundaries have the same environment, the chain codes are different due to local changes. A critical preprocessing is required for the robust operation of the chain code algorithm. Objects with a large surface will have a long circumference. Noisy regions can be found on the surrounding. They reduce the generalization capability of the chain code. After finding the noisy boundary, the spline algorithms may be used to fit smooth curve, or morphological operations may be preferred on the binary objects before acquiring of boundary. Another point to note is rotation normalization. In this way, unknowable problem of starting point can be overcome.

Region: Area is one of the other geometric properties that describe an object. Objects that are figuratively similar to each other in the image can be classified according to their area. But this information on its own is not always sufficient. In addition, it is necessary to obtain values such as minor and major axis lengths, solidity and eccentricity of the object.

Some metrics have been defined to reveal the differences of objects from known geometric shapes. The first of the area-based shape descriptors is the net area and the convex area. The net area is calculated as the number of pixels filled by the object. Convex area represents the area of the convex hull surrounding the object. Another metric is compactness. It is defined as the ratio of the area value calculated on the object to the circle with the same radius. If this value is 1 then the object is a complete circle. If there are indented and protruding places on the surface of the object, this value will be between (0, 1). Eccentricity is defined as the ratio of the minor axis to the major axis. This value is between (0, 1). If it is close to 1, it means the object has high circularity. Another descriptor acquired through the field is solidity. Here the real area is obtained by the ratio of the convex area. All these values can give an idea about the geometric properties of the object.

4.2 Textural Features

Textural features are features that find patterns in an image, not only an object in the image. In order for images with a common pattern to be detected in the same class, the images must be close to each other in the feature space. At this point, when making global image classification of feature descriptors, it is expected that the intra-class variance is low and the inter-class variance is high. In this section, a few algorithms that are frequently used in pattern analysis will be presented. First of them is Histogram of Gradient (HoG) algorithm.

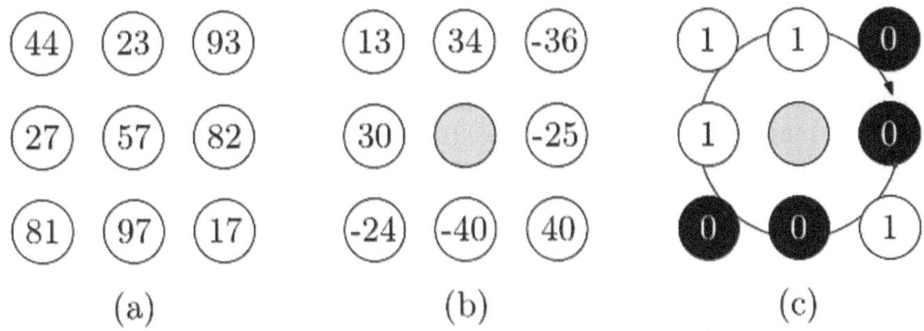

Fig. 39 Working schema of LBP algorithm. (a) gray-level intensity, (b) difference of center pixel with neighbors, (c) extraction of binary pattern

Histogram of Gradients: Although the origins of the HOG algorithm go back to the 1980s [26], its popularity is the work of and on pedestrian detection in images in 2005 [27]. After this study, the HoG algorithm has been applied to many various image processing problems [28, 29].

The essence of the algorithm is based on the idea that boundaries provide the discriminative of objects from each other. When determining the boundaries, different amounts of components appear in the x and y-directions. As a result, gradients of different angles and magnitudes occur. In this algorithm, the pattern information can be acquired by using histogram of orientations.

Linear Binary Pattern: LBP [30, 31] produces a binary sequence by comparing a pixel with its neighboring pixels. Its value determines the local pattern. The histogram of the pattern values is used as a visual descriptor. It is an efficient algorithm that can be extracted quite fast. Its discriminative side and low computational complexity have made it applicable to many problems. It is a suitable algorithm for systems that will work in real time. Another aspect of the algorithm is that it is robust against enlightenment problems that are frequently encountered in real life. It can successfully represent tissues illuminated under different conditions.

As a first step, the LBP algorithm compares the center pixel with its neighbors at a given radius. In Fig. 39, it is selected as 1. The difference pattern between them is evaluated. Then, a binary pattern is created from these values. As an example, $b = [0, 1, 0, 0, 1, 1, 1, 0]$ is obtained from Fig. 39c. The decimal value of this binary number represents the relation of the bunch of pixels. In the image processed in blocks, the histograms of these values form the feature vector.

5 Example: Predicting the Number of Glass Sheets Using Image Analysis

Image processing has become an increasingly common tool in business, driven by recent advances in AI. Thanks to computer vision (CV) techniques, mundane tasks can be performed efficiently by machines. In this way, it provides person-day savings for companies. Inventory management system [32], product control and damage detection [33, 34], autonomous driving [35], fire detection [36] and health-care industries [22, 37] are the areas where IP is frequently preferred and increasingly widespread. In this context, a case study about how IP algorithms find a solution to an inventory management problem is presented below.

One of the problems in the glass manufacturing industry is estimating accurately the number of glass sheets lined up at the delivery stage. The proposed algorithm for count prediction presented below depends on a frequency analysis-based approach.

Once the glass bundles are captured with a camera, a region of interest (RoI) is manually cropped and is converted to gray-level format (cf. Fig. 40).

Glass types can be different. Samples of two different types are given in Fig. 41. Glass No. 1 and 2 are normal glass, while 3 and 4 are examples of mirrored glass which

Fig. 40 Whole glass bundles (a) and gray-level cropped RoI image (b) captured by camera

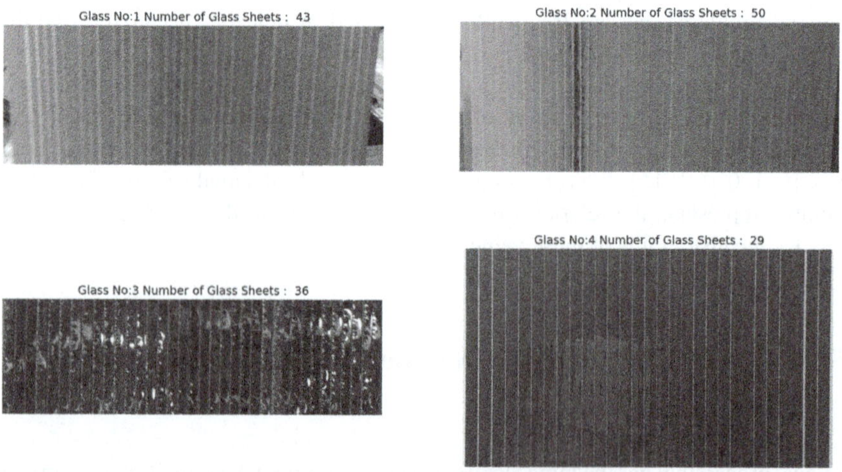

Fig. 41 RoI samples and sheet amounts of two different glass types

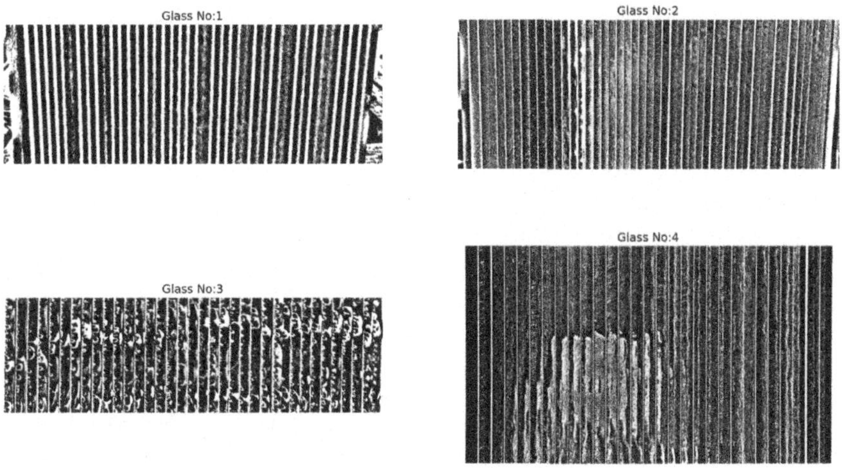

Fig. 42 Binary images obtained by applying adaptive thresholding

is a very challenging example of this problem. Since they reflect the cameraman, glass edges are distorted. The reflection case of cameraman can be seen in No. 4. The actual sheet amounts in the bundle are also shown in Fig. 41.

To obtain edge information, thresholding is applied to the gray-level RoIs. Since the illumination is heterogeneously distributed in the images, a global thresholding does not give satisfactory results here. Separate threshold values for local regions should be determined to get rid of this problem. Hence, adaptive thresholding was applied to the images, and Fig. 42 is obtained. Consequently, edges in the images become apparent.

After this process, pixel intensity values are accumulated column-wise, from top to bottom, resulting in vertical projection profile signals (cf. Fig. 43). These profile signals contain spatial period information of the glass array in the horizontal direction. However, it is not possible to estimate the horizontal period value from such a noisy signal in the discrete-index domain. Therefore a frequency analysis-based approach is adopted.

To obtain an accurate estimate, the signal is processed further, initially making it a zero-mean one. Then, in order to eliminate the discontinuities at the ends of the signal, it is multiplied by the Blackman window, and thus the damped signals were obtained. In order to increase the spectral resolution, zero padding was applied to the right and left sides as a quarter of the length of the signal (cf. Fig. 44).

After preprocessing, Power Spectrum (PS) was obtained over FT, and this spectrum was normalized by using (15).

$$X[k] = FFT\{x[n]\} \quad PS[k] = \sum_{k=0}^{N_s-1} |X[k]|^2 \quad PS_N[k] = \frac{PS[k]}{\max\{PS[k]\}} \tag{15}$$

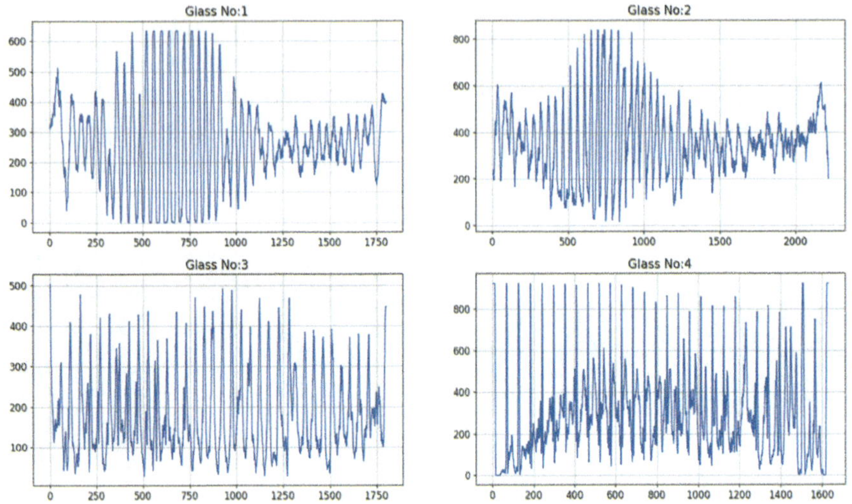

Fig. 43 Profile signals extracted from binary images. Horizontal and vertical axes denote the x spatial axis, and the total foreground pixel amount in the y-direction (downward), respectively

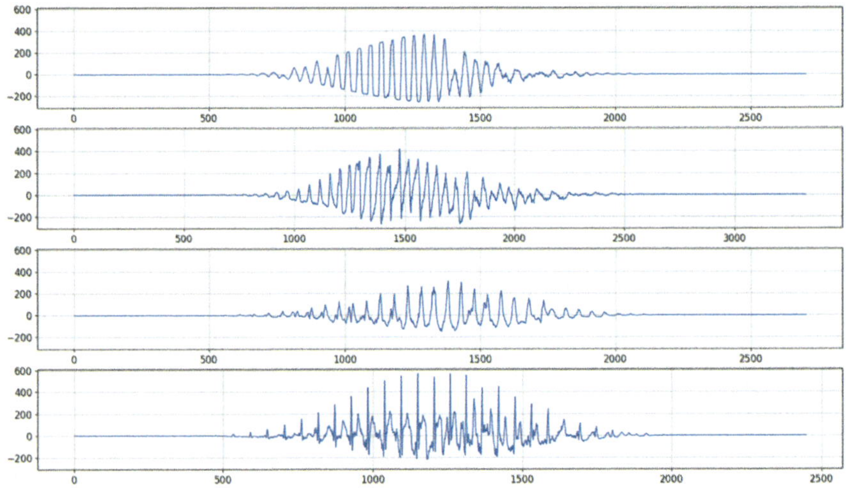

Fig. 44 Preprocessed profile signals corresponding to glasses from 1 (top) to 4 (bottom). Horizontal and vertical axes denote the x spatial axis, and zero-centered values of signals, respectively

Herein, $X[k]$, $PS[k]$, $PS_N[k]$ and represent FT of profile signal ($x[n]$), PS and normalized PS of $x[n]$. As can be seen in Fig. 45, there is a peak of the PS signal. The frequency value above this peak is related to the period of the glass array. By dividing the period value obtained from here by the length of the profile signal, the number of glasses can be predicted.

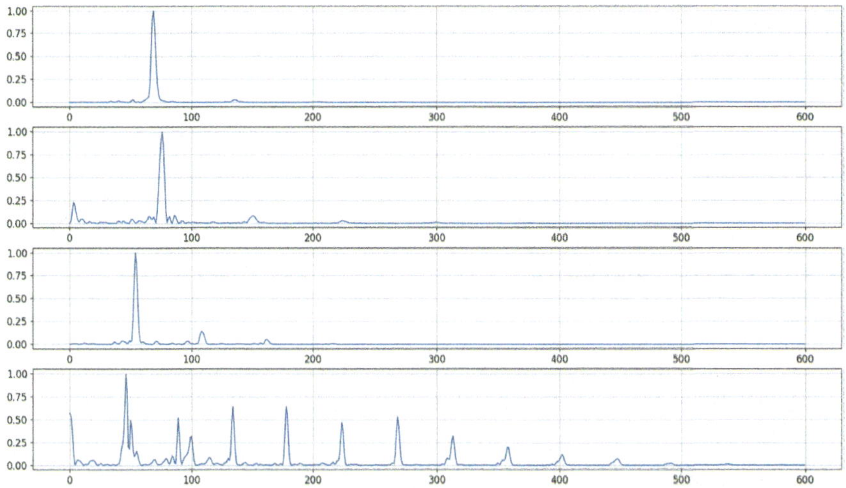

Fig. 45 Normalized power spectrum of signals. Peak value is related with glass array period. Horizontal and vertical axes denote the relative frequency indices and magnitude of $PS_N[k]$, respectively

Glass numbers estimated using this algorithm and their actual values are given in Fig. 46. The reason for differences in Glass No. 1 is that the glass image is not captured from a perfectly aligned vertical angle. In such a case, the length of the profile signal turns out to be longer than the actual glass array, and this causes the prediction to be overestimated. Glass No. 2 and 3 were correctly predicted. In No. 4,

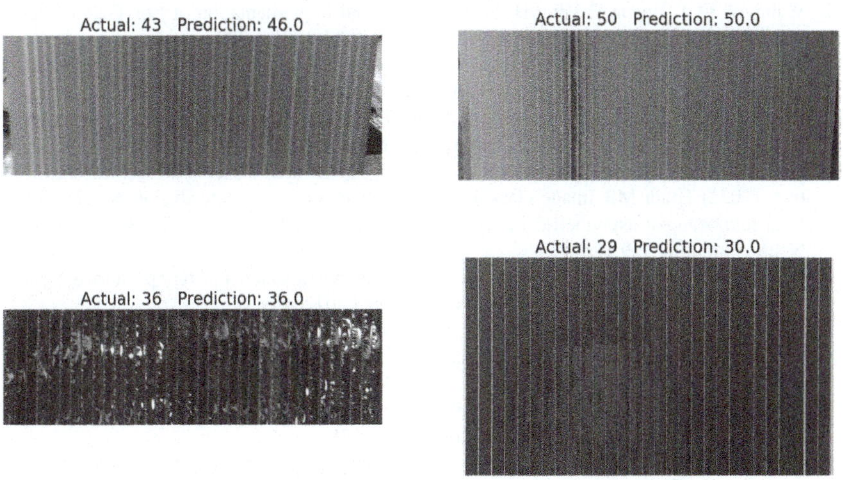

Fig. 46 Comparison of the values obtained by the algorithm with the actual values

as can be seen from the relevant profile signal, a noisy signal is extracted due to the reflective feature of the surface. Despite this, a near-accurate estimate could be made.

References

1. Dean J (2014) Big data, data mining, and machine learning: value creation for business leaders and practitioners. Wiley & Sons
2. Bose I, Mahapatra RK (2001) Business data mining—a machine learning perspective. Inf Manage 39(3):211–225
3. Litjens G, Kooi T, Bejnordi BE, Setio AAA, Ciompi F, Ghafoorian M, Sánchez CI (2017) A survey on deep learning in medical image analysis. Med Image Anal 42:60–88
4. Zhu XX, Tuia D, Mou L, Xia GS, Zhang L, Xu F, Fraundorfer F (2017) Deep learning in remote sensing: a comprehensive review and list of resources. IEEE Geosci Rem Sens Mag 5(4):8–36
5. Gonzalez RC, Woods RE (2018) Digital image processing. Pearson
6. Solomon C, Breckon T (2011) Fundamentals of digital image processing: a practical approach with examples in Matlab. Wiley & Sons
7. Shrivakshan GT (2013) An analysis of SOBEL and GABOR image filters for identifying fish. In: 2013 International conference on pattern recognition, informatics and mobile engineering. IEEE, pp 115–119
8. Ahmed AS (2018) Comparative study among Sobel, Prewitt and Canny edge detection operators used in image processing. J Theor Appl Inf Technol 96(19):6517–6525
9. Shrivakshan GT, Chandrasekar C (2012) A comparison of various edge detection techniques used in image processing. Int J Comput Sci Issues (IJCSI) 9(5):269
10. Meer P, Baugher ES, Rosenfeld A (1987) Frequency domain analysis and synthesis of image pyramid generating kernels. IEEE Trans Pattern Anal Mach Intell 4:512–522
11. Feifei S, Xuemeng Z, Guoyu W (2011) An approach for underwater image denoising via wavelet decomposition and high-pass filter. In: 2011 Fourth international conference on intelligent computation technology and automation, vol 2. IEEE, pp 417–420
12. Wang W, Li J, Huang F, Feng H (2008) Design and implementation of Log-Gabor filter in fingerprint image enhancement. Pattern Recogn Lett 29(3):301–308
13. Kamarainen JK, Kyrki V, Kalviainen H (2006) Invariance properties of Gabor filter-based features-overview and applications. IEEE Trans Image Process 15(5):1088–1099
14. Dora L, Agrawal S, Panda R, Abraham A (2017) An evolutionary single Gabor kernel based filter approach to face recognition. Eng Appl Artif Intell 62:286–301
15. Poloni KM, de Oliveira IAD, Tam R, Ferrari RJ, Alzheimer's Disease Neuroimaging Initiative. (2021) Brain MR image classification for Alzheimer's disease diagnosis using structural hippocampal asymmetrical attributes from directional 3-D log-Gabor filter responses. Neurocomputing 419:126–135
16. Karo NNB, Sari AY, Aziza N, Putra HK (2019) The enhancement of fingerprint images using Gabor filter. J Phys Conf Ser IOP Publishing 1196(1):012045
17. Niu X, Wu X, Xie P, Pan L (2014) A time-frequency analysis of event-related desynchronization/synchronization based on Gabor filter. In: Proceeding of the 11th World congress on intelligent control and automation, pp 5179–5184
18. Hsieh I, Saberi K (2016) Imperfect pitch: Gabor's uncertainty principle and the pitch of extremely brief sounds. Psychon Bull Rev 23:163–171
19. Sagiv C, Sochen NA, Zeevi YY (2006) The Uncertainty principle: group theoretic approach, possible minimizers and scale-space properties. J Math Imag Vis 26:149–166
20. Kartsov SK, Kupriyanov DY, Polyakov YA, Zykov AN (2020) Non-local means denoising algorithm based on local binary patterns. In: Computer vision in control systems, vol 6. Springer, Cham, pp 153–164

21. Zou J, Shen M, Zhang Y, Li H, Liu G, Ding S (2018) Total variation denoising with non-convex regularizers. IEEE Access 7:4422–4431
22. Albayrak A, Akhan AU, Calik N, Capar A, Bilgin G, Toreyin BU, Durak-Ata L (2021) A whole-slide image grading benchmark and tissue classification for cervical cancer precursor lesions with inter-observer variability. Med Biol Eng Comput 59(7):1545–1561
23. Hass G, Simon P, Kashef R (2020) Business applications for current developments in big data clustering: an overview. In: 2020 IEEE international conference on industrial engineering and engineering management (IEEM), IEEE, pp 195–199
24. Zhuang K, Wu S, Gao X (2018) Auto insurance business analytics approach for customer segmentation using multiple mixed-type data clustering algorithms. Tehnički vjesnik 25(6):1783–1791
25. Azmi AN, Nasien D, Omar FS (2017) Biometric signature verification system based on freeman chain code and k-nearest neighbor. Multimedia Tools Appl 76(14):15341–15355
26. McConnell RK (1986) U.S. Patent No. 4,567,610. U.S. Patent and Trademark Office, Washington, DC
27. Dalal N, Triggs B (2005) Histograms of oriented gradients for human detection. In: 2005 IEEE computer society conference on computer vision and pattern recognition (CVPR'05), vol 1. IEEE, pp 886–893
28. Folador JP, Rosebrock A, Pereira AA, Vieira MF, de Oliveira Andrade A (2019) Classification of handwritten drawings of people with Parkinson's disease by using histograms of oriented gradients and the random forest classifier. In: Latin American conference on biomedical engineering, Springer, Cham, pp 334–343
29. Kar NB, Babu KS, Jena SK (2017) Face expression recognition using histograms of oriented gradients with reduced features. In: Proceedings of international conference on computer vision and image processing, Springer, Singapore, pp 209–219
30. Ojala T, Pietikäinen M, Harwood D (1996) A comparative study of texture measures with classification based on featured distributions. Pattern Recogn 29(1):51–59
31. Wang X, Han TX, Yan S (2009) An HOG-LBP human detector with partial occlusion handling. In: 2009 IEEE 12th international conference on computer vision, IEEE, pp 32–39
32. Rahmadya B, Sun R, Takeda S, Kagoshima K, Umehira M (2020) A framework to determine secure distances for either drones or robots based inventory management systems. IEEE Access 8:170153–170161
33. Yi G, Yi L, Li J (2022) Collaborative design method for product quality based on neural network. In: Proceedings of 2021 Chinese intelligent systems conference, Springer, Singapore, pp 22–30
34. Honarvar F, Varvani-Farahani A (2020) A review of ultrasonic testing applications in additive manufacturing: defect evaluation, material characterization, and process control. Ultrasonics 106227
35. Kukkala VK, Tunnell J, Pasricha S, Bradley T (2018) Advanced driver-assistance systems: a path toward autonomous vehicles. IEEE Consum Electron Mag 7(5):18–25
36. Cetin AE, Merci B, Günay O, Töreyin BU, Verstockt S (2016) Methods and techniques for fire detection: signal, image and video processing perspectives. Academic Press
37. Dastres R, Soori M (2021) Advanced image processing systems. Int J Imag Robot 21(1):27–44

Nurullah Calik received his M.Sc. and Ph.D. degree in Electronics and Communication Engineering from the Yıldız Technical University, Turkey, in 2013 and 2019, respectively. He worked as a postdoctoral researcher at the Informatics Institute, Istanbul Technical University. He is currently an Assistant Professor with the Department of Biomedical Engineering, Istanbul Medeniyet University, Turkey. His main research areas are large-scale data analysis, signal and image processing and AI applications in engineering. His research interests include especially biomedical image analysis, deep learning regression and optimization.

Behçet Uğur Töreyin received the B.S. degree from the Middle East Technical University, Ankara, Turkey, in 2001, and the M.S. and Ph.D. degrees from Bilkent University, Ankara, in 2003 and 2009, respectively, all in electrical and electronics engineering. He is currently a Professor with the Informatics Institute, Istanbul Technical University. His research interests include signal processing and pattern recognition with applications to computational intelligence.

Prescriptive Analytics: Optimization and Modeling

Nursah Alkan, Kenan Menguc, and Özgür Kabak

Prescriptive analytics, a type of complex business analytics, aims to suggest the best among various decision options to benefit from the predicted future using large amounts of data. In this process, prescriptive analytics combines the output of predictive analytics and uses artificial intelligence, optimization algorithms, and expert systems to provide adaptive, automated, constrained, time-bound, and optimal decisions, thus having the potential to bring the greatest intelligence and value to businesses [1]. An important element for prescriptive analytics is to provide a decision proposal using mathematical models on a realistic problem under consideration. For example, modeling many complex real-life problems such as deciding which product should be produced to maximize the profits of the businesses, deciding which warehouses should serve which customers, which warehouses should be kept open to establish the lowest-cost supply chain network, and deciding the lowest-cost vehicle route while supplying the products of the business is not something to be done manually. Here, the application of prescriptive analytics will enable businesses to make the best decision that will support them in getting the desired output in line with the target determined in the considered problems. There are also many classes of models depending on the considered problem and often many specific techniques for solving each.

Unlike the predictive and descriptive analytics, prescriptive analytics provides decision makers with an action plan. Prescriptive analytics not only answers why, how, when, and what questions but also generates recommendations on how to act under given circumstances. Prescriptive analytics allows to find the best or closest to the best moves for a situation with constraints and goals. Usually, it is the final phase of the business analytics application that also benefits from the results of descriptive and predictive analytics tools.

N. Alkan · K. Menguc · Ö. Kabak (✉)
Department of Industrial Engineering, Faculty of Management, Istanbul Technical University, 34367, Macka Istanbul, Turkey
e-mail: kabak@itu.edu.tr

A. Ustundag et al. (eds.), *Business Analytics for Professionals*, Springer Series in Advanced Manufacturing, https://doi.org/10.1007/978-3-030-93823-9_9

Prescriptive analytics methods can be classified into six main categories: probabilistic models, machine learning/data mining, mathematical programming, evolutionary computation, simulation, and logic-based models [1]. Among these categories, mathematical programming models have an important place. Mathematical programming (or mathematical optimization or optimization) is finding the best decision with respect to some criteria, from a set of available alternatives that are defined by some rules, limitations, and constraints. Linear programming, integer programming, nonlinear programming, fuzzy programming, and stochastic programming are among the well-known mathematical programs. In this chapter, the basic mathematical programming tools, linear programming and integer programming, are introduced.

1 Linear Programming

Linear programming (LP) is the basic approach for solving optimization problems. Objective function and the constraints are defined using linear equations. It can be solved by an efficient algorithm called simplex algorithm. Thus, it has been applied to many diverse industries such as manufacturing, assembly lines, finance, education, and transportation. Modeling in LP also serves as an introduction to other optimization tools such as integer programming and nonlinear programming.

In this section, basic concepts of LP are introduced on an example, and two interesting examples are provided.

Glass Production Example:

ATK-Brown produces handmade glass ornaments. The company's portfolio includes three products as follows: spherical glass, glass bird, and plate. A spherical glass sells for $35. The total cost of producing one piece of spherical glass is $30. A glass bird and a plate sell for $50, and $25 and costs $42 and $21, respectively.

The basic raw material of the glass used for the products is a special kind of sand. The sand is processed in an oven before being shaped by the operators. For the production of spherical glass, two units of sand are required. It is processed in the oven for 0.1 h and shaped by an operator in 0.2 h. For a glass bird, the required amount of sand is 1 unit, and oven and operator hours are 0.2 and 0.5. For a plate, the required amount of sand is three units, and oven and operator hours are 0.3 and 0.4.

ATK-Brown has 400 units of available sand for this week's production. Oven can operate 40 h in a week. The company employs two operators, each working 40 h a week. ATK promises 50 pieces of glass birds to one of their customers. Summary of the given information is given in Table 1.

ATK-Brown wants to maximize this week's profit (revenues costs). Formulate a mathematical program for ATK-Brown to accomplish their goal.

Table 1 Glass production example

	Sales price ($)	Cost ($)	Sand (units)	Oven (h)	Operator (h)
Spherical glass	35	30	2	0.1	0.2
Glass bird	50	42	1	0.2	0.5
Plate	25	21	3	0.3	0.4
Available amount			400	40	80

LP Formulation

While formulating a mathematical program, or an LP in particular, the focus is on three factors: decision variables, objective function, and constraints.

Decision Variables:

Decision variables describe the decisions to be made. In the glass production example, it has to be decided how many spherical glasses, glass birds, and plates will be manufactured in this week. Therefore, the following decision variables are defined:

x_1 : Number of spherical glasses produced this week
x_2 : Number of glass birds produced this week
x_3 : Number of plates produced this week

Objective Function:

In any mathematical programming problem, the aim is to minimize (usually cost) or maximize (usually profit or revenue) a function of the decision variables. The objective function is the function to be minimized or maximized. For the given example, ATK-Brown wants to maximize the total weekly profit. The profit is composed of profits coming from the three products. For each product, the profit can be calculated by multiplying unit profit with number of products produced. The unit profit of each product can be calculated as the difference of sales price and cost. For spherical glass, the profit is $5 ($35–$30). For a glass bird and a plate, the unit profit values are $8 ($50–$42) and $4 ($25–$21), respectively. Therefore, the objective function for the problem can be stated as follows:

$$\text{Max } Z = 5x_1 + 8x_2 + 4x_3$$

Here, Z represents the total profit. The coefficients of the decision variables (e.g., 5, 8, and 4) are called objective function coefficients.

Constraints:

In any decision problem, there are restrictions, rules, and limits while trying to reach the optimum decision. In the glass production example, it is not possible to produce 1 million units of the products because of limits for raw materials and resources. Therefore, it must define constraints to introduce limits or other rules as functions of decision variables.

In the glass production example, there are limits in sand, oven capacity, and operator hours. For the available sand, 400 units can be used in this week's production. Total sand used in the production is the sum of sand used in the products. For a product, the required sand is calculated by multiplying unit sand usage and number of the product. For instance, for the spherical glass, $2x_1$ units of sand are used. The total amount of sand used for the whole production is $2x_1 + x_2 + 3x_3$. This amount must not exceed 400 units. The related constraint can be written as follows:

$$2x_1 + x_2 + 3x_3 \leq 400$$

For oven capacity, a similar constraint can be added:

$$0.1x_1 + 0.2x_2 + 0.3x_3 \leq 40$$

For operator hours, the available hours are not provided directly. One operator works 40 h per week. Thus, two operators can work 80 h in total. Total operator hours used for the products should not exceed 80:

$$0.2x_1 + 0.5x_2 + 0.4x_3 \leq 80$$

Related to the order of a customer, ATK must produce at least 50 glass birds:

$$x_2 \geq 50$$

After adding the constraints related to the nature of the problem, sign restrictions of the decision variables are written. Sign restrictions show if the decision variables are nonnegative or can take positive or negative values. In most of the LP formulations, the decision variables are nonnegative. In the glass production example, as number of production amounts cannot be negative, the decision variables are defined as nonnegative:

$$x_1, x_2, x_3 \geq 0$$

In an LP formulation, the constraints can be $\geq$ (greater than or equal to), $\leq$ (less than or equal to), or $=$ (equal to). Any constraint can be organized as the decision variables are left-hand side of the constraint sign and a scalar value on the right-hand side. The value on the right is called right-hand side value. The coefficients of the decision variables in the constraints are called technology coefficients.

The final LP for ATK-Brown is given below:

$$\text{Max } Z = 5x_1 + 8x_2 + 4x_3$$

Subject to (s.t.):

$$2x_1 + x_2 + 3x_3 \leq 400 \text{ (Sand Constraint)}$$

$$0.1x_1 + 0.2x_2 + 0.3x_3 \leq 40 \text{ (Oven Constraint)}$$

$$0.2x_1 + 0.5x_2 + 0.4x_3 \leq 80 \text{ (Operator Constraint)}$$

$$x_2 \geq 50 \text{ (Customer Order)}$$

$$x_1, x_2, x_3 \geq 0 \text{ (Sign Restriction)}$$

When this LP is solved, $x_1 = 150$, $x_2 = 100$, $x_3 = 0$, and $Z = 1550$ are obtained. ATK-Brown is recommended to produce 150 spherical glasses and 100 glass birds that will result with a \$1550 profit. All available sand and operator hours are used in this production plan. 35 h of oven is required for the production; thus, the oven will have 5 h of free time.

Assumptions of LP

As can be seen in the example, the fundamental property of the LP formulation is that the objective function is a linear function, and all constraints are linear inequalities or equalities. The function, inequalities, and equalities are summation of decision variables multiplied by coefficients. This implies two assumptions: proportionality and additivity. Each decision variable's contribution to the objective function is proportional to the value of the decision variable. In this respect, nonlinear equations such as x^2 or $\log(x)$ are not acceptable in LP formulations. Each decision variable's contribution to the objective function is independent from the other decision variables. Thus, multiplication or division of decision variables is not used in LPs. If these assumptions are violated in a formulation, it is suggested to use nonlinear programming.

Another important assumption of LP is divisibility. Decision variables in LP formulations are defined as continuous variables, that is, decision variables can get fractional values. For instance, in the glass production example, it is acceptable to produce 75.2 spherical glasses or 42.7 glass birds. This result can be interpreted as 75 spherical glasses, and 20% of a spherical glass will be produced in the planning period. If this assumption is not acceptable in a problem, then it is suggested to use integer programming (IP). IP is explained in Sect. 2.

The last assumption of LP is certainty assumption. It implies that all parameters of an LP (objective function coefficients, right-hand side values, and technology coefficients) are known with certainty. If any of the parameters are not known exactly, or cannot be assumed as exact values, it is suggested to use fuzzy or stochastic programming.

Under all these assumptions, one can question the use of LP in real-life problems. Actually, it has applied in many real-life problems. Besides, it has an efficient solution algorithm—the simplex algorithm—that makes it preferable to nonlinear programming, integer programming, fuzzy programming, and stochastic programming in which the solutions of the models are much more complicated.

Solution of LPs

As mentioned above, LPs can be solved by the simplex algorithm efficiently. There are also other solution approaches such as graphical solution for the problems with two variables and interior point methods. Many software packages are developed to solve LP models. See https://en.wikipedia.org/wiki/Linear_programming#Solvers_and_scripting_(programming)_languages for details. Pyomo programming language is used in this chapter to solve the formulated models. Pyomo is a collection of Python software packages for formulating optimization models.

1.1 Diet Problem

Diabetes or diabetes mellitus is a disease that occurs as a result of insufficient insulin secretion or insufficient insulin use of pancreas. A nutrition program is prepared for Type 2 diabetes patients as given in Table 1. Seven food options (oat, egg, milk, etc.) are determined, and their nutrition factors (energy, protein, magnesium, etc.) are measured. For a patient, the minimum and maximum mineral and nutritional values are defined as given in the last two rows of Table 2. For example, patient should get at least 600 kcal of energy and at most 4 mg of sugar. There is no upper limit for energy and no lower limit for sugar. Nutrition factors are mutually exclusive except one instance. Each 100 mg. magnesium reduces the cholesterol by 5 mg.

Suppose a patient wants to find out which foods he has to take. Formulate an LP to prepare diet program for the patient that provides the necessary mineral and nutrient intake with the minimum cost.

LP Formulation

Indices:

i : foods, $i = 1, 2, \ldots, 7$. (Corresponding foods are given in Table 2.)

Parameters:

c_i : *The cost considered for each unit of food* i.
e_i : *The amount of energy for each unit of food* i.
p_i : *The amount of protein for each unit of food* i.
m_i : *The amount of magnesium for each unit of food* i.
s_i : *The amount of sugar for each unit of food* i.
o_i : *The amount of oil for each unit of food* i.
h_i : *The amount of cholesterol for each unit of food.*
g_i : *The amount of Omega-3 (mg) for each unit of food.*

Table 2 Nutrition program

	Values for each unit of food intake							
	Energy (Kcal)	Protein (mg)	Magnesium (mg)	Sugar (mg)	Oil (mg)	Cholesterol (mg)	Omega-3 (mg)	Price per unit
Oat (100 gr) ($i = 1$)	351	11.4	131	0.3	5.8			20$
Egg (One unit) ($i = 2$)	140	13.2	13		9.7	360		2$
Milk (1 lt) ($i = 3$)	64	3.2	9	3.9	3.6	8		1$
Walnut kernel (100 gr) $i = 4$	679	14.5	165		64.8		6.5	14$
Salmon (100 gr) ($i = 5$)	149	25.6	27		4.4	53	2.5	35$
Tuna fish (100 gr) ($i = 6$)	116	25.5	18		0.85	47	1	12$
Flaxseed (100 gr) ($i = 7$)	454	18.2	379		33.2		17	4$
Minimum value	600	55	90				1	
Maximum value				4	65	800		

Decision Variables:

x_i: Amount of unit of food that will be included in the diet.

Objective Function:

$$\text{Min}z : \sum_{i=1}^{7} c_i x_i \tag{1}$$

Subject to:

$$\sum_{i=1}^{I} e_i x_i \geq 600 \tag{2}$$

$$\sum_{i=1}^{I} p_i x_i \geq 55 \tag{3}$$

$$\sum_{i=1}^{I} m_i x_i \geq 90 \tag{4}$$

$$\sum_{i=1}^{I} s_i x_i \leq 4 \tag{5}$$

$$\sum_{i=1}^{I} o_i x_i \leq 65 \tag{6}$$

$$\sum_{i=1}^{I} h_i x_i - 0.05 \sum_{i=1}^{I} g_i x_i \leq 800 \tag{7}$$

$$\sum_{i=1}^{I} g_i x_i \geq 1 \tag{8}$$

$$x_i \geq 0 \tag{9}$$

The objective function given in Eq. (1) minimizes the total cost of food that will be included in the diet. Equations (2), (3), and (4) indicate the minimum amount of energy, protein, and magnesium to be taken, respectively. Equations (5) and (6) indicate the maximum amount of sugar, oil to be taken, respectively. In Eq. (7), the maximum cholesterol to be taken is determined by considering its interaction with the amount of magnesium taken. The constraint in Eq. (8) determines the lower limit for the amount of Omega-3. Sign restrictions given in Eq. (9) ensure that the decision variables do not take negative values.

Solution

The solution given by the Pyomo library to the diet problem is $x_2 = 0.1806$ and $x_7 = 2.8910$. It indicates that the patient should take 0,1806 unit of egg and 289,10 gr flaxseeds in his diet. This solution corresponds to the following nutritional values.

- Total energy intake is 1337.80 kcal
- Total of protein intake 55 mg
- Total of magnesium intake 1098.05 mg
- Total of sugar intake 0.0 mg
- Total of oil intake 65.00 mg
- Total of cholesterol intake 65.00 mg
- Total of Omega-3 intake 49.15 mg
- Total cost of diet program is 11.93 dollars.

1.2 Police Station Problem

The chief officer of the Macka Police Station will plan the shifts of the officers. Each officer works for 5 days per week in consecutive days and takes a break in the following two days. The crime rate fluctuates with the day of week, so the number of the police officers required each day depends on which day of the week it is: Monday, 15; Tuesday, 18; Wednesday, 19; Thursday, 11; Friday, 21; Saturday, 16; Sunday, 11. The chief wants to schedule police officers to minimize the total number of the officers. Formulate an LP that will accomplish this goal.

LP Formulation

Decision Variables:

$X_1 =$ Number of officers works in days Monday to Friday (Shift 1)
$X_2 =$ Number of officers works in days Tuesday to Saturday (Shift 2)
$X_3 =$ Number of officers works in days Wednesday to Sunday (Shift 3)
$X_4 =$ Number of officers works in days Thursday to Monday (Shift 4)
$X_5 =$ Number of officers works in days Friday to Tuesday (Shift 5)
$X_6 =$ Number of officers works in days Saturday to Wednesday (Shift 6)
$X_7 =$ Number of officers works in days Sunday to Thursday (Shift 7)

Objective Function:

$$\text{Min} = X_1 + X_2 + X_3 + X_4 + X_5 + X_6 + X_7 \tag{10}$$

Subject to:

$$X_1 + X_4 + X_5 + X_6 + X_7 \geq 15 \tag{11}$$

$$X_1 + X_2 + X_5 + X_6 + X_7 \geq 18 \tag{12}$$

$$X_1 + X_2 + X_3 + X_6 + X_7 \geq 19 \tag{13}$$

$$X_1 + X_2 + X_3 + X_4 + X_7 \geq 11 \tag{14}$$

$$X_1 + X_2 + X_3 + X_4 + X_5 \geq 21 \tag{15}$$

$$X_2 + X_3 + X_4 + X_5 + X_6 \geq 16 \tag{16}$$

$$X_3 + X_4 + X_5 + X_6 + X_7 \geq 11 \tag{17}$$

$$X_1, X_2, X_3, X_4, X_5, X_6, X_7 \geq 0 \tag{18}$$

where the total number of police officers working at the Macka Police Station is represented by Eq. (10). Equation (11) shows that the total number of police officers working for all variables except X_2 and X_3, which includes officers not working on Monday, must be equal to or greater than the number of police officers required on Monday. Equation (12) shows that the total number of police officers working for all variables except X_3 and X_4, which includes officers not working on Tuesday, must be equal to or greater than the number of police officers required on Tuesday. Equation (13) shows that the total number of police officers working for all variables except X_4 and X_5, which includes officers not working on Wednesday, must be equal to or greater than the number of police officers required on Wednesday. Equation (14) shows that the total number of police officers working for all variables except X_5 and X_6, which includes officers not working on Thursday, must be equal to or greater than the number of police officers required on Thursday. Equation (15) shows that the total number of police officers working for all variables except X_6 and X_7, which includes officers not working on Friday, must be equal to or greater than the number of police officers required on Friday. Equation (16) shows that the total number of police officers working for all variables except X_7 and X_1, which includes officers not working on Saturday, must be equal to or greater than the number of police officers required on Saturday. Equation (17) shows that the total number of police officers working for all variables except X_1 and X_2, which includes officers not working on Sunday, must be equal to or greater than the number of police officers required on Sunday. Equation (18) shows the sign constraint where all decision variables are nonnegative values.

Solution

The results obtained after solving the model using the Python Pyomo library are as follows:

$$\text{Minimum number of officer} = 23.667$$
$$X_1 = 7.667, X_2 = 5, X_3 = 3.667, X_4 = 2, X_5 = 2.667, X_6 = 2.667, X_7 = 0$$

According to the results obtained, the number of officers working in some of the shifts is not integer values. Since the decision variables are continuous variables in the LPs, it is assumed and acceptable to get such non-integer values. If the number of police officers working in the shifts is desired to be integer, the model should be set up and solved using integer programming. Notice that integer-valued solution for this problem and other LPs cannot be found by rounding the resulting non-integer values.

2 Integer Programming

In LP formulation, it is assumed that the decision variables take continuous values. In an optimal solution of an LP, decision variables can be fractional non-integer numbers. While this is acceptable in some situations, in many cases it is not. In many real-life problems, non-integer solutions are not applicable. For instance, in the police station example in Sect. 1.2, it is not possible to assign 7.667 officers to the first shift. In a project selection problem, it may not be possible to select half of a project. In a course timetabling problem, it is not possible to assign half of the lecturers to courses or a very small percentage of the course in a certain hour. The number of examples can be increased for many problems in real life.

On the other hand, there may be conditional restrictions in the problem. For instance, while making a production plan, there may be fixed investment costs such as assembly line construction cost and machine purchasing cost to enable the manufacturing of a product. One restriction would be valid only in some specific situations. For instance, if more than ten units of a product are purchased, its price drops by 10%. If one product is purchased, another one cannot be purchased.

For all the above-given situations and much more, decision variables should be defined to take only integer or binary (i.e., 0 or 1) values. The mathematical program used for this purpose is called integer program (IP), and the subject of solving such programs is called integer programming.

An IP in which all variables are required to be integers is called a pure IP problem. If some variables are restricted to be integer and some are not, then the problem is a mixed IP problem. The case where the integer variables are restricted to be 0 or 1 comes up surprisingly often. Such problems are called pure (mixed) 0–1 programming problems or pure (mixed) binary IP problems.

Formulation of IPs is the same as LP formulation except integer or binary variables are indicated separately. See an example of a mixed IP below:

$$\text{Max } Z = x_1 + x_2 + 2x_3$$

Subject to:

$$x_1 + 2x_2 - 3x_3 \leq 9$$

$$x_1 - 10x_4 \leq 0$$

$$2x_1 - x_2 + x_3 \geq 5$$

$$x_1, x_3 \geq 0 \text{ and integer}, \quad x_2 \geq 0, \quad x_4 \in \{0, 1\}$$

In the given model, x_1 and x_3 are nonnegative integer variables that can take values $0, 1, 2, 3, \ldots$; x_2 is a nonnegative continuous variable, and x_4 is a binary variable that can take 0 or 1.

In this chapter, initially, basic information for IP modeling is explained with small examples, and then two comprehensive examples are given in Sects. 2.1 and 2.2.

Real Estate Investment Example

AEK real estate company owns houses all over Istanbul from which they earn rental income. At the moment, they have 7 million dollars to invest in four possible houses. The costs of houses are 2.5, 3.5, 2, and 1.5 million dollars, respectively. Expected rental incomes of the houses are 18, 20, 12, and 10 thousand dollars per month, respectively. Formulate an IP to maximize the monthly income of the company from the houses that they will invest with their limited budget.

IP Formulation:

Decision variables are defined as x_i that indicates whether AEK purchases house i $(i = 1, 2, 3, 4)$ or not. It is not possible to buy a percentage from a house or more than one from a specific house; therefore, x_i can be 1 or 0 and is a binary variable. x_i will be 1 if AEK purchases house i and 0 if they do not.

The IP can be formulated as follows:

$$\text{Max } Z = 18x_1 + 20x_2 + 12x_3 + 10x_4$$

Subject to:

$$2.5x_1 + 3.5x_2 + 2x_3 + 1.5x_4 \leq 7$$

$$x_i \in \{0, 1\} \quad i = 1, 2, 3, 4$$

When this IP is solved, the result will be $x_1 = 0, x_2 = 1, x_3 = 1, x_4 = 1, Z = 42$. The model suggests that all of the budgets should be spent on houses 2, 3, and 4; and the company will get 42 thousand dollars rent per month in the coming years.

This type of a problem where only one constraint exists and the decision variables are binary is the famous Knapsack problem. The idea is to fill a capacitated knapsack with items according to their weights and values to maximize the total value of the selected items.

In an IP formulation, it is possible to formulate logical rules. For instance, consider the following conditions in the real estate investment example:

- If house 1 is purchased, house 2 must also be purchased (Condition 1).
- If house 2 is purchased, house 4 cannot be purchased (Condition 2).
- Either house 2 or house 4 must be purchased (Condition 3).

As the state of the purchase of a house is defined as binary variable, the above-given restrictions can be added to the IP formula by the following constraints:

- $x_2 \geq x_1$ (Condition 1)
- $1 - x_4 \geq x_2$ or $x_2 + x_4 \leq 1$ (Condition 2)
- $x_2 + x_4 \geq 1$ (Condition 3).

Fixed Charge Problems

In some problems, initial flat costs occur in an activity regardless of the level of the activity. For example, suppose that a special machine needs to be rented to produce a particular product. If the company decides to produce the product, there will be machinery rental cost (F) in addition to the variable cost of the product (c). In a fixed charge problem, total cost of an activity is defined as follows:

$$C(x) = \begin{Bmatrix} F + cx & x > 0 \\ 0 & x = 0 \end{Bmatrix} \tag{19}$$

where x is level of the activity, F is the fixed cost, and c is the variable cost.

In order to formulate fixed charges in IP, a binary decision variable is defined to indicate the level of activity is more than 0. For the cost function in Eq. (19), the following constraints are added to the model:

$$C(x) = Fy + cx \tag{20}$$

$$x \leq My \tag{21}$$

$$y \in \{0, 1\} \tag{22}$$

where M is a sufficient large positive number that can be calculated as the high possible value of x according to the other constraints. By the use of Eq. (21), y will be equal to 1 if $x > 0$, and fixed cost F will be added to the total cost. When $x = 0$, y will be free and can take 0 or 1. During the optimization process, while the solver minimizing the total cost, y will tend to be equal to 0.

Either–Or Constraints

In some situations, satisfying one of the given set of constraints will be enough for meeting a restriction. Suppose at least one of the following constraints must be satisfied:

$$f(x_1, x_1, \ldots, x_n) \leq 0 \tag{23}$$

$$g(x_1, x_1, \ldots, x_n) \leq 0 \tag{24}$$

The following either–or constraints are added to the model:

$$f(x_1, x_1, \ldots, x_n) \leq My \tag{25}$$

$$g(x_1, x_1, \ldots, x_n) \leq M(1 - y) \tag{26}$$

$$y \in \{0, 1\} \tag{27}$$

where M is a sufficiently large positive number. The idea here is that for binary values of y, one of the constraints becomes active. When $y = 0$, Eq. (25) becomes $f(x_1, x_1, \ldots, x_n) \leq 0$, and constraint given in Eq. (23) becomes active. Equation (26) becomes $g(x_1, x_1, \ldots, x_n) \leq M$ that indicates that constraint in Eq. (24) may not be satisfied. When $y = 1$, $g(x_1, x_1, \ldots, x_n) \leq 0$ (Eq. (24)) is satisfied, and the other may not be satisfied.

If–Then Constraints

In some situations, a constraint becomes active when another constraint is satisfied. Suppose it is wanted to ensure that a constraint $f(x_1, x_1, \ldots, x_n) > 0$ implies the constraint $g(x_1, x_1, \ldots, x_n)0$.

The following constraints are added in the formulation:

$$-g(x_1, x_1, \ldots, x_n) \leq My \tag{28}$$

$$f(x_1, x_1, \ldots, x_n) \leq M(1 - y) \tag{29}$$

$$y \in \{0, 1\} \tag{30}$$

where M is a sufficiently large positive number. According to this formulation, if $f(x_1, x_1, \ldots, x_n) > 0$, Eq. (29) forces y to be equal to 0, and when $y = 0$, $g(x_1, x_1, \ldots, x_n)0$, constraint becomes active by Eq. (28). $f(x_1, x_1, \ldots, x_n) > 0$ is not satisfied, y can be 0 or 1, and $g(x_1, x_1, \ldots, x_n)0$ can be satisfied or not.

In the following subsections, two examples are provided to show how the given formulation tricks are applied.

2.1 Example: Production Planning

ATK-White has plans to produce five new products. In Table 3, the initial investment costs required to produce the products and the net profit values of the products have been given. ATK-White considers the following conditions:

- Up to three products will be produced.

Table 9.3 Information for the products

	Products				
	1	2	3	4	5
Investment cost	15,000$	35,000$	50,000$	25,000$	40,000$
Unit net profit	50$	80$	70$	85$	90$
Required labor hours per product	5	3	7	4	5
Required raw material per product	4	4	5	2	3

- In order to produce product 3 or product 4, at least one of product 1 or product 2 must be produced.
- Required labor hours and raw material for the products are given in Table 3. The company has 7000 working hours and 7000 units of raw materials during the planning period.
- In accordance with the agreement of the company with a customer, 500 units of the first and third products in total or 750 units of the second and fourth products in total should be sent to the customer.
- If the fifth product is produced more than 500, the fourth product must be produced at most 500.

Under given conditions, formulate an integer program to maximize the total profit of ATK-White.

IP Formulation

Decision Variables:

$$X_i = \text{Production quantity for product } i, \quad \text{for } i = 1, 2, 3, 4, 5$$

$$Y_i = \begin{cases} 1, \text{ if investment is made in product } i \\ 0, \text{ otherwise} \end{cases} \quad \text{for } i = 1, 2, 3, 4, 5$$

Objective Function:

$$\text{Max } Z = 50X_1 + 80X_2 + 70X_3 + 85X_4 + 90X_5 - 15,000Y_1$$
$$35,000Y_2 - 50,000Y_3 - 25,000Y_4 - 40,000Y_5 \quad (31)$$

Equation (31) shows the objective function to maximize the total net profit of ATK-White.

Constraints:

The labor hours and raw material constraints of ATK-White are indicated as given Eqs. (32) and (33), respectively.

$$5X_1 + 3X_2 + 7X_3 + 4X_4 + 5X_5 \leq 7000 \quad (32)$$

$$4X_1 + 4X_2 + 5X_3 + 2X_4 + 3X_5 \leq 7000 \tag{33}$$

The following constraints are added to provide $Y_i = 1$ when $X_i > 0$.

$$X_i \leq M_i Y_i \quad for\ i = 1, 2, \ldots, 5 \tag{34}$$

where M_i values are sufficiently big numbers and can be calculated by using the other constraints. For instance, for M_1 that is related to product 1, considering the labor (Eq. (32)) and raw material constraints (Eq. (33)), a maximum of 1400 units of product 1 can be produced that is $M_1 = \min(7000/5, 7000/4) = 1400$. Therefore, the constraint to be added for product 1 does not create an unnecessary limit on X_1.

The other M_i values can be found as follows:

$$M_2 = \min(7000/3, 7000/4) = 1750 \text{ for product 2,}$$
$$M_3 = \min(7000/7, 7000/5) = 1000 \text{ for product 3,}$$
$$M_4 = \min(7000/4,\ 7000/2) = 1750 \text{ for product 4,}$$
$$M_5 = \min(7000/5,\ 7000/3) = 1400 \text{ for product 5,}$$

According to the agreement made with the customer, either 500 units of the first and third products in total or 750 units of the second and fourth products in total must be ordered. These conditions are as shown in Eqs. (35) and (36).

$$X_1 + X_3 \geq 500 \tag{35}$$

$$X_2 + X_4 \geq 750 \tag{36}$$

Since at least one condition is desired to be satisfied, these conditions must be represented using either–or constraints.

For $f(x) = 500 - X_1 - X_3$ and $g(x) = 750 - X_2 - X_4$.

The following constraints are added to the model to ensure at least one of the constraints in Eqs. (35) and (36) are satisfied.

$$500 - X_1 - X_3 \leq M_6 K_1 \tag{37}$$

$$750 - X_2 - X_4 \leq M_7(1 - K_1) \tag{38}$$

$$K_1 = 0\ or\ 1 \tag{39}$$

where M_6 and M_7 must be at least 500 and 750 to meet order constraints, respectively.

If more than 500 of the fifth product is produced, 500 of the fourth product must be produced at most. In order to fulfill these conditions, it is necessary to use the if–then rule. If the condition $f(x) > 0$ is met, then the condition $g(x) \geq 0$ must also be met, or if the condition $f(x) > 0$ is not met, then it does not matter whether the condition $g(x) \geq 0$ is met or not.

If $X_5 > 500$, then $X_4 \leq 500$.

For $f(x) = X_5 - 500$ and $g(x) = 500 - X_4$.

We can ensure this condition is ensured by adding the following pair of linear constraints:

$$X_4 - 500 \leq M_8 K_2 \tag{40}$$

$$X_5 - 500 \leq M_9(1 - K_2) \tag{41}$$

$$K_2 = 0 \text{ } or \text{ } 1 \tag{42}$$

where $M_8 = \min(7000/4, 7000/2) = 1750$ and $M_9 = \min(7000/5, 7000/3) = 1400$.

Up to three products will be produced. The following constraint ensures this condition:

$$Y_1 + Y_2 + Y_3 + Y_4 + Y_5 \leq 3 \tag{43}$$

At least one of product 1 or product 2 must be produced in order to produce product 3 or product 4, respectively. The following constraints can be used to satisfy this restriction.

$$Y_3 \leq Y_1 + Y_2 \tag{44}$$

$$Y_4 \leq Y_1 + Y_2 \tag{45}$$

The sign and integer constraints are indicated as in Eqs. (46) and (47), respectively.

$$X_1, X_2, X_3, X_4, X_5 \geq 0 \tag{46}$$

$$Y_1, Y_2, Y_3, Y_4, Y_5 = 0 \text{ } or \text{ } 1 \tag{47}$$

As a result, the following IP is formulated for the given problem.

$$\begin{aligned} \text{Max } Z = {} & 50X_1 + 80X_2 + 70X_3 + 85X_4 + 90X_5 - 15,000Y_1 \\ & - 35,000Y_2 - 50,000Y_3 - 25,000Y_4 - 40,000Y_5 \end{aligned}$$

Subject to:

$$5X_1 + 3X_2 + 7X_3 + 4X_4 + 5X_5 \leq 7000$$

$$4X_1 + 4X_2 + 5X_3 + 2X_4 + 3X_5 \leq 7000$$

$$X_1 \leq 1400Y_1$$

$$X_2 \leq 1750Y_2$$

$$X_3 \leq 1000Y_3$$

$$X_4 \leq 1750Y_4$$

$$X_5 \leq 1000Y_1$$

$$500 - X_1 - X_3 \leq 500K_1$$

$$750 - X_2 - X_4 \leq 750(1 - K_1)$$

$$X_4 - 500 \leq 1750K_2$$

$$X_5 - 500 \leq 1400(1 - K_2)$$

$$Y_1 + Y_2 + Y_3 + Y_4 + Y_5 \leq 3$$

$$Y_3 \leq Y_1 + Y_2$$

$$Y_4 \leq Y_1 + Y_2$$

$$X_1, X_2, X_3, X_4, X_5 \geq 0$$

$$Y_1, Y_2, Y_3, Y_4, Y_5 \in \{0, 1\}$$

$$K_1, K_2 \in \{0, 1\}$$

Solution

The results obtained after solving the model using the Pyomo library are as follows:

$$\text{Max } Z = 111500, X_2 = 1400, X_4 = 700, Y_2 = 1, Y_4 = 1, K_1 = 1, K_2 = 1$$

According to the results obtained, ATK-White will achieve the maximum profit if it produces 1400 units of the second product and 700 units of the fourth product. In this situation, the company will use all 7000 labor hours and all 7000 units of raw materials during the planning period. In addition, regarding the agreement of the company with a customer, which required a total of at least 500 units of the first and third products or 750 units of the second and fourth products, the need can be met from production of 2100 units of the second and fourth products. Furthermore, since the production of the fifth product is not realized, the situation that a maximum of 500 units of the fourth product should be produced has not been met. Moreover, since the second product is produced among the first and the second products, fourth product could be produced.

2.2 *Example: Warehouse Location*

A cargo company wants to locate warehouses for their operations in the Asian side of Istanbul. A warehouse costs $30,000. The company can open one warehouse in a district and can serve that district and neighbor districts. Please see Fig. 1 for the districts in the Asian side of Istanbul. For instance, if a warehouse is located in Beykoz, then it can serve to Uskudar, Umraniye, Çekmekoy, Sile, and Beykoz. The company should provide service to all districts. The aim of the company is to find the minimum cost of locating warehouses to serve all the districts. Please formulate and solve an IP for this purpose.

IP Formulation

Notice that this problem is an example of a set covering problem.

Index:

i : *Districts where a warehouse can be located* (i = 1, 2, ...,13)

Decision Variables:

x_i : Binary variable to show if a warehouse is located at district *I* (*i* = *1, 2, ...,13*). See IDs of the districts in Table 4.

$$x_i = \begin{cases} 1 & \text{If aware house is located at district } i \\ 0 & \text{otherwise} \end{cases}$$

Parameters:

c_i : *Cost of locating a warehouse in district i.*

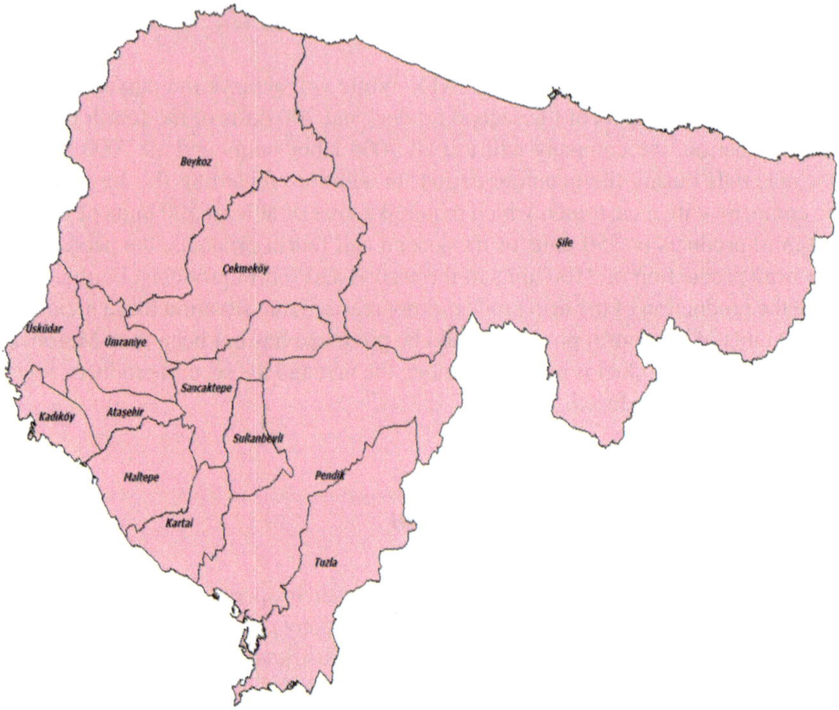

Fig. 1 Map of Asian side in Istanbul

In the given problem, $c_i = 30,000$ for all districts.

Objective Function:

$$\min \ Z : \sum_{i=1}^{13} c_i x_i \tag{48}$$

Subject to:

$$x_1 + x_4 + x_6 + x_8 + x_{12} + x_{13} \geq 1 \tag{49}$$

$$x_2 + x_3 + x_9 + x_{12} + x_{13} \geq 1 \tag{50}$$

$$x_2 + x_3 + x_7 + x_8 + x_9 + x_{12} \geq 1 \tag{51}$$

$$x_1 + x_4 + x_6 + x_8 + x_{13} \geq 1 \tag{52}$$

$$x_1 + x_4 + x_5 + x_6 + x_8 + x_9 + x_{11} \geq 1 \tag{53}$$

$$x_3 + x_5 + x_7 + x_8 + x_9 + x_{10} + x_{11} \geq 1 \tag{54}$$

$$x_1 + x_3 + x_5 + x_6 + x_7 + x_8 + x_{10} + x_{12} \geq 1 \tag{55}$$

$$x_2 + x_3 + x_7 + x_9 \geq 1 \tag{56}$$

$$x_5 + x_7 + x_8 + x_{10} \geq 1 \tag{57}$$

$$x_7 + x_{11} \geq 1 \tag{58}$$

$$x_1 + x_2 + x_3 + x_8 + x_{12} + x_{13} \geq 1 \tag{59}$$

$$x_1 + x_2 + x_4 + x_{12} + x_{13} \geq 1 \tag{60}$$

$$x_i \in \{0, 1\} \quad i = 1, 2, \ldots, 13 \tag{61}$$

All constraints between Eqs. (49) and (60) represent the situation, where each of the 14 locations receives warehouse service from at least one location. The last constraint ensures that the decision variables are binary. Notice that instead of writing the constraints given Eqs. (49)–(60) one by one for each district, the following closed form can be used by defining a parameter n_{ij} as a binary parameter. n_{ij} gets a value of 1 in Eq. (62) if districts i and j are neighbors (See Table 4 for the values of n_{ij}).

$$\sum_{j=1}^{13} n_{ij}x_j \geq 1 \quad i = 1, 2, \ldots, 13 \tag{62}$$

Solution

The given model is solved using the Pyomo library. x_7, x_8, and x_{13} values are found to be 1. The model allowed to locate three warehouses in locations. The total cost of building warehouses is \$90,000 dollars. Figure 2 shows the places where the warehouse will be located (Pendik, Sancaktepe, Uskudar). As can be seen in Fig. 2, all districts have at least one warehouse located in itself or its neighbor.

Table 4 Neighborhood relations of districts on the Asian side of Istanbul

ID		1	2	3	4	5	6	7	8	9	10	11	12	13
1	Atasehir	1			1		1		1				1	1
2	Beykoz		1	1						1			1	1
3	Çekmekoy		1	1				1	1	1			1	
4	Kadıkoy	1			1		1							1
5	Kartal					1	1	1	1		1			
6	Maltepe	1			1	1	1		1					
7	Pendik			1		1		1	1	1	1	1		
8	Sancaktepe	1		1		1	1	1	1		1		1	
9	Sile		1	1				1		1				
10	Sultanbeyli					1		1	1		1			
11	Tuzla							1				1		
12	Umraniye	1	1	1					1				1	1
13	Uskudar	1	1		1								1	1

2.3 *Example: Revised Warehouse Location*

Consider the Warehouse location problem in Sect. 2.2. Suppose the cargo company has an option to build two types of warehouses. A small size warehouse costs $30,000 and can serve up to three districts. A large size warehouse costs $40,000 and can serve up to five districts. The rule for serving its own district and neighbor district is also valid in this case. Each of the districts should be served by at least one warehouse.

For this aim, the cargo company wants to determine the types of warehouses and places to be located with the least cost. Formulate an IP to help the company.

IP Formulation

Indices:

i, j : Districts where the warehouse can be located. $i, j = 1, 2, \ldots, 13$.
k : Type of warehouse. $k = 1, 2$.

Parameters:

c_k : *Cost of warehouse type k.*
n_{ij} : Neighborhood status of district i and district j.

Decision Variables:

x_{ij} : Binary variable to show if warehouse located in district i serves to district j

$$x_{ij} = \begin{cases} 1 & \text{If warehouse located in district } i \text{ serves to district } j \\ 0 & \text{otherwise} \end{cases}$$

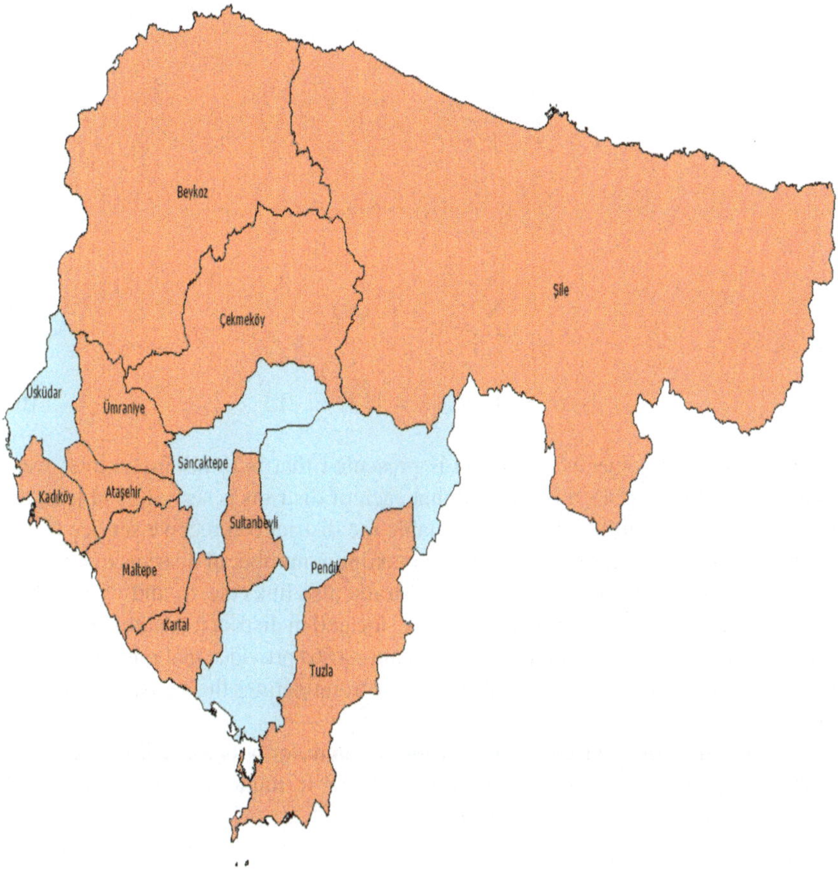

Fig. 2 Solution for the cargo company

y_{ik} : *Binary variable to show if a k-type warehouse is located in the district I*

$$y_{ik} = \begin{cases} 1 \text{ If a } k \text{ - type warehouse is located in district } i \\ 0 \quad \text{otherwise} \end{cases}$$

Objective Function:

$$\min Z : \sum_{i=1}^{I} \sum_{j=1}^{J} \sum_{k=1}^{K} c_k y_{ik} \tag{63}$$

Subject to:

$$\sum_{k=1}^{K}\sum_{i=1}^{I} n_{ij}x_{ij} \geq 1 \quad \forall \ j \tag{64}$$

$$\sum_{j=1}^{J} x_{ij} \leq 3y_{i1} + 5y_{i2} \quad \forall \ i \tag{65}$$

$$\sum_{k=1}^{K} y_{ik} \leq 1 \quad \forall \ i \tag{66}$$

$$x_{ik} = 0, 1 \qquad i = 1, 2 \ldots, 13 \tag{67}$$

In Eq. (63), the objective function is presented that is to minimize total warehouse costs. By Eq. (64), it is ensured that each of districts is served by at least one warehouse. x_{ij} is multiplied by n_{ij} to enable the district can receive service from a neighbor district. Equation (65) limits the maximum number of districts to be served from small and large warehouses. This constraint also links the x_{ij} and y_{ik} variables, that is if a district is served from a warehouse located in district i, a warehouse (small or large) should be located in district i. Equation (66) provides that only one type of warehouse is opened in a district. The last constraint defines the decision variable as binary.

Notice that in the revised version of the problem, only one condition related to types of warehouses is changed; however, the IP formula including the decision variable definitions changes significantly.

Solution

According to the solution gathered by the Pyomo library, $y_{5,2}$, $y_{7,1}$, and $y_{13,2}$ values are found to be 1. Moreover, $x_{5,5}, x_{5,6}, x_{5,7}, x_{5,8}, x_{5,10}, x_{7,3}, x_{7,9}, x_{7,11}, x_{5,5}, x_{13,1}, x_{13,2}, x_{13,4}, x_{13,12}$, and $x_{13,13}$ are equal to 1.

These results indicate that three warehouses, two large and one small, will be built with the total cost of $110,000. Figure 3 shows the places where the warehouse will be located.

A large size warehouse will be located in Kartal, and it will serve to Kartal, Sancaktepe, Sultanbeyli, Maltepe, and Pendik districts.

A small size warehouse will be located in Pendik, and it will serve to Cekmekoy, Tuzla, and Sile districts.

A large size warehouse will be located in Uskudar, and it will serve to Uskudar, Kadikoy, Atasehir, Umraniye, and Beykoz districts.

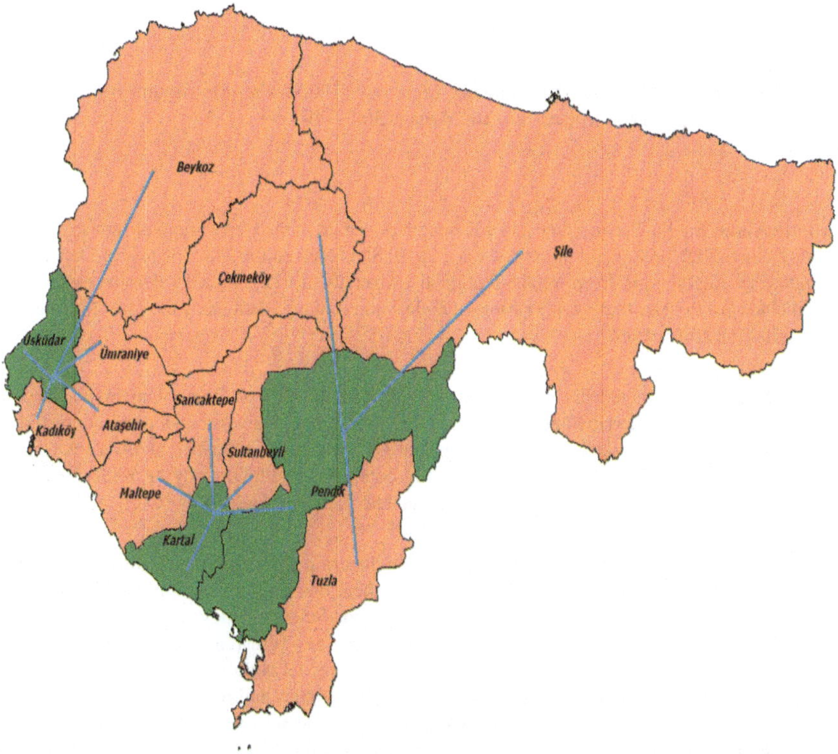

Fig. 3 Solution map of Asian side in Istanbul alternative question

It is interesting to find that the warehouse located in Pendik does not serve Pendik. This may be because the transportation costs are not considered in the proposed model.

In this section, two prescriptive analytics methods, namely LP and IP, are introduced. Notice that there are many other methods in the literature and applied in industry to assist the decision makers to make rational decisions based. To give the idea of LP and IP, several examples are provided. There is no specific algorithm to formulate mathematical programs. Mathematical programming formulation skills are gathered through experiences in diverse areas of applications. The readers are advised to review more examples from the literature in order to get advanced in LP and IP modeling.

Reference

1. Lepenioti K, Bousdekis A, Apostolou D, Mentzas G (2020) Prescriptive analytics: literature review and research challenges. Int J Inf Manage 50:57–70

Nursah Alkan is a Ph.D. candidate and a Research Assistant in the Department of Industrial Engineering at Istanbul Technical University since 2019. She received the MSc. degree in Department of Industrial Engineering from Yildiz Technical University, Turkey, in 2019. Her research interests include fuzzy sets, multi-criteria/objective decision making, data analysis, machine learning and digital transformation.

Kenan Menguc completed his undergraduate education in 2016 at Doğuş University, Department of Industrial Engineering and his M.Sc. in 2018 at Istanbul University, Department of Industrial Engineering. He worked as a lecturer in the Department of Logistics at Beykent University between 2018 and 2020. While he currently conducts his Ph.D. study at Yıldız Technical University, he works as a research assistant at Istanbul Technical University.

Özgür Kabak is a full professor at Industrial Engineering Department of Istanbul Technical University (ITU-IE). Before lecturing in ITU-IE, he spent one year at Belgium Nuclear Research Centre for a post-doc research that was granted by Belgian Science Policy (BELSPO). He teaches undergraduate and graduate education courses in operations research, group decision-making and logistics management. His research explores how to model complex systems such as socio-economic systems, productions systems, supply chains, logistics and transportation systems, etc. and how to make decisions in group decision making environments and incomplete information situations. He has published in indexed journals including European Journal of Operations Research, Information Fusion, Transport Policy, Knowledge-based Systems, IEEE Transactions on Knowledge and Data Engineering, and Socio-Economic Planning Sciences, etc. His work has been featured in international conferences and meetings including MCDM, FLINS, WCTR and EURO-k conference series.

Big Data Management and Technologies

Altan Cakir

Abstract We create and collect more data today than we have in the past. All this data comes from and is reviewed from different sources, including social media platforms, our phones and computers, healthcare gadgets and wearable technology, scientific instruments, financial institutions, the manufacturing industry, news channels and more. When these small and wide data are analyzed, it offers businesses the opportunity to take quick action in business-development processes (B2B, B2C), gain a different perspective, and better understand applications, creating new opportunities. While changing their sales and marketing strategies, businesses are now able to manage the data they collect in real time to transform themselves, to record them in a healthy way, to analyze and evaluate data-based processes, and to determine their digital transformation roadmaps, their interactions with their customers, sectoral diffraction, application and advanced analyses. They want to accelerate the transformation processes within the technology triangle. As a result, big data technologies and applications are at the center of everything and becomes an important application for digital transformation.

1 Introduction

Today we create and collect more data than we have in the past. All data is produced from different sources, including social media platforms, our phones and computers, healthcare gadgets and wearable technology, scientific instruments, financial institutions, the manufacturing industry, news channels, and more [1]. When these heterogeneous data are analyzed, it offers businesses the opportunity to take quick action in business-development processes (B2B, B2C), gain a different perspective, and better understand applications, creating new opportunities. While changing their sales and

A. Cakir (✉)
Department of Physics Engineering, Faculty of Science and Letters, Istanbul Technical University, Maslak, Istanbul, Turkey
e-mail: altan.cakir@itu.edu.tr

Artificial Intelligence and Data Science Research and Application Center, Istanbul Technical University Artificial Intelligence, Istanbul, Turkey

A. Ustundag et al. (eds.), *Business Analytics for Professionals*, Springer Series in Advanced Manufacturing, https://doi.org/10.1007/978-3-030-93823-9_10

marketing strategies, businesses are now able to manage the data they collect in real-time to transform themselves, to record them in a healthy way, to analyze and evaluate data-based processes, and to determine their digital transformation roadmaps, their interactions with their customers, sectoral diffraction, application and analysis. They want to accelerate the transformation processes within the technology triangle for information and knowledge management. Thus, big data is at the center of everything and becomes an important application for digital transformation and technological development of the companies [2].

Digital transformation and big data management help companies embrace change and stay competitive in an increasingly digital world. The value of big data in digital technologies comes from its ability to combine both in an organization's efforts to both digitize and automate its business operations. In this section, big data solutions that increase efficiency, encourage innovation and support new business models in different industry segments will be demonstrated and discussed in the context of big data management and technological evolution.

2 Big Data Technologies

Big data and technologies are becoming a topic that has started to gain importance in the last decade. The amount of data collected (health sector, social media, smart cities, agricultural practices, finance, education, etc.) have reached a very large scale. Establishing a data-based information network needed specifically for business intelligence, especially for scientific developments, is possible by processing data with different structures produced from various sources.

In this context, by the end of the 1950s, when the computer was invented, computer technology has progressed rapidly in parallel with the software technologies that enable the hardware elements that make up the computer to communicate within themselves, and today it has become a topic that has begun to affect many areas significantly. A process in which an ecosystem is created that works in harmony with the systematic implementation of decision-making mechanisms based on a certain logic, algorithms, and the configuration of computer operating systems that we all actively experience, are actively used in many areas today. The topics that are important for big data emerge as searching, identifying, storing, distributing, analyzing, and visualizing data, respectively. Therefore, the technological developments emerged in the last 15 years deeply affect the science and business world. For this reason, the following main topics are:

1. The use and impact of computer and IoT hardware technologies in big data
2. Development of software technologies, compatibility with operating systems and use of open source code
3. Artificial Intelligence technologies, algorithms, big data analytics

will be discussed in a completely active-application oriented structure centered on the introductory applications of big data in terms of today's emerging technological developments.

Although each of the related topic titles contains theoretical, technical and applied developments that contain a lot of details and developments in their own way, in this subsection, references will be provided, specific to a use that exists only when there is interest, and the details of the subject and the mathematical infrastructure are referenced as needed. More will be left to the reader, which will be briefly summarized in the information section. Rather than being a long narrative section describing historical development processes, this chapter, written in an application setup from start to finish, will mainly discuss the use of open source software and the readers will be guided by the book's author in a reasoned manner. A terminological introduction will be made in order for the readers of the book to follow the relevant developments specific to each subject title, and the trends of these subjects will be discussed as of today. All book content has been prepared within the scope of the Istanbul Technical University Big Data and Business Analytics graduate program, and has been created within the framework of all the courses and assignments given in this program.

3 The Big Data Idea: A Methodology of Information Acquisition

The way the human mind thinks in the face of all kinds of situations, analyzes and separates data, associates it with a similar and/or opposite event in the subconscious and makes a decision can actually be associated with the emergence of the methodology of acquiring information today. The emergence of a predictive idea based on a question asked here, a hypothesis, making predictions on it and testing it basically emerges with the idea of bringing the big data idea together. The flow of data created and collected over different sources, for example, provides an empirical observation to form a judgment with the specific reasons for the observation and to understand the implications by bringing together the relevant data. For example, over an image where hundreds of people swimming at the same time can be watched at the same time on the beach, it can be predicted under which conditions people exhibit risky behaviors, by evaluating their age groups, demographic characteristics, distance from the beach, weather conditions, and environmental effects. This example can be extended in terms of large scale manufacturing, e-commerce, and finance applications.

In this case, together with the previously experienced information (referring the structured data), the evaluation of the situations obtained from the events that are taking place at the moment (the situations arising from human behavior such as weather, boat and/or speedboat—unstructured data) data will provide very important inputs for the understanding of the relevant empirical observation. Continuously gathering and examining this data flow will enable early intervention in case of a possible dangerous situation and at the same time prevent the occurrence of an undesirable situation. At this stage, the volume and speed of the collected data stream and the reliability of the received data are important for understanding the situation. Here, the fact that the data is constantly changing, and at the same time, bringing together the weather and other human factors, and examining the whole data flow together, actually gives us a simple idea about big data.

In a study conducted in 2014, the average daily Internet data flow was determined as 1826 Petabytes (PB) [1], and this increased to 2.5 quintillion (!) bytes, or 2.5 Exabytes (EB), in 2018 with the advancing technologies reached [3]. Previously, it was observed that the amount of data produced according to the research, whose estimate was announced by the International Data Institution, doubled every year. However, it has been stated that 90% of the amount of data produced all over the world has been produced in the last few years, and Google is currently processing 40,000 searches per second, or 3.5 billion searches per day [2–4]. Facebook users upload 300 million photos, make 510,000 comments, and share 293,000 status updates every day [4]. In the face of these striking examples, without further examples, it would not be wrong to say that the amount of data produced daily has reached quite shocking levels as of today. As a result, at this level, understanding and analyzing data sets and extracting scientific and/or industrially useful information emerge as a very important topic in this decade.

With advanced data analysis techniques, it is possible to transform big data into intelligent data for the purpose of obtaining important information over large data sets [4–6]. Thus, these processed intelligent data structures provide ease of use in scientific and industrial frameworks, and improve decision-making skills by increasing the ease of taking action. For example, in the healthcare sector, analytical studies can be performed to include all the records of the relevant patient over large datasets (electronic health records and data provided by Clinical decision systems) and the demographic structure similarity of the relevant patient in a feasible and affordable framework. It is expected that similar diagnosis and treatment practices should be compared and the situation will make negative and positive behavior patterns predictable. In this case, it is possible to make quick decisions on all recorded data in accordance with the instantaneous data flow and to know the expectations against the treatment type clearly. In this case, it is very difficult to implement the five characteristic definitions included in big data definition with traditional analysis techniques, high volume, low uncertainty, high data flow rate, multivariate variability, high accuracy [7]. In addition, the characteristics of big data have become increasingly diverse over time. These are, respectively, variability, data flow resistance (viscosity), validity, viability, and ethic (virtue). Various artificial intelligence techniques, machine learning, natural language processing, cognitive intelligence and data mining are structures designed on the basis of more sensitive, more accurate and rapid examination of data structures with complexity on a large scale.

The purpose of these advanced analysis techniques is to extract information from large data sets, discover hidden structures and extract unknown correlations [8]. For example, relying on a patient's previous knowledge to detect a contagious disease that may have a high initial impact will enable the disease to be treated and/or a more optimal treatment plan to be implemented [9]. In addition, it can benefit from complex simulations on a large scale to improve decision-making abilities in risky business decisions (entering a new market and/or launching a new product) [7–10].

The first thing perceiving through this simple example is to show that the definition of big data is basically 3 main topics.

Considering the definition of big data within the framework of data volume, a few petabyte levels from a few terabytes (TB) levels are generally considered as large data volumes today. This situation changes every year and it is observed that it is in an increasing trend. However, today, a few gigabytes (GB) of data volume is not considered big data. In principle, this amount of data, which can be processed efficiently and quickly by a well-equipped computer, is generally not suitable for big data applications.

Youtube, Facebook, etc. within the framework of various sampling. Companies when the data streams produced by research institutions such as CERN, NASA, are examined, it is seen that the data volume is created in a very large (100 TB–500 EB) range, and the understanding of this data is an important scientific and business intelligence development. It is known to be used in breeding strategies. With the increase in data, the importance of hard drive technologies where this data will be stored should be evaluated separately. Evaluating hardware processes in addition to software technologies is also important for big data applications. In particular, AWS, etc., which is taken as a service without setting up infrastructure. In services, choosing the most efficient equipment for work or research becomes an important topic.

Applications that enter our daily lives in parallel with the developing software and hardware technologies show significant transitions at certain time intervals, and the development of these related processes can be understood with their development. This situation will be discussed in detail in the second part.

4 Big Data Management and Business Applications

Modern business world requires using information as the important resource in increasing competition (Fig. 1). Organizational success can only be achieved by managing information, transforming it to knowledge and using the knowledge [11]. In Digital Transformation age, generating data-based strategies will provide optimizing the performance of the companies by collecting and analyzing data through the product life cycle [12].

Processing data is not a new issue for the companies. Companies have systems for storing and processing the data in order to create value for various business purposes. On the other hand, most of them have traditional data warehouse technology that is not suitable for extremely high speed data from various sources [8]. IoT enables the enlargement of collectible data, big data technologies provide data to be processed and transformed into knowledge. Big data generally described as technology that enables to obtain information from high speed data from different sources

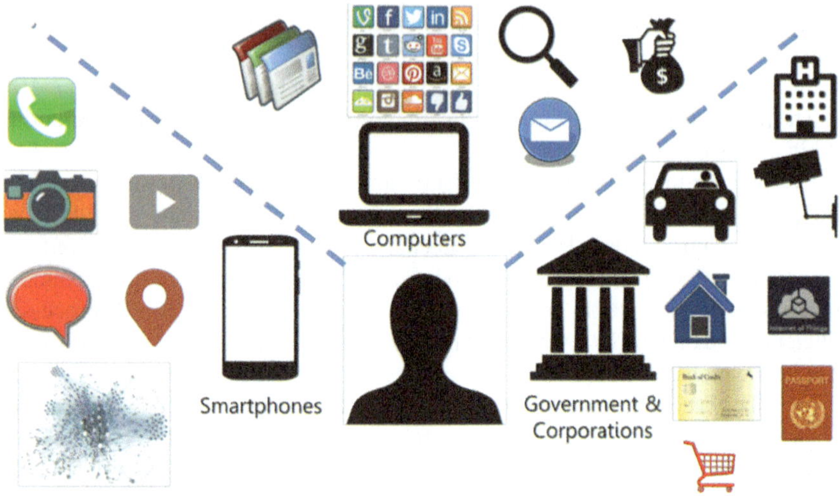

Fig. 1 Heterogenous data flow from various sources

by applying statistics, mathematics, econometric, simulations, optimizations, and/or other techniques to support decision-making processes [12]. Therefore, big data technologies and deployment of data lake are considered in companies now instead of data warehouse solutions. Managing all types of data, studying specific analytical studies by multiple users, providing the required data lineage back to source systems and enabling users to access the analytical contents in a self-serve manner are the necessities of deploying the data lake [10].

Advanced data analysis techniques, such as machine learning and deep learning, can be used to transform big data into usable content for the purposes of obtaining critical information in the business [8]. In this manner, refined and optimized data provides useful information and improves decision-making qualities for wide range of applications. A common example can be given health sector: analytical outputs upon these datasets may enable faster decisions for health care practitioners to deliver effective and affordable solutions. This is a very practical operational help in comparison to relying on evidence provided with strictly localized or current data. Big data analysis is difficult to handle using traditional data analytics [13] as they can lose effectiveness due to the five V's characteristics of big data: high volume, low veracity, high velocity, high variety, and high value. Moreover, many other characteristics exist for big data such as variability, viscosity, validity, and viability [7]. Several artificial intelligence (AI) techniques, such as machine learning (ML), natural language processing (NLP), computational intelligence (CI), and data mining were designed to provide big data analytic solutions as they can be faster, more accurate, and more precise for massive volumes of data [14]. The aim of these advanced analytic techniques is to discover information, hidden patterns, and unknown correlations in massive datasets [14]. For instance, a detailed analysis of clinical trial or patient data could lead to the detection or prevention of many diseases at an early

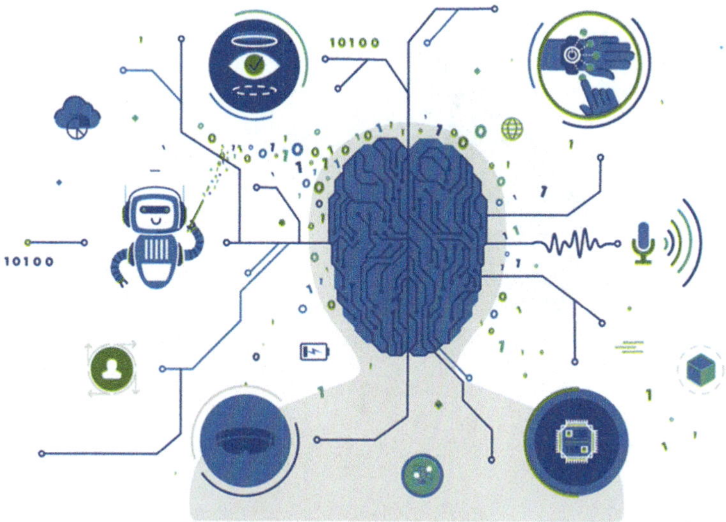

Fig. 2 Smart decision-making algorithms with various data formats and applications in the context of AI. https://www.educba.com/future-of-artificial-intelligence/

stage, thereby enabling either a precaution or more effective treatment plan [7, 8]. Additionally, risky business decisions (e.g., entering a new market or launching a new product) can get better options from simulations that have better decision-making skills [10] (Fig. 2).

5 Big Data Characteristics

In May 2011, big data term was established as the next frontier for business, innovation, and competition. In 2018, the number of Internet users grew 7.5% from 2016 to over 3.7 billion people [3] and extended the large scale orders with COVID-19 in worldwide. In 2001, the usual characteristics of big data were accumulated with three V's (Volume, Velocity, and Variety) [14, 15]. Similarly, it is extended by International Data Corporation to four V's (Volume, Variety, Velocity, and Value) in 2011 [15]. In 2012, Veracity was introduced as a fifth characteristic of big data [16–18] in the context of advanced analytics features. While many other V's exist [7], the five most common characteristics of big data, as next illustrated in Fig. 3 are focused in this study. On the other hand, Virtue term (the ethics of big data usage)—need to be addressed in light of all the regulations for data privacy and compliance—is deserves to be cited.

The first of all, *Volume*, a unique characteristic term for big data, refers to the massive amount of data generated in time and applies to the data size and scale of a corresponding dataset. It is hard to describe a universal threshold for big data volume

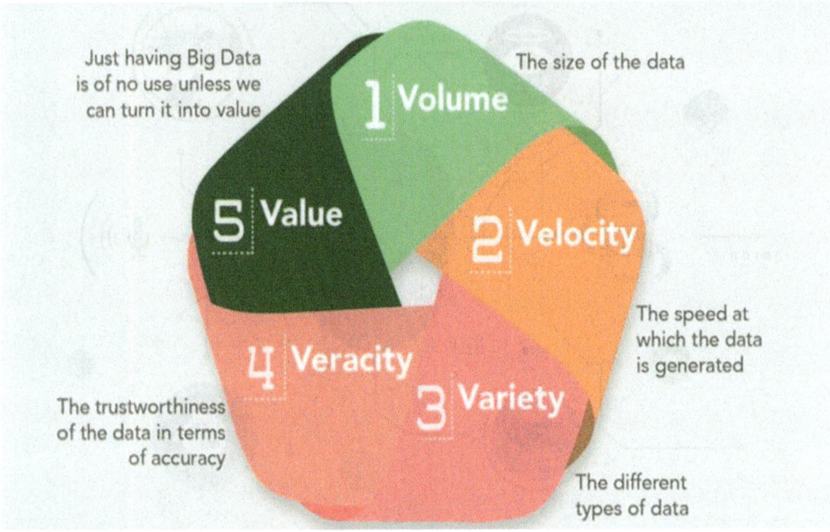

Fig. 3 Common big data characteristics

(i.e., what constitutes a "big dataset") since the time and type of data can influence its definition [18]. Currently, datasets that reside in the Exabyte (EB) or ZB ranges (for many companies a couple of hundred TB's) are generally considered as big data [18–20], however various orders still exist for datasets in smaller size ranges. For example, Walmart collects more than 2 PB from over a million customers every hour [19]. These numbers of data size can introduce scalability and applicability problems (e.g., a database or analytics tools may not be able to accommodate infinitely large datasets). Many existing data analysis techniques are not designed for large scale databases and can fall short when trying to scan and understand the data at scale.

Velocity, the second term, refers to the speed (contains of batch, near real-time, real-time, and streaming processing) of data processing, emphasizing that the data processed must meet the speed with which the data is produced [11]. For example, Internet of Things (IoT) devices, i.e., health, manufacturing, etc., continuously produce large amounts of sensor data. If the device causes delays or latency in processing the data and sending the results to clinicians may result in unexpected problems for operations (e.g., a pacemaker that reports emergencies to a doctor or facility) [18]. In the same manner, devices in the cyber-physical systems often rely on real-time nature enforcing strict timing standards on execution, and as such, may encounter similar issues when data provided from a big data application fails to be delivered on time.

The third term, *Variety*, refers to the various forms of data in a corresponding dataset types including structured data, semi-structured data, and unstructured data. Structured data (e.g., stored in a relational database) is mostly well-organized and easily sorted, but unstructured data (e.g., text and social media content) is unorganized and difficult to analyze directly. Semi-structured data (e.g., NoSQL databases) contains different elements to separate tags in operations [20], but manipulating this structure is completed to the database user. These manipulations can manifest when converting between these data types (e.g., from unstructured to structured data), in representing data of mixed data types, and in changes to the underlying structure of the dataset at run time. From the point of view of variety term, traditional techniques for analytics operations and algorithms face challenges for handling multi-modal or high dimensional data, incomplete and noisy data. As is known, data mining techniques are designed to consider well-formatted structured data, which may not be able to deal with incomplete and/or different formats of input data formats.

Advanced analyses of various forms of data can be challenging, as the data under observation comes from heterogeneous input sources with representations. For example, daily operations of company databases are negatively influenced by inconsistent, incomplete, and noisy data. Therefore, a number of data preprocessing techniques, including data cleaning, data manipulation, data integrating, and data transforming used to remove these issues from data [21]. Data manipulation techniques address data quality and noise problems resulting from variety in big data operations. Removing unnecessary information (i.e., cleaning, refinement, manipulation) during the analysis process can significantly enhance the performance of data analysis. Furthermore, data refinement for error detection and correction is facilitated by identifying and eliminating mislabelled training samples, ideally resulting in an improvement in classification accuracy in machine learning and deep learning methods [22].

Veracity represents the correctness of the data (e.g., quality level data collection or imprecise data). For example, technology companies estimate that poor data quality costs the economy more than a trillion $ per year [16]. Data quality, in general, can be inconsistent, noisy, ambiguous, or incomplete, therefore data veracity can be categorized as good, bad, and undefined. Due to the diverse sources and heterogeneous data formats, accuracy, and trust become more difficult to establish in big data analytics. For example, a person, who represents a company operation, may use Twitter to share official corporate information but at other times use the same account to express personal opinions, causing problems for refinement to work on the Twitter dataset. Furthermore, when analyzing millions of health care or manufacturing machines records to determine or detect disease trends and predictive maintenance operations, for instance to mitigate an outbreak that could impact many people, any ambiguities or inconsistencies in the dataset can interfere or decrease the precision of the analytics process [16].

Value represents the context and usefulness of data for decision-making, whereas the prior V's focus more on representing challenges in big data. As shown in Fig. 4, there are five key elements in the Value Chain. For example, big companies, such as Amazon, Google, Facebook, have boosted the value of big data via analytics in their diverse product segments. Amazon uses large datasets of users and their interest to provide product insight, thereby increasing sales and user participation. Google uses location data fromAndroid mobile phones users to improve location services in Google Maps operations. Facebook collects users' activities to provide targeted advertising and friend suggestions. These three big companies are getting

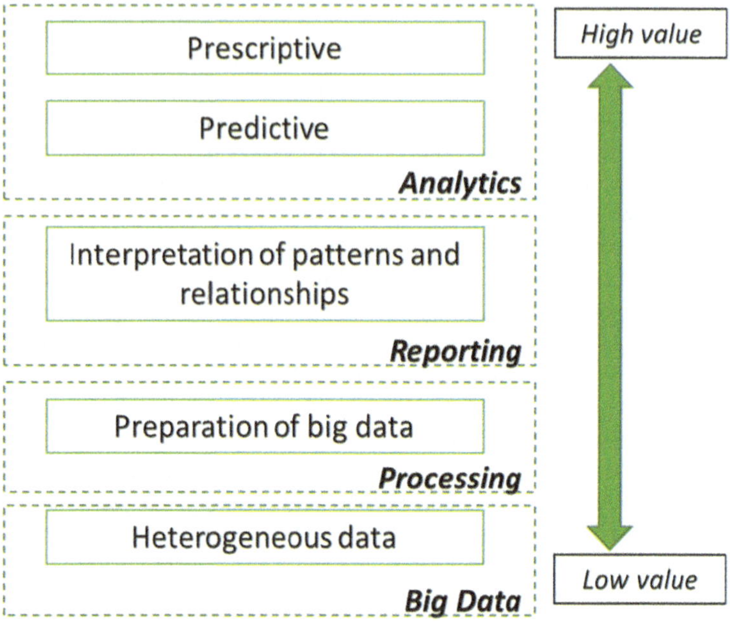

Fig. 4 *Value* represents the context and usefulness of data for decision-making

massive scale by examining these individual datasets of raw data and retrieving useful information to make better business decisions [23].

6 Big Data Analytics

Big data analytics refers to the process of large scale datasets to retrieve patterns, unknown correlations, market trends, user behaviors, and other valuable information at scale [24]. With the general framework of the big data's five V, analysis techniques needed to be reconsidered to get over their limitations on processing in terms of time limitations. Business revenues for utilizing big data are growing in the modern world of digital data and corresponding technologies. The global annual growth rate of big data technologies and services increased about 36% between 2014 and 2019, with the global income for big data and business analytics anticipated to increase more than 60% [25].

Several AI techniques and advanced algorithms together with big data analytics (i.e., ML, data mining, NLP, and CI) and potential applications such as parallelization, divide-and-conquer, transfer learning, sampling, granular computing, feature selection [13], and instance selection can transform big bulk problems to easy-applicable problems and can be used to make effective decisions, reduce costs, and enable more operational knowledge management.

In the context of big data analytics, parallelization reduces computation time by splitting large datasets, as a result smaller problems, into smaller instances of itself and performing the smaller tasks at the same time (e.g., distributed computing for sharing cores, GPU's or/and processors). Parallelization does not decrease the amount of work performed but rather reduces computation time for real-time nature as the small tasks are completed at the same point in time instead of one after another sequentially [26].

This splitting strategy plays an important role in processing big data. Splitting and distribution of the work consist of three phases: (1) reduce the problem size into several smaller parts, (2) complete the smaller sets, where the solving of each small problem contributes to the solving of the large part, and (3) incorporate the solutions of the smaller problems into one large scale. For many years distributed strategy has been used in large scale databases to manipulate data in groups rather than all the data at once [27].

Incremental learning, which is dedicated to streaming data, is trained only with new data rather than only training with existing data. Due to time dependent incremental adjustment of parameters in the data the process is for the learning algorithm over time according to each new input data and each input is used for training only once [26]. That is the strong part of the incremental learning, especially time dependent business processes.

Sampling method is generally used as a data reduction technique for big data analytics for extracting patterns in large data sets by choosing, manipulating, and analyzing a subset of the data [26]. Many researches indicate that obtaining effective results using sampling methods depends on the data sampling conditions/criteria with respect to corresponding use-case [26].

Feature selection is a critical term of analytical investigation, which is a conventional approach to handle big data with the purpose of targeting a subset of indirect features for an aggregate but more precise data representation [28, 29]. Feature selection is a very useful strategy in data science for analyzing high-scale data [28].

7 Architectural Designs

7.1 Big Data Architecture for Manufacturing Processes: An Intelligent Maintenance

Manufacturing industries, technical machine/equipment providers have been working with a growing interest in implementing artificial intelligence-based maintenance solutions [30–32]. This is due to various reasons, such as digital transformation, changing or distributed design of automation systems (new generation of manufacturing execution systems), advanced data acquisition techniques, transformation and storage, as well as the recent development of machine/deep learning algorithms integrated machines. The need for end-to-end solutions for data-driven decision systems enabling intelligent solutions extends over many different fields of industrial applications. This application provides end-to-end solutions, which are expected to integrate all of the necessary building blocks, starting from data acquisition and optimization together with distributed storage nature, onto real-time streaming and data warehousing, through data processing, analytics and visualization of condition monitoring data up to smart manufacturing decision support.

The field of smart data-driven business applications for intelligent manufacturing ranges from simple condition monitoring, especially health data-taking/data acquisition through fault identification, up to fault detection and diagnosis. In this manner, the terms "Predictive/Preventive Maintenance" are getting a difficult application for factories due to a very high variability of operating conditions with respect to the lack of availability of segmented data which is representative of all of these conditions for operations. However, rule-based automation/database oriented solutions is inadequate to find out faults or degradation in complex systems.

Therefore, the development, especially software tracks possible open source solutions, of an end-to-end architecture that enables highly effective solutions of business-specific data analysis as well as general insight of overall operations, requires the collaboration of domain experts, big data engineer experts, and data science experts. The major challenge is how to determine a big data architecture for intelligent solutions, such as Kappa architecture or Lambda architecture that enables efficient searchable and real-time processing as well as data science applications to see the health of the operation, identify useful/harmful life of critical components during the operation, including finance, supply chain and ERP.

In the following use-case an example for such an end-to-end intelligent manufacturing system, which has been developed for industrial plastic injection manufacturing machines is presented. The example of big data architecture enables efficient data storage, data streaming, query processing and data analytics for various data-driven business intelligence applications.

The use-case makes the following example in detail:

A detailed manufacturing application of an end-to-end IoT solution for predictive maintenance is presented. In this respect the following terms will be discussed with respect to use-case: big data architecture, storage design, analytic pipeline or machine learning applications [32–36]. The example that is proposed here can be used as reference for similar IoT-based manufacturing systems in diverse application fields. In addition, similar approach can be also used for wider applications such as finance, e-commerce.

A scalable big data analytics pipeline based on a lambda architecture, which integrates heterogeneous platforms/services across entire systems is presented in Fig. 5.

The application enables near real-time and/or discrete interval-based machine manufacturing company in Turkey. A detailed evaluation of different big data system operations (software/hardware) such as distributed storage and computing to enable efficient, multi-dimensional processing is performed. The main challenges and

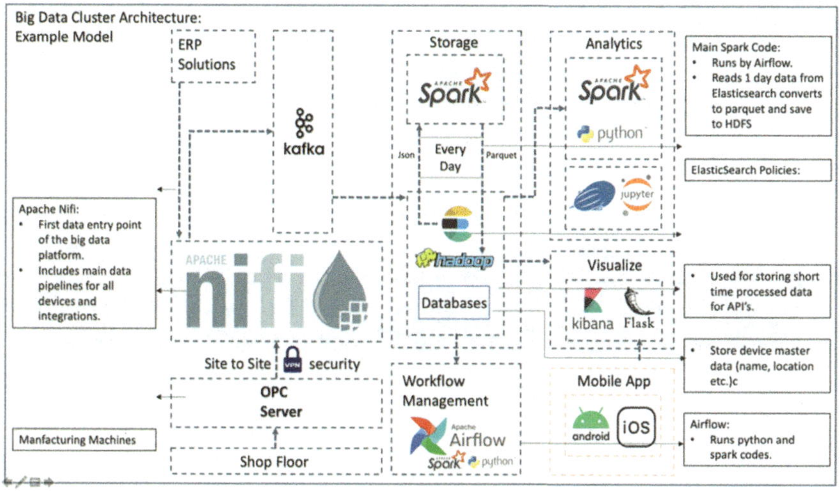

Fig. 5 Big data cluster architecture for manufacturing applications

lessons learned from implementing the sensor data analytics pipeline within an industrial setting using state-of-the-art big data applications, such as open source development framework is shown. Finally, the possibility of time series integrated energy consumption rate together with auto-encoder prediction technique is discussed in the operation. In particular, both statistical methods and deep learning are applied in a real-time nature.

7.2 *Big Data Infrastructure Design and Definitions—Fundamentals*

Big data cluster may build on $2n + 1$ (redundant approach) physics or virtual machines with various technical features. The machines in the cluster totally have certain number of CPUs, in case of necessity GPU's, large volume of RAM, and storage, especially SSD type of disk space. The distribution of the hardware was made by considering the requirements of the big data system tools to be installed and the amount of data to be processed at a certain time interval. In this context, a namenode and a secondary namenode are used primarily to ensure high availability in the Hadoop file system. The task of the secondary namenode is to keep the replica of the system records and to ensure that the system continues without interruption in case a problem occurs in the namenode. After HDFS and MapReduce configurations were made, YARN and SPARK installed on all nodes where the HDFS system exists. In this way, it is aimed to use the power in the entire cluster in the big data analytics to be made over in-memory-computation model aka Apache Spark approach. Example model for the applications can be shown as an example in Fig. 6.

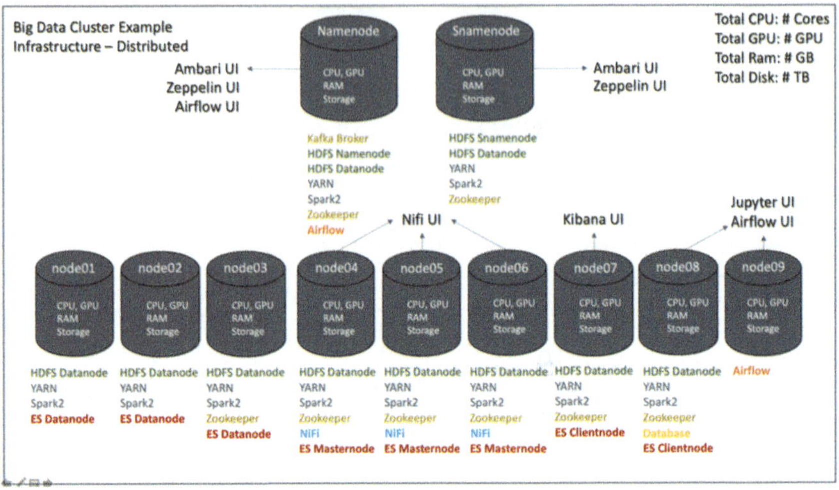

Fig. 6 Big data cluster example with potential hardware solutions

7.3 *Big Data Architectural Components*

AMBARI

Big data platforms consist of many open source software working synchronously on a distributed architecture. General perspectives on these systems that it is not easy to install and maintain these open source components individually. In other words, it is difficult to keep these systems alive in the production environment and to follow the problems in the systems. The hortonworks data platform was developed in 2011 by the YAHOO team responsible for the HADOOP Project. This platform provides many facilities for installing and managing big data systems over a single interface (includes software version control and maintenance) which is called as Ambari. Thanks to Ambari, big data system components can be installed easily and effectively on the cluster and at the same time, it is possible to monitor the problems in the system with monitoring features.

Example big data cluster in Fig. 6 is designed by installing HDFS, MapReduce, YARN, Spark, Zeppelin, Kafka and Hive software by using Ambari interface. The example system is based on eleven nodes and a secondary master node is added to ensure sustainability. A hadoop file system, which can be accessed through the Ambari interface, has been installed, allowing easy access to the folders stored in hdfs. Thanks to the Ambari interface, upload and download operations can be performed on the file system established. Necessary configurations of the specified software are made through the interface. After installation phase, the Ambari is also used to monitor the system status, and information about the number of nodes, storage space, network and CPU usage and memory usage were visualized via the system monitoring dashboard.

Apache NIFI

Apache NiFi was formerly developed by the National Security Agent (NSA) and is based on open source Niagara Files software as part of the technology transfer program in 2014. It is an open software for automating and monitoring network data movement between heterogeneous systems. Niagara files (project) aimed to address the issues of data flow in the context of security, interactivity, fault tolerance, scalability, routing, data transformation, conversion, encryption, governance, provenance, buffering with back-pressure, prioritized queuing, clustering, and many more [37]. It can analyze different data formats and protocols and operates on Java Virtual Machine [37, 38]. NiFi implements the concept of Flow Based Programming (FBP) as a real-time ETL for collecting large amounts of data in a distributed environment. Generally, ETL means to extract data from the same model or other model to a data warehouse or data lake in the case, transform the data to save the data in appropriate format or structure for inquiry or analysis. In ETL or data in motion, the major concern that how to secure, track and manipulate the data flow. Therefore, Apache NiFi, which is the data automation platform, is capable of handling milions/bilions of records to ingest and enrich data with near real-time in distributed environment. In addition, Apache NiFi has been able to utilize user friendly web UI with a lot of ready-use processors. Processors is doing some combination of data routing, transformation, or mediation between systems by accessing the incoming flow files and its contents. The Processor can also operate independently without the flow files [38]. In big data approach, there is structured data from SQL Server and unstructured data from MQTT server and this occupation is another issue to handle by dataflow management platform. Apache NiFi supports dataflow management for both structured and unstructured data, powered by separation of metadata and payload. Schema is not required, one can use it as optional if the operation is required for schema.

Apache Kafka

Apache Kafka is a scalable, elastic, fault tolerant and secure manner publish-subscribe messaging system and used for building real-time data pipelines. Kafka can be deployed on bare-metal hardware, virtual machines, and containers, and on-premises as well as in the cloud [39]. Kafka consists of several capabilities; therefore, it can be implemented for end-to-end event streaming. These capabilities are ordered below:

1. To publish (write) and subscribe to (read) streams of events, including continuous import/export of data from other systems.
2. To store streams of events durably and reliably.
3. To process streams of events as they occur or retrospectively.

At LinkedIn, Apache Kafka supports dozens of subscribing systems, and delivers more than 55 billion messages to consumers daily. According to Kafka approach, a distributed messaging system that it is specially designed for high volume of log events processing. It also provides integrated distributed support and can scale out.

Kafka achieves much higher throughput than conventional messaging systems (such as ActiveMQ and RabbitMQ) [40].

According recent analyses show integrating with tens or hundreds of thousands IoT devices and processing the data can be processed in real-time nature without any latency. The corresponding systems generally include Kubernetes, Kafka, MQTT and TensorFlow in a scalable, cloud-native infrastructure to integrate and analyze sensor data from IoT integrated devices in real-time. A proposed solution shows the system performance of Apache Kafka is utilized for stream processing. The result showed that Apache Kafka achieved a higher scalability and faster responses as well as cost reduction compared to traditional systems.

ElasticSearch

Elasticsearch is an open source, real-time and distributed search and analytics engine based on the Lucene library, released in February 2010 [41, 42]. It is used to store, search and analyze big data. Elasticsearch stores data as Javascript Object Notation (JSON) documents, and retrieves whatever document with query. It is generally used by big companies such as Netflix, Linkedin to web search, log analysis, and big data analytics [42]. In order to understand the working principle of Elasticsearch, the terms index, shard, node, replica, cluster should be known. Index is similar to database in a relational database. Elasticsearch stores data in one or more indexes and reads the data from indexes. Each index must have a unique name and must be created in lowercase letters. Every index is made up of one or more shards [43, 44].

Each Elasticsearch server is called a node, and is similar to a traditional database system. An ES node can be one or more of any 3 of the master node, data node, and client node. Elasticsearch is a distributed system, so can be run on multiple nodes. This collaboration is called a cluster and works like a single server [42]. Elasticsearch cluster allows to process big data easily and quickly. Cluster provides us to access data due to its nature even if some of the nodes are not available [41]. Another feature of the cluster structure is the shards. Shards are smaller pieces of the index, in other words, the data is horizontally split. Each shard is distributed to a different server, thus when a query is sent to the data, the data is quickly retrieved from different shards. Finally, the replica is an exact copy of the shards. Replica is created in order to data is not lost. If one of the shards lost, the data can be accessed from the replica of shard.

In order for Elasticsearch to work fast, the number of shards, nodes and number of replicas should be optimized. The correct cluster setup should be created according to the needs.

Kibana

Kibana is an open source interactive visualization interface and analysis tool for data in Elasticsearch. It is an Elastic plugin. Kibana offers the opportunity to visualize data with different types of graphs such as maps, charts, bars, tables, and pies. [45]. Kibana allows real-time monitoring with the power of Elasticsearch. Users can create dashboards by editing the generated graphics [46]. Kibana presents all created visualizations, analyses and time series to the user with dashboards. It shows the whole picture to the user along with the filters. In the picture, different graphics are shown on the dashboard (Fig. 7).

Besides visualizations, Kibana also offers: Time series: Perform advanced time series analysis on Elasticsearch data with time series interface [46]. Machine learning: Kibana detects anomalies in your Elasticsearch data with unsupervised machine learning models and identifies features. Graphs and networks: Kibana reveals relationships in your data and takes search engine capabilities and combines them with graphics.

To visualize data in Kibana, one needs to create an index pattern in the Kibana application (Fig. 8). Kibana provides a web interface for the management of indexes. These indexes should be created for data analysis and visualization [47]. When

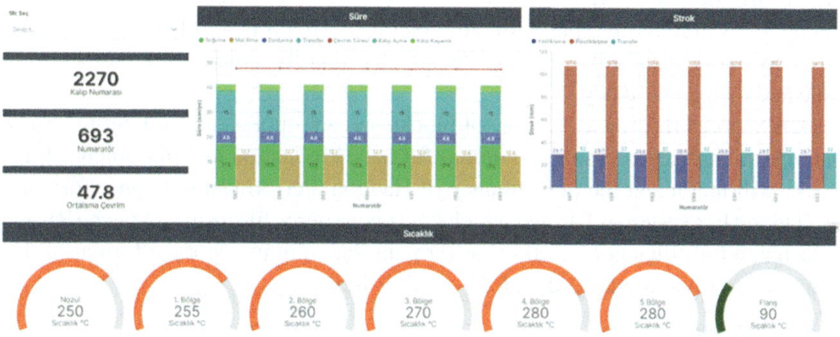

Fig. 7 Kibana dashboard example in the big data cluster data processing

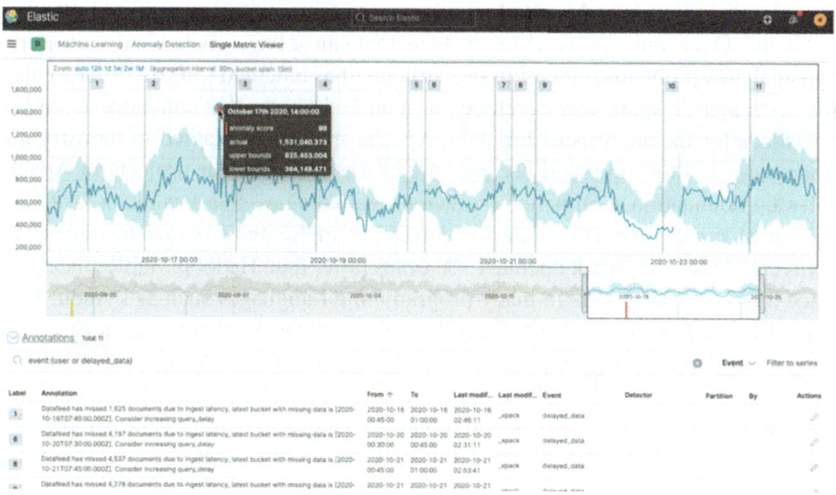

Fig. 8 Kibana time series integrated machine learning deployment example for continues data flow real-time prediction

Elasticsearch is installed, users can be created, defined roles and created space, thanks to the Xpack that comes by default. Objects and dashboards recorded with Kibana spaces can be organized by categorizing them into separate spaces. It offers fields that different users can enter. And also, a user can be created and a role is assigned to authorize certain spaces and the data with the corresponding index is figured out.

Apache Hadoop

Manufacturing data consists of structured, semi-structured and unstructured data. High volume of data is stored in HDFS for batch analysis. Hadoop Distributed File System (HDFS) is one of the key elements in Hadoop Framework and it is an open source implementation of the Google File System (GFS) for storing data. HDFS is a large distributed file system designed to be deployed on low-cost hardware [48, 49] and provides high throughput with fault tolerance. HDFS is widely used in research for its open source and advanced architecture. It primarily settles the massive capacity issue and data consistency.

HDFS stores records as a series of blocks and are duplicated for fault tolerance. Data partitioning, processing, columnar formats, replication and placement of data blocks will increment the performance of Hadoop. Performant columnar format in the Hadoop Ecosystem is Parquet which is an open source file format. Apache Parquet is designed for efficient as well as performant flat columnar storage format of data compared to row-based files like CSV or TSV files [49].

Apache Spark

With the increase in data volumes, the users' need for systems that can perform calculations in a distributed structure has increased. Previously, specialized systems were developed such as MapReduce, Pregel, and Storm which were able to do a single job. Over time, the need for systems that can solve complex and various big data analytics problems with all in one structure has emerged [50]. In order to meet this need, apache spark was developed as a unified engine that can work at various workloads for the analysis of big data. Apache spark was adapted to industry and academia in a short time and became one of the most popular projects in Apache Software Foundation. Compared to previous technologies, Spark is much faster and has a much more compact structure. For example, in the study on logistic regression, it was concluded that spark operates 100 times faster than Hadoop MapReduce [51]. Since Spark provides APIs in many programming Languages such as Python, Java, Scala, R, it makes complex distributed big data analysis operations easier.

Spark provides many impressive high-level tools that can be used in cluster computing such as spark SQL for running SQL queries on spark, MLlib to build machine learning models, Spark Streaming to handle streaming data, GraphX for graph processing [52]. Having these libraries, which have separate very important functions, made spark a general big data processing engine. These tools are developed and reproduced in each version of spark to keep it up-to-date and functional.

A spark application runs on five main components. These components are a driver program, a cluster manager, workers, executors and tasks as shown in Fig. 1. Driver program defines the calculation to be performed using spark as a library [51]. Spark context connects to the cluster manager that will make the necessary resource allocations. This cluster manager can be the manager of the spark itself, as well as managers such as Mesos, YARN or Kubernetes. After connecting cluster manager, spark will have executors on the worker nodes that will process the data and perform the calculations [53].

Apache Airflow

Apache Airflow is an Apache Software Foundation project, originally developed by Airbnb and released in 2016. Apache Airflow is a python-based fast- running, easy-to-integrate and customizable platform to programmatically author, schedule and monitor workflows as Directed Acyclic Graphs (DAGs) of tasks [54]. In this way, both graphically observing and creating workflows create a facilitating effect for processes. At the same time, the created tasks do not need to share resources with each other, so it is possible to run independently from each other. It supports the establishment of workable, repeatable structures of work flow studies. General architecture of Airflow includes three main components which are webserver, scheduler and worker in the context of design.

Workflows are becoming increasingly important to automate more complex tasks in the software development lifecycle, for instance Oozie, Airflow or Azkaban that allow to author workflows that schedule jobs sequentially, in parallel or based on conditions and triggers [54]. Regarding the research and modular needs, Airflow is selected for the pipeline workflow engine. Airflow is very modular and provides many pre-built interfaces (Hooks) to common clouds and database systems such as Amazon S3, Google Cloud, or HDFS among others, and has a modular execution engine for computational tasks (Operators) [55–59].

8 Summary

This part has discussed how big data can impact business applications and technical perspectives, both in terms of analytics and the dataset itself. The aim is to discuss the fundamentals with open source big data analytics techniques, how smart decision-making strategies can impact such various applications, and examine the open issues that remain in business. For each common technique, introductory level relevant research has been summarized to aid others in this community when developing their own techniques. The issues surrounding the five V's of big data have been discussed, however many other V's exist. In terms of existing research, much focus has been provided on volume, variety, velocity, and veracity of data, with less available work in value (e.g., data related to corporate interests and decision-making in specific domains).

In addition to business perspective, this section has reviewed numerous techniques on big data analytics and the impact of big data analytics with one of the use-cases, manufacturing, in terms of application. Figures 5 and 6 summarizes these findings and explain general perspective of lambda architecture (12 nodes in total) including the real-time processing nature. First, each component is categorized as either distributed nature, stream processing, batch processing, and ML-based analytics layers. Figure 7 illustrates how IoT and big data applications in manufacturing impacts on business, both in terms of application later in the operation and the technique itself. Finally, the final part, Fig. 8, summarizes proposed mitigation strategies for each applications challenge and also analytics pipeline. For example, the predictive maintenance example and real-time energy prediction and optimization processes illustrate one possibility for analytics pipeline to be introduced in ML via time series integrated predictive modeling. One approach to extend this specific form of application Fis to use an active learning technique that uses a subset of the data chosen to be the most significant, thereby countering the problem of continuous available training data, which obviously big data approaches help.

Note that each big data characteristic is explained separately. However, combining one or more big data characteristics will occur exponentially more applications, thus requiring even further explanation for use-cases.

References

1. Forbes MB (2018) How much data do we create every day? https://www.forbes.com/sites/bernardmarr/2018/05/21/how-much-data-do-we-create-every-day-the-mind-blowing-stats-everyone-should-read/#4146a89b60ba. Accessed on 28 Nov 2019
2. Ionescu A (2018) Processing petabytes of data in seconds with databricks delta. https://databricks.com/blog/2018/07/31/processing-petabytes-of-data-in-seconds-with-databricks-delta.html. Accessed 28 Nov 2019
3. McAfee A, Brynjolfsson E, Davenport TH, Patil DJ, Barton D (2012) Big data: the management revolution. Harvard Bus Rev 90(10):60–68
4. Pouyanfar S, Yang Y, Chen SC, Shyu ML, Iyengar SS (2018) Multimedia big data analytics: a survey. ACM Comput Surv 51(1):1–34
5. Zephoria. Digital Marketing. The top 20 valuable Facebook statistics https://zephoria.com/top-15-valuable-facebook-statistics/. Accessed Nov 2018
6. Iafrate F (2014) A journey from big data to smart data. In: Digital enterprise design and management. Cham, Springer, pp 25–33
7. Borne K (2014) Top 10 big data challenges a serious look at 10 big data v's. https://mapr.com/blog/top-10-big-data-challenges-serious-look-10-big-data-vs. Accessed 11 Apr 2014
8. Lenk A, Bonorden L, Hellmanns A, Roedder N, Jaehnichen S (2015) Towards a taxonomy of standards in smart data. IEEE international conference on big data (Big Data). Piscataway, IEEE, pp 1749–1754
9. Tsai CW, Lai CF, Chao HC, Vasilakos AV (2015) Big data analytics: a survey. J Big Dat 2(1):1–32
10. Manyika J, Chui M, Brown B, Bughin J, Dobbs R, Roxburgh C, Byers AH (2011) Big data: the next frontier for innovation, competition, and productivity. McKinsey Global Institute
11. Chen M, Mao S, Liu Y (2014) Big data: a survey. Mobile Netw Appl 19(2):171–209
12. Iafrate F (2014) A journey from big data to smart data. Digital enterprise design and management. Springer, Cham, pp 25–33
13. Ma C, Zhang HH, Wang X (2014) Machine learning for big data analytics in plants. Trends Plant Sci 19(12):798–808

14. Saidulu D, Sasikala R (2017) Machine learning and statistical approaches for big data: issues, challenges and research directions. Int J Appl Eng Res 12(21):11691–11699
15. Laney D (2001) 3D data management: controlling data volume, velocity and variety. META Group Res Note 6(70):1
16. Jain A (2017) The 5 Vs of big data. IBM Watson Health Perspectives. https://www.ibm.com/blogs/watson-health/the-5-vs-of-big-data/. Accessed 30 May 2017
17. IBM big data and analytics hub. Extracting Business Value from the 4 V's of Big Data (2016) http://www.ibmbigdatahub.com/infographic/extracting-business-value-4-vs-big-data
18. Gandomi A, Haider M (2015) Beyond the hype: big data concepts, methods, and analytics. Int J Inf Manage 35(2):137–144
19. Marr B (2017) Really big data at Walmart: real-time insights from their 40+ Petabyte data cloud. https://www.forbes.com/sites/bernardmarr/2017/01/23/really-big-data-at-walmart-real-time-insights-from-their-40-petabyte-data-cloud/#2a0c16916c10
20. Pokorný J, Škoda P, Zelinka I, Bednárek D, Zavoral F, Kruliš M, Šaloun P (2015) Big data movement: a challenge in data processing. In: Big data in complex systems. Cham, Springer, pp 29–69
21. Han J, Pei J, Kamber M (2011) Data mining: concepts and techniques. Elsevier, Amsterdam
22. Xiong H, Pandey G, Steinbach M, Kumar V (2006) Enhancing data analysis with noise removal. IEEE Trans Knowl Data Eng 18(3):304–319
23. Court D (2015) Getting big impact from big data. McKinsey Q 1:52–60
24. Golchha N (2015) Big data—the information revolution. IJAR 1(12):791–794
25. Khan M, Ayyoob M (2018) Big data analytics evaluation. Int J Eng Res Comput Sci Eng (IJERCSE) 5(2):25–38
26. Wang X, He Y (2016) Learning from uncertainty for big data: future analytical challenges and strategies. IEEE Syst Man Cybern Mag 2(2):26–31
27. Jordan MI (2012) Divide-and-conquer and statistical inference for big data. In: Proceedings of the 18th ACM SIGKDD international conference on knowledge discovery and data mining. ACM, New York, p 4
28. Guyon I, Weston J, Barnhill S, Vapnik V (2002) Gene selection for cancer classification using support vector machines. Mach Learn 46(1–3):389–422
29. Liu H, Motoda H (eds) (2007) Computational methods of feature selection. CRC Press, Boca Raton
30. Belhadi A, Zkik K, Cherrafi, A, Yusof SM, fezazi SE (2019) Understanding the capabilities of big data analytics for manufacturing process: insights from literature review and multiple case study. Comput Ind Eng 137. https://doi.org/10.1016/j.cie.2019.106099
31. IBM Analytics: IBM Industry Model support for a data lake architecture (2016). https://www.ibm.com/downloads/cas/DNKPJ80Q. Accessed 20 Feb 2021
32. Tao F, Qi Q, Liu A, Kusiak A (2018) Data-driven smart manufacturing. J Manuf Syst 48:157–169. https://doi.org/10.1016/j.jmsy.2018.01.006
33. Shao G, Jain S, Shin SJ (2014) Data analytics using simulation for smart manufacturing. Proc Winter Simul Conf. https://doi.org/10.1109/WSC.2014.7020063
34. Syafrudin M, Fitriyani NL, Li D, Alfian G, Rhee J, Kang YS (2017) An open source-based real-time data processing architecture framework for manufacturing sustainability. Sustainability 9(11). https://doi.org/10.3390/su9112139
35. Dai HN, Wang H, Xu G, Wan J (2019) Big data analytics for manufacturing internet of things: opportunities, challenges and enabling technologies. Enterprise Inf Syst 14(6). https://doi.org/10.1080/17517575.2019.1633689
36. Wilcox T, Jin N, Flach P, Thumim J (2019) A big data platform for smart meter data analytics. Comput Ind 105:250–259. https://doi.org/10.1016/j.compind.2018.12.010
37. https://en.wikipedia.org/wiki/ApacheNiFi. Accessed 20 Feb 2021
38. Pandya A, Kostakos P, Mehmood H, Cortes M (2019) Privacy preserving sentiment analysis on multiple edge data streams with apache nifi. In: Proceedings of European intelligence and security informatics conference (EISIC). https://doi.org/10.1109/EISIC49498.2019.9108851

39. Kreps J, Narkhede N, Rao J (2011) Kafka: a distributed messaging system for log processing. In: Proceedings of the NetDB, Athens, Greece
40. https://kafka.apache.org. Accessed 20 Feb 2021
41. Mu C, Zhao J, Yang G, Zhang J, Yan Z (2018) Towards practical visual search engine within elasticsearch. 1806.08896
42. Srivastava A, Miller D (2019) Elasticsearch 7 quick start guide. Packt Publishing
43. Ku´c R, Rogozin´ski M (2015) Mastering Elasticsearch. Packt Publishing
44. Kuc R, Rogozinski M (2014) Elasticsearch Server. Packt Publishing
45. Srivastava A, Azarmi B (2019) Learning Kibana 7: build powerful elastic dashboards with Kibana's data. Packt Publishing
46. Build visualizations simply and intuitively. https://www.elastic.co/kibana. Accessed 20 Feb 2021
47. Flask Web Development, one drop at a time. http://flask.pocoo.org Accessed 20 Feb 2021
48. Erraissi A, Belangour A, Tragha A (2017) A big data hadoop building blocks comparative study. Int J Comput Trends Technol 48(1):36–40. https://doi.org/10.14445/22312803/IJCTT-V48P109
49. Chiary MR, Anand R (2015) Hadoop cluster on linode using ambari for improving task assignment scheme running in the clouds. Int J Comput Sci Inf Technol 6(1):586–589
50. Salloum S, Dautov R, Chen X, Peng PX, Huang JZ (2016) Big data analytics on apache spark. Int J Data Sci Anal 1:145–164
51. Shoro AG, Soomro TR (2015) Big data analysis: Apache spark perspective. Global J Comput Sci Technol 15(1)
52. Armbrust M, Fox A, Griffith R, Joseph AD, Katz RH, Konwinski A, Zaharia M (2009) Above the clouds: a berkeley view of cloud computing. Technical Report UCB/EECS-2009-28, vol. 17. EECS Department, University of California, Berkeley
53. https://databricks.com. Accessed 20 Feb 2021
54. Beauchemin M Airflow: a workflow management platform. https://medium.com/airbnb-engineering/airflow-aworkflow-management-platform-46318b977fd8. Accessed 20 Feb 2021
55. Gantz J, Reinsel D (2011) Extracting value from chaos. IDC iview 1142:1–12
56. https://nifi.apache.org. Accessed 20 Feb 2021
57. Samal B, Panda M (2017) Real time product feedback review and analysis using apache technologies and nosql database. Int J Eng Comput Sci 6(10):22551–22558. https://doi.org/10.18535/ijecs/v6i10.04
58. Soner K, Upadhyay H (2014) A survey: Ddos attack on internet of things. Int J Eng Res Dev 10(11):58–63
59. What is Elasticsearch. https://www.elastic.co/guide/en/elasticsearch/reference/master/elasticsearch-intro.html. Accessed 20 Feb 2021

Altan Cakir received his M.Sc. degree in theoretical particle physics from Izmir Institute of Technology in 2006 and then went straight to graduate school at the Karlsruhe Institute of Technology, Germany. During his Ph.D., he was responsible for a scientific research based on new physics searches in the CMS detector at the Large Hadron Collider (LHC) at European Nuclear Research Laboratory (CERN). Thereafter he was granted as a post-doctoral research fellow at Deutsche's Elektronen-Synchrotron (DESY), where he spent more than five years, and then he got his present a full professor position at Istanbul Technical University (ITU), Istanbul, Turkey. Currently, Altan Cakir is a group leader of ITU-CMS group at CERN and leading a data analysis group at the CMS detector. His group's expertise is focused around machine learning techniques in large scale data analysis. Prof. Cakir has been a board member of the ITU Artificial Intelligence and Data Science Research and Applications Center since 2019 and he is a co-president of Turkish Artificial Intelligence Platform (http://www.ai.org.tr) since the beginning of 2021. He is the founder of Parton Big Data Analytics and Consulting Inc. operating in ITU Teknokent in Istanbul.

Business Applications

Supply Chain Analytics

Yakup Turgut, Kenan Menguc, Nursah Alkan, Yildiz Kose, Seval Ata, Sule Itir Satoglu, and Ozgur Kabak

1 Introduction

Supply chain management (SCM) is the process of managing the people, resources, activities, and technology involved in manufacturing and delivering a product or service in order to reduce costs and avoid shortages. Supply chain analytics is the term that refers to the analytical decision-making processes using huge amount of data generated through the supply chain. Analytics in the supply chain is a critical component of SCM. Descriptive, predictive, and prescriptive analytics should be combined to optimize supply chain planning processes. In today's business, strategic, tactical, and operational decisions can be automated by using advanced technologies such as optimization, AI, and RPA. These technologies are utilized to increase operational efficiency, save costs, and avoid shortages. This chapter covers strategic and operational supply chain decisions and presents sample analytics models and solutions for demand forecasting, network optimization, inventory replenishment, and transportation planning for different case studies.

Y. Turgut · K. Menguc · N. Alkan · Y. Kose · S. Ata · S. I. Satoglu (✉) · O. Kabak
Department of Industrial Engineering, Faculty of Management, Istanbul Technical University, 34367 Macka, Istanbul, Turkey
e-mail: onbaslis@itu.edu.tr

Y. Turgut
e-mail: turgut16@itu.edu.tr

A. Ustundag et al. (eds.), *Business Analytics for Professionals*, Springer Series in Advanced Manufacturing, https://doi.org/10.1007/978-3-030-93823-9_11

2 Demand Forecasting

Problem Definition

Demand forecasting is the prediction of a possible future demand in a specified time period for an organization's products or services. This term is often used alternatively with *demand planning;* however, demand planning starts with the forecast but then takes other activities into consideration like distribution and storage. Demand forecasting is important because it allows a business to control inventory levels, price the products properly, and allocate the resources efficiently. The accurate business decisions related to production, inventory, personnel, and facilities depend on the accuracy of demand forecast.

There are three main forecast horizons:

Short-term forecast horizon is usually from 1 month to 1 year. However, weekly and daily forecasts are also necessary for several business decisions. Short-term forecasting is mostly used for operational decisions, such as scheduling production activities, managing purchasing, controlling inventory, formulating pricing policy, and developing suitable sales strategy.

Medium-term forecast horizon is from 3 months to 3 years. Generally, sales and production planning, budgeting decisions are taken accordingly.

Long-term forecast horizon is from 1 year to 3+ years. It is usually used for strategic decisions like new product planning, supply chain network optimization.

The importance of forecast accuracy is essential to a company's operational success. Short-term forecasts tend to be more accurate than longer-term forecasts, as the uncertainty level of input variables is higher in long term. Long-term forecasts deal with issues that require investment and support strategic decisions.

Quantitative and qualitative methods and techniques are used for long- and short-term forecasting (Fig. 11.1). There are four types of qualitative techniques:

Executive opinion: Group of managers meet, and they estimate demand by working together and combining managerial experience with statistical models.

Salesforce opinion: Salespeople convey predictions about the region they are responsible for.

Market survey: Interviews and surveys are conducted to understand preferences of customers and to estimate demand.

Delphi method: A consensus is provided among a group of specialists.

There are two fundamental approaches in quantitative techniques. The first approach is the times series model which examines historical data patterns and attempts to forecast the future using the underlying patterns in the data. In this method, the factors affecting the demand are not considered as it is assumed that factors influencing past and present will continue to influence in future. There are several techniques used for time series modeling:

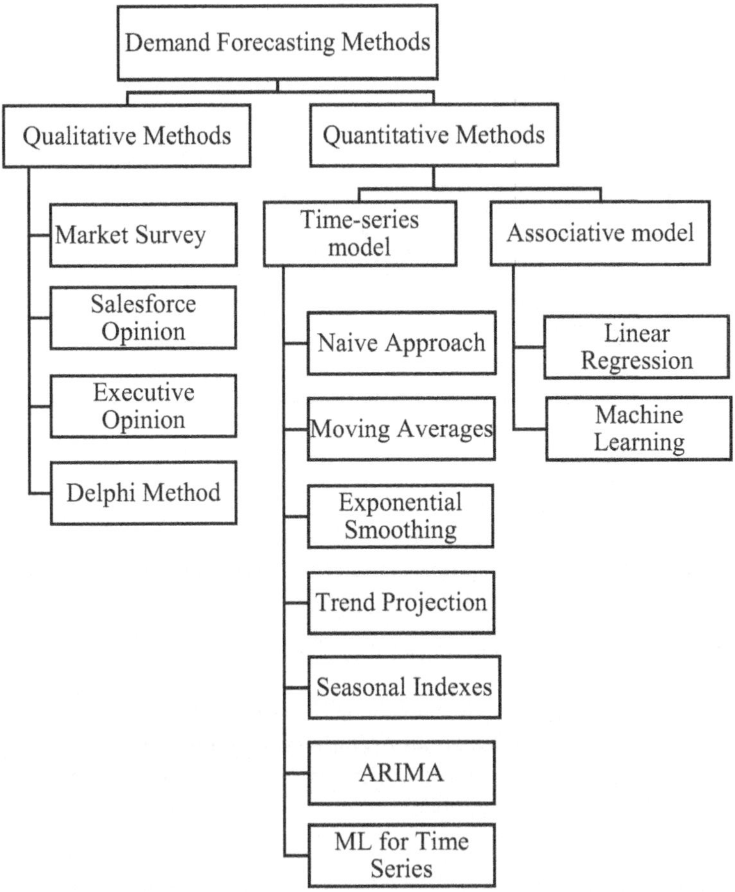

Fig. 11.1 Demand forecasting methods

Naive approach uses actual value of the previous period as a forecast.

Moving average uses an average of a specified number of the past data. As observations receive the same weight in the simple moving average, they receive different weights in the weighted moving average.

Exponential smoothing uses a weighted average technique with weights decreasing exponentially as data gets older.

Trend projection uses the least-squares method to fit data with linear function.

Seasonal indexes are used for adjusting the forecast by considering any seasonal patterns in the data.

ARIMA is autoregressive integrated moving average predicting the future values on the basis of the data's own lags and its lagged errors.

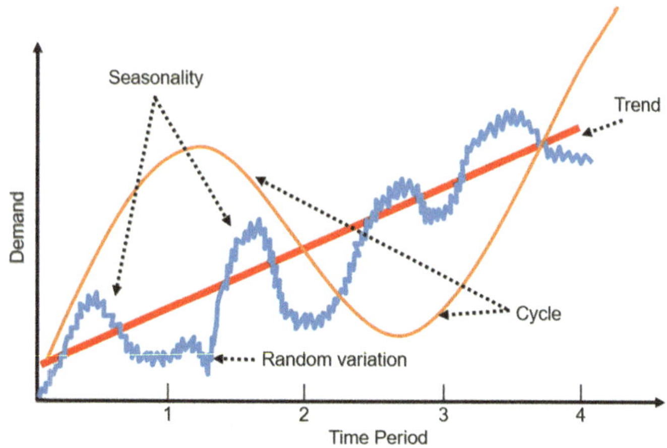

Fig. 11.2 Time series components

Machine Learning techniques are also used for univariate time series data.

There are four types of time series components (Fig. 11.2).

- Trend indicates either upward or downward changes in data values over a length of time.
- Seasonality represents periodic fluctuations in a short to intermediate time period.
- Cycles exhibit periodic fluctuations over a very long time period.
- Random variations correspond to the erratic movements that appear irregularly and generally during short periods.

The second approach for quantitative techniques is associative model (often called causal model). It assumes that the variable being forecasted is related to other variables in the environment. Several factors are used for demand forecasting including psychological, sociological, competitive, economic, market, environmental conditions, etc. Linear regression models and machine learning techniques can be both employed in this multivariate approach.

Case Study: A Fast-Food Restaurant Company

Fast-food company ABC has six restaurants in a district. Customers place orders on its Web site or mobile application. The company managers would like to optimize the number of carriers for each hour. Therefore, they should predict the hourly demand first. The list of variables used to predict demand is given (Table 11.1).

Initially, hourly demand forecasts are obtained using the LSTM method based on the data of the previous 12 h. The error reduction graph for the train and test sets of the LSTM model is shown in Fig. 11.3.

Six other machine learning techniques, namely MLP, ridge regression, support vector machine (SVR), decision tree, random forest, XGB and LGB are also included to forecast. As stated in Table 11.2, mean absolute percentage errors (MAE) of the

Table 11.1 List of variables affecting demand

Variable name	Variable description
WorkingDay	Workday information
ClosedRestaurant	Number of closed restaurants on a specific hour
WeatherCode	General weather condition forecast (rainy, sunny, cloudy, etc.)
Day	Day of month
ReligiousHoliday	Is there a religious holiday on that day
Event	Is there any event on that day
Hour	Hour of day
Humidity	Weather humidity forecast
Rain	Rain forecast
Snow	Snow forecast
FootballMatch	Is there any football match on that day
Month	Month of year
Ramadan	Is the day in Ramadan
Holiday	Is there a holiday on that day
ExamDay	Is it an exam day for students
WinterHoliday	Is there a semester holiday on that day
Temperature	Temperature forecast
UnexpectedEvent	Is there any other event on that day
WeekDay	Is it a weekday
WindDirection	Wind direction forecast
Wind	Wind speed forecast
SummerHoliday	Is there a summer holiday on that day
Year	Year
NewYear	Is there a Christmas on that day

methods are obtained in the test data. Finally, it can be said that deep learning method (LSTM) and ensemble methods (LGB, random forest, and XGBoost) are superior to others.

3 Inventory Management

Problem Definition

It is critical for any business to maintain an adequate inventory of required items in the required quantity and at the appropriate time. From this vantage point, inventory management in any business entails addressing the following three questions:

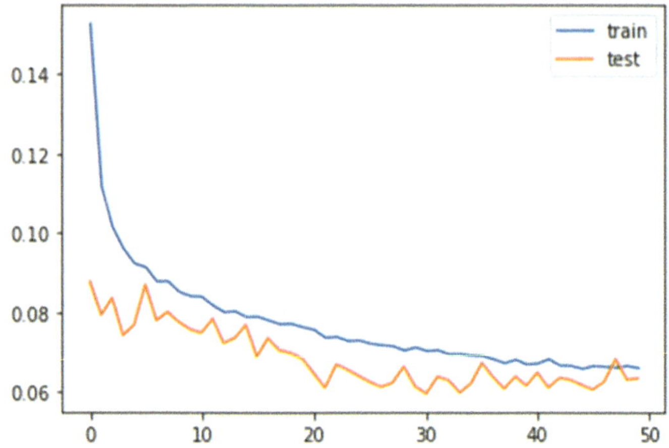

Fig. 11.3 Mean absolute errors of train and test sets for LSTM iterations

Table 11.2 Forecast accuracy

Method	MAE
LSTM	2.03
MLP	2.77
Ridge	3.22
SVR	3.96
Decision tree	2.44
Random forest	2.02
XGBoost	2.09
LGB	1.97

- How often should an item's inventory status be monitored?
- When is the best time to put a replenishment order?
- What is the optimal order size?

Inventory managers can determine the most optimal or near-optimal responses to these questions through the use of prescriptive analytics methods. The aim of inventory is to achieve adequate customer service while maintaining fair inventory costs. The total cost of inventory management includes the following basic costs: 1) **unit cost**: this is the price that suppliers charge for one unit or the cost to organize the purchase of one unit; (2) **ordering cost**: this is the costs associated with generating and processing an order to a supplier; (3) **inventory holding cost**: this is the cost of stocking one unit of an item for a specified time period; (4) **inventory shortage cost**: if a business runs out of an item and there is consumer demand, a shortage occurs, resulting in a cost [1]. The simplest example would be to lose profit from a lost sale by a business.

Table 11.3 Characteristics of inventory problem/models

Dimension	Values
Demand	Constant, deterministic, stochastic
Lead time	0, >0, stochastic
Time horizon	Finite, infinite
Products	Single item, multiple item
Capacity	Limited order, unlimited order
Customer service	Inventory shortage allowed, not allowed

The inventory models developed to address the aforementioned concerns are organized around the dimensions listed in Table 11.3 [2]. The bold values in column "Values" represent the case study's dimension values.

Inventory systems are also classified into two broad categories: continuous review and periodic review. This classification defines whether inventory levels are monitored continuously or periodically. An order will be placed in continuous review as soon as the inventory falls to the prescribed reorder point. In a periodic review, the inventory level, on the other hand, is examined at discrete intervals, for example at the end of each week, and ordering decisions will only be made at such times even though the inventory level falls below the reorder point between previous and current reviews.

Case Study: An Online Food Ordering Company

Online food ordering is simply the act of placing an order with a Web site or other application. The largest food delivery platform X recently joined the internet market with their new initiative Y. Y delivers market-level services to its customers in 10 min or less. Frozen foods, fruits and vegetables, hygiene products, and baby food delivery, and diaper delivery are just a few of the hundreds of thousands of market items that can be ordered through Y. Inventory management decisions they make to maintain service quality and economic profitability are critical. Because they provide a diverse range of product services, some of which involve deterioration-prone products. Additionally, a lack of stock may result in the loss of a customer or a loss of sales. Figure 11.4 depicts a typical cycle of an item's stock level over time. In the case study, determining the item's order points and order size for the online food delivery company's beverage product are focused. Assumptions for the case include the following:

1. The duration of each period in the planning horizon is known. Each period is equivalent to one week, and lead time is not taken into account.
2. At the beginning of the planning horizon, demand is estimated.
3. The item's price remains constant over the entire planning horizon.
4. The total quantity of required items can be ordered over a period of time.
5. The inventory manager must determine the order quantity at the beginning of each week. The lot-sizing models discussed in the next section are used to assist him in making the decision.
6. The inventory levels are reviewed periodically, i.e., weekly.

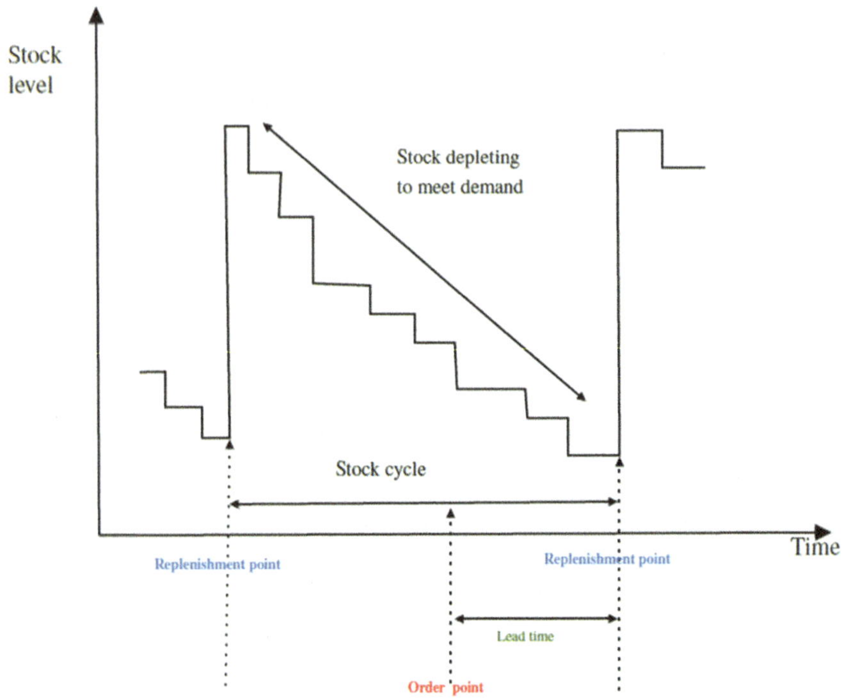

Fig. 11.4 Stock levels within a normal cycle

Prior to addressing these issues, the item's demand for the planning periods should be estimated. The data set contains 25 weekly sales records for the item, of which 21 to forecast demand for the item over the next six weeks will be used.

Table 11.4 details several features as well as handcrafted features. Prophet library, an open-source library developed by Facebook to automatically predict uniform time series data, is used to forecast the demand over the planning horizon. The forecasting results are illustrated in Fig. 11.5.

After estimating demand for the next six weeks, the weekly order size is determined. Subsequently, the order size decisions based on actual demand over these weeks are evaluated.

Model

The notation and description of the model are shown below:

$t = 1, 2, \ldots, T$, where T is the planning horizon

I_t: is the inventory at the end of time t

I_0: is the initial inventory at the beginning of planning period

D_t: the actual demand in period t

Table 11.4 Specifics of the features that were used to estimate demand

Variable name	Variable description
Week	The week number of the year
Year	Year
Price	Price of one unit of beverage for that week
Demand	Demand of beverage for that week
Percent	How many days free product campaign was made for that week
MinTemp	Minimum temperature for that week
MaxTemp	Maximum temperature for that week
Precipitation	Total precipitation for that week
NextWeekPrice	Price of one unit of beverage for next week
NextWeekFree	How many days free product campaign was made for next week
NextWeekPercent	How many days discounted product campaign was made for next week
Holidays	Number of holidays for that week
SchoolHolidays	Number of school holidays for that week
NextHoliday	Number of holidays for next week

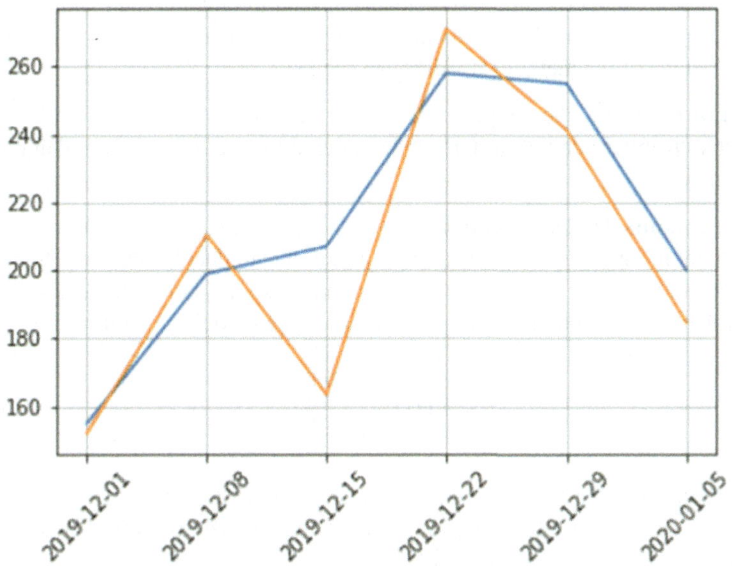

Fig. 11.5 Actual and forecasted demand plot (blue line indicates the actual demand, orange line indicates the forecasted demand)

R_t: the number of items arrived in period t

P_t: is the price of one item in period t

O_t: the ordering cost (setup cost) in period t

H_t: the holding cost of one item during period t

S_t: the storage cost of one item during period t

TC: the total costs related to inventory management decisions.

$$\min TC = \sum_{t=1}^{T} H_t I_t + P_t R_t + S_t \max(0, D_t - (R_t - I_{t-1}) + O_t Y_t \tag{11.1}$$

where Y_t is a binary variable that takes the value one if an item is being ordered and zero otherwise. In this case, the decision-maker must determine the order size (or lot size) for each period t while taking the total cost of these decisions into account. To keep things simple, only the ordering and holding costs are included when comparing the performance of the following lot-sizing techniques. Additionally, it is assumed that the holding and ordering costs remain constant throughout the planning period. Before applying the methods, quickly review the fundamentals of the techniques. For the details of the methods, readers can see [3].

LoL (Lot for Lot): The simplest lot-sizing rule is known as lot-for-lot. This rule states that the total number of items to be ordered in a period equals to that period's requirement.

EOQ (Economic Order Quantity): The economic order quantity (EOQ) is a widely used model in inventory control settings. The aim is to decide when and how much inventory should be replenished in order to minimize the sum of the costs (holding cost, ordering cost, and unit cost) per unit time. The model is as follows:

$$EOQ = \sqrt{\frac{2DS}{H}} \tag{11.2}$$

where D indicates the annual demand, S indicates the ordering cost (setup cost), and H indicates the holding (carrying) cost on an annual basis per unit. The order size for each period is indicated by the EOQ.

Fixed Order Quantity (FOQ) It specifies an arbitrary number of units to be ordered each time. For instance, economic order quantity can be used as a fixed order quantity for each period. In the case, economic order quantity as a fixed order quantity per period is used.

Part Period Balancing (PPB) PPB aims to balance the inventory-carrying cost and setup cost. The objective is to keep the inventory holding costs close to the ordering costs as possible. The following describes how the technique works:

Let $d_1, d_2, d_3, \ldots, d_n$ be the demand for an item over a period of n periods:

O: be the ordering cost

H: be the holding cost for an item per period

$C(j)$: be the total holding cost over j period

$$C(j) = Hd_2 + 2Hd_3 + \cdots + (j-1)Hd_j \tag{11.3}$$

$y_1, y_2, y_3, \ldots, y_n$ be the order size over a period of n periods.
The steps of the method are as follows:

Start with $t^* = 1$.

Step I: Start the calculation from period t^* to the next period. Calculate $C(t*), C(t*+1), \ldots, C(j)$ until $C(j) > O/H$

Step II: If $C(j) > O/H$ set $y_{t*} = d_{t*} + d_{t*+1} + \cdots + d_{j-1}$

Step III: Go to Step I and start over period j.

Step IV: Repeat step (I–III) until $y_1 + y_2 + \cdots + y_n >= d_1 + d_2 + \cdots + d_n$.

Silver–Meal Heuristic: It is a heuristic technique for lot sizing. The following describes how the technique works:
Let

$d_1, d_2, d_3, \ldots, d_n$ be the demand for an item over a period of n periods

O: be the ordering cost

H: be the holding cost for an item per period

$C(j)$ be the average holding cost and setup cost over j period

$$C(j) = (O + Hd_2 + 2Hd_3 + \cdots + (j-1)Hd_j)/j \tag{11.4}$$

$y_1, y_2, y_3, \ldots, y_n$ be the order size over a period of n periods.
The steps of the method are as follows:

Start with t* = 1.

Step I: Start the calculation from period t* to next period. Calculate $C(t*), C(t*+1), \ldots, C(j)$ until $C(j) > C(j-1)$

Step II: Stop the calculation when $C(j) > C(j-1)$ and set $y_{t*} = d_{t*} + d_{t*+1} + \cdots + d_{j-1}$

Step III: Go to Step I and start over period j.

Step IV: Repeat step (I–III) until $y_1 + y_2 + \cdots + y_n >= d_1 + d_2 + \cdots + d_n$.

Wagner–Whitin Algorithm: It ensures the optimal (minimizing total cost of replenishment and holding cost) selection of replenishment quantities under some assumptions. Wagner–Whitin algorithm provides and optimal solution for finite, discrete time inventory management problems with no shortage [4]. The algorithm minimizes the summation of ordering cost, holding cost, and purchasing cost. The following describes how the algorithm works:

Let $C(t)$ will be the optimal cost from period 1 to period t.

Minimum cost from periods l through $s - l$ can be defined as $F(s - l)$.

Let denote the K_t^s as the minimum cost up to time t when the demand for period t is ordered at period s.

Assume v is the period when the last order is placed.

The following steps are performed to find the minimum total cost.

Step 1: Assign $F(1) = O_1$, and $v = 1$

Step 2: In the second step determine if the demand for period $t = 2$ should be satisfied from first period or second period as follows:

$$F(2) = \min\{O_1 + O_2, O_1 + hd_2\} = \min\{F(1) + K_2^2, K_2^1\} \tag{11.5}$$

$$K_2^1 = F(1) + hd_2 \tag{11.6}$$

$$K_t^s = F(s) + hd_{s+1} + 2hd_{s+2} + \cdots + (t - s)hd_t \tag{11.7}$$

Step 3: change $v = 2$ to if $O_1 + O_2 < O_1 + hd_2$ otherwise v remains unchanged.

Step 4: for the t-period problem, determine the optimum value of v by minimizing

$$F(t) = \min\{K_t^v, K_t^{v+1}, \ldots, K_t^{t-1}, K_t^t\} \tag{11.8}$$

Solution and Analysis

The following are the model inputs in the case: The demand for the next six weeks is forecast to be 152, 210, 163, 271, 241, and 184, respectively. The holding cost is \$0.5 per unit, and the order cost is \$100 per order.

Results of the above-mentioned methods for the case are presented in Tables 11.5, 11.6, 11.7, 11.8, and 11.9 for each method. The overall inventory costs for each method are shown in Table 11.10. The results indicate that among all methods, the Silver–Meal heuristic and Wagner–Whitin algorithm have the lowest total cost. When the procedures of the methods are considered, it is expected that Wagner–Whitin will produce the best result, as it is a dynamic programming algorithm for finding the optimal solution. On the other hand, among all methods, the EOQ method used in FOQ yields the worst results. This is due to the change in demand over a period of weeks. Erroneous assumptions of steady demand and steady sales in the EOQ model leave you unable to account for shifts in demand around holidays or seasonal events.

Table 11.5 Fixed order quantity solution summary

	Period	Beginning inventory	Demand	Orders	Ending inventory	Holding cost	Order cost
0	1	0.0	152	285.0	133.0	0.0	100
1	2	133.0	210	285.0	208.0	66.5	100
2	3	208.0	163	0.0	45.0	104.0	0
3	4	45.0	271	285.0	59.0	22.5	100
4	5	59.0	241	285.0	103.0	29.5	100
5	6	103.0	184	285.0	204.0	51.5	100

Table 11.6 Lot for lot solution summary

	Period	Beginning inventory	Demand	Orders	Ending inventory	Holding cost	Order cost
0	1	0.0	152	152.0	0.0	0.0	100
1	2	0.0	210	210.0	0.0	0.0	100
2	3	0.0	163	163.0	0.0	0.0	100
3	4	0.0	271	271.0	0.0	0.0	100
4	5	0.0	241	241.0	0.0	0.0	100
5	6	0.0	184	184.0	0.0	0.0	100

Table 11.7 PPB solution summary

	Period	Beginning inventory	Demand	Orders	Ending inventory	Holding cost	Order cost
0	1	0.0	152	525.0	373.0	0.0	100
1	2	373.0	210	0.0	163.0	186.5	0
2	3	163.0	163	0.0	0.0	81.5	0
3	4	0.0	271	696.0	425.0	0.0	100
4	5	425.0	241	0.0	184.0	212.5	0
5	6	184.0	184	0.0	0.0	92.0	0

Table 11.8 Silver–Meal heuristic solution summary

	Period	Beginning inventory	Demand	Orders	Ending inventory	Holding cost	Order cost
0	1	0.0	152	152.0	0.0	0.0	100
1	2	0.0	210	373.0	163.0	0.0	100
2	3	163.0	163	0.0	0.0	81.5	0
3	4	0.0	271	271.0	0.0	0.0	100
4	5	0.0	241	425.0	184.0	0.0	100
5	6	184.0	184	0.0	0.0	92.0	0

Table 11.9 Wagner–Whitin solution summary

	Period	Beginning inventory	Demand	Orders	Ending inventory	Holding cost	Order cost
0	1	0.0	152	152.0	0.0	0.0	100
1	2	0.0	210	373.0	163.0	0.0	100
2	3	163.0	163	0.0	0.0	81.5	0
3	4	0.0	271	271.0	0.0	0.0	100
4	5	0.0	241	425.0	184.0	0.0	100
5	6	184.0	184	0.0	0.0	92.0	0

Table 11.10 Total inventory costs for various models

Model	Total cost (TC)
LoL	$600
Fixed order quantity	$774
Part period balancing	$772.5
Silver–Meal heuristic	$573.5
Wagner–Whitin algorithm	$573.5

4 Network Optimization

Supply chain is composed of facilities and movement of goods among the facilities. Considering all movements through the facilities using the transportation routes, supply chain is a network, instead of a "chain." In order to make various supply chain design, planning and operation decisions, network optimization techniques are used.

Problem Definition

A supply chain is a set of relationships and links that involve various participants and carry out a series of activities in transporting physical goods or services from a point of origin to a point of consumption. The supply chain, which consists of all parties directly or indirectly involved in fulfilling customer demand, includes not only manufacturers and suppliers, but also shippers, wholesalers, warehouses, retailers, distribution centers, and even customers themselves [5]. By the effective management of the supply chain, the values that make up the entire performance of the enterprise such as improved production capacity, market sensitivity, and customer/supplier relations can be increased [6].

One of the most important issues in supply chain management is to design and plan the overall architecture of the supply chain network. A holistic approach is required for the supply chain to develop strategies and design processes by maximizing total supply chain added value and/or minimizing the total supply chain cost.

In general, the supply chain network structure can also be shown as given in Fig. 11.6. In order to design a supply chain network specific for a product (or a product

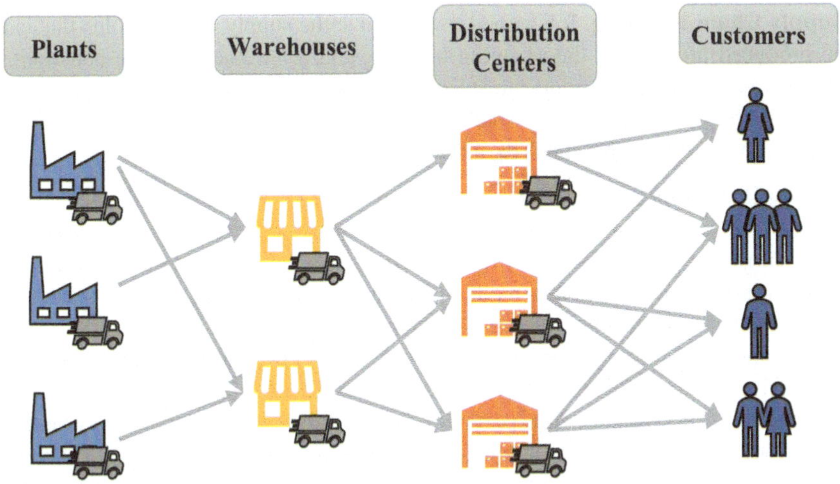

Fig. 11.6 Supply chain network design

family), the tiers in the network and transportation means among the tiers are specified. Managers must make two important decisions when designing a distribution network:

1. Will the product(s) be picked up from a predetermined site or delivered to the customer's location?
2. Will the product(s) pass through an intermediate location (or intermediary)?

Depending on the answers to these two questions, the sector in which the company is located and company's strategic priorities, one of six different distribution network designs can be used when moving the products from the factory(s) to the customer(s) [5]:

- Manufacturer storage with direct shipping,
- Manufacturer storage with direct shipping and in-transit merge,
- Distributor storage with carrier delivery,
- Distributor storage with last-mile delivery,
- Manufacturer/distributor storage with customer pickup,
- Retail storage with customer pickup.

Successful supply chain management requires many decisions to be addressed regarding information flow, products, and funds. These decisions are divided into three categories or phases, depending on the frequency of each decision and the time frame over which a decision phase has an impact [5].

Supply Chain Strategy or Design: At this stage, a company is concerned with making decisions about how to structure its supply chain. For this, it is decided what the configuration of the chain will be, how the resources will be allocated, and which processes each stage will perform.

Supply Chain Planning: It is aimed to maximize the supply chain surplus that can be produced throughout the planning horizon, taking into account the constraints identified at the strategic or design stage. Planning includes making decisions about which markets to source from which locations, outsourcing production, inventory policies to follow, timing and size of marketing, and price promotions.

Supply Chain Operation: At this stage where decisions about individual customer orders are taken, it is aimed to handle incoming customer orders in the best possible way. During this phase, inventory or production is allocated to individual orders, an order is assigned to a particular mode of shipment and shipment, delivery schedules of trucks, and replenishment orders are determined by firms.

To understand how companies can improve supply chain performance in terms of responsiveness and efficiency, cross-functional drivers of supply chain performance such as facilities, inventory, transportation, information, sourcing, and pricing need to be considered. To improve the supply chain surplus and the firm's financial structure, drivers need to be configured to achieve the desired level of responsiveness at the minimum possible cost [5].

Facilities: Decisions about the location, role, flexibility, and capacity of production sites and inventory storage facilities play an important role in the performance of the supply chain.

Inventory: Adjusting or changing inventory policies, taking into account questions such as what kind of inventory should be stocked at each stage of the supply chain, and how much inventory should be kept in raw materials, semifinished products or finished goods can significantly change the efficiency and responsiveness of the supply chain [7].

Transportation: Various transportation options need to be considered in order to move inventory from one point to another in the supply chain. These transportation options can be in the form of a single type of transportation or in the form of a combination of many modes and routes [7].

Information: It directly affects each of the other drivers in the supply chain as it encompasses information and data about facilities, customers, shipping, prices, costs, and inventory throughout the supply chain. Therefore, it is the biggest driver affecting supply chain performance [5].

Sourcing: It includes decisions about who will perform a particular supply chain activity, such as production, storage, transportation, or information management. These are the decisions that include which functions the firm performs and which functions to outsource [5].

To determine the applicability of decision alternatives, supply chain network design should be evaluated under factors such as capacity, flow, service compliance, and demand scope/amount. It is aimed to maximize or minimize the objective functions created based on the decisions to be taken, by evaluating them within the framework of the constraints that define their applicability. As a result, optimization models such as mixed-integer linear programming, stochastic modeling, uncertainty modeling, bi-level optimization are used among the most commonly used methods to give the best solution by examining all possible solutions [5].

Supply chains need to not only reduce operating costs, but also offer value-added products and services to meet the needs of internal and external consumers. The ability of a business to accurately identify the needs of its final and internal customers also indicates the effectiveness of industrial and personal sales in businesses. Quantitative demand forecasting methods such as time series and regression analysis and qualitative demand forecasting methods such as survey and Delphi are used to accurately predict and meet customer needs [5, 6].

The total cost within a supply chain network is the sum of inventory, transportation, and facility costs. As the number of facilities increases, as shown in Fig. 11.7, the total cost first decreases and then increases. Each firm should have a number of facilities that will minimize its total costs. If firms want to further reduce customer response time to increase customer satisfaction, they may have to increase the number of facilities beyond the point that minimizes their total cost. In this case, facilities can be added beyond the point of minimizing costs if it is concluded that the increase in revenues from better response outweighs the increased costs due to additional facilities [5].

Case Study: An Automotive Spare Part Producing Company

ABC company produces automotive spare parts. It has four different plants at different locations for the production of 2 different products, product1 and product2. The products manufactured in a plant must be conveyed to the appropriate warehouses and distribution locations, respectively, in appropriate quantities and under capacity constraints, before reaching the customers. Customer demands are met by the whole operation. For this, the company has 5 different warehouses, 6 different distribution centers, and serves 22 customers. Flow of the products is as follows:

$$\text{Plants} \rightarrow \text{Warehouses} \rightarrow \text{Distribution Centers} \rightarrow \text{Customers}$$

Fig. 11.7 Variation in total cost and response time with number of facilities

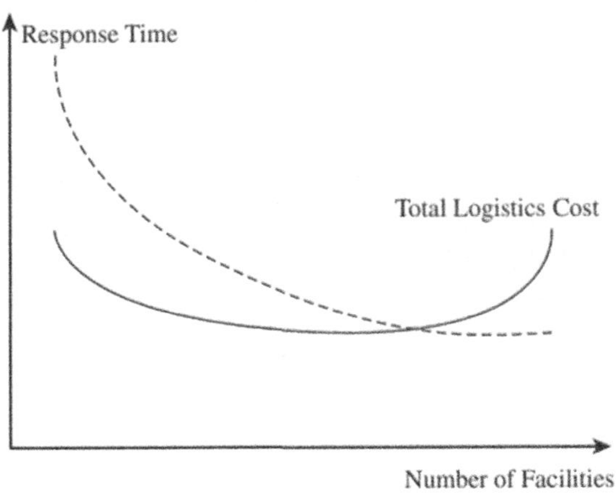

The firm aims to meet customer demands and to design a supply chain network that minimizes the total cost of the operation under recourse restrictions. Assumptions related to the case include:

- Transportation between facilities has different unit transportation costs.
- There is only one route option between facilities.
- Each of the warehouses and distribution centers has different operating costs.
- Each warehouse and distribution center has a different capacity.
- Each customer can demand two products at the same time in different demand quantities.
- The total number of distribution centers to be opened cannot be greater than the determined upper limit.
- The total number of warehouses cannot be greater than a determined upper limit.
- Each customer receives service from only one distribution center.

In the supply chain network in model, the capacities of warehouses and DCs are as given in Table 11.11. The operation costs of warehouses and DCs are presented in Table 11.12. Only unit transportation costs from plants to warehouses are given as in Table 11.13 to illustrate input parameters due to space constraint.

Model

An integer programming model is formulated to make network decisions described in the case. The decision variables, parameters, objective function, and constraints of the model are presented as follows, respectively.

Table 11.11 Capacities of warehouses and DCs

Warehouses	Capacity	DCs	Capacity
W1	100	DC1	100
W2	100	DC2	150
W3	150	DC3	150
W4	200	DC4	200
W5	150	DC5	150
		DC6	150

Table 11.12 Operation costs of warehouses and DCs

Warehouses	Operation costs	DCs	Operation costs
W1	400,000	DC1	450,000
W2	600,000	DC2	400,000
W3	600,000	DC3	500,000
W4	800,000	DC4	650,000
W5	400,000	DC5	450,000
		DC6	500,000

Table 11.13 Unit transportation costs from plants to warehouses

Plants/warehouses	W1	W2	W3	W4	W5
P1	2890	21,191	47,084	41,301	13,331
P2	46,480	51,280	65,115	63,263	35,518
P3	57,479	56,837	62,242	68,933	46,517
P4	41,641	36,803	35,102	40,129	37,786

Decision Variables:

Continuous variables:

X_{jki}: number of product i to be shipped from plant j to warehouse store k

Y_{kli}: number of product i to be shipped from warehouse k to distribution center (DC)l.

Z_{lmi}: number of product i to be shipped from DC l to customer m.

Binary variables:

$$A_l = \begin{cases} 1, if DClisopen \\ 0, otherwise \end{cases}$$

$$W_k = \begin{cases} 1, ifwarehousekisopen \\ 0, otherwise \end{cases}$$

$$B_{lm} = \begin{cases} 1, if DClservestothecustomerm \\ 0, otherwise \end{cases}$$

Parameters:

T_k : Capacity of warehouse k

D_l : Capacity of DC l

S_{ji} : Supply limit of plant j for product i

d_{mi} : Demand amount of product i for customer m

e_l : Operation cost of DC l

f_k : Operation cost of warehouse k

C_{jk} : Unit transportation cost from plant j to warehouse k

C_{kl} : Unit transportation cost from warehouse k to DC l

C_{lm} : Unit transportation cost from DC l to customer m

K: Maximum number of warehouses in operation

L : Maximum number of DCs in operations.

Objective Function:

$$\text{Min} = \sum_j \sum_k \sum_i X_{jki} C_{jk} + \sum_k \sum_l \sum_i Y_{kli} C_{kl} + \sum_l \sum_m \sum_i Z_{lmi} C_{lm} + \sum_k f_k W_k + \sum_l e_l A_l \tag{11.9}$$

Subject to:

$$\sum_k X_{jki} \leq S_{ji} \forall j \text{and} i \tag{11.10}$$

$$\sum_j \sum_i X_{jki} \leq T_k W_k \forall k \tag{11.11}$$

$$\sum_l \sum_i Y_{kli} \leq T_k W_k \forall k \tag{11.12}$$

$$\sum_k \sum_i Y_{kli} \leq D_l A_l \forall l \tag{11.13}$$

$$\sum_m \sum_i Z_{lmi} \leq D_l A_l \forall l \tag{11.14}$$

$$\sum_l A_l \leq L \tag{11.15}$$

$$\sum_k W_k \leq K \tag{11.16}$$

$$\sum_j X_{jki} = \sum_l Y_{kli} \forall k \text{and} i \tag{11.17}$$

$$\sum_m Z_{lmi} = \sum_k Y_{kli} \forall l \text{and} i \tag{11.18}$$

$$\sum_l B_{lm} = 1 \forall m \tag{11.19}$$

$$Z_{lmi} = B_{lm} d_{mi} \forall l, m \text{and} i \tag{11.20}$$

$$A_l, W_k, B_{lm} = 0 \text{or} 1 \forall k, l \text{and} m \tag{11.21}$$

$$X_{jki}, Y_{kli}, Z_{lmi} \geq 0 \forall j, k, l, m \text{and} i \tag{11.22}$$

where the total cost of supply chain network is represented by Eq. (11.9). Equation (11.10) shows the constraint that the quantity sent from the plant to the warehouse cannot be greater than the capacity of the plant. Equation (11.11) shows the constraint that the quantity coming from the plant to the warehouse cannot be greater than the capacity of the warehouse. Equation (11.12) presents the constraint that the quantity sent from the warehouse to the DC cannot be greater than the capacity of the warehouse. Equation (11.13) represents the constraint that the quantity coming from the warehouse to the DC cannot be greater than the capacity of the DC. Equation (11.14) presents the constraint that the quantity sent from the DC to the customer cannot be greater than the capacity of the DC. Equation (11.15) shows the constraint that the total number of distribution centers to be opened cannot be greater than the determined distribution center upper limit L. Equation (11.16) shows the constraint that the total number of warehouses to be opened cannot be greater than the determined warehouse upper limit K. Equation (11.17) presents the constraint that the quantity sent from the plant to the warehouse should be equal to the quantity sent from the warehouse to the DC. Equation (11.18) represents the constraint that the quantity sent from the DC to the customer should be equal to the quantity sent from the warehouse to the DC. Equation (11.19) presents the constraint that a DC should be assigned to a customer. Equation (11.20) shows the constraint that the quantity sent from the DC to the customer should be equal to the demand of customers. Equation (11.21) shows the constraint that binary restrictions while Eq. (11.22) represents the sign restrictions.

In constraints (11.11) and (11.12), there are checks about the capacity and determine whether the warehouse will be opened or not. If desired, the model can be solved by removing one of the constraints (11.11) or (11.12). Similarly, the same is true for constraints (11.13) and (11.14). It will be sufficient to use a constraint for the proposed model.

Solution and Analysis

The parameter values of the model based on capacity, operation costs, and unit transportation costs have been presented in Tables 11.11, 11.12, and 11.13, respectively. Here, as an example, only the parameter values of unit transportation costs from plants to warehouses are given. Other transportation costs can be found in the code file. In addition, the values of parameters K and L, which show the maximum number of warehouses and DCs in operation, are used in the model as 5 and 6, respectively. The established model has been solved by using the Python-pyomo library with the given parameter values. Considering the objective function and constraints in the developed model, the total cost has been found to be 77,738,950 and the results obtained for the amount of flow between the facilities have been presented in Tables 11.14, 11.15, and 11.16, respectively. The supply chain design network that emerged as a result of the developed model is given in Fig. 11.8. Besides, the results of transportation and operation costs at each stage are shown in Fig. 11.9.

For the given case, an integer program has been used to find the optimal solution. The results showed that the highest cost was from the warehouse to the DC and the highest operation cost belonged to the warehouses. Besides, it has been seen

Table 11.14 Product and flow quantity from plant to warehouses

Plant	Warehouse	Product	Quantity
P1	W1	Product1	100
P1	W2	Product1	55
P1	W5	Product1	45
P2	W4	Product1	110
P2	W5	Product1	25
P3	W5	Product2	50
P4	W2	Product2	45
P4	W3	Product2	150
P4	W4	Product2	75
P4	W5	Product2	30

Table 11.15 Product and flow quantity from warehouse to DC

Warehouse	DC	Product	Quantity
W1	DC4	Product1	70
W1	DC5	Product1	30
W2	DC1	Product1	45
W2	DC4	Product1	10
W2	DC4	Product2	45
W3	DC1	Product2	50
W3	DC4	Product2	20
W3	DC5	Product2	80
W4	DC3	Product1	110
W4	DC3	Product2	40
W4	DC4	Product2	35
W5	DC4	Product1	15
W5	DC6	Product1	55
W5	DC6	Product2	80

that a large part of the total costs consisted of transportation costs. Each customer was served by only one distribution center, and all customer needs have been met. Furthermore, the results showed the service of DC2 was shut down while other DCs were kept open for minimum supply chain network cost.

5 Route Planning and Transportation

In supply chains, the movement of the goods between the facilities such as raw material sources, production plants, warehouses, and retail stores is realized by the

Table 11.16 Product and flow quantity from DC to customer

DC	Customer	Product	Quantity	DC	Customer	Product	Quantity
DC1	C4	Product2	25	DC4	C10	Product1	15
DC1	C16	Product1	25	DC4	C10	Product2	15
DC1	C16	Product2	10	DC4	C11	Product1	20
DC1	C17	Product1	20	DC4	C12	Product1	15
DC1	C17	Product2	15	DC4	C12	Product2	20
DC3	C2	Product1	50	DC4	C21	Product1	10
DC3	C2	Product2	10	DC4	C21	Product2	20
DC3	C3	Product1	20	DC5	C8	Product1	15
DC3	C7	Product1	10	DC5	C8	Product2	35
DC3	C7	Product2	10	DC5	C13	Product1	5
DC3	C14	Product1	30	DC5	C13	Product2	35
DC3	C15	Product2	20	DC5	C19	Product1	10
DC4	C1	Product1	15	DC5	C19	Product2	10
DC4	C1	Product2	35	DC6	C6	Product1	40
DC4	C5	Product1	20	DC6	C9	Product2	40
DC4	C5	Product2	10	DC6	C18	Product2	40
				DC6	C22	Product1	15

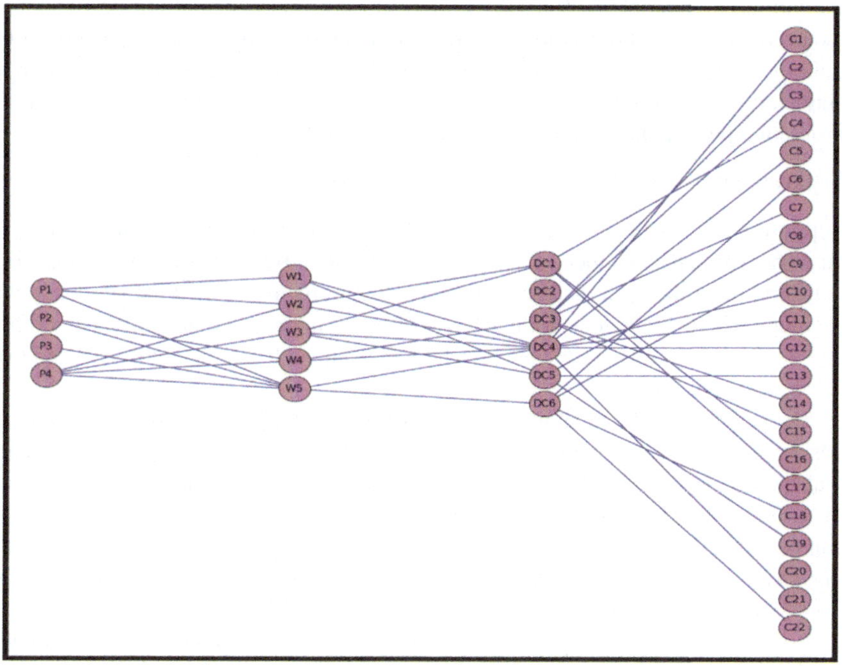

Fig. 11.8 Supply chain design network of the proposed model

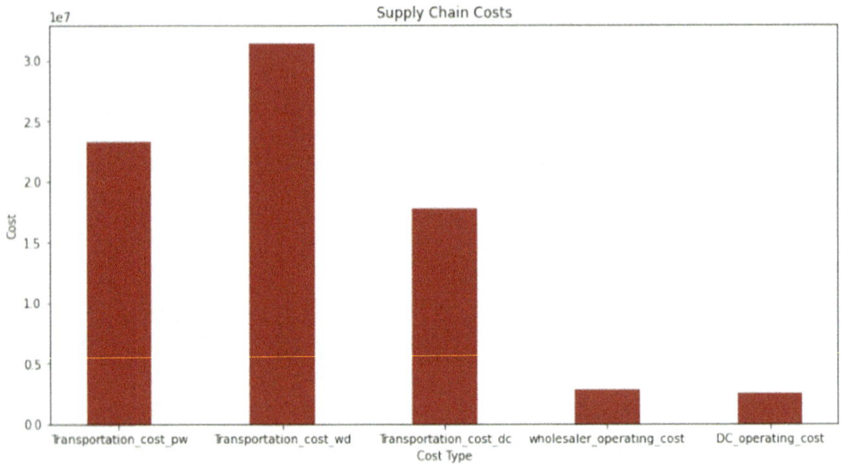

Fig. 11.9 Supply chain transportation and operation costs

transportation services. There are several transportation problems in the supply chain management. Among many others, three of them in this chapter will be handled. When origin and destination points are separate and the aim is to find the cheapest route from origin to the destination point in the given network, the problem is called the shortest path problem. When there are multiple origins and multiple destinations, the problem is called transportation problem in the operations research context. On the other hand, when one or several vehicles visit several locations starting from a single origin, the problem is called vehicle routing problem. In this subsection, these problem types are introduced and a case is presented.

Separate and Single Origin and Destination Points: Shortest Path problem

Suppose a truck will travel from a production plant located at a certain point to a customer located at another point using the network of roads and highways. The aim is to minimize the total distance traveled. This problem is called the shortest path problem. The vehicle does not have to visit every node in the network. The origin and destination points are the points where the vehicle has to be at the start and the finish, respectively.

Shortest path problem allows us to decide the shortest path or route over (or least cost) a large number of nodes between a starting node and an ending node. It is seen that the shortest time options in transportation with the navigation tools are used in our daily life. If the maps are concretized, every street intersection on highways can be defined as a node. And in this case, the shortest path can sometimes be the distance of the road, sometimes the duration of the trip, sometimes the fuel cost or the ratios of these parameters (distance/time).

Linear programming (LP) formulation of the shortest path problem is given below. In the LP model, i and j are indices for the nodes in the network S, and (i, j) represents the arcs in the network. x_{ij} is a binary variable that gets 1 if arc (i, j) is on the shortest route and 0, otherwise. c_{ij} is a parameter that shows the cost (distance, time, etc.) of using arc (i, j).

Objection function:

$$\min \sum_{(i,j)\in S}^{M} c_{ij}x_{ij} \tag{11.23}$$

Subject to:

$$\sum_{j,(1,j)\in S}^{M} x_{1j} = 1 \tag{11.24}$$

$$\sum_{j,(i,j)\in S}^{M} x_{ij} - \sum_{k,(k,i)\in S}^{M} x_{ki} = 0i = 2, 3, \ldots, m - 1 \tag{11.25}$$

$$\sum_{i,(i,m)\in S}^{M} x_{im} = 1 \tag{11.26}$$

$$x_{ij} \in \{0, 1\} for(i, j) \in S \tag{11.27}$$

The objective function given in Eq. (11.23) aims to minimize the total cost of the selected route. Equation (11.24) indicates that the route has a starting point, while Eq. (11.25) indicates that the route has an end point. In addition, Eq. (11.25) ensures that the route does not end between the start and end points. The last constraint ensures that the variables take binary values in the model [8].

There are several well-known algorithms to solve the shortest path problem, for instance, Dijkstra, dynamic programming, labeling, and Bellman–Ford algorithms. Dijkstra's algorithm is a labeling method, in which the nodes are given temporary and permanent labels. In dynamic programming, the shortest path is found by evaluating the routes from the start point to the end point or in the opposite direction. Processing time is high for large-sized problems. The Bellman–Ford algorithm computed shortest paths from a single source vertex to all of the other vertices in a weighted directed graph.

Multiple Origin and Destination Points: Transportation Problem

Multiple Origin and Destination points are the case where there is more than one starting point and more than one destination. Suppose a company has several production plants manufacturing the same product with certain capacities and several customers to be served with certain demands. The unit transportation costs between the production plants and customers are known. The problem is to distribute the capacities of the plants to meet the demands of the customers with minimum cost.

In the transportation problem, there are supply nodes such as production plants and demand nodes such as customers. Supply nodes have capacities that cannot be exceeded, and demand nodes have demands that should be met.

The LP formulation for the transportation problem is presented below. In the model, i is an index for supply points and j is an index for demand points. m and n are the number of supply and demand points, respectively. x_{ij} is continuous decision variable that shows the amount of product transported from supply point i to demand point j. c_{ij} is a parameter that shows the unit transportation cost from supply point i to demand point j. s_i is a parameter that shows supply amount of supply point i, and d_j is parameter that shows demand of demand point j.

Objection function:

$$\min \sum_{i=1}^{m} \sum_{j=1}^{n} c_{ij} x_{ij} \tag{11.28}$$

Subject to:

$$\sum_{j=1}^{n} x_{ij} \leq s_i (i = 1, 2, \ldots, m) \tag{11.29}$$

$$\sum_{i=1}^{m} x_{ij} \geq d_j (j = 1, 2, \ldots, n) \tag{11.30}$$

$$x_{ij} \geq 0 (i = 1, 2, \ldots, m)(j = 1, 2, \ldots, n) \tag{11.31}$$

The objective function given in Eq. (11.28) aims to minimize the total cost of the selected distribution plan. Equation (11.29) ensures that no more than supply resources are distributed, and Eq. (11.30) ensures that customer demands are met. In addition, the constraint given in Eq. (11.31) ensures that the amounts to be distributed do not take negative values. The network representation of a generic transportation problem is given in Fig. 11.10.

In order to solve the transportation problem, transportation simplex algorithm can be used. This algorithm depends on the duality theory and is the basic version of network simplex algorithms.

Vehicle Routing Problem

Suppose that a retail company will distribute goods from a central depot to several retail stores located at different points in the network. The stores have demands to be filled. The trucks have limited capacities. The aim is to plan routes for one or several trucks to visit the stores under given restrictions such as truck capacities, time intervals for the visits, driving time limits, etc. This problem is called vehicle routing problem (VRP).

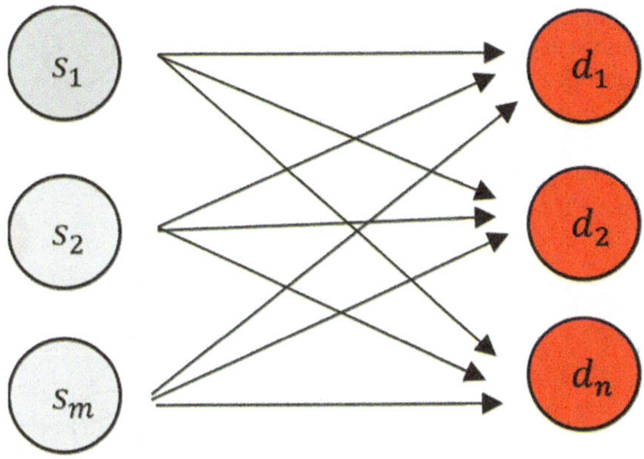

Fig. 11.10 Transportation network

In VRP, the aim is to find the best vehicle route(s) to serve a set of orders from customers. The routes start and end at the origin. The best route is sometimes the least cost, sometimes the least distance, sometimes the least travel time. Vehicles can deliver the products they receive from the warehouse to the customers, as well as pick up the product from any node and deliver it to any node. In Fig. 11.11, a representation of the VRP is given where the red point is the origin and two routes are generated to serve 10 points in the network.

In Fig. 11.12, triangles represent product collection points and circles represent delivery points. The red circle is both the starting and ending point of both routes. The situations where vehicles both deliver and collect products are included in current vehicle routing problems.

Some properties of VRP variants are listed below:

- VRP problems produce solutions for different situations, for example, for single vehicle or multi-vehicle deployment situations.
- There may be different vehicle capacities, such as weight, cubic feet, floor space, and value in the problem type.
- The objective function may consist of different costs or benefits such as minimum route distance, minimum number of vehicles and minimum number of trips, minimum driving time, or maximum profit.

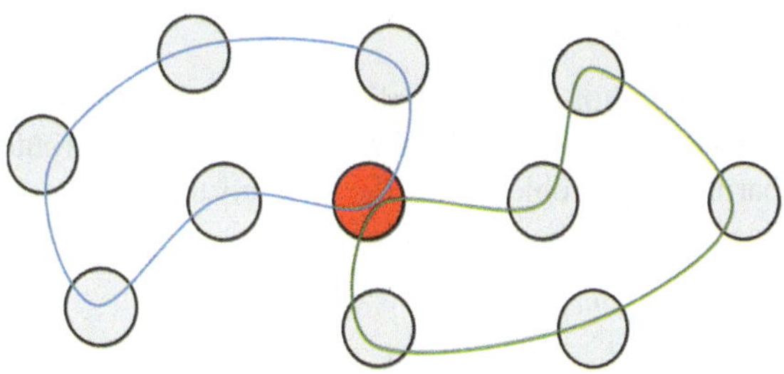

Fig. 11.11 Vehicle routing network

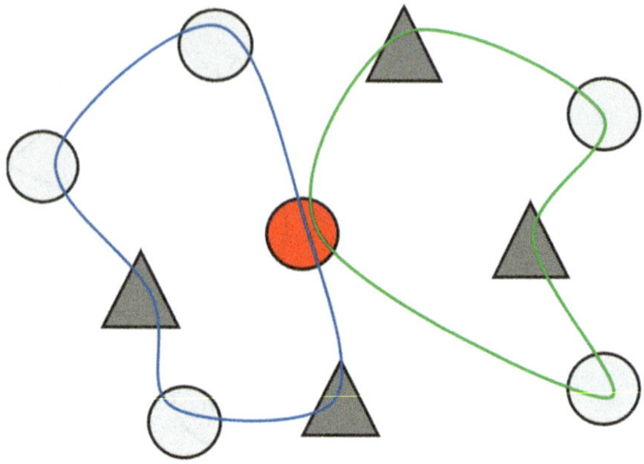

Fig. 11.12 Multi-vehicle routing and scheduling network

- Each variable can have different costs, such as fixed charge, variable costs per loaded mile, per empty mile, cost per stop, loading and unloading cost.
- There may be customer constraints such as customer priority or customer importance weight.

There are many solution algorithms for VRP problems. Some basic methods are explained below:

Sweep Algorithm

The sweep algorithm was first used by Gillett and Miller [9]. Suppose that the coordinates of the distribution center and distribution points are known. Therefore, the angles between the distribution center and the distribution point are calculated. Routes are created in a way that does not exceed capacity value for each demand amount of demand point. As soon as the capacity exceeds, new routes are created for the remaining demand points. Figure 11.13 indicates the angle of two demand points.

Savings Method

The points that offer the greatest savings when combined on the same route are those that are farthest from the depot and that are closest to each other. If the route meets the capacity constraint, these demand points must be within the same route.

Petal Algorithms

The algorithm is a different version of the sweep algorithm. The algorithm creates petal-shaped subsets that form the routes. These routes (petals) are solved like a set partitioning problem, and the least costly routes are selected. The following IP is used to select the optimum petals.

Objection function:

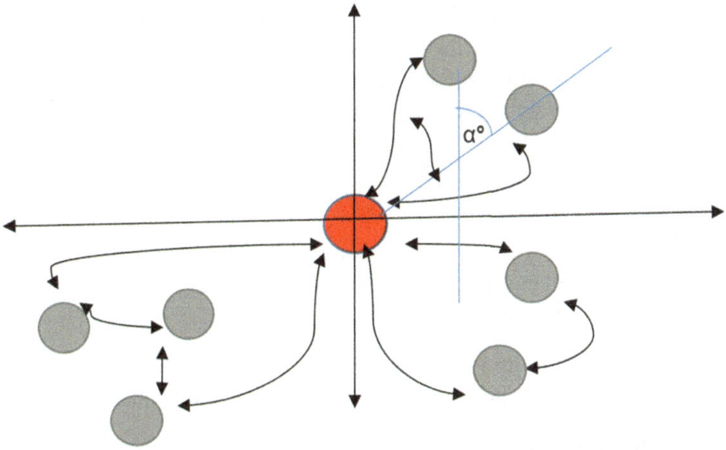

Fig. 11.13 Sweep algorithm

$$min \sum_{k \in S}^{K} c_k x_k \tag{11.32}$$

Subject to:

$$\sum_{k \in S}^{K} a_{ki} x_k = 1 \forall i \tag{11.33}$$

$$x_k = 0 or 1 k \in S \tag{11.34}$$

S indicates the set of the routes. x_k is a binary variable for determining which route will be in the solution set. a_{ki} is a binary parameter that shows the demand point i belongs to route k. c_k is the cost of route k.

The objective function given in Eq. (11.32) aims to minimize the total cost of the selected route. Equation (11.33) indicates that only a single demand point belongs to a route, and Eq. (11.34) indicates that the decision variables are binary.

Case Study

ATK-Cargo has a main distribution center in Uşak to serve the customers located in the Aegean region in Turkey. The company first carries out the customer cargo to province centers via trucks and then delivers the intra-city shipments via small vehicles. The company purchased two trucks for delivering customer cargoes from main station to other cities in the Aegean region. Each truck is to return to the main station (Uşak) after delivering cargoes. The company cannot load more than 30 tons or 80 cubic meters for each truck as per the legislation. The trucks are expected to distribute the freights to all the locations in the Aegean region. The load of each truck can only differ up to 4 tons. In addition, there may be a difference of up to 10 cubic meters between the volumes of the loads included in each truck. The company wants to distribute the loads evenly on the trucks. Therefore, the deviation of each truck's load in terms of both weight and volume from the average loads (per truck) can be at most 10%.

- Aegean region in Turkey and the location of the distribution center in the region are shown in Fig. 11.14.
- The distances (km) between the cities are given in Table 11.17.
- Each truck has a fuel cost of 30 cents for 1 km. The cargoes of each province are given in Table 11.18.

The company wants to find which city should be assigned to which truck and the routes of the trucks. Please help the company to answer these questions.

Solution to the Case Study

The problem is a VRP where two different capacity restrictions exist (i.e., weight and volume). Unlike the classical VRP problems, there is a restriction for distributing the load evenly to the trucks.

Fig. 11.14 Location of distribution center in the Aegean region

Table 11.17 Possible distances between provinces

	Uşak	Afyon	Aydın	Balıkesir	Bursa	Çanakkale	Denizli	Eskişehir	İzmir	Kütahya	Manisa	Muğla
Uşak	0	114	155	153	176	262	95	191	191	72	165	185
Afyon	114	0	261	245	214	354	169	116	306	117	280	268
Aydın	155	261	0	188	279	257	109	346	87	212	92	84
Balıkesir	153	245	188	0	114	109	216	262	146	130	114	264
Bursa	176	214	279	114	0	182	267	180	257	109	224	342
Çanakkale	262	354	257	109	182	0	314	356	183	237	165	341
Denizli	95	169	109	216	267	314	0	273	184	167	172	99
Eskişehir	191	116	346	262	180	356	273	0	368	143	336	371
İzmir	191	306	87	146	257	183	184	368	0	224	33	170
Kütahya	72	117	212	130	109	237	167	143	224	0	192	255
Manisa	165	280	92	114	224	165	172	336	33	192	0	177
Muğla	185	268	84	264	342	341	99	371	170	255	177	0

Table 11.18 Weight and volume of province cargoes

i	1	2	3	4	5	6	7	8	9	10	11	12
	Uşak	Afyon	Aydin	Balikesir	Bursa	Canakkale	Denizli	Eskisehir	Izmir	Kutahya	Manisa	Mugla
Cargo weight/tons	–	2	4	3	6	5	3	6	8	4	5	6
Cargo volume/m3	–	9	12	5	15	16	7	19	22	13	12	15

In order to solve such problems, integer programs (IP) can be used. Although IP solution of VRP problems is NP-Complete, it is efficient to use it in given problem as the problem size is relatively small (12 nodes and 2 trucks). The IP formulation of the given problem is presented as follows:

Indices:

i, *j*, *v*: cities in the Aegean region ($i, j, v = 1, 2, \ldots, 12$)

k: trucks ($k = 1, 2$).

Parameters:

p_j: Cargo weight of city *j*.

c_j: Cargo volume of city *j*.

d_{ij}: Distance between city *i* and city *j* (km).

Sdw: Allowable deviation for a weight of a truck from average weight (in percentages-%10).

Sdv: Allowable deviation for volume of a truck from average volume (in percentages-%10).

w: Average weight.

v: Average volume.

b: maximum load-carrying capacity (30 ton).

s: maximum transport volume (80 cubic meter).

Decision variable:

x_{ijk}: Whether or not the cities *i* and *j* are in the *k* truck route (binary variable).

u_{ij}: Sub-tour variable (continuous variable).

Objection function:

$$\min z : \sum_{i=1}^{I} \sum_{j=1}^{J} \sum_{k=1, k \neq l}^{K} x_{ivk} d_{ij} \tag{11.35}$$

Subject to:

$$\sum_{i=1}^{I} \sum\nolimits_{j=1}^{J} p_j x_{ijk} \leq w(1 + Sdw) \forall k \tag{11.36}$$

$$\sum_{i=1}^{I} \sum\nolimits_{j=1}^{J} p_j x_{ijk} \geq w(1 - Sdw) \forall k \tag{11.37}$$

$$\sum_{i=1}^{I}\sum\nolimits_{j=1}^{J} c_j x_{ijk} \le w(1 + Sdv)\forall k \tag{11.38}$$

$$\sum_{i=1}^{I}\sum\nolimits_{j=1}^{J} c_j x_{ijk} \ge w(1 - Sdv)\forall k \tag{11.39}$$

$$\sum_{i=1}^{I}\sum\nolimits_{j=1}^{J} p_j x_{ijk} \le b \tag{11.40}$$

$$\sum_{i=1}^{I}\sum\nolimits_{j=1}^{J} c_j x_{ijk} \le s \tag{11.41}$$

$$\sum_{i=2}^{K} x_{i1k} = 1\forall k \tag{11.42}$$

$$\sum_{j=2}^{K} x_{1jk} = 1\forall k \tag{11.43}$$

$$\sum_{k=1}^{K}\sum\nolimits_{i=1}^{N} x_{ijk} = 1\forall j j = 2, 3, 4 \ldots N \tag{11.44}$$

$$\sum_{k=1}^{K}\sum\nolimits_{j=1}^{N} x_{ijk} = 1\forall i i = 2, 3, 4 \ldots N \tag{11.45}$$

$$\sum_{i=1}^{I}\sum_{j=1}^{J}\sum_{k=1,k\ne l}^{K} x_{ivk} - \sum_{i=1}^{I}\sum_{j=1}^{J}\sum_{k=1,k\ne l}^{K} x_{vjk} = 0\forall k, v, v = 2, 3, 4 \ldots N \tag{11.46}$$

$$u_{ij} \ge 1\forall i, j \tag{11.47}$$

$$u_{ij} \le n\forall i, j \tag{11.48}$$

$$u_{ij} - u_{ji} + 1 \le (n - 1)\left(1 - x_{ijk}\right)\forall i, j, k \tag{11.49}$$

$$x_{ijk} = 0, 1\ u_{ij} \ge 0 \tag{11.50}$$

In Eq. (11.35), the cost of energy has been aimed to be minimized for the route of each truck to be formed. Equations (11.36) to (11.37) provide a balance of 10% deviation in the load and volume of the trucks. Equations (11.38) and (11.39) provide the legal weight and volume constraints in the loading. Equations from (11.40) to (11.46) enable visits of the vehicles to the points starting and ending at point 1.

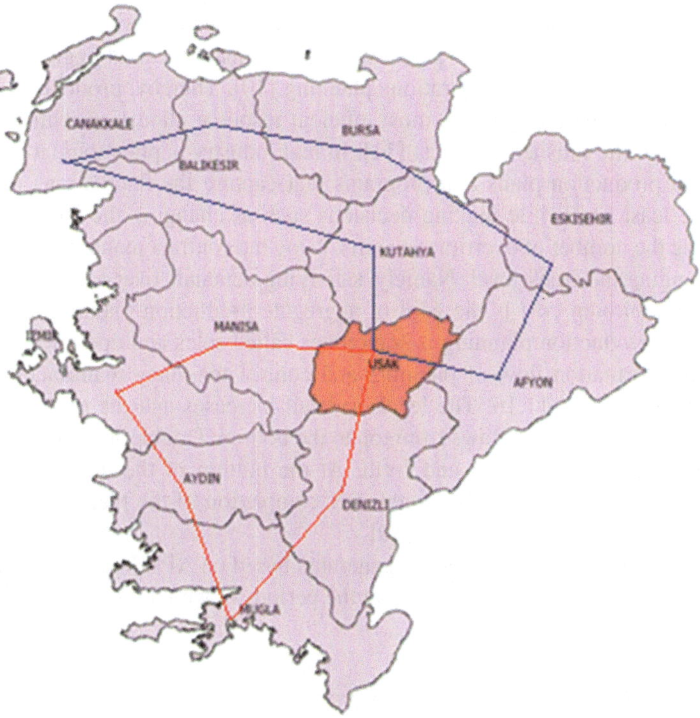

Fig. 11.15 Two different truck routes for the case study

Possible sub-tours are barred in Eqs. (11.47), (11.48), and (11.49). Equation (11.50) provides the definition of the decision variables and ensures non-negativity.

Result

The problem was solved in the Pyomo library and the following results were obtained. Figure 11.15 shows the solution for the case study.

- A route of the first vehicle: Uşak-Kütahya-Çanakkale-Balıkesir-Bursa-Eskişehir Afyon- Uşak.
- A route of the second vehicle: Uşak- Manisa- İzmir- Aydın- Muğla- Denizli-Uşak.
- Optimal route cost: 38,040 dollars.

6 Aggregate Production Planning

Problem Definition

Strategic and business plans tend to be very general in coordinating and timing the resource needs and action plans of many key functions of the firms. However, more detailed planning of resources, including the type and amount of the resources, and the timing of resources, is performed by the function called aggregate production planning or sales and operations planning [10]. Thereby, production planning is essential for determining the most efficient resource allocation while satisfying demands for the finished products [11]. In real industry applications, it is intended to obtain production plans and programs that capture the fluctuations in demand with the least cost while making decisions such as changing the workforce level, changing the number of overtime or shifts, using the contract manufacturing facility, and changing the stock level. Namely, satisfying demands over a planning horizon with the minimum cost is the goal of aggregate production planning (APP) [12]. Aggregate production planning is sometimes called sales and operations planning. General information flow of planning and control for many manufacturing firms is illustrated in Fig. 11.16. The level of detail increases and the planning horizon decreases as the diagram flows from top to the bottom. Double-headed arrows mean that information flows back and forth. At the bottom of the diagram, execution activities control the actual activities after completion of the planning section and commencement of the production [10].

Master production schedules are generated based on APP. Master schedule planning yields the production quantities in the periods in which production takes place

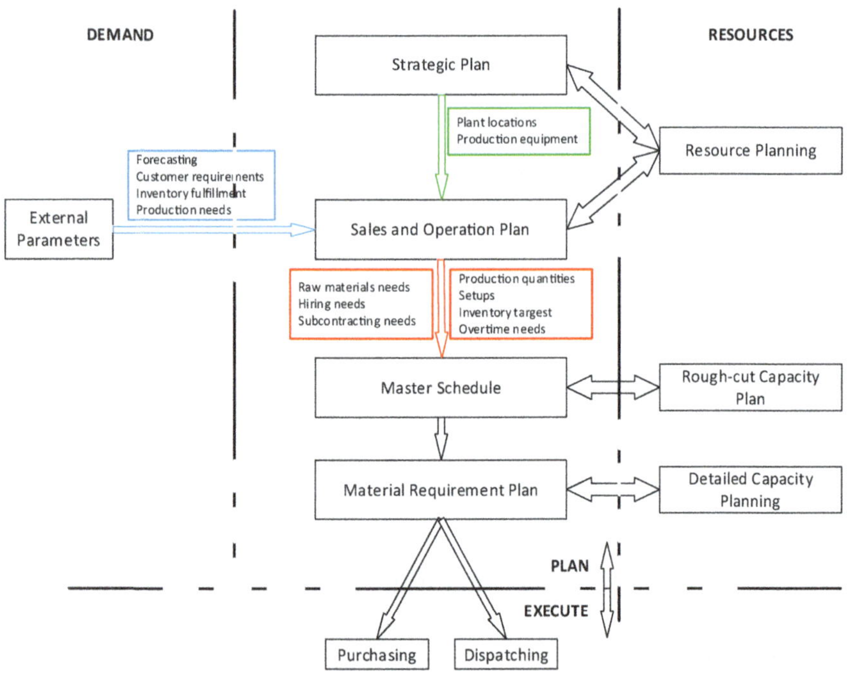

Fig. 11.16 Information flow of planning and control activities

for each product type according to the forecasted quantities in the APP and the finalized (firm) orders. In a more detailed production schedule, on the other hand, the date on which the production of each part will begin, the machines on which the operations will be carried out, the priority of processing each part and the completion times of the operations are specified.

In general, there are three basic strategies in preparing an aggregate production plan with an approach that integrates sales, production, and inventory functions:

Chase: The production rate monitors the changes in demand very closely, and therefore, the stock level is close to zero. However, resources altering are performed, including internal strategies such as hiring or firing, overtime, and subcontracting.

Level: The demand is tried to be met during the planning period by keeping the parameters of production rate and workforce level constant.

Combination: It is a strategy created by combining the first two strategies to balance their drawbacks and capture their advantages. The graphical representation of these three strategies is given in Fig. 11.17.

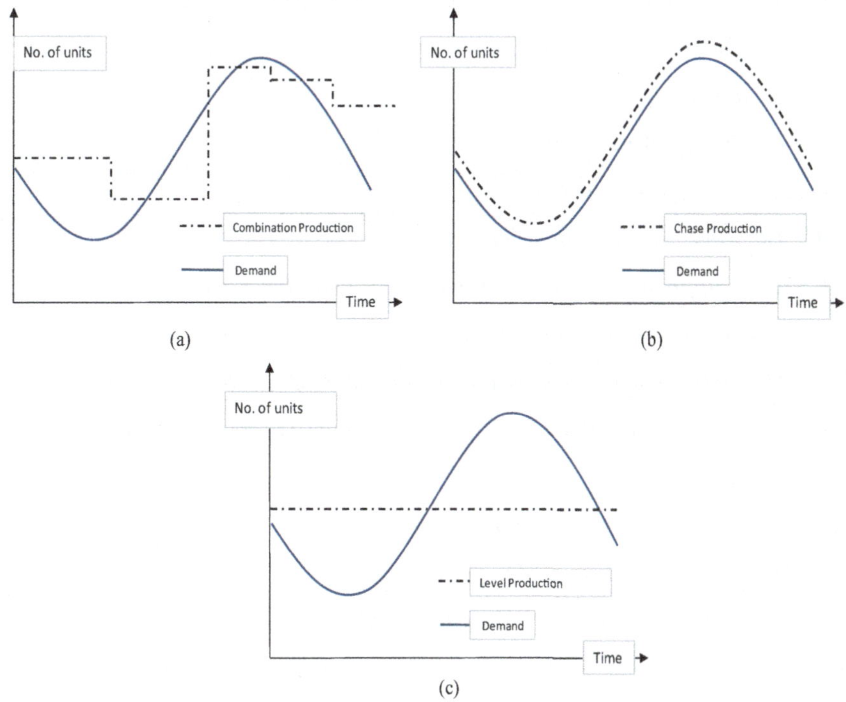

Fig. 11.17 Demand patterns of **a** combination, **b** chase and **c** level strategy

Model

The parameters and variables of the production planning case are introduced as follows.

Parameters:

$t : 1, 2, \ldots, T$, *where T is the planning horizon.* ST is the set of periods.

$i : 1, 2, \ldots, I$, *where I is the number of product families.* SI is the set of product families.

R_i : *The number of units for product i that one worker can produce in a period in regular time.*

O_i : *The number of units for product i that one worker can produce in a period in vertime.*

IW : *The initial workforce level.*

II_i : *The initial inventory level for product i.*

DW : *The desired workforce level at the end of the planning horizon.*

DI_i : *The desired inventory level at the end of the planning horizon for product i.*

D_{it} : *Demand for product i in period t.*

CH : *The cost of hiring a worker.*

CF : *The cost of firing a worker.*

CW : *The wage cost of a worker in a period.*

CR_i : *The unit production cost for product i per period in regular time.*

CO_i : *The unit production cost for product i per period in overtime.*

CS_i : *The unit production cost for product i per period in subcontracting.*

CK_i : *The cost per period of carrying one unit of inventory for product i.*

Decision variables:

H_t : *The number of workers hired in period t.*

F_t : *The number of workers fired in period t.*

W_t : *The number of workers employed in period t.*

r_{it} : *The number of units produced for product i on regular time in period t.*

o_{it} : *The number of units produced for product i on overtime in period t.*

k_{it} : *The number of units stored for product i in inventory at the end of period t.*

Objective function:

$$TC = \min\sum_{t=1}^{T} CH * H_t + CF * F_t + CW * W_t$$
$$+\sum_{i=1}^{I}\sum_{t=1}^{T}(CR_i * r_{it} + CO_i * o_{it} + CS_i * s_{it} + CK_i * k_{it}) \quad (11.51)$$

Constraints:

$$\left(k_{i(t-1)} + r_{it} + o_{it} + s_{it} - D_{it}\right) = k_{it}, \forall t \in ST : t > 1, \forall i \in SI \quad (11.52)$$

$$(II_i + r_{i1} + o_{i1} + s_{i1} - D_{i1}) = k_{i1}, \forall i \in SI \quad (11.52')$$

$$\sum_{i=1}^{I} \frac{r_{it}}{R_i} \le W_t, \forall t \in ST \quad (11.53)$$

$$\sum_{i=1}^{I} \frac{o_{it}}{O_i} \le W_t, \forall t \in ST \quad (11.54)$$

$$\sum_{i=1}^{I} s_{it} \le 1000, \forall t \in ST \quad (11.55)$$

$$W_{(t-1)} + H_t - F_t = W_t, \forall t \in ST : t > 1 \quad (11.56)$$

$$IW + H_1 - F_1 = W_1 (11.56')$$

$$W_T = DW, \forall t \in ST \quad (11.57)$$

$$k_{i0} = II_i, \forall t \in ST, \forall i \in SI \quad (11.58)$$

$$k_{iT} = DI_i, \forall t \in ST, \forall i \in SI \quad (11.59)$$

$$H_t, F_t, r_{it}, o_{it}, s_{it} = 0, t = 0, \forall i \in SI \quad (11.60)$$

$$H_t, F_t, W_t, r_{it}, o_{it}, s_{it}, k_{it} \ge 0, \forall t \in ST, \forall i \in SI \quad (11.61)$$

Objective function minimizes the total cost of hiring, firing, wages, regular time, overtime, subcontracting, and inventory for T periods. Constraints (11.52) and (11.52') indicate the inventory balance relationships in each period and for each product. Constraint (11.53) denotes that production in regular time does not exceed the maximum allowed production quantity of regular time. Likewise, constraint (11.54) ensures that overtime production quantity does not exceed the maximum capacity of overtime. Constraint (11.55) ensures that the total amount of subcontracting in a period is not greater than 1000 units. Constraints (11.56) and (11.56') state the workforce balance relationship in each period. Constraints (11.57)–(11.60) are initializing conditions for workforce level and inventory level. Constraint (11.61) shows the sign restrictions.

In addition to the mathematical model, two general strategies of sales and operations planning are implemented. Chase and level strategies are given as pseudocode in Figs. 11.18 and 11.19, respectively. These heuristics are valid for a single product family.

Case Study: A Bicycle Production Company

A bicycle production company aims to optimize its aggregate planning to minimize costs. Three products are manufactured in the company. The details of the parameters utilized in the case are presented in Table 11.19. The aggregate demand forecast is shown in Table 11.20.

Assumptions for the case include the following:

- Parameters are deterministic and products' demands are known.
- Cost values are constant over the entire planning horizon.
- As the number of operators increases, the capacity increases linearly.
- Demands of the customers should be met.

Compute production levels in a period ($Production_t$) and safety stock level (SS_t)

$$Production_t = D_t + SS_t$$

$SS_t = D_t * s$, where s is the safety stock percentage

If over time is allowed, then

For $t=1$ **to** T, **compute**

$$k_t = SS_t$$
$$W_t = Production_t/(\#ofworkdays_t * (R+O))$$
$$H_t - F_t = W_t - W_{t-1}$$
$$r_t = R * W_t * (\#ofworkdays_t)$$
$$o_t = O * W_t * (\#ofworkdays_t)$$
$$TC_t = CH * H_t + CF * F_t + CW * W_t + CR * r_t + CK * k_t + CO * o_t,\ t=1,\ldots,T$$

Next t **compute** total cost with over time ($TC_{chase\ OT}$)

$$TC_{chase\_OT} = \sum_{t=1}^{T} TC_t,$$

If overtime is not allowed, **then compute**

$$k_t = SS_t$$
$$W_t = Production_t/(\#ofworkdays * (R))$$
$$H_t - F_t = W_t - W_{t-1}$$
$$r_t = R * W_t * (\#ofworkdays_t)$$
$$TC_t = CH * H_t + CF * F_t + CW * W_t + CR * r_t + CK * k_t,\ t=1,\ldots,T$$

Next t **compute** total cost without over time ($TC_{chase\ RT}$)

$$TC_{chase\_RT} = \sum_{t=1}^{T} TC_t,$$

If $TC_{chase\_RT} > TC_{chase\_OT}$ **then** select TC_{chase_OT}

Else select TC_{chase_RT}

Fig. 11.18 Chase aggregate production planning heuristic

Compute total number of workdays (*TW*), total forecasted demand of product family (*TFD*) and daily production rate (*DPR*)

$$TW = \sum_{t=1}^{T} \#\_of\_Workdays_t,$$
$$TFD = \sum_{t=1}^{T} D_t,$$
$$DPR = (TFD - Initial\_inventory)/TW$$
$$Production_t = DPR * \#\_of\_Workdays_t$$

If overtime is allowed, **then compute** the number of workers in period t (W_t), the number of hired workers in period 1 (F_1), the number of fired workers in period 1 (H_1), regular time production in period t (r_t), over time production in period t (o_t)

$$W_1 = W_t = DPR\ /\ (R + O)$$
$$H_1 = (DPR/(R + O)) - IW$$
$$F_1 = IW - (DPR/(R + O))$$

Note: R and O indicate the number of product units that one worker can produce in a period on regular and over time, respectively.

$$r_t = W_t * R * \#\_of\_Workdays_t$$
$$o_t = W_t * O * \#\_of\_Workdays_t$$

For $t=1$ **to** T, **compute** inventory level for each period (k_t), and total cost for each period (TC_t)

$$k_t = k_{t-1} + r_t + o_t - D_t + B_t$$
$$TC_1 = CH * H_1 + CF * F_1 + CW * W_1 + CR * r_1 + CO * o_1 + CK * k_1 + CB * B_1$$
$$TC_t = CW * W_t + CR * r_t + CO * o_t + CK * k_t + CB * B_t\ ,\ t=2,\ldots,T$$

Note: B_t and CB indicate the backlog amount in period t and the cost of backlog, respectively.

Next t **compute** total cost with over time (TC_{wo})

$$TC_{wo} = \sum_{t=1}^{T} TC_t$$

If overtime is not allowed, **then compute**

$$W_1 = W_t = DPR\ /\ R$$
$$H_1 = (DPR/(R)) - IW$$
$$F_1 = IW - (DPR/(R))$$
$$r_t = W_t * R * \#\_of\_Workdays_t$$
$$o_t = 0$$

For $t=1$ **to** T, **compute** inventory level for each period (k_t), total cost for each period (TC_t)

$$k_t = k_{t-1} + r_t - D_t + B_t$$
$$TC_1 = CH * H_1 + CF * F_1 + CW * W_1 + CR * r_1 + CK * k_1 + CB * B_1$$
$$TC_t = CW * W_t + CR * r_t + CK * k_t + CB * B_t\ ,\ t=2,\ldots,T$$

Note: B_t and CB indicate the backlog amount in period t and the cost of backlog, respectively.

Next t **compute** total cost without over time (TC_{no})

$$TC_{no} = \sum_{t=1}^{T} TC_t$$

If $TC_{wo} > TC_{no}$ **then** select TC_{no}

Else select TC_{wo}

Fig. 11.19 Level aggregate production planning heuristic

Results and Discussion

The mathematical model was solved by CPLEX solver for the case explained above. The solution yielded the production quantities in regular time, overtime as well as subcontracting, the amount to be stored, the number of workers fired or hired in each period, as shown in Table 11.21. The total cost was found as US$109,210,929.

Table 11.19 Characteristics of the production system

Parameters	Value	Notation
The number of units that one worker can produce in a period on regular time for product i	10, 8, and 9 units	R_i
The number of units that a worker can produce in a period on overtime for product i	8, 6, and 7 units	O_i
The initial workforce level	100 workers	IW
The desired workforce level at the end of the planning horizon	140 workers	DW
The initial inventory level for product i	1000, 1150, and 1800 unit	II_i
The desired inventory level at the end of the planning horizon for product i	0 unit for all product type	DI_i
Unit production cost per period on regular time for product i	30, 35, and 20 \$/hour	CR_i
Unit production cost per period in overtime for product i	40, 44, and 38 \$/hour	CO_i
Unit production cost per period in subcontracting for product i	500, 550 and 440 \$/hour	CS_i
The cost of hiring a worker	3500\$/worker	CH
The cost of firing a worker	1000\$/worker	CF
The wage cost of a worker in a period	4000\$/worker	CW
Cost per period of holding one unit of inventory for product i	15, 20, and 13 \$/unit	CK_i

Table 11.20 Demand forecast of three products

Product 1	Period	D_{it}	Period	D_{it}	Period	D_{it}
	1	1000	6	1000	11	4400
	2	3300	7	1500	12	6000
	3	5800	8	2400	13	5500
	4	3200	9	3000	14	3000
	5	2200	10	3500	15	2200
Product 2	1	2000	6	1000	11	2700
	2	3600	7	1600	12	5000
	3	4200	8	3500	13	5700
	4	2300	9	3400	14	4500
	5	3300	10	3500	15	5600
Product 3	1	1500	6	2000	11	1800
	2	2500	7	1700	12	5000
	3	1600	8	2000	13	5700
	4	4200	9	3000	14	1000
	5	2000	10	3700	15	1600

Table 11.21 Optimal results of the model

	Product 1					Product 2					Product 3				
t	D_{it}	r_{it}	o_{it}	s_{it}	k_{it}	D_{it}	r_{it}	o_{it}	o_{it}	k_{it}	D_{it}	r_{it}	o_{it}	s_{it}	k_{it}
0					1000					1150					1800
1	1000	0	724	0	724	2000	800	3	47	0	1500	0	63	953	1316
2	3300	0	3896	0	1320	3600	3512	0	88	0	2500	432	0	912	160
3	5800	0	4480	0	0	4200	3200	0	1000	0	1600	1440	0	0	0
4	3200	0	3200	0	0	2300	1896	0	404	0	4200	2907	1120	596	423
5	2200	0	2200	0	0	3300	2304	0	997	1	2000	2448	1995	3	2869
6	1000	0	1000	0	0	1000	0	0	1000	1	2000	5040	3045	0	8954
7	1500	0	1504	0	4	1600	600	0	1000	1	1700	4365	2604	0	14,223
8	2400	0	2396	0	0	3500	2496	3	1000	0	2000	2232	1820	0	16,275
9	3000	0	4480	0	1480	3400	2400	0	1000	0	3000	2340	0	0	15,615
10	3500	0	4480	0	2460	3500	2504	0	1000	4	3700	2223	0	0	14,138
11	4400	0	4480	0	2540	2700	3936	0	1000	2240	1800	612	0	0	12,950
12	6000	0	4480	0	1020	5000	4480	0	1000	2720	5000	0	0	0	7950
13	5500	0	4480	0	0	5700	4480	0	1000	2500	5700	0	0	0	2250
14	3000	0	4080	0	1080	4500	4480	0	1000	3480	1000	0	350	0	1600
15	2200	0	1120	0	0	5600	1120	0	1000	0	1600	0	0	0	0

Figure 11.20 presents comparisons among production levels for the product families. Accordingly, product families 2 and 3 are produced during regular time and by subcontracting. However, product family 1 is totally produced during overtime. In other words, there are no regular time and subcontracting production for product family 1 and no overtime production for product family 2.

Figure 11.21 indicates the number of hired and fired workers for each period. Production starts with 100 workers. Afterward, 387 workers are hired in the second period and 73 workers are hired in the third period, 420 workers are fired in the last period, reaching the desired number of workers of 140.

In addition to the solution of the model, the level and chase heuristics are employed for solving the case, only by considering Product Family-1, as the heuristics are designed for a single product family. The codes of the heuristics are formulated using Python and developed in a notebook, then run the codes of the heuristic algorithms. In the level aggregate production planning heuristic, the workforce is constant throughout the periods, while in the chase strategy it is adjusted according to the production amount. The proposed heuristic algorithms for the level and chase strategies handle two different scenarios. In one scenario, overtime will be allowed, while in the other, production will be covered entirely during regular time. It is assumed that there is a single product family production. The results are given in Table 11.22.

The results indicated the total cost improvement for the scenario where overtime is not allowed in the level strategy heuristic (12.7%) with reference to a scenario with overtime. However, the best decision is allowing overtime in terms of total cost

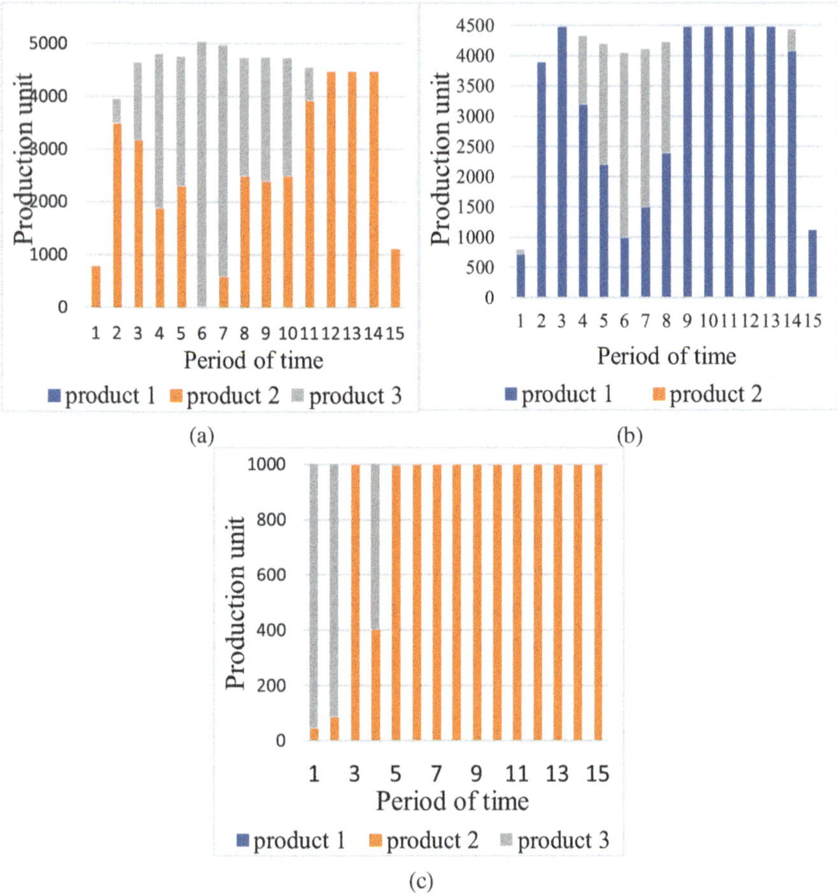

Fig. 11.20 Production levels of regular time (**a**), overtime (**b**), and subcontracting (**c**)

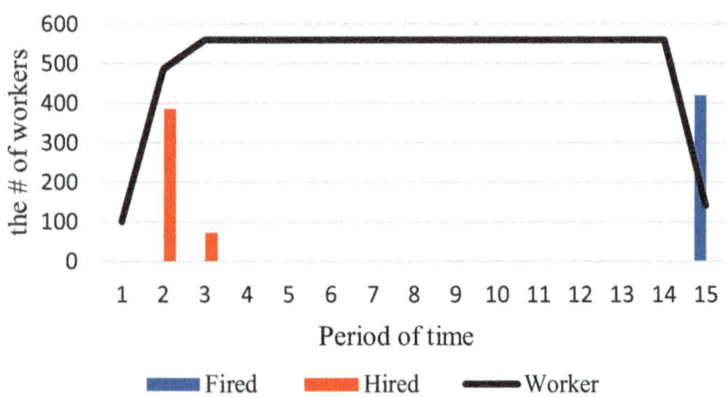

Fig. 11.21 Workforce level of plant

Table 11.22 Results of the proposed heuristics

Strategies	Scenario	Total cost
Level	With overtime	$3,646,630
	Without overtime	$3,183,250
Chase	With overtime	$2,428,860
	Without overtime	$3,550,060

improvement (31.58%) for the chase strategy production plan. Besides, the results reveal that the chase strategy algorithm has the lowest total cost in the case where overtime is allowed.

References

1. Silver EA, Pyke DF, Peterson R (1998) Inventory management and production planning and scheduling, vol 3. Wiley, New York
2. Prasad S (1994) Classification of inventory models and systems. Int J Prod Econ 34(2):209–222
3. Heizer J (2016) Operations management, 11/e. Pearson Education India
4. Suwondo E, Yuliando H (2012) Dynamic lot-sizing problems: a review on model and efficient algorithm. Agroind J 1(1):36
5. Chopra S, Meindl P (2013) Supply chain management strategy, planning, and operation, US: Pearson Education
6. Crandall R, Crandall W, Chen C (2015) Principles of supply chain management, Taylor & Francis Group
7. Hugos M (2011) Essentials of supply chain management. Wiley, Canada
8. Bang-Jensen J, Gutin GZ (2008) Digraphs: theory, algorithms and applications. Springer Science & Business Media
9. Gillett BE, Miller LR (1974) A heuristic algorithm for the vehicle-dispatch problem. Oper Res 22(2):340–349
10. Chapman SN (2006) The fundamentals of production planning and control. Upper Saddle River, NJ: Pearson/Prentice Hall
11. Pereira DF, Oliveira JF, Carravilla MA (2020) Tactical sales and operations planning: a holistic framework and a literature review of decision making models. Int J Prod Econ 228:107695
12. Hafezalkotob A, Chaharbaghi S, Lakeh TM (2019) Cooperative aggregate production planning: a game theory approach. J Ind Eng Int 15(1):19–37

Yakup Turgut was born in Turkey, in 1993. He received a B.E. degree in industri-al engineering from Yıldız Technical University, Turkey, in 2015, and a master's degree from the Istanbul Technical University, Turkey, in 2018. He is a Ph.D. candidate in industrial engineering at Istanbul Technical University. His research interests are simulation modeling, mathematical modeling, and artificial intelli-gence. His email address is turgut16@itu.edu.tr.

Kenan Menguc completed his undergraduate education in 2016 at Doğuş University, Department of Industrial Engineering and his M.Sc. in 2018 at Istanbul University, Department of Industrial Engineering. He worked as a lecturer in the Department of Logistics at Beykent University between 2018 and 2020. While he currently conducts his Ph.D. study at Yıldız Technical University, he works as a research assistant at Istanbul Technical University.

Nursah Alkan is a Ph.D. candidate and a Research Assistant in the Department of Industrial Engineering at Istanbul Technical University since 2019. She re-ceived the M.Sc. degree in Department of Industrial Engineering from Yildiz Technical University, Turkey, in 2019. Her research interests include fuzzy sets, multi-criteria/objective decision making, data analysis, machine learning and digital transformation.

Yildiz Kose graduated from Industrial Engineering Department of Gazi Universi-ty, Ankara, in 2014. She received her M.Sc. degree from Industrial Engineering Department of Karadeniz Technical University, Trabzon, in 2017. Since 2018, she has been a Ph.D. candidate and research assistant at Istanbul Technical University. Her current research interests are production planning, assembly line and lean production systems.

Seval Ata is a Research Assistant in the Department of Industrial Engineering at Istanbul Technical University and a Ph.D. Candidate in the Industrial Engineering department at Boğaziçi University. She received her BS and MS degrees in Industrial Engineering from Boğaziçi University in 2015 and 2018, respectively. Her re-search interests include stochastic modeling of hybrid manufacturing systems, optimal production decision mechanisms and currently, she is mainly focusing on adaptive production control mechanisms with the aim of increasing environmental sustainability.

Sule Itir Satoglu earned her undergraduate degree from Yildiz Technical University, Mechanical Engineering Department, then her master's and Ph.D. degrees from Istanbul Technical University, Industrial Engineering Department, in 2002 and 2008, respectively. She has been working as faculty member of Istanbul Technical University, Industrial Engineering Department, since 2009. Her research interests include supply chain management, sustainability, humanitarian logistics and lean production systems.

Ozgur Kabak is an associate professor at Industrial Engineering Department of Istanbul Technical University (ITU-IE). Before lecturing in ITU-IE, he spent one year at Belgium Nuclear Research Centre for a post-doc research that was granted by Belgian Science Policy (BELSPO). He teaches undergraduate and graduate education courses in operations research, group decision-making and logistics management. His research explores how to model complex systems such as socio-economic systems, productions systems, supply chains, logistics and transportation systems, etc. and how to make decisions in group decision making envi-ronments and incomplete information situations. He has published in indexed journals including European Journal of Operations Research, Information Fusion, Transport Policy, Knowledge-based Systems, IEEE Transactions on Knowledge and Data Engineering, and Socio-Economic Planning Sciences, etc. His work has been featured in international conferences and meetings including MCDM, FLINS, WCTR and EURO-k conference series.

CRM and Marketing Analytics

Sultan Ceren Oner, Yusuf Isik, Abdullah Emin Kazdaloglu, Mirac Murat, Tolga Ahmet Kalayci, Kubra Cetin Yildiz, Aycan Pekpazar, Mahmut Sami Sivri, Nevcihan Toraman, Basar Oztaysi, Umut Asan, and Cigdem Altin Gumussoy

1 Introduction

Customer relationship management (CRM) and marketing analytics is a combination of techniques, technologies, and strategies that serve to create and deliver value to profitable customers. It involves internal business processes and functions (such as marketing, sales) as well as external influences (such as competitors). CRM systems collect, analyze, and model information about customers using data science methods at all stages of their life cycle to establish and maintain long-term profitable relationships and create loyal customers. Attracting and retaining profitable customers, preventing loss of customers as well as reactivating passive customers are main functions of CRM systems. Further applications of CRM and marketing analytics can be summarized as follows. Adopting analytical methods in pricing allows examining past purchases and thereby forecasting demand, which leads to optimizing profits under limited resources. Firms use web analytics to observe, model, and predict behaviors of website users, which helps to detect potential customers of new products, develop personalized offers, and optimize web usage. Text mining enables analysis of unstructured text data such as customer complaints and suggestions to extract useful, interesting, and hidden knowledge. Empowered by supervised and unsupervised learning algorithms, recommendation systems generate useful data for analyzing customer desires and data-driven suggestions.

S. C. Oner
Saray, Ahmet Tevfik Ileri Street 10B, 34768 Umraniye, Istanbul, Turkey

Y. Isik · A. E. Kazdaloglu · M. Murat · T. A. Kalayci · K. C. Yildiz · A. Pekpazar · M. S. Sivri · N. Toraman · B. Oztaysi · U. Asan · C. A. Gumussoy (✉)
Department of Industrial Engineering, Faculty of Management, Istanbul Technical University, 34367 Macka, Istanbul, Turkey
e-mail: altinci@itu.edu.tr

A. Ustundag et al. (eds.), *Business Analytics for Professionals*, Springer Series in Advanced Manufacturing, https://doi.org/10.1007/978-3-030-93823-9_12

In this chapter, five cases are presented including revenue management and pricing, customer churn analysis, social media and web analytics, complaint analysis, and recommendation systems.

2 Revenue Management and Pricing

Revenue management, which emerged in the USA in the late 1970s and was initially compatible only with the airline concept but is now applied in many sectors such as airline, hotel, health, printing, publishing, railway, telecommunications, car rental, e-commerce, also plays a very important role in supply chains. In addition, in the presence of factors such as whether the products sold have seasonal characteristics, whether they are sold wholesale and spot, whether they have different values in different segments, and whether they tend to deteriorate or be defective, they have a significant share in terms of profitability of companies [1].

The primary goal of pricing and revenue management in the supply chains of companies is to optimize profits. While performing this optimization, it is necessary to consider the balance between supply and demand [2]. From the point of view of companies, although adjusting the size of the companies' supply chain assets is the first method that comes to mind for revenue management, companies today use pricing as the primary tool. When considered in a broader framework, it was proven that profit margins can be increased more effectively in managerial approaches in which pricing and revenue management are considered together in the supply chains of companies [3]. The information learned through revenue management indicates that pricing should be used first in keeping the balance between supply and demand for any company. Yet, after this balance is maintained, it is suggested that various financial investments should be made and/or some asset items should be eliminated. In short, pricing is considered to be a factor that increases profits by adjusting the balance of supply and demand. Revenue management, on the other hand, is defined as a practice in which pricing is used to maximize profits in the supply chain.

The major purpose of revenue management, which takes a qualitative approach to pricing and inventory management, is to maximize the profit that the company can achieve through demand forecasting of the market. Adopting analytical methods about where and when demand will come, revenue management examines past purchases and customer behavior. In this way, it optimizes the supply chain processes and, as a result, promotes the customer experience by using the pricing factor. Moreover, it was underlined that revenue management strategies yield better results on products that are prone to price fluctuations due to the pricing dimension of revenue management. For instance, it is emphasized that the revenue management process should be considered for products with seasonal effects or for products sold only at certain times of the year [4]. In summary, pricing and revenue management should be implemented by companies to optimize profits when operating with limited assets, such as capacity and inventory in a supply chain.

2.1 Problem Definition

In the last few decades, the increasing competitive environment due to globalization has made revenue management an important part of companies, which they see as a solution to sustain their profitability. It is a fact that revenue management encourages innovative research methods for firms to develop optimal price strategies for their products [5]. It was originally introduced only for the airline industry, but has been adopted in many different industries. In this way, revenue management systems have been developed over the years by following multi-criteria pricing policies in industries. Thus, it is known that different companies significantly increase their revenues annually thanks to revenue management. However, it is observed that decision makers in firms often tend to set the prices of products lower than the theoretical optimum price. Therefore, it is possible that the company cannot obtain the potential maximum income that can be elicited from the products. Therefore, various mathematical models and revenue optimization model should be adopted.

There are several challenges in revenue management and pricing. The first problem that comes to mind is which model to choose while predicting and which parameters have a role in pricing. Hence, the model should include a holistic approach through various parameters such as exchange rate, previous sales amounts and revenues, past sales amounts and revenues of competitors, promotion and/or discount of the product. In order to eliminate all these problems, it is obvious that companies should look for solid, robust, and dynamic solutions for product pricing while managing their revenue. With this point of view, necessary parameters should be obtained by making databases with dense information content more understandable. Various machine learning algorithms and product-based sales prediction and revenue optimization should be made by means of these parameters. In this context, answers will be sought to the following main issues:

- Which machine learning method should be used in forecasting?
- Which variables should be considered when setting up the model?
- How should the company's revenue corresponding to the optimal price on a product basis be calculated?

2.2 Case Study: A White Goods and Technology Company (Company A)

Founded in 1986, Company A is a company with more than 30 thousand employees worldwide, with 26 production facilities in ten countries in total in Asia, Europe, and Africa, and sales and marketing offices in 39 countries. Company A provides services in nearly 150 countries, both online and offline, and its headquarter is in country B. Company A, whose main focus is to produce products such as white goods, electronics, small household goods, and air conditioners, sold 55 thousand products on a global scale in 2020 and earned a total income of $ 4.5 million from these products. In addition, Company A also sells through its own website. In 2020, the website was visited by more than 400 million unique users, it receives approximately 3 million clicks on a monthly basis, and over 10 thousand transactions are made. Considering online shopping alone, Company A sold approximately 15,000 products in 2020, generating over $1 million in revenue from these products.

In this case study, the main focus will be the 17-month sales data of the X product produced by company A for the years 2019 and 2020. The dataset includes basic sales information of product X, information about competitors of company A, discount information, and various economic indicators. All information in the dataset has the following assumptions and operational financial environment:

- The numerical information in the dataset has not been intentionally or unintentionally distorted, and this data was audited by an independent auditor.
- In terms of product X, sales in company A's online and offline stores are evaluated together.
- It is assumed that government policies have no and will not have an impact on product X produced by company A.
- All aspects of the X product's technical specifications are not covered by the case study.

The sales department of company A wants to develop a revenue optimization model based on the number of sales of X product. In this context, a machine learning model will be established by considering the dataset of product X, and the discounted price of product X will be obtained that maximizes revenue. Table 1 lists the attributes to be used in the dataset. By using StatsModel and Sklearn open libraries, revenue optimization of product X will be made.

2.3 Model

The notations and explanations of the model are shown below:

$f(x)$: Prediction of number of sales.

ρ: Distribution of average selling price.

c: Profitability.

$$\text{Income} = f(x) \times \rho \tag{1}$$

$$\text{Optimum Revenue: } \max[f(x) \times \rho] \tag{2}$$

$$\text{Profit} = f(x) \times \rho \times c \tag{3}$$

Table 1 Specifics of the features that were used to maximize revenue

Variable name	Variable description
Year	Production year
Month	Production month
Recommended price	Recommended price of current month
Avg sales price	Average sales price of current month
Total sold	Total sales of current month
Sold without discount	Total sales per month w/o discount
Discount	Discount (1) or not (0)
Promotion	Promotion (1) or not (0)
Competitor_price	Selling price of competitors in the current month
Previous avg sales price	Average sales price of last month
Previous total sold	Total sales of last month
Two previous avg sales price	Average sales price of two months before
Two previous total sold	Total sales of two months before
Previous competitor_price	Selling price of competitors in last month

$$\text{Optimum Profit: } \max[f(x) \times \rho \times c] \tag{4}$$

Parameters indicated with the notation 'x' in the prediction of the number of sales are product price, competitor prices, market information, economic indicators, and discount information. In this model, the aim is to create the function that gives the average monthly sales price using the discounted sales price of the X product. Then, using the parameters determined with the 'x' notation, a function that calculates the predicted number of sales on a monthly basis is created. Finally, the revenue item, which will maximize the monthly total revenue, is calculated by the monthly average sales price and the total number of sales. Thus, the predicted number of sales and estimated sales revenue are calculated by taking the possible discounted prices into account. These accounts are compared with the actual total sales and revenue items. The inconsistency is evaluated with the MAPE parameter, which calculates the prediction accuracy.

2.4 Solutions and Analysis

The model inputs of the case study are as follows: The data of product *X* for the months of April, June, and July 2020 was considered. The average price function was created by linear regression using the discounted price of these months. The coefficients of the regression equation and the R2 value are as follows (Table 2):

The actual monthly total sales values of April, June, and July were compared with the estimated monthly sales amounts of the actual discounted prices (Table 3), and the accuracy of the sales figures was found. Here, the accuracy is used as a measure of how much the estimated number of sales deviates from the actual number of sales in percentage terms. For instance, in July, the difference between the actual total sales and the predicted total sales is 216 units. 216 (difference)/6636 (actual total sales) = 2.95%. This indicates that the accuracy rate in July equals ~97%.

Afterward, revenue optimization was made separately for April, June, and July of 2020. By testing potential discounted prices between 50 and 450 TL for each month (by increasing 10 TL in each trial), the average selling price was found with the regression equation whose parameters were given above, and the possible income calculation was made using this price. The results for April, June, and July, respectively, are given in the following tables. Also discounted price and predicted revenue of months are illustrated in Figs. 1, 2, and 3. The curve plotted in Figs. 1, 2, and 3 for April, June, and July, respectively, shows the predicted revenue value for different discounted prices. In other words, the curve in these figures represents the simulation of possible income for different discounted prices (Tables 4, 5, 6, 7, 8 and 9).

The actual values and the estimated values of the model for the revenue-maximizing discounted prices in April, June, and July are shown in Table 10.

Table 2 Parameters of the regression equation

Coefficient	Value
B0	53.014
B1	0.898
R2	0.994

Table 3 Actual total sales versus predicted total sales

Month	Actual total sales	Predicted total sales	Accuracy (%)
April 2020	15.935	15.948	99
June 2020	14.924	14.498	97
July 2020	6.636	6.832	97
Average			98

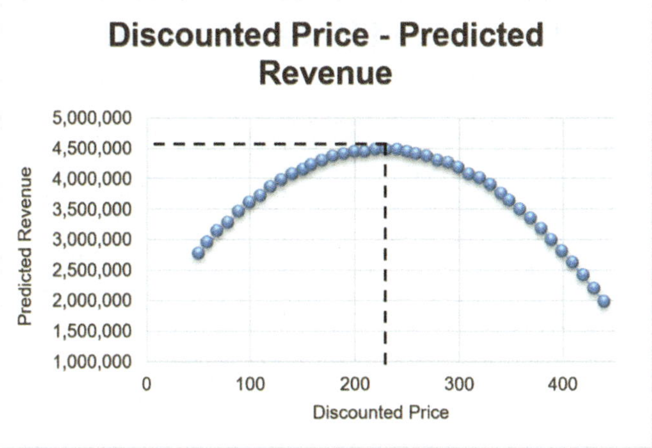

Fig. 1 Discounted price and predicted revenue of April 2020

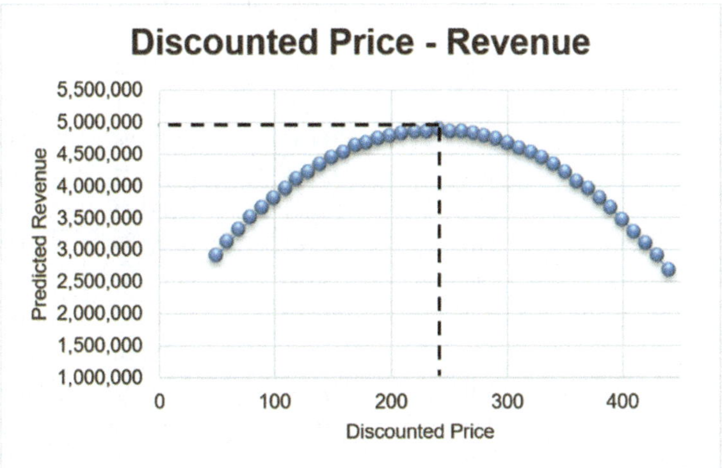

Fig. 2 Discounted price and predicted revenue of June 2020

The tables above were created for 3 months by using the elements mentioned in the model section. It is seen that the accuracy of the model created in this case study, which aims to find the optimum revenue point, is quite high.

Fig. 3 Discounted price and predicted revenue of July 2020

Table 4 Optimal revenue of April 2020

Potential sales price	Potential average sales price	Predicted total sold	Predicted revenue
160	196.71	21,431.44	4,215,831
170	205.69	20,822.21	4,282,994
180	214.67	20,212.98	4,339,214
190	223.66	19,603.75	4,384,491
200	232.64	18,994.52	4,418,825
210	241.62	18,385.29	4,442,216
220	250.60	17,776.06	4,454,664
230	259.58	17,166.83	4,456,168
240	268.56	16,557.60	4,446,730
250	277.54	15,948.37	4,426,348
260	286.52	15,339.14	4,395,024
270	295.50	14,729.91	4,352,756
280	304.49	14,120.68	4,299,545
290	313.47	13,511.45	4,235,391
300	322.45	12,902.22	4,160,294
310	331.43	12,292.99	4,074,254

Table 5 Actual revenue of April 2020

Actual discounted price	Actual average sales price	Actual total sold	Actual revenue
249	279.03	15,935	4,446,380

Table 6 Optimal revenue of June 2020

Potential sales price	Potential average sales price	Predicted total sold	Predicted revenue
160	196.71	23,027.61	4,529,818
170	205.69	22,418.38	4,611,317
180	214.67	21,809.15	4,681,873
190	223.66	21,199.92	4,741,485
200	232.64	20,590.69	4,790,155
210	241.62	19,981.46	4,827,881
220	250.60	19,372.23	4,854,664
230	259.58	18,763.01	4,870,504
240	268.56	18,153.78	4,875,401
250	277.54	17,544.55	4,869,355
260	286.52	16,935.32	4,852,366
270	295.50	16,326.09	4,824,433
280	304.49	15,716.86	4,785,558
290	313.47	15,107.63	4,735,739
300	322.45	14,498.40	4,674,978
310	331.43	13,889.17	4,603,273

Table 7 Actual revenue of June 2020

Actual discounted price	Actual average sales price	Actual total sold	Actual revenue
299	317.17	14,924	4,733,471

2.5 Conclusion

Thanks to the model within the scope of this case study, it is possible to predict the number of sales according to different discount setups and to determine the optimum discount budget. In addition, this study can be developed with differentiations/constraints such as determining the optimum revenue point according to the minimum profit constraint (e.g., profit > 30%). The effect of the price and discount of the substitute products on the sales of the product *X* and modeling the cannibalization effect of the lower–upper category products can be tried. Furthermore, by considering other competitors in the industry, the revenue model can be simulated depending on the changes in the prices of the competitors' substitute products. For all these reasons, this model can be considered a sufficient starting point. Thanks to this model alone, the ideal discounted price and predicted revenue for product *X* of company A were clearly demonstrated.

Table 8 Optimal revenue of July 2020

Potential sales price	Potential average sales price	Predicted total sold	Predicted revenue
160	196.71	19,016.29	3,740,742
170	205.69	18,407.06	3,786,215
180	214.67	17,797.83	3,820,744
190	223.66	17,188.60	3,844,331
200	232.64	16,579.37	3,856,974
210	241.62	15,970.14	3,858,674
220	250.60	15,360.92	3,849,431
230	259.58	14,751.69	3,829,245
240	268.56	14,142.46	3,798,116
250	277.54	13,533.23	3,756,044
260	286.52	12,924.00	3,703,029
270	295.50	12,314.77	3,639,070
280	304.49	11,705.54	3,564,169
290	313.47	11,096.31	3,478,324
300	322.45	10,487.08	3,381,536
310	331.43	9877.85	3,273,805

Table 9 Actual revenue of July 2020

Actual discounted price	Actual average sales price	Actual total sold	Actual revenue
359	377.42	6636	2,504,602

Table 10 Actual versus predicted of all months

Month	Discounted price	Actual average sales price	Actual total sold	Actual revenue	Predicted average sales price	Predicted total sold	Predicted revenue
April	230	279.03	15,935	4,446,380	259.58	17,166.83	4,456,168
June	240	317.17	14,924	4,733,471	268.56	18,153.78	4,875,401
July	210	377.42	6636	2,504,602	241.61	15,970.14	3,858,674

3 Customer Churn Analysis

3.1 Customer Relationship Management and Customer Churn

Customer relationship management (CRM) is a combination of techniques, technologies, and strategies that serve to create and deliver value to profitable customers [6]. It involves internal business processes and functions (such as marketing, sales) as well as external influences (such as competitors). CRM systems collect and analyze information about customers at all stages of their life cycle to establish and maintain long-term profitable relationships and create loyal customers [7, 8].

Customer-centric organizations exploit CRM to enhance competitive advantage of the organization. In such organizations, CRM aims to attract and retain profitable customers, prevent loss of customers as well as reactivate passive customers. In line with these purposes, the main applications of CRM can be classified under the following three categories: (i) customer acquisition, (ii) customer retention and churn, and (iii) customer win-back [9]. All these applications correspond to the stages that any customer goes through in his/her relationship with a company, i.e., customer life cycle. Therefore, CRM allows managing the customer life cycle across all these stages more efficiently and effectively:

Customer acquisition: It is the first stage of the customer life cycle. In this stage, while acquiring customers to do businesses, the questions to be answered are 'which potential customers should be targeted?', 'how should they be approached?', and 'what should they be offered?' [10]. The right CRM strategies in this step aim at acquiring profitable customers.

Customer retention and churn: Customer retention involves the company's skills and efforts to keep their current customers. In other words, the core purpose of CRM strategies related to retention is to maintain the continuity and quality of relationships with profitable customers. Actions and strategies that improve customer retention play also a key role in creating loyal customers and, thereby, growing the existing customer base [8]. However, it is also critical for organizations to analyze and understand the loss of customers (i.e., customer churn) and avoid current and potential mistakes [11].

Customer win-back: This stage involves winning customers back who have once terminated their relationship with the company. The CRM strategies for win-back differ from the acquisition in that lost customers have certain experience and their own judgment about being in a relationship with that company's products and services [9].

Several studies in the literature have reported that efforts to keep existing customers are considerably less costly than acquiring new customers [11, 12] among others. Therefore, being much less hard and costly than customer acquisition, customer retention is critical to survive in a highly competitive environment. At the retention stage, the aim is to keep a high proportion of profitable customers by reducing customer defections. Customer defection refers to switching merchandisers or terminating the relationship and is also known as customer churn or attrition.

There will always be some customer churn while trying to engage current customers to continue buying products or services from the company. Therefore, customer churn prediction is critical in understanding whether a company is retaining customers [13]. It helps companies to identify customers who are likely to terminate the relationship, to predict when they are likely to quit and what the drivers are that help explain why a customer is likely to leave the company [7, 9].

Customer churn might be voluntary, unavoidable, or involuntary [14]. In voluntary churn, customers prefer leaving the company because they are not satisfied with the product/service, other companies provide better products/services, or they no longer need it. The customer's decision relies on his/her experience. Unavoidable churn (also known as incidental churn) occurs when the customer permanently leaves the market, for example, when he/she dies or migrates. In involuntary churn, on the other hand, the relationship with the customer ends when without the customer's decision. For example, when the customer fails to pay his/her bill, the customer is asked by the company to leave. Involuntarily churned customers are likely to continue their relationship with the current brand or company [7].

Especially under intense competition, customer churn analysis may help to adjust the firm's marketing strategy and minimize operational costs. Hence, both in theory and practice, churn analysis is widely performed in markets where customers might switch to competitors easily, such as telecommunications, insurance, credit card and financial services, and online gaming industry[15]. The studies in the literature are also rich in methods used to model and predict churn behavior. Table 11 provides examples of recent contributions to the customer churn literature.

3.2 Case Study: A Telecommunication Industry

The telecommunication industry renders services to communicate in the form of words, voice, audio, or video both on global and local scales whether over the phone or the Internet, through airwaves or cables, over wires or wirelessly. Focus on improving driving forces, i.e., to present quicker and clearer data services, to provide multi-application usage, and to increase connectivity makes the competition within the sector challenging [25]. Therefore, to compete effectively, improving customer experience and service as well as developing new technologies is highly critical. To do this, it is essential to understand the behavior of telecom customers [26].

In this case study, the churn behavior of telecom customers is predicted. Here, *X* is a telecommunication company providing its customers with various services. *X* wants to identify the customers who are likely to terminate the relationship or switch to any other provider. To do this, the company will analyze a dataset of 3333 records [27] involving features related to customers' service usage (see Table 12).

Table 11 Summary of current studies focusing on customer churn prediction

Author(s)	Summary	Methods
Zhang et al. [16]	People's watching and spending habits were analyzed in order not to lose tracking of customers	Logistic regression
Mandak et al. [17]	Demographic and service usage variables were used to predict customer churns and find the factors influencing them in European telecommunications providers	Logistic regression
Khamlichi et al. [18]	Several data mining methods were used, and a final model by hybridization of all used models was set as a best of all	Hybrid model (Logistic Reg. SVM, random forest, decision tree, XGBoost, K-nearest neighbor)
Karvana et al. [19]	To offer incentives to survive, potential churn customers were predicted using past data and previous behavior	Support vector machine (SVM)
Jafar et al. [20]	A prediction model that helps telecom operators to predict customers who are most likely to churn was developed by building churn models with data mining techniques	Decision tree, random forest, GBM tree, XGBboost
Adnan et al. [21]	The focus is on determining the effectiveness of the factors, i.e., lower and upper distance between the samples. A novel solution is proposed for the telecommunication sector showing the hidden factors considered for predicting the customer churn	Manhattan distance formula, Naïve Bayes (NB), Bayesian binomial test
Pamina et al. [22]	Three prominent classifiers were compared to enhance customer churn prediction accuracy. Defining the attributes was focused on higher churn cognition with XGBoost classifier	Random forest, K-nearest neighbors, XGBoost
Amin et al. [23]	The study addresses the gap not having enough historical data. A cross-company data was used in the context of just-in-time for addressing customer churn prediction problems in the telecom sector	Feature selection via mRMR, SVM, K-nearest neighbors, neural network
Caigny et al. [24]	The inclusion of textual data in a CCP model was investigated to improve its predictive performance	Text mining, convolutional neural networks

Table 12 Predictor features in the dataset [27]

Feature	Data type	Definition
State	String	2-letter code of the US state of customer residence
Account_length	Numerical	Number of months the customer has been with the current telco provider
Area_code	String	'Area_code_AAA' where AAA = 3 digit area code
İnternational_plan	Yes or No	The customer has international plan
Voice_mail_plan	Yes or No	The customer has voice mail plan
Number_vmail_messages	Numerical	Number of voice mail messages
Total_day_minutes	Numerical	Total minutes of day calls
Total_day_calls	Numerical	Total number of day calls
Total_day_charge	Numerical	Total charge of day calls
Total_eve_minutes	Numerical	Total minutes of evening calls
Total_eve_calls	Numerical	Total number of evening calls
Total_eve_charge	Numerical	Total charge of evening calls
Total_night_minutes	Numerical	Total minutes of night calls
Total_night_calls	Numerical	Total number of night calls
Total_night_charge	Numerical	Total charge of night calls
Total_intl_minutes	Numerical	Total minutes of international calls
Total_intl_calls	Numerical	Total number of international calls
Total_intl_charge	Numerical	Total charge of international calls
Number_customer_service_calls	Numerical	Number of calls to customer service
Churn	True or False	Customer churn—target variable

The analysis to be performed by the company to model its customers' behavior is an example of supervised machine learning. Supervised learning grounds on algorithms that develop general hypotheses with reasoning based on externally supplied instances. Thus, the hypotheses can constitute a comprehensive model of the distribution of churned customers considering predictor features given in Table 12. The company decides to develop a classification algorithm, because, with this model, it will be able to categorize its customers into churn and non-churn classes.

The churn behavior of the company's customers will be modeled using four different classification algorithms, including artificial intelligence applications along with basic statistical methods.

3.2.1 Data Examination

The first and fundamental step is data examination. The raw dataset has 3333 instances with 20 features. To test the accuracy of the models to be developed, the raw data is divided into two parts with an 80/20 ratio preserving the distribution of the classes. Thus, the training dataset has 2666 instances, the test dataset has 667 instances, and 15% of each set represents instances for churned customers.

There is no missing value in the raw data. However, the three categorical features, 'State', 'International plan', and 'Voice mail plan', need to be encoded to allow numerical analysis. The features 'International plan' and 'Voice mail plan' have a sequence of labels with values 'Yes' and 'No', and the feature 'State' has 51 unique labels. Using one-hot encoding, a representation of categorical variables as binary vectors, the feature of 'State' is represented by a 51-sized vector, and the features 'International plan' and 'Voice mail plan' are transformed into vectors with two elements. After encoding the features, the final dataset comprises 71 features.

3.2.2 Performance Criteria for Comparison

The model development process needs performance evaluation measures to evaluate a model and compare it with others. In order to measure the classification performance in classification problems, several functions, known as loss, score, and utility, can be used. Before the model development, the basic performance criteria for model evaluation will be introduced.

The most common criterion is the accuracy by which the mean of the proportion of correct predictions concerning actual labels specific to the classes is represented. The prediction accuracy of a model over n_{samples} is defined as

$$\text{accuracy}\left(y, \hat{y}\right) = \frac{1}{n\_\ \text{samples}} \sum_{i=0}^{n_{\text{samples}}-1} 1\left(y_i = \hat{y}_i\right) \tag{5}$$

where $\hat{y}_i$ and y are the predicted and true values corresponding to i-th sample and $1(x)$ is indicator function, which takes its value if x is realized [28].

Different performance measures are also used to compare the performances of models from different perspectives. For example, a classifier has several capabilities, such as determining positive samples and accurately classifying negative samples, and there are relevant different measures. These are *precision*, the ratio of positive identifications to the samples that are actually positive, and *recall*, the ratio of samples that are actually positive to samples identified as positive [29]. For more details, the reader should refer to Part I of this book. There is also a general measure to interpret precision and recall simultaneously: F-score. F-score is the way of combining precision and recall into a weighted harmonic mean and is calculated as expressed by the following equation [28].

$$F_\beta = \left(1 + \beta^2\right) \frac{\text{precison} * \text{recall}}{\left(\beta^2 * \text{precision}\right) + \text{recall}} \tag{6}$$

when $\beta = 1$, F_β is equal to F_1 and the influence of precision and recall in f-score becomes equal.

A graphical illustration of a classification model's performance is a curve called receiver operating characteristic (ROC), which plots a classifier's performance at varied discrimination thresholds. It is a plot displaying the proportion of sensitivity and specificity. Curve information summarized in one number is revealed by computing the area under the ROC curve (abbreviated to AUC).

Another tool for the performance evaluation is the confusion matrix. This matrix illustrates the accuracy of the classification by a matrix with true classes in rows and predicted classes in columns.

3.2.3 Model Development

In this case study, logistic regression, multilayer perceptron (MLP), random forest, and gradient boosting trees are used as classifiers which are explained in detail in Chap. 3.

3.2.4 Comparison of the Models

For each technique listed below, the classifiers are trained on the training dataset. To form a judgment about the generalization ability of each model and to compare them for the best model, the test dataset is used. The default initialization parameters and the training and test performances of the models are given in Tables 13, 14, 15, and 16.

Considering the unbalanced distribution of positive and negative labels (churn and non-churn), it seems that logistic regression results are not better than a default non-churn choice in terms of test accuracy provided in Table 17. In addition, neural network predictions are similar compared to logistic regression in terms of accuracy. Random forest predictions are the best among the algorithms used. The high number of one-hot encoded features and the power of ensemble prediction methods are some of the main reasons for this difference. Another ensemble method GMB's predictions are very close to random forest prediction.

Table 13 Parameters of logistic regression classifier

Parameter	Value
Penalty	'12'
Tol	0.0001
C	1.0
Solver	'lbfgs'
Fit_intercept	True

Table 14 Parameters of MLP classifier

Parameter	Value
Activation	'Relu'
Solver	'Adam'
Alpha	1e-5
Learning_rate_init	0.01
Hidden_layer_sizes	(5,5)
Max_iter	1000
Learning_rate	'Constant'
Momentum	0.9

Table 15 Parameters of random forest classifier

Parameter	Value
N_estimators	150
Max_depth	15
Criterion	'Gini'
Min_samples_split	2
Min_samples_leaf	1

Table 16 Parameters of gradient boosting trees classifier

Parameter	Value
N_estimators	150
Learning_rate	0.1
Loss	'Deviance'
Criterion	'Friedman_mse'
Min_samples_split	2
Min_samples_leaf	1

Table 17 Prediction performance of the developed models

	Accuracy			
Model	Training	Test	Test AUC	Test F_1
Logistic regression	0.8709	0.8545	0.8141	0.3021
Neural network classifier	0.8653	0.8545	0.7883	0.4260
Random forest	0.9857	0.9430	0.9217	0.7564
Gradient boosting trees	0.9609	0.9385	0.9189	0.7484

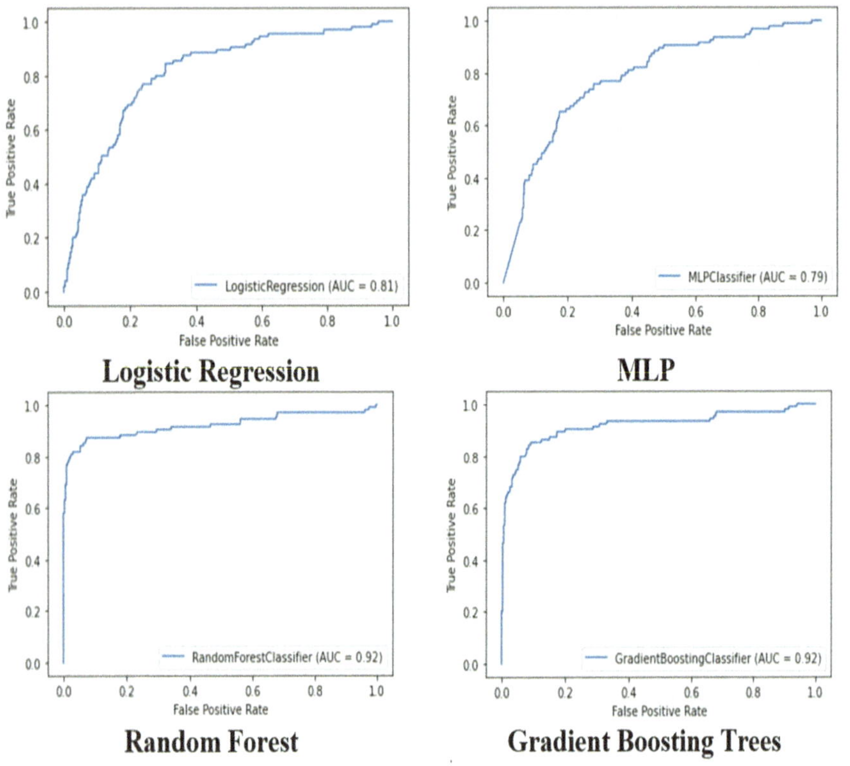

Fig. 4 ROC curves of classifiers for test dataset

The ROC curves in Fig. 4 also show the superiority of the ensemble models developed for predicting customers who will churn. The AUC with varied classification thresholds is higher for two ensemble models than logistic regression and MLP classifiers.

Finally, Fig. 5 illustrates the confusion matrix of the models developed for the test dataset. Accordingly, with the best prediction model developed, at least 62% of the 95 customers who have churned can now be accurately identified.

4 Social Media and Web Analytics

Internet usage has increased rapidly, approximately 5 times in the last 15 years. The number of users increased from 1.023 billion in 2006 to 5.16 billion users in 2021, as well as 65% of the world population [30]. In addition, this rise was followed by the usage of social media from 0.97 billion to 3.7 billion active users in the last decade [31].

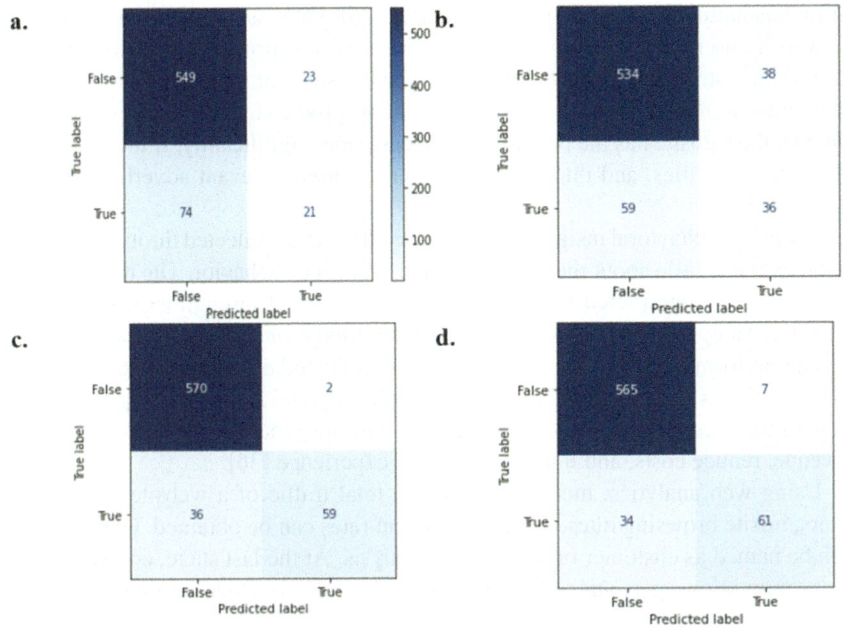

Fig. 5 Confusion matrices of classifiers (**a.** logistic regression, **b.** MLP, **c.** random forest, and **d.** gradient boosting trees) for test dataset

Companies are adapting themselves to these changes by using social technologies such as marketing through social media networks and online platforms since they realized the potential of the new marketing era. Companies can achieve up to a 25% productivity increase of their highly skilled employees by using social networks [32]. Also, about three-fourths of the surveyed companies are practicing social technologies, and 90% of them are recognizing the benefits of using this technology. When dealing with analytics of this new business area, it would be convenient to split it into two fragments of social media and web analytics and point them out in detail.

Social media, which has emerged as a different form of human communication, has also affected the company–customer relationship. The way to grow a business rapidly is now through social media and networking, which make it less painful to reach large audiences [33]. Almost all companies take part in social media to make their brand perceptions effective. In addition to being an effective and affordable marketing method, social media [34] can also provide an environment for rapid reactions or feedback of customers. However, it can be easily out of control in some negative situations. In social media, the data can be collected through the application programming interfaces (APIs) of some social media networks or by web crawling or scrapping. From this data, one can obtain users' comments, feelings and opinions about a brand or a product by performing sentiment analysis.

Another term to mention is web analytics which is defined as follows: 'the measurement, collection analysis, and reporting of Internet data for the purpose of understanding and optimizing web usage' [35]. One can perceive the behaviors of website users, detect the potential customers of new products, and also enhance the existing structure of the website by following steps of web analytics. From the marketing point of view, suggesting an appropriate product to the customer at the right time on the website has the potential to increase sales significantly. If the customer's character, activities, and interests can be known, more relevant advertising can be conducted.

Likewise, behavioral insights can be derived from data collected through websites, but it explains little about the motivation and process of behavior. The missing parts are tried to be interpreted by utilizing the data gathered through experiments and surveys. Analyzing process starts with the determination of a goal, according to a certain performance criterion, then new data is collected after adjustments are made in line with this goal, and finally, it is checked for progress. Although the performance criteria vary according to the type of website, they are generally designed to increase revenue, reduce costs, and improve customer experience [36].

Using web analytics, metrics such as the total traffic of a website, location of users, in-site browsing rates, and abandonment rates can be obtained. Observing all can be named as customer online activity analysis. At the last stage, companies can make promotions, personalized offers, and product recommendations on top of these basic analyses.

4.1 Problem Definition

In today's data-driven and complex world, companies have difficulties connecting with all customers in large, broad, or diverse markets, and customers expect to be understood and communicated with a personal perspective. This problem occurs in various areas such as communication, new product development, and service-level determination. To overcome this problem, identifying and uniquely satisfying the right market segments is proposed [37]. The process of market segmentation can be defined as to divide a market into well-defined slices and each slice consists of a group of customers who share a similar set of needs or characteristics [38]. In the literature, the most popular ways of marketing segmentation are as follows: geographic, demographic, psychographic, and behavioral segmentation.

- Demographic segmentation uses data such as age, gender, and income to create market segments. Demographic segmentation is highly beneficial for markets where customer demographics are associated with customer needs and wants. However, the association between demographics and customer wants may not be valid for most markets. In other terms, two different customers of the same age have the same gender, income may have different buying behaviors, and demographic segmentation may not show the difference.
- Psychographic segmentation aims to segment the market by customers' lifestyles, attitudes, and aspirations. The variables used in psychographic segmentation

include interests, activities, opinions, and values. This type of segmentation is generally used for strengthening brand identity and creating an emotional connection.
- Behavioral segmentation uses customers' product or service-related behaviors data. The source data for this kind of segmentation includes benefits sought, usage rate, brand loyalty, user status, readiness to buy, and occasions. Behavioral segmentation can be very beneficial for companies that provide membership-type relationships to customers, such as banks and telecommunications providers. The disadvantage of this approach is that the data used includes only the company's data. The user's usage data in other companies is ignored.
- Need-based segmentation groups users according to their wants and needs. In this sense, the expectations of the customers in the same group from the product or service will be similar. Although it is the most suitable method for basic marketing studies such as developing new products, deciding on product features, and pricing, it is difficult to access data on customer requests and needs. For this reason, data obtained from methods such as surveys or focus groups is used in this type of segmentation.

Social media and the web provide novel application perspectives on customer segmentation. Both channels provide customer data such as web browsing, search, product view, product sell, engaging content, and friends' network. The data gathered from these channels enables us to build new customer segments.

4.2 Case Study: Customer Segmentation Based on Page View Data

On-demand 'ultrafast' delivery service companies aim to deliver the ordered products to customers in a very short period. As the market is growing and competition is high, a company aims to build a segmentation model based on page view data. By doing this, the company wants to create market segments based on customers' both behaviors and needs. In order to accomplish such a study, list of variables given in Table 18 is needed.

In web analytics, identification of the customer is an important issue, and generally, cookies are utilized. To this end, each visitor is assigned a unique number which is stored in the cookies. This value is represented as customer ID. In the second or later

Table 18 Input data of the segmentation model

Variable name	Variable description
Customer ID	The unique ID of a customer
Category ID	The ID which represents a category
Visits	Total number of page visits of a specific customer to a specific category

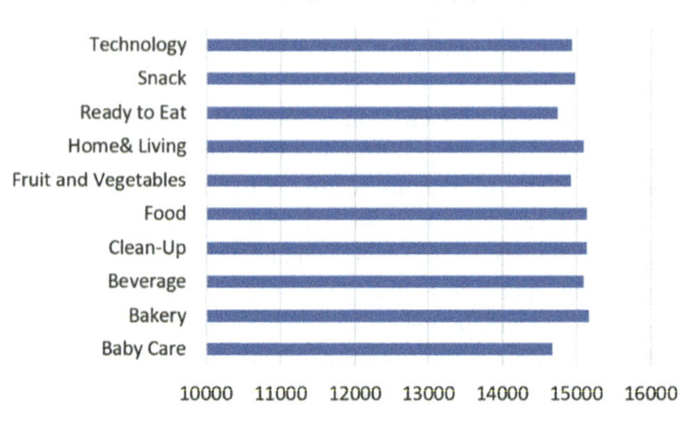

Fig. 6 Categories and the number of visits in each category

visits, the unique ID stored in the cookie is used to identify the user. On the other hand, the data gathered from cookies is vague since the cookies are not transferred from different hardware of the user, or similarly, the user can erase the cookies from his/her hardware. In such cases, the same user can have more than one unique ID.

Another data related with e-commerce is the category ID. Each product is listed under at least one category. The related products are listed under the category page, and visitors browse the products. Since various products are served to the visitors, using product visits as a source for segmentation has some drawbacks. Visitor patterns can be undetected, or the segments cannot be well formed because of sparse data. Thus, category ID variable is used for segmentation. The number of visits is a metric used to measure the total number of times a user navigates to a category page or a product page that is listed under that category. Since the raw visit data can be very huge, a summarized version of visit data is used here.

For this case, 149,919 visit data of 3000 customers is used. The categories and number of visits in each category are shown in Fig. 6.

4.3 *Model and Application*

In this application, *k*-means clustering method, which is explained in Chap. 10, is used for segmentation. *K*-means algorithm is a method of vector quantization that aims to partition *n* observations into *k* clusters in which each observation belongs to the cluster with the nearest, and the application steps can be summarized as in the following.

Step 1. Determine the data which will be included in the study.

Step 2. Standardize the data so that they become unitless. Since the algorithm is based on distances between points, the units of the data may distort the formation of the clusters. To get better results, the data should be standardized.

Step 3. Apply the k-means algorithm for different k values. In the following, steps of the k-means algorithm are given.

Step 3.a. Determine the value of k.

Step 3.b. Select k random points from the data as centroids: While there are different initiation approaches, the classical k-means algorithm starts with randomly selecting k points.

Step 3.c. Assign all the points to the closest cluster centroid: In this approach, segments are created by assigning each data point to the nearest centroid.

For every i, set

$$c^{(i)} := \arg\min_{j} \left\| x^{(i)} - \mu_j \right\| \tag{10}$$

where $x^{(i)}$ represents each data point and μ_j represents the centroid values. The distance between each data point and centroids is calculated by using Euclidian distance.

Step 3.d. Recompute the centroids of newly formed clusters: As data points are assigned to their new clusters, the center points of the segments may change. So, in this step, the new centroids are calculated.

For each j, set

$$\mu^{(i)} := \frac{\sum_{i=1}^{m} 1\{c^{(i)} = j\}x^{(i)}}{\sum_{i=1}^{m} 1\{c^{(i)} = j\}} \tag{11}$$

Step 3.e. Check the stopping criteria: A criteria should be defined for the algorithm to stop. Basically, the algorithm stops when there are no changes in the segments or the iteration reaches to a certain threshold value.

Step 3.f. Repeat Steps 3.c, Steps 3.d, and Step 3.e: If the algorithm does not stop, the next step is to turn back to Step 3.c.

Step 4. Calculate the performance scores for different k values and select the most suitable k value. Silhouette score is used as a performance score in clustering. The calculation is as follows:

$$\text{Silhouette Score} = \frac{(b - a)}{\max(a, b)} \tag{12}$$

where a is the average intra-cluster distance, i.e., the average distance between each point within a cluster. b is the average inter-cluster distance, i.e., the average distance between all clusters.

Step 5. Prepare the centroid table for the segments created for the selected k value.

Step 6. Interpret the centroid table and define different segments in the market.

4.4 Solution and Analysis

The data used for the application of customer segmentation based on page view data is an anonymized and simplified version of a real-world data. A sample view of the raw data used for the application is given in Table 19.

The raw data is later converted to tabular form as given in Table 20. In Table 20, each cell represents the number of visit data of a customer to a category mentioned in the associated column.

The next step is to standardize the column values, and the standardized values are shown in Table 21. In this step, outliers are eliminated. A regular way of eliminating the outliers is to eliminate the users which have standardized values outside the [−3, 3] interval.

Table 19 Sample raw data

Customer ID	Category	Number of visits
2614	C1	4
3731	C5	8
2028	C1	3
1026	C10	15
1823	C6	3
3735	C4	13
2050	C1	3

Table 20 Tabular form of the raw data

	C1	C2	C3	C4	C5	C6	C7	C8	C9	C10
1000	2	0	1	4	4	2	9	5	5	13
1001	1	1	1	4	3	5	8	4	2	10
1002	4	2	2	3	1	1	9	4	3	7
1003	5	2	2	12	8	14	2	2	5	1
1004	..	...	..	..	..	..	..	..	..	..

Table 21 Scaled data

	C1	C2	C3	C4	C5	C6	C7	C8	C9	C10
1000	−0.76	−1.29	−1.02	−0.27	−0.27	−0.77	1.02	0.02	0.00	2.06
1001	−1.03	−1.03	−1.02	−0.27	−0.53	0.01	0.77	−0.24	−0.76	1.29
1002	−0.24	−0.78	−0.77	−0.54	−1.04	−1.03	1.02	−0.24	−0.51	0.52
1003	0.03	−0.78	−0.77	1.83	0.76	2.34	−0.78	−0.75	0.00	−1.02
1004	..	...	..	..	..	..	..	..	..	..

After this step, data is ready for k-means algorithm, and the algorithm is applied with a different number of clusters. The silhouette scores related to each different cluster number are calculated to show the performance of clustering. The resulting silhouette numbers are given in Fig. 7.

The results reveal that $k = 4$ show the highest silhouette values which means the best mathematical clusters are obtained when four clusters are formed. However, still, the formed clusters should be examined. One way of examining the clusters is the number of customers in each segment, and Fig. 8 shows the number of customers

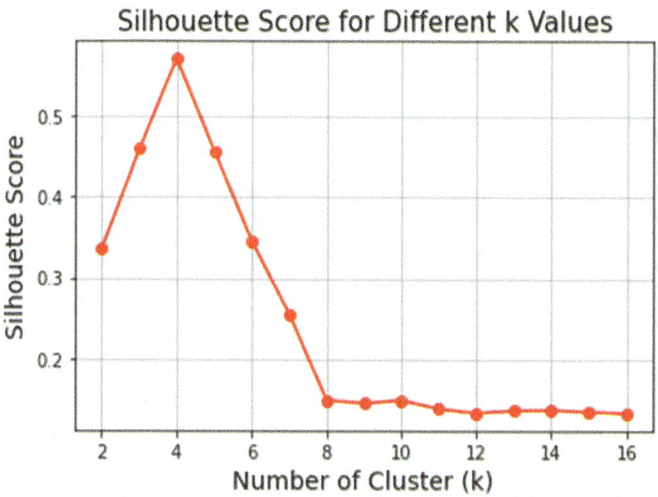

Fig. 7 Silhouette values for different numbers of clusters

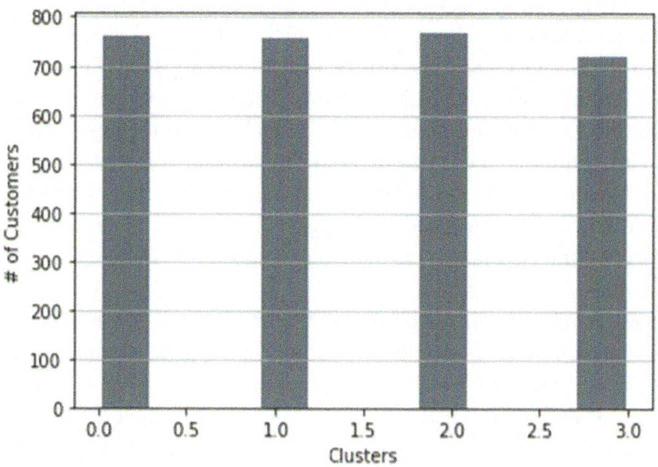

Fig. 8 Number of customers in each segment

Table 22 Centroid table with original data

	C 1	C2	C3	C4	C5	C6	C7	C8	C9	C10
Cluster 1	3.03	2.99	2.90	3.03	3.10	2.92	11.01	2.94	2.91	10.94
Cluster 2	2.99	3.01	2.96	10.82	10.96	10.86	3.01	3.01	2.91	2.93
Cluster 3	3.01	11.02	11.03	3.18	2.99	2.94	3.03	2.98	10.97	2.95
Cluster 4	10.90	2.99	3.05	3.04	3.02	3.07	3.01	11.09	2.99	3.04

Table 23 Centroid table with scaled data

	C 1	C2	C3	C4	C5	C6	C7	C8	C9	C10
Cluster 1	−0.49	−0.52	−0.54	−0.53	−0.50	−0.53	1.54	−0.51	−0.53	1.53
Cluster 2	−0.50	−0.52	−0.53	1.52	1.53	1.53	−0.52	−0.49	−0.53	−0.52
Cluster 3	−0.50	1.52	1.52	−0.49	−0.53	−0.53	−0.52	−0.50	1.52	−0.52
Cluster 4	1.58	−0.52	−0.50	−0.53	−0.52	−0.49	−0.52	1.59	−0.51	−0.50

in each segment when four is selected as the parameter. As the number of customers in each group is similar, the clustering is suitable from this perspective.

Another perspective to validate the results is to check the centroid tables and interpret the results with market experts. The centroid tables show the center point of each segment so they are used to understand the characteristics of each segment. The centroid tables can be formed by using original data or scaled data. The centroid table with original data is given in Table 22, and the centroid table with scaled data is given in Table 23.

According to the centroid table, the customers in Cluster 1 are interested in Category 7 (Home&Living) and Category 10 (Technology). It can be seen from Table 22 as the numbers associated with Category 7 and Category 10 are more than 10. In Table 23, the same values are around 1.5 which means the page visits are higher than the mean. Similarly, it can be observed that customers in Cluster 2 are related to Category 4 (Cleanup), Category 5 (Food), and Category 6 (Fruit and Vegetables). The customers in Cluster 3 are related to Category 2 (Bakery), Category 3 (Beverage), and category 9. Finally, Cluster 4 is related to Category 1 (Baby Care) and Category 8 (Ready to Eat).

From these results we can conclude that Cluster 1 shows "HomeTech" customers who care about technology and their life at home. In the same manner, Cluster 2 can be defined as "Everyday Users" who buys nearly all of the daily needs from the system. Cluster 3 can be defined as FastFooders since their focus is on the product which is ready to consume quickly. And finally, Cluster 4 can be defined as BigFamily, which refers to couples with babies.

5 Complaint Analysis

Analyzing customer complaints is very important for service quality management in order to prevent customer dissatisfaction and loss of loyalty. Elimination of customer dissatisfaction begins with identifying the weak and strong aspects of the service. This useful information is then used to redesign and improve the service [39]. Analyzing customer reviews is a good and frequently used tool to monitor and increase customer satisfaction. Analysis of customer reviews is better than many other methods because it can reveal the positive and negative aspects of the service or product. It is an important opportunity for companies to take immediate action if there is dissatisfaction with the relevant product or service. In addition, the knowledge gained in this way is highly reliable as customers voluntarily write reviews. Therefore, examining the reviews of real customers is a good source of information for companies while monitoring customer complaints and developing methods to increase satisfaction [40].

The reason for this study is that companies are now starting to give importance to analyzing their customers' opinions with text analysis in order to better understand them and thus to improve their products or services. In this direction, the studies in the literature were examined, and the techniques and usage areas used in the analysis of customer reviews were internalized. Afterward, an exemplary customer review analysis study was carried out in the field of e-commerce, where the analysis of customer reviews is much more critical than in other areas.

5.1 Problem Definition

Electronic commerce (e-commerce) is one of the most important building blocks for many retail companies, and e-commerce sales in the retail sector are expected to reach 4.5 trillion dollars by the end of 2021 [41]. The e-commerce industry, which has become so widespread, has started to be an element of choice compared to traditional commerce, and thanks to this awareness, companies seek to set and design their services to boost customer experience. Hence, it is an undeniable fact that e-commerce companies focus on the provision of innovative and competitive advantages to their customers and try to differentiate from other companies.

Despite all this penetration rate and success of e-commerce, compared to the traditional retail sector, e-commerce does not provide any staff or salesperson to help customers, so customers make purchase decisions based on the comments and reviews made on e-commerce portals. The fact that e-commerce websites contain textual reviews of a large number of customers and that these comments include a wide range of subjective evaluations of products and product features affect customer purchase decisions as a starting point for prospective customers. With e-WoM, which is an electronic word of mouth, the characteristics and uses of products are transferred to new customers through informal channels, and it plays a major role in the customer purchase decision process. However, a large number of reviews made about the products make it difficult for customers to notice and interpret those reviews that are useful to them. At the same time, the existence of some conflicting reviews makes the customer purchase decision more insensible. Therefore, it is not always possible to match the appropriate product with the right customer. Furthermore, hundreds of comments for the same product cause a cognitive load on customers' minds.

It is crystal clear that companies should adopt various innovative solutions in order to eliminate this problem. With this perspective, e-commerce companies need to apply text analytics methods and adopt these method-based systems to provide better service quality. In this context, this study aims to understand how product comments that are made by customers help recommend a product to someone else.

5.2 *Case Study: Clothing Product-Based E-Commerce Website*

Company *M* provides services on an online platform and focuses on fashion products. Company *M* has an average of 4 million members and sells over 35,000 products annually. Monthly, roughly 1 million clicks are received, and approximately 10,000 transactions are made. Moreover, *M* company has 350 offline stores in country *K*, and more than 25,000 products are sold in these stores on an annual basis.

In this case study, the main focus will be on the women's clothing products sold by company *M* over 1 year. In 1 year, more than 23,000 women's clothing products are sold through the online platforms of *M* company. These products are intended for all women over the age of 18 and categorically consist of intimate, dresses, bottoms, tops, jackets, and trend products. On the online platform of *M* company, customers, who have purchased the products before, write a review text about the products, the products are rated, and it is projected whether the products will be recommended to others. In this way, foresight is shown for future new customers with the potential to purchase these products. Company *M* has the following assumptions and operational environment:

- Feedback and comments for purchases in *K* company's offline stores do not affect comments on the online platform.
- The seasonal changes in the demands of the products or the shopping made by women from different age groups do not distort the ratings of the products and the comments about the products.
- Shopping for products other than women's clothing is out of context, and there is not any significant relationship between these products and women's products.
- The discounts or campaigns on the products (such as the third product free or discounted if two products are purchased) do not affect the ratings and reviews of the products.

Table 24 Variables in the dataset

Variable name	Description
Clothing ID	Unique codes assigned to products
Age	Reviewers age
Title	The title of the review
Review text	Review body
Rating	Product score given by the customer (1: Worst,…,5: Best)
Recommended IND	The field where customers say 1 for the product they recommend and 0, otherwise
Positive feedback count	The number of other customers who found this review positive
Division name	Name of the product high-level division
Department name	Name of the product department
Class name	Name of the product class

Company *M* wants to directly assist its customers in its online stores as well as its offline stores. While there is a staff in offline stores, this is not available in online stores, and customers have trouble finding ones that will work for them among thousands of reviews. Therefore, Company *M* is aware of the need to develop a system that analyzes customer reviews and outputs whether customers recommend that product to someone else. Hence, in the context of this case study, dataset containing customer reviews will be processed to reveal various prediction models about whether customers recommend the product or not, and the model performances will be compared. Information about the variables in the dataset is given in Table 24.

5.3 Model

The data goes through certain basic steps in the analysis process. It is helpful to identify and follow these basic steps before implementing a data analysis project. In this study, data analysis was carried out by following the steps shown in Fig. 9. In addition, each step will be examined in this section before proceeding to the implementation phase.

Step 1: Define the target of your analysis.

When data analysis is done with the right data and methods, it can prevent many mistakes in the critical decisions of companies. For this, the first step is to reveal what the company is aiming for with this analysis.

Step 2: Collect the customer review data.

Collecting as much data as possible from different sources allows data analysis to produce more accurate results. Although different methods are used to collect data,

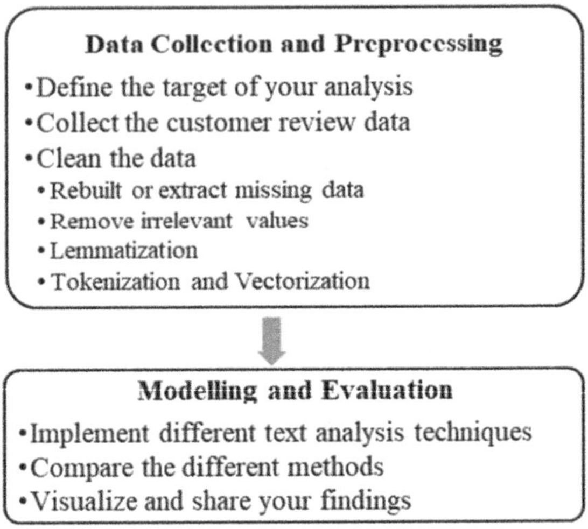

Fig. 9 Basic steps in the data analysis process

the most frequently used ones today include computers, social media and blogs, forum sites, mobile applications, and websites.

Step 3: Clean the data.

Rebuilt or Extract Missing Data

Most of the real-world datasets contain missing values. It is rare to find a real-world dataset without missing values. Examining missing values is important because leaving missing values intact can affect the analysis. Missing values in the dataset can be filled with different methods, or these rows can be removed from the data altogether. In this study, missing values in the features that are used in the analysis were removed from the data.

Furthermore, processes can be made to strengthen the meaning of relationships in the data such as creating a new variable and merging some variables. The significance of the model can be increased with these operations. In this study, a filtering process was performed on the ratings given by the customers in order to reduce the contradictions in the data.

Remove irrelevant values

Stop words and punctuations are often removed from the text before training deep learning and machine learning models because they are abundant and do not contain useful information that can be used for classification and thus can adversely affect analysis results. This process aims to focus on more important information and thus reduce the training time by reducing the data size. Different options can be used to remove stop words and punctuations in the Python programming language.

Lemmatizing

When dealing with text processing, specialists are more interested in the root form of the word than the suffix. Stemming is one of the techniques used in text analysis to ensure that variants of words are not left out. Stemming algorithms attempt to truncate the beginning or end of the word by considering the list of common prefixes and suffixes. There are various stemming algorithms developed to reduce words to their root form. Unlike stemming, lemmatization uses morphological analysis of the word to remove inflectional suffixes and return words to dictionary form. It also helps to match synonyms using a thesaurus [42].

Text Tokenization and Vectorization

In order to perform machine learning algorithms on text datasets, the dataset is needed to transform into vector forms. This process is a kind of feature extraction for free texts for machine learning models. There are many different techniques for the vectorization process of converting text into a numerical representation. In natural language processing, n-grams are a contiguous array of n elements generated from a given sample of text where the items can be characters or words and n can be any numbers like 1, 2, 3, etc. [43].

- N-grams are useful to create features from text corpus for machine learning algorithms like SVM, Naive Bayes, etc.
- N-grams are useful for creating capabilities like autocorrect, autocompletion of sentences, text summarization, speech recognition, etc.

An *n*-gram vector represents text as a collection of unique n-grams [44]. These *n*-grams are then converted to numeric vectors that machine learning applications can handle. After indexes are assigned to the *n*-grams, different methods such as one-hot encoding, count encoding, and TF-IDF encoding can be used for vectorization.

Step 4: Implement different text analytics methods.

Logistic regression, support vector machines, gradient boosting trees, and random forest are all useful classification methods for prediction.

Step 5: Compare the different methods.

Performance is often estimated with one-dimensional indicators such as accuracy, precision, recall, or F-score. In the analysis part of this study, these three performance indicators will be considered. These performance indicators will be explained with the confusion matrix [45].

5.4 Analysis and Results

The purpose of this study is to analyze old customer reviews, so to predict whether a new customer will recommend the product to someone else based on the review. Therefore, recommendation classification will be made, which determines whether a review text recommends the product under review.

For this study, the dataset of Women's Clothing E-Commerce Reviews is utilized [46]. This dataset consists of reviews of real customers. For analysis, the dataset was anonymized by removing the customer names and changing the company name to 'retailer'. This dataset consists of 23,486 rows and ten feature variables, but in this study, the entire dataset was not used, only the features of 'Rating', 'Review Text', 'RecommendedIND', and 'ClassName' were used in the analysis, and unnecessary features were eliminated from the data.

It was mentioned that the missing values in the dataset could be filled using different methods or that these rows could be completely removed from the data. When the missing values in this study were analyzed as percentages, it was observed that 3.59% of the 'ReviewText' and 0.05% of the 'ClassName' contained missing data, and there was no missing data in the 'Rating' and 'RecommendedIND' variables. Since the 'ClassName' variable will not be used exactly in the analysis, it will only be used in the interpretation phase, and the missing values of this variable have not been dealt with. Accordingly, only the missing values in the 'ReviewText' variable were taken into account, and since it was difficult to fill in the missing data in the string variable, these lines were completely removed from the data.

Furthermore, the lengths of the 'ReviewText' according to the target 'RecommendedIND' were analyzed. It is expected that the averages of the text lengths with 0 (not recommend) and 1 (recommend) values are close to each other and that there are no extreme values in the minimum and maximum values. According to the results, the average text length of the comments with 'RecommendedIND' 0 is 318.33, the minimum value is 20, and the maximum value is 508; while the average length of the comments with 1 is 306.64, the minimum value is 9 and the maximum value is 508.

Accordingly, as can be seen in Fig. 10, the lengths of the recommended and non-recommended comments were close to each other on average. This indicates that the length of comments will not cause any bias in the analysis. Considering the minimum lengths, it was concluded that the comment lengths are good enough and do not need to be removed. Then, the number of comments was examined based on other variables, and general information about the data was obtained.

In Fig. 11, the number of comments is analyzed based on rating. Accordingly, it is seen that the comments with a rating of 5 have the highest number of comments, while those with a rating of 1 have the least number, and as the rating increases, the number of comments also increases.

As seen from Fig. 12, the number of comments with a RecommendedIND of 1 is considerably higher than those with 0.

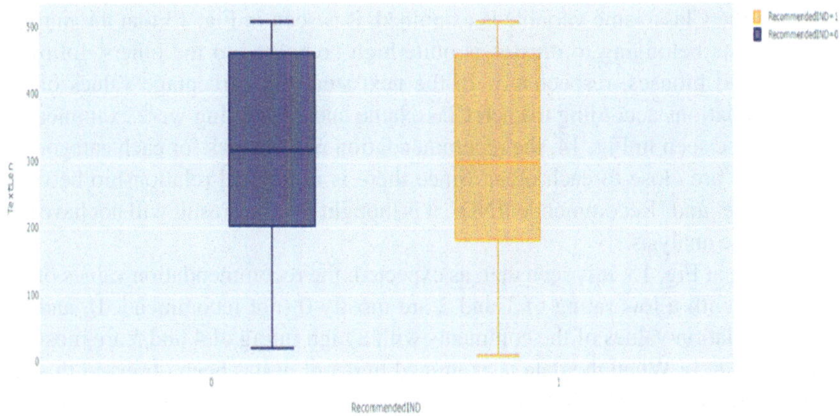

Fig. 10 Boxplot to visualize the length of comments by recommendation value

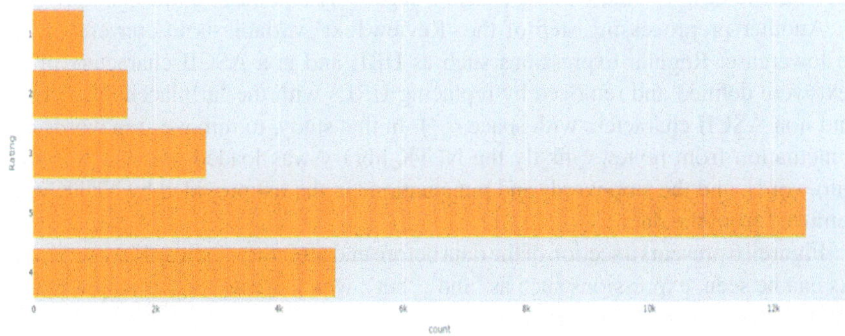

Fig. 11 Boxplot visualizing the distribution of the number of comments by rating

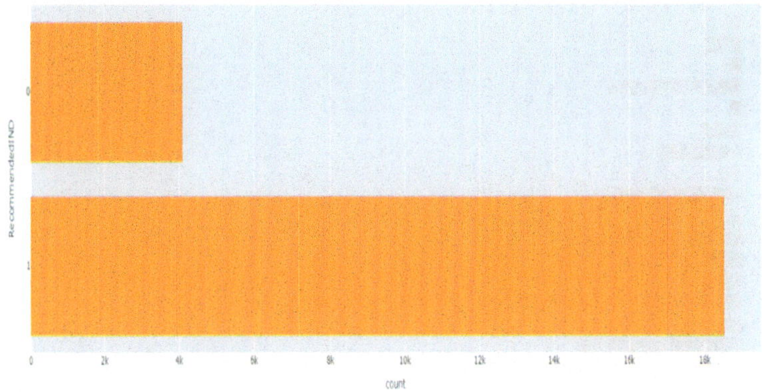

Fig. 12 Boxplot visualizing the distribution of the number of comments by RecommendedIND

When the ClassName variable is examined, it is seen in Fig. 13 that the number of comments belonging to dresses is quite high compared to the others, followed by knits and blouses, respectively. In the next step, the percentage values of the recommendations according to each ClassName and each rating were examined.

As can be seen in Fig. 14, the recommendation percentages for each category in ClassName are close to each other. Since there is no logical relationship between 'ClassName' and 'RecommendedIND', it is thought that this result will not have any effect on the analysis.

Looking at Fig. 15, it is seen that, as expected, the recommendation values of the comments with a low rating of 1 and 2 are mostly 0 (not recommended), and the recommendation values of the comments with a high rating of 4 and 5 are mostly 1 (recommended). When the data is examined in detail, it has been observed that the reviews with a rating of three do not have a single negative or positive tendency and are abstaining. At this stage, reviews with a rating of three were excluded from the data. It is predicted that including abstaining reviews in the analysis of whether to recommend or not will affect the analysis negatively and may be misleading.

Another preprocessing step of the 'ReviewText' variable is to convert all text to lowercase. Regular expressions such as URL and non-ASCII characters in the text were defined and removed by replacing URLs with the 'urlplaceholder' token and non-ASCII characters with space (' '). In this study, to remove stop words and punctuation from reviews, firstly the NLTK library was loaded, the text was split into words, and the stop words and punctuations in the list provided by NLTK were removed from the data.

Figure 16 presents a section of the data before and after the stop words are removed. As can be seen, expressions such as 'and', 'but', which are defined as stop words in English, have been cleared from the data.

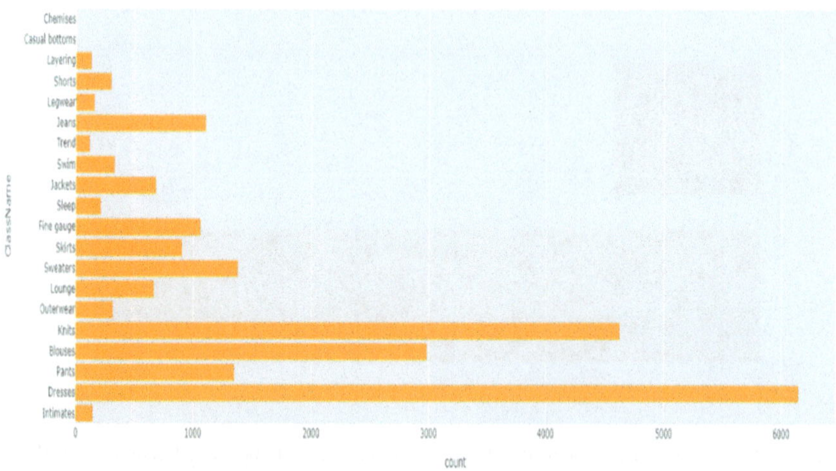

Fig. 13 Boxplot visualizing the distribution of the number of comments by ClassName

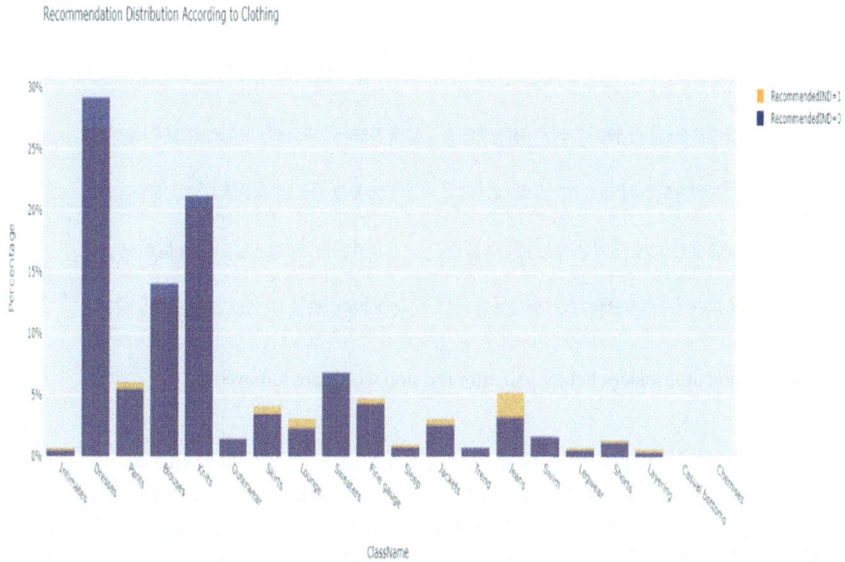

Fig. 14 Recommendation distribution according to clothing

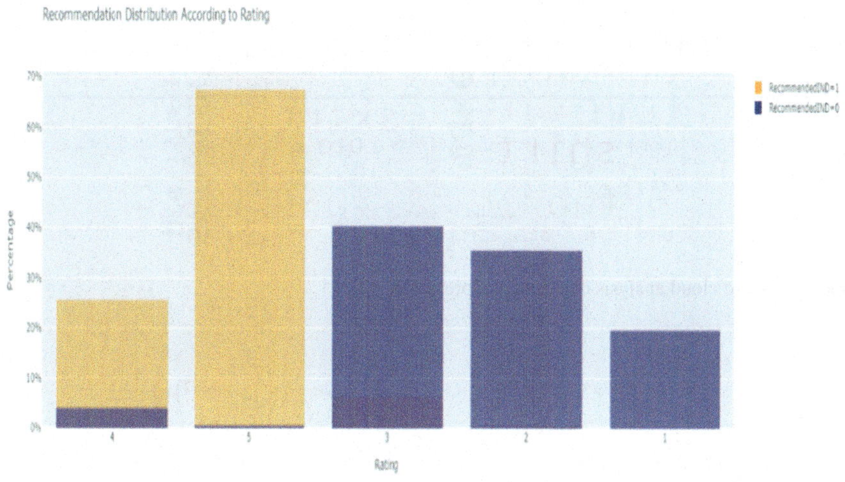

Fig. 15 Recommendation distribution according to rating

Word cloud analysis, which helps to analyze the text quickly and visualize the keywords, was applied separately for positive and negative expressions. It is concluded that the larger a word in the visual, the more common that word is in the data. This type of visualization can be used to explore text and identify key points in the reporting phase.

ReviewText1	ReviewText2
absolutely wonderful silky and sexy and comfo...	absolutely wonderful silky sexy comfortable
love this dress its sooo pretty i happened t...	love dress sooo pretty happened find store im ...
i love love love this jumpsuit its fun flirty ...	love love love jumpsuit fun flirty fabulous ev...
this shirt is very flattering to all due to th...	shirt flattering due adjustable front tie perf...
i love tracy reese dresses but this one is not...	love tracy reese dresses one petite 5 feet tal...

Fig. 16 Comment instances before and after the stop words are removed

In the analysis of positive expressions seen in Fig. 17, there are words such as love, comfortable, and pretty among the most common and prominent words.

In the analysis of negative expressions seen in Fig. 18, there are words such as small, didn't buy, and return among the most common and prominent words. Some

Fig. 17 Word cloud analysis of positive expressions

Fig. 18 Word cloud analysis of negative expressions

Fig. 19 Vectorized tokens format

Unigrams tokens:

Flattering	I	love	this	dress

Bigrams tokens:

Flattering, I	I love	love this	this dress

expressions, such as dress, which are very common in both analyses, show that the number of comments about them is high in the data, and these words are not associated with negative or positive tendencies.

Before starting the model predictions, the customer reviews in the given dataset are transformed into vector form with 'countvectorizer' *n*_gram range which is applied as (1, 2) means unigrams and bigrams for creating a feature for models.

For instance, if the following review is vectorized with unigrams and bigrams,

Flattering, I love this dress.

The results will be viewed vectorized tokens format as (Fig. 19).

Then, these tokens are converted into numeric values by existence in all sentences of the dataset.

The separation procedure of train and test data is the process of dividing a dataset into two subsets at a given ratio. While the training subset is used to train the model, the test subset given to the model as input is used to make predictions and compare these predictions with expected values. In this study, a train/test split of (75–25%) was applied. Finally, the different models were run, and the results were obtained as seen in Table 25.

In terms of accuracy, recall, and F-score, logistic regression gave the best results compared to other algorithms. On the other hand, in terms of precision, SVM is better than logistic regression and the other models.

As can be seen from the ROC curves shown in Figs. 20 and 21, logistic regression and SVM gave better results than gradient boosting trees and random forest. These ROC curves are closer to the upper left corner, so the area under the curves as desired is noticeably larger than the other two methods. This result, therefore, supports the above performance indicators. It is seen that predicted probabilities are mostly shifted toward positive interpretation in gradient boosting trees and random forest models.

Table 25 Performance measurements of all methods

Model	Accuracy	Precision	Recall	F-Score
Logistic regression	0.929	0.865	0.780	0.815
Support vector machines	0.924	0.887	0.729	0.781
Gradient boosted tree	0.897	0.770	0.696	0.724
Random forest	0.876	0.438	0.500	0.467

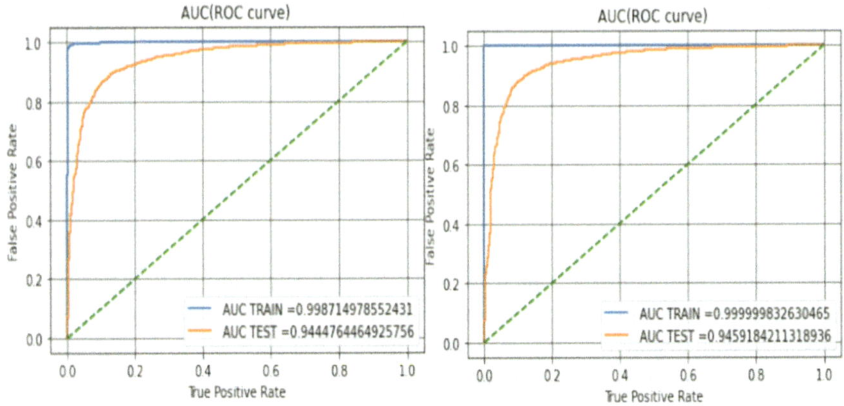

Fig. 20 ROC curves of logistic regression and SVM, respectively

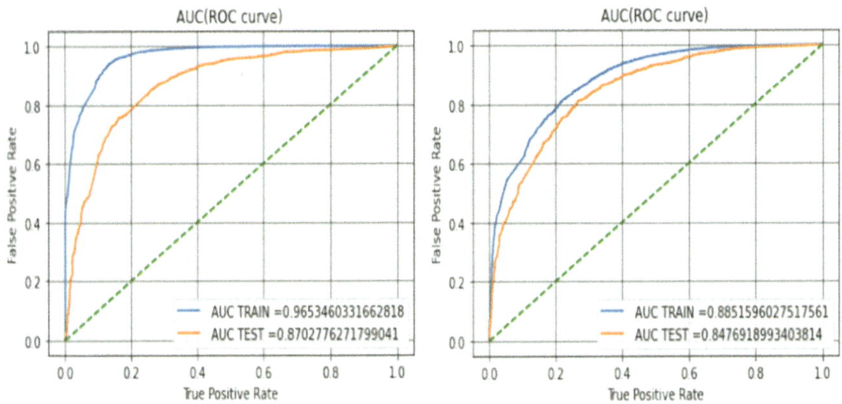

Fig. 21 ROC curves of gradient boosting trees and random forest, respectively

Due to the small sample size in the dataset and the unbalanced distribution of the target variable, it has been concluded that although these algorithms produce the strongest results in the literature, they are not suitable models for this case. With these models, the results tend toward the excess value of the target variable (Recommended_IND = 1).

With the model to be established in this study, it is provided that getting preliminary insight about the products by predicting whether customers who have already bought that product will recommend the product to another customer based on their reviews.

At the end of the model prediction trials, the best fit model is selected for the given case and its dataset. Then some validation tests are conducted to confirm that the prediction model has good results on unseen new data. In most cases, an out-of-time dataset is used for the validation analysis. Since there was not enough new data for this case, a few artificial interpretations were produced, and the prediction results of the models were simply examined. Two unseen reviews prediction results are given below:

I loved this classic jacket. It really suits me. I like it

Built Models' prediction probability results on given review:

Logistic Regression Prediction: [0.930]

SVM Prediction: [0.850]

Gradient Boosting Trees Prediction: [0.959]

Random Forest Prediction: [0.878]

Pants are really oversized! I didn't like it. It was too big, and not appropriate with instruction. I never bought again

Built Models' prediction probability results on given review:

Logistic Regression Prediction: [0.405]

SVM Prediction: [0.128]

Gradient Boosting Trees Prediction: [0.960]

Random Forest Prediction: [0.874]

Predictions are calculated as the probability of recommendation of a customer. If the predicted probability is close to 1, it means that the customer is more willing to recommend the product with his/her review. If the predicted probability is close to 0, it means that the customer is closer to not recommending the product with his/her review. As it is seen in the above examples, probability results, logistic regression, and SVM predictors have better results on reviews. On the other hand, it gives a better idea about models to validate with the use of bulk unseen new review data.

In order to get better prediction results from the given case,

- The more feature engineering processes can be applied, and review text can be strengthened by new variables.
- Dataset sample size can be increased.
- With more hyperparameter tuning and different predictors, models that fit better to the dataset can be obtained.

6 Recommendation Systems

Recommendation systems have been widely penetrated to all parts of life and have gained significant attention both for practitioners and academicians. Utilizing past behavioral patterns of end users, a recommender system specifies items or contents that would be interesting for a user. In this aspect, association rule mining enables gathering the rules for preferable items according to the sequence of past behavior of the user. Thus, recommendation systems can be supported by using these rules to find out what users might like and what could be shown that is not submitted before. This study provides the following: (i) The way of association rule mining can be used to improve state-of the-art collaborative filtering recommender systems; (ii) association rule mining can be used as an alternative way for current state-of-the-art recommendation approaches especially in terms of collaborative filtering approaches and (iii) a case study to prove the way for implementation of association rule mining. The case study is conducted using MovieLens 100 k dataset, and results are reported for building the infrastructure of recommendation systems.

6.1 Problem Background

6.1.1 Motivation

Recommender systems try to make suggestions in accordance with a potential or specific interest of a user with respect to the user's individual preferences. In addition to that, recommendation systems have changed the way to find specific items for users among a large number of potential items or content. Such recommender systems are the focus of current interest especially for a part of electronic commerce research. As realized, recommendation systems are widely applied to the fields such as book, e-learning, music, movie, advertisement, and touristic area recommendations that have extensive variety of implementation domains and further application opportunities [47].

From the perspective of association rule extraction, the motivation is to extract meaningful patterns from massive amounts of transactional data. For instance, sales data can be evaluated in terms of which items purchased on a per-transaction basis considering the sequential way. The main aim is to discover 'connected' items according to associative relations between each item and occurrence across transactions. Here, one can cope with the sequence of the batches or focus on the items that are borrowed, bought, viewed, or liked together by users. In addition, extracted association rule includes two Boolean suggestions that if the left-hand side is true, then the right-hand side will also be true. On the other hand, if probabilistic rule is used, right-hand side will be true with a predetermined probability, and one could conclude that the left-hand side is also true [48].

In association rule mining, support and confidence levels are comprised of two main indicators: (i) Support (s) is the percentage of transactions that contain both item A and item B. (ii) Confidence (c) is the percentage of transactions containing item A that also contain item B. To evaluate rule mining model, lift is a critical indicator for the evaluation of the strength of the relationship between item A and item B. In other words, if lift is calculated as greater than 1, positive association is captured; on the other hand, if lift appeared less than 1, negative association is observed [49].

Association rule mining could support recommendation systems in the following way: (i) figure out all the important association rules, (ii) setting a predictive model based on these association rules, and (iii) recommend to users the top-n items that are connected with the items they rated or viewed. In this respect, transactions should be converted as users' ratings with respect to transaction types of frequent itemsets. Regarding this, the main step is mining associations to transactions. To prepare relevant recommendations to each user, an adequate number of rules for both users and items should be covered. This extraction relies on pre-defining the minimum support and minimum confidence of rule mining process. For instance, if the minimum support and confidence are high, rules for recommendation could not be discovered. In contrast, if minimum support and confidence are low, meaningless rules are gathered that could increase computational effort for recommendation systems. Considering this situation, minimum support and minimum confidence should be balanced and determined properly according to capturing users' behavior changes to find out desired range for the number of rules [50].

To achieve a well-designed recommendation system, the practitioners should better consider the most important factor items and user-based similarities. According to recent studies, gathering the similarities between items could assist the limitation of data sparsity and cold start problems in recommendation environment [51]. Using associated items to extract alternative items is invaluable information that may support recommendation system to enhance its decision-making capabilities regarding chances in user attitudes.

This study provides a transition of association rule mining and recommendation systems. For case study, MovieLens 100 k data that includes 943 users and 1682 movies as unique items is examined, and mapping of items as integers to reduce the computational effort is conducted. The connection of collaborative filtering with respect to MovieLens data is shown with an experiment to provide the improvements in recommendation system.

This part of this chapter is organized as follows: A brief introduction and theoretical structure of association rule mining and recommendation systems are presented in Sect. 2. In Sect. 3, proposed methodology for the implementation of association rule mining and recommendation system is explained. A case study considering grocery transaction data is deeply examined, and associated rules are used in recommendation system in Sect. 4. Finally, conclusions and future suggestions are discussed.

6.1.2 Theoretical Background

As mentioned before, association rule mining consists of three main stages: (i) frequent itemset generation, (ii) sequential pattern mining, and (iii) structural pattern determination. Here, the focus of these studies is to extract necessary knowledge from a huge amount of transactional data. These applications are generally conducted on sales transactional data to find out meaningful relationships between items. Hence, a brief literature review is presented on how association mining is working and adapted to support recommendation systems to provide preliminary information on the topic.

With the help of increasing potential of computational efficiency, association rule mining aims to find relationships between attributes of a database. For instance, finding out useful rules to present customer behavior enables specification of customer needs and assists the implementation of personalized recommendation systems. Different from market basket analysis, medical diagnosis to extract illness and treat patterns and also analyzing protein sequences can be given as an example of other application fields of association rule mining [52].

A frequent itemset is defined as a pattern that is determined with a specific threshold, named as minimum support that is the most applied approach in association rule mining field. Naturally, frequent itemset mining intends to determine patterns such as classifiers, correlations, clusters, association rules, and sequences. It aims to optimize the extraction of patterns in a specific dataset. The methods for utilizing frequent itemset can be described as (i) pruning approaches to reduce the number of candidates, (ii) direct hashing and pruning method to reduce number of transactions, and (iii) reducing the number of pairwise comparisons. Note that Table 26 defines frequent pattern mining approaches.

As seen from Table 26, discrimination of the methods highly depends on dataset configuration: Here, datasets can be organized in two approaches: horizontal approaches such as hash tree and prefix tree include transactions with row-based appearing. On the contrary, vertical design presents items with columns while presenting exchanges of items. In addition to all these approaches, transaction reduction, dataset partitioning, and sampling could be applied for executing some of the

Table 26 Comparison of commonly used methods for frequent itemset mining [52]

Methods	Data configuration	Description
Apriori	Hash tree based (horizontal)	Eliminating some of the itemsets for less execution time
Eclat	Vertical	The algorithm extracts the elements using a single depth that executes the search on dataset in advance
P mine	Horizontal	Mining with parallel disk on multi-core processor
FP growth	Tree based (frequent pattern tree)	Preserving the association information of all itemsets
COFI	Horizontal	Mining with pruning method that uses relatively small trees from FP tree
TM (transaction mapping)	Vertical	Transactions are listed as transaction intervals and search are adapted with lexicographic tree
SSR	Horizontal	Generating subtree for each frequent item and then generating candidates in batch from this subtree

transactions later. For instance, parallel frequent pattern mining enables the segmentation of algorithms to handle computational dependencies that computational time is linear. On the other hand, it requires extensive scale data processing.

Apart from frequent itemset generation, sequential pattern mining is another method for a sliding time window as time constraints. Sequential pattern mining utilizes the downward closure approach of sequential patterns and conducts a multiple pass, candidate generate-and-test approach. As another method, structured pattern mining conducts sophisticated patterns including trees, lattices, and graphs for modeling complex structures such as text retrieval and video understanding. Here, graph-based illustration of the problem and greedy search methods is widely applied to discover best frequent patterns [53].

In this respect, the most important point to use association rule mining is supporting candidate generation especially for item-based collaborative filtering to extract similar items or groups. This task is mandatory to the evaluation of user item interactions. Therefore, association rule mining supported recommendation system examples are given in Table 27.

As seen from Table 27, association rule mining has been adapted widely for improving recommendation performance in various fields such as stock market recommendation, disaster management, book recommendation, and web personalization. Also, recommendation performance could increase by utilizing association rule mining. For instance, association rule mining is conducted for coping with cold start problem about the issue of recommendation system that could not extract any suggestions for new users or items caused by insufficient information. This particular case is generally faced in collaborative filtering. In addition to that case, similarity detection of transactions is also provided with clustering techniques.

To sum up, this study aims to present association rule mining adapted movie recommendation system to increase the effectiveness of top k recommendations. For this purpose, user similarity is detected from k-nearest neighborhood (k-NN), and item similarity is calculated with Jaccard index. After user item similarities are gathered, association rule mining is gathered for target users, and rules are given in descending order. According to these rules, similar items are recommended to user groups. This approach enables improvement of stability and decreases the effect of cold start problem which would state the computational effectiveness.

6.2 Problem Definition

Recommendation system aims to identify several points for making suggestions to each user:

- How to rank items for each user?
- How to group each user and item?
- How to select candidate items for each user group?
- What is the cost if we show ith item rather than jth item?

Table 27 Association rule mining supported recommendation system examples

Problem statement	Author(s)	Approach	Specifications
Adaptive websites and web personalization systems	[57]	Frequent itemset generation	Session and transaction identification enabled for personalized systems
Rule extraction	[55]	Personalized rule extraction for each user	Association between rules and items
	[58]	Top N rule extraction with diversified confidence and drank levels	Diversity improvement for ranking tasks
Web mining/collaborative filtering	[59]	Associative classification	Adapting multiclass to whole systems to prevent cold start problem
Disaster management	[51]	A priori algorithm	Identification of risk factors and handle cold start problem
Stock market recommender systems	[60]	Fuzzy rule generation	Fuzzy and time lag-based predictions for relationships between stocks
Increasing the performance of recommendation system	[61]	A priori and *k*-nn comparison	Robustness and stability study for collaborative filtering
	[62]	Clustering enhanced association rule mining	User profiling is supported with clustering
Book recommendation	[63]	Frequent itemset generation	Content-based filtering with rule extraction considering table of contents
Product profitability	[64]	Frequent itemset for prediction of transactional profitability	Cross-selling profitability is supported with collaborative filtering

- How to predict missing values in ratings?

To have proper response to that questions, recommendation system definition could help to give these questions:

Let C be the set of all users and S be the set of all possible items that can be recommended. A utility function U indicates the satisfaction of a specific item s in S to user c in C. In addition to that, utility directly depends on best similar user and item matching. Ordered set of items are gathered at the end of the extraction process. Note that the set of feasible items can be enormous, hundreds or even millions of transactions.

Table 28 Sample transactions [55]

Transaction ID	Items
1	A, B, C
2	A, D
3	A, B, C, E
4	B, E, F

6.2.1 Preliminaries

Association rule mining can be defined as the extraction process of meaningful rules considering the correlation between attributes in transactional datasets. The formal problem can be given as follows: Let *I* present the set of n number of items, *D* denote the group of transaction records, and *T* denote the group of items as a subset of *I*. Note that every transaction is a list of items for a specific identifier of user. This transaction includes a subset of items defined as *A*. Association rule assists the mapping of *A* to another subset (*B*) with a support s. The support formula is given in the following:

$$\text{Support } (AB) = \frac{\text{Support sum of } A \text{ and } B}{\text{Total records in database } D} \quad (14)$$

The rule of this mapping process $A \rightarrow B$ also has a confidence level c that include A also include B.

$$\text{Confidence } \{A|B\} = \frac{\text{Support } (AB)}{\text{Support } (A)} \quad (15)$$

The problem of association rule is to present whole relationship as $A \rightarrow B$ with a minimum support to give rule frequency and confidence to measure degree of correlation between items in database.

A sample transaction is given for better understanding in Table 28. For the association rule given as $A \rightarrow B, C$ confidence is 0.66, and minimum support is 0.50.

To sum up, association rule mining problem can be described as extracting all related rules with user specified minimum support and confidence level. The problem actually covers detection of similar user behavior for the purpose of recommending items to specific target users.

6.2.2 Proposed Approach

Association rule mining-based recommendation system includes two stages: (ii) profile matching that contains the detection of target profile, *p* is compared to each profile case, *pc*, to select *n* most similar cases. (ii) The uncovered items in target profile within selected cases are listed considering the relevance of the target user, and the r most common items are given as recommendations.

Profile determination: The profile similarity is calculated as the weighted sum of the similarities between the items in the target group and main profile case. Note that if there is a direct dependency between an item in the source, $\boldsymbol{pc_i}$, and the target, $\boldsymbol{p_j}$, then maximal similarity is achieved. On the other hand, direct dependencies could not be reached easily because of the rare ratings, and in that case, similarity of the source profile item $\boldsymbol{pc_i}$ is calculated as taking the mean similarity between this item and the $\boldsymbol{n}$ most similar items in the target profile $\boldsymbol{p_j}$.

$$\begin{aligned} \mathrm{Sim}_{\mathrm{profile}}(p, pc, r) &= \sum_{pc_i \in pc} w_i \mathrm{Sim}_{\mathrm{item}}(p, pc_i, r) \\ \mathrm{Sim}_{\mathrm{Item}}(p, pc_i, r) &= 1 \quad \text{if} \quad p_j = pc_i \\ &= \frac{\sum_{j=1..n} \mathrm{sim}\left(p'_j, pc_i\right)}{n} \end{aligned} \tag{16}$$

Here, the formula does not directly depend on one to one correspondence appearance. If there is no such correspondence, similarity between target and profile cases can also be generated.

Ranking task: After the connection between target and profile cases is adapted, item recommendations can be given in lists named as ranking. In ranking tasks, much more importance could be given to items that have higher similarity. On the other hand, a balance between diversity and popular items should be provided. Generally, items should be recommended to similar profile groups compared with target group. Thus, relevance should be measured as [56] mentioned in their paper. Relevance enables the comparison of an item pc_i to profile case pc considering target group p. If the set of retrieved groups are given as PC' which is the subset of profile cases PC, relevancy is calculated as follows:

$$\mathrm{Rel}(pc_i, p, n) = \mathrm{Sim}_{\mathrm{Item}}(pc_i, p, n). \frac{|PC'|}{|PC|}. \sum_{pc \in PC'} \mathrm{Sim}_{\mathrm{profile}}(pc, p) \tag{17}$$

6.2.3 Association Rule Mining to Item-Based Recommendation System

The first step of the proposed recommendation system relies on user collaborative filtering to capture user similarities. This step includes the calculation of nearest neighbors of the target user using K-nearest neighbor (KNN) algorithm. Similarity is calculated from Pearson correlation coefficient on the item ratings given by the user. Note that items which are rated by a user assigned as a transaction to the related user. Secondly, the nearest neighbors' transactions are merged as a unique transaction of the target user. Therefore, transactions of the users are collected in transaction database.

Secondly, items should be grouped for the further implementation of recommendation systems. Thus, similarity of the items is calculated using Jaccard index and similarity degrees form similarity matrix as the initial step of item-based filtering. Finally, association rules are gathered from frequent itemset generation using target user, minimum support, and minimum confidence. To provide the generation of an adequate number of rules, support threshold is settled to facilitate the decision of minimum support level. After association rules are obtained by using minimum support level, they are ranked according to confidence levels. To provide recommendations, association rule results are added into the proposed item list whose preceding is totally the same as input item list. Note that, if there is an inadequate number of recommendations extracted from association rules, empty places are assigned with the similarity values calculated for item-based filtering. In other words, more similar items are added in the suggested item list as a result of association rule mining.

6.3 Case Study: Movie Recommendation System

6.3.1 Dataset

A well-known public dataset, MovieLens 100 K dataset [54], has been considered for the setting of this hybrid approach. The data was collected through the MovieLens website (movielens.umn.edu) during the seven-month period from September 19, 1997, to April 22, 1998. The dataset contains 100,000 ratings given from 1 to 5 for 943 users and 1682 movies. Each user has rated at least 20 movies. Dataset has movies in the following categories: 'Action', 'Adventure', 'Animation', 'Children's', 'Comedy', 'Crime', 'Documentary', 'Drama', 'Fantasy', 'Film-Noir', 'Horror', 'Musical', 'Mystery', 'Romance', 'Sci-Fi', 'Thriller', 'War', and 'Western'. One of the descriptive information can be given as the most rated movies as 'Top 25 rated movies' as seen from Fig. 22. Age distribution of the users who gave ratings is also given in Fig. 23.

List of the main features are also seen in Table 29.

6.3.2 Application

The proposed methodology aims to achieve several points to make proper recommendations:

- To ensure both diversity and specification of the movies, in other words, not to present similar movies to a particular user group.
- Aggregation of association rules to recommendation systems to handle computational efficiency in matching process for large amounts of data.
- Contribution of minimum support level and confidence to provide necessary amount of rule extraction.

After data cleaning to eliminate enormous number of rates, preprocessing to gather binary reviews is achieved using the following pseudocode:

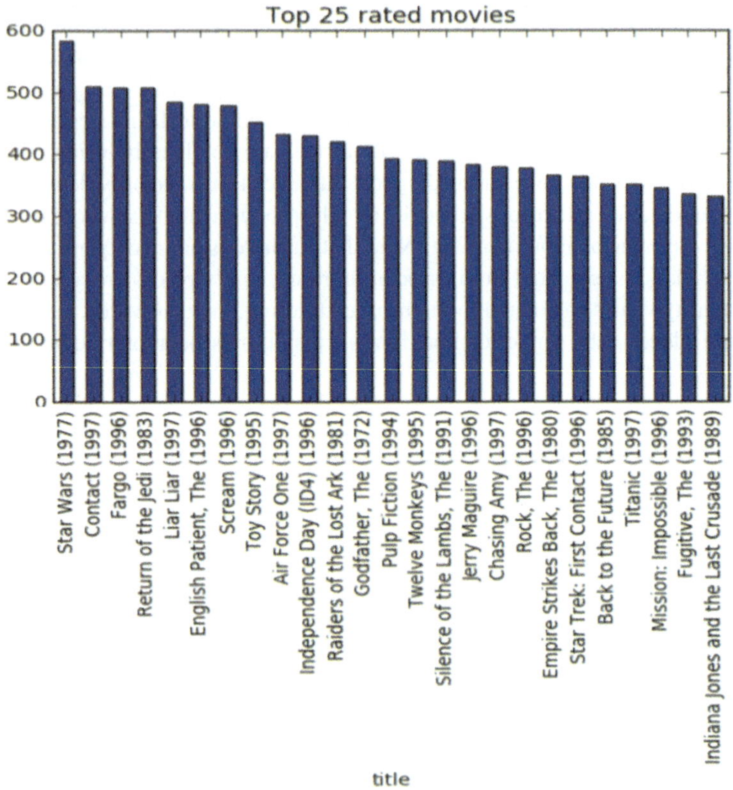

Fig. 22 Top 25 rated movies in MovieLens 100 K dataset

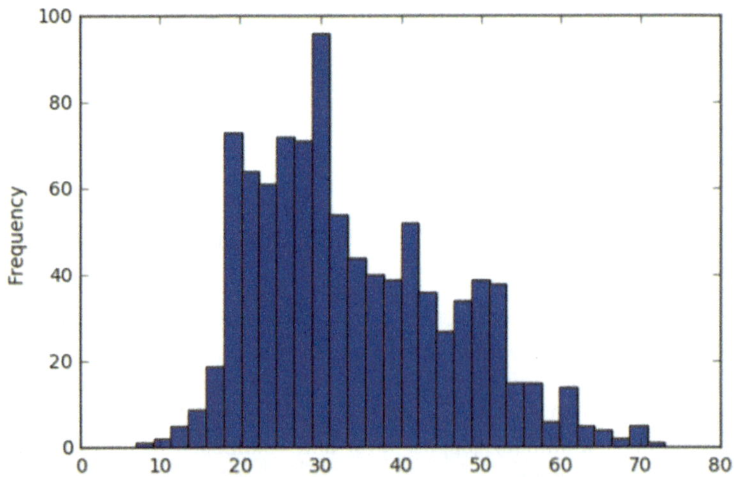

Fig. 23 Age distribution of the users

Table 29 List of features

Feature name	Description
User ID	Unique ID of users
Item ID	Unique ID for items
Rating	Rating of each movie
Timestamp	Unix seconds
Movie ID	Unique ID for movies
Movie title	Name of the movie
Release date	Date of movie release
Video release date	Video release time of a unique movie
IMDB URL	Path of IMDB page
Movie types	Type of the film: unknown|Action |Adventure |Animation |Children's |Comedy |Crime |Documentary |Drama |Fantasy |Film-Noir |Horror |Musical |Mystery |Romance |Sci-Fi |Thriller |War |Western|
Age	User age
Gender	User gender
Occupation	User's occupation

```
Verbose=False
def get_abs_file_path(file_dir, fn):
    cur_dir = os.path.abspath(os.curdir)
    return os.path.normpath(os.path.join(cur_dir, "..", file_dir, fn))
r_fn = get_abs_file_path("data", "c.csv")
w_fn = get_abs_file_path("data", "ratings_c.csv")
with open(w_fn, 'w') as output:
    wr = csv.writer(output)
    with open(r_fn, 'rt') as input:
        header = True
        for row in csv.reader(input):
            if verbose:
                print(row)
            if header:
                wr.writerow(row[:2])
                header = False
            else:
           if rating >= 4:
              wr.writerow(row[:2])
```

In addition to that, pairs of co-occurrences are extracted using the pseudocode:

```
method Map_Users(userid u, movieid m, rating i):
    Emit(userid u, m)
class Reducer:
    method Reduce_Users(userid u, movieids[m1, m2, ...]):
    Movies_reviewed = new Array
    for all movieid m in movieids[m1, m2, ...] do:
        Append(Movies_reviewed, m)
    for all movieid i in Movies_reviewed do:
        for all movieid j not i in Movies_reviewed do:
            Emit(pair(movieid i, movieid j), 1)
class Reducer:
    method Reduce_Cooccurrence_Count(pair p, counts [c1, c2, ...])
    cooccurence_sum = 0
    for all count c in counts [c1, c2, ...] do:
        cooccurence_sum = cooccurence_sum + c
    Emit(pair p, cooccurrence_sum)
    method Map(user u, movies_reviewed m)
    for all reviews i in movies_reviewed m do:
        for all user v not user u do:
            if user v has review i do:
                for all review j not review i in movies_reviewed
n do:
                    Emit(pair(i, j), 1)
class Reducer
    method Reduce(pair p, counts [c1, c2, ...])
    cooccurence_sum = 0
    for all count c in counts [c1, c2, ...] do:
        cooccurence_sum = cooccurence_sum + c
    Emit(pair p, cooccurrence_sum)
```

Once similar users are gathered, forming the user item matrix is provided from the following pseudocode:

```
#Create a user-rating matrix
data_matrix = np.zeros((info['Counts'][0], info['Counts'][1]))
for line in data.itertuples():
    data_matrix[line[1]-1, line[2]-1] = line[3]
#Calculate user similarity matrix based on cosine similarity of ratings
user_similarity = pairwise_distances(data_matrix, metric='cosine')
#Create dictionary of each user and user with most similar preferences
sim_user= {}
for i in range(943):
    sim_user[i+1] = [np.argmax(user_similarity[i])+1]
#Save the results in a text file
fout = "MostSimUser.txt"
fo = open(fout, "w")
for k, v in sim_user.items():
    fo.write(str(k) + ' >>> '+ str(v) + '\n\n')
fo.close()
```

The output of user to user similarity is given for five users in the following:

1 >>> [341]
2 >>> [172]
3 >>> [67]
4 >>> [67]
5 >>> [33]

In similar manner, genre similarities are calculated as seen from Fig. 24.

Finally, the recommended top three films with respect to each user group are given in Table 30. Note that the proposed system presents additional alternatives to each user group from item (genre) similarities list as seen in Fig. 25.

7 Conclusion

This study has configured a recommendation approach by using association rule mining. The system had been tested by existing data in terms of coverage. Best results have been determined for MovieLens 100 k dataset with 65.4% coverage. Processing time was 654 min for this dataset.

```
unknown >>> [('Action', 0), ('Adventure', 0)]
Action >>> [('Adventure', 10305), ('Thriller', 10100)]
Adventure >>> [('Action', 10305), ('Sci-Fi', 5135)]
Animation >>> [("Children's", 2932), ('Musical', 1868)]
Children's >>> [('Comedy', 3122), ('Animation', 2932)]
Comedy >>> [('Romance', 7542), ('Drama', 4117)]
Crime >>> [('Drama', 4067), ('Thriller', 3157)]
Documentary >>> [('Drama', 57), ('War', 53)]
Drama >>> [('Romance', 8017), ('Thriller', 5419)]
Fantasy >>> [("Children's", 807), ('Sci-Fi', 656)]
Film-Noir >>> [('Thriller', 974), ('Mystery', 766)]
Horror >>> [('Thriller', 1811), ('Action', 1285)]
Musical >>> [("Children's", 2492), ('Animation', 1868)]
Mystery >>> [('Thriller', 3270), ('Drama', 1343)]
Romance >>> [('Drama', 8017), ('Comedy', 7542)]
Sci-Fi >>> [('Action', 7968), ('Adventure', 5135)]
Thriller >>> [('Action', 10100), ('Drama', 5419)]
War >>> [('Drama', 5162), ('Action', 4527)]
Western >>> [('Action', 848), ('Comedy', 568)]
```

Fig. 24 Item similarities from association rule mining for recommendation systems

Table 30 Recommendation list for each user group

Occupation	Movie title
Administrator	When We Kings (1996)
Administrator	The Winter Guest (1997)
Administrator	World of Apu, The (Apu Sansar) (1959)
Artist	The 39 Steps (1935)
Artist	The Adventures of Pinocchio (1996)
Artist	Anne Frank Remembered (1995)
Doctor	Kiss the Conqueror (1997)
Doctor	Kull the Conqueror (1997)
Doctor	Leave It to Beaver (1997)
Educator	Incognito (1997)
Educator	Infinity (1996)
Educator	Kaspar Hauser (1993)

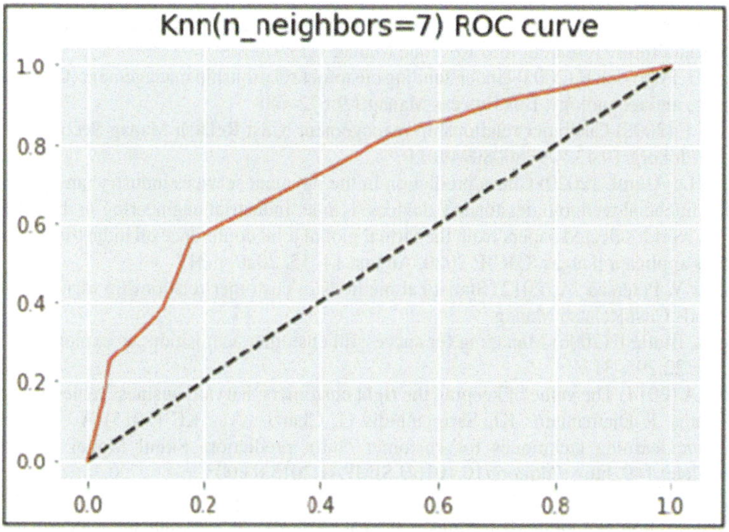

Fig. 25 User and item grouping with respect to *k-nn*

The experiments verify the assumption that a limited number of rules are adequate for making recommendations to a user in a considerable time. An enormous number of rules are not required and might cause complex matching problems in recommendation systems. In addition to that case, the minimum confidence level of rules has a substantial effect on the performance as expected. The reason for that case is that the confidence level of a rule accounts for the average precision that recommend movies in a specific group of users.

In the approach, association rule mining is conducted to extract the collaborative information.

Besides that, association rules can also be assisted to gather content or text-based information. For future direction, using both content and collaborative filtering-based methodology can be better to build hybrid recommendation systems.

References

1. Supply Chain Management (2021) SCM—pricing & revenue. TutorialsPoint, https://www.tutorialspoint.com/supply_chain_management/supply_chain_management_pricing_and_revenue.htm. Access: 06 Sep 2021
2. Cross RG, Higbie JA, Cross ZN (2011) Milestones in the application of analytical pricing and revenue management. J Revenue Pricing Manag 10(1):8–18
3. Rockton Software (2021) How to efficiently manage pricing and revenue in a supply chain. Rockton software, https://www.rocktonsoftware.com/how-to-efficiently-manage-pricing-and-revenue-in-a-supply-chain/. Access: 6 Sep 2021
4. Black J (2019) Revenue management: definition and dynamic pricing. Prisync, https://prisync.com/blog/revenue-management-dynamic-pricing/. Access: 6 Sep 2021

5. Chiang WC, Chen JC, Xu X (2007) An overview of research on revenue management: current issues and future research. Int J Revenue Manag 1(1):97–128
6. Chen IJ, Popovich K (2003) Understanding customer relationship management (CRM): people, process and technology. Bus Process Manag J 9:672–688
7. Buttle F (2008) Customer relationship management. Cust Relatsh Manag Second Ed 1–500. https://doi.org/10.4324/9780080949611
8. Aslan D, Asan U (2020) Churn prediction in the payment services industry: an application at token financial technologies for IoT devices. Içinde: industrial engineering in the internet-of-things world: selected papers from the virtual global joint conference on industrial engineering and its application areas, GJCIE 2020, August 14–15. 2020. p 317
9. Kumar V, Petersen JA (2012) Statistical methods in customer relationship management. Stat Methods Cust Relatsh Manag
10. Ang L, Buttle F (2006) Managing for successful customer acquisition: an exploration. J Mark Manag 22:295–317
11. Gallo A (2014) The value of keeping the right customers, harvard business review, October 29
12. Vafeiadis T, Diamantaras KI, Sarigiannidis G, Chatzisavvas KC (2015) A comparison of machine learning techniques for customer churn prediction. Simul Model Pract Theory Complete: 1–9. https://doi.org/10.1016/J.SIMPAT.2015.03.003
13. Gordini N, Veglio V (2017) Customers churn prediction and marketing retention strategies. An application of support vector machines based on the AUC parameter-selection technique in B2B e-commerce industry. Ind Mark Manag 62:100–107
14. Yang LS, Chiu C (2006) Knowledge discovery on customer churn prediction. Içinde: procedings of the 10th WSEAS international conference on applied mathematics, Dallas, Texas, USA, November 1–3, 2006, pp 523
15. Amin A, Anwar S, Adnan A, Nawaz M, Alawfi K, Hussain A, Huang K (2017) Customer churn prediction in the telecommunication sector using a rough set approach. Neurocomputing 237:242–254. https://doi.org/10.1016/J.NEUCOM.2016.12.009
16. Bing-Zhang H, Yue W, Li-Ming Z, Dong-Lai Z, Ao-Ran X (2018) Customer churn prediction in Chinese traditional broadcasting industry: a positive analysis. Int Conf Manag Sci Eng—Annu Conf Proc 2017-August. 596–605. https://doi.org/10.1109/ICMSE.2017.8574436
17. Mandák J, Hančlová J (2019) Use of logistic regression for understanding and prediction of customer churn in telecommunications
18. Khamlichi FI, Zaim D, Khalifa K (2019) A new model based on global hybridization of machine learning techniques for "customer churn prediction". 2019 3rd Int Conf Intell Comput Data Sci ICDS 2019. https://doi.org/10.1109/ICDS47004.2019.8942240
19. Karvana KGM, Yazid S, Syalim A, Mursanto P (2019) Customer churn analysis and prediction using data mining models in banking industry. 2019 Int Work Big Data Inf Secur IWBIS 2019:33–38. https://doi.org/10.1109/IWBIS.2019.8935884
20. Ahmad AK, Jafar A, Aljoumaa K (2019) Customer churn prediction in telecom using machine learning in big data platform. J Big Data 61(6):1–24. https://doi.org/10.1186/S40537-019-0191-6
21. Amin A, Al-Obeidat F, Shah B, Adnan A, Loo J, Anwar S (2019) Customer churn prediction in telecommunication industry using data certainty. J Bus Res 94:290–301. https://doi.org/10.1016/J.JBUSRES.2018.03.003
22. Pamina J, Raja B, SathyaBama S, Sruthi MS, Vj A (2019) An effective classifier for predicting churn in telecommunication. Jour Adv Res Dyn Control Syst 11:221–229
23. Amin A, Al-Obeidat F, Shah B, Al Tae M, Khan C, Durrani HUR, Anwar S (2017) Just-in-time customer churn prediction in the telecommunication sector. J Supercomput 766(76):3924–3948. https://doi.org/10.1007/S11227-017-2149-9
24. De Caigny A, Coussement K, De Bock KW, Lessmann S (2020) Incorporating textual information in customer churn prediction models based on a convolutional neural network. Int J Forecast 36:1563–1578. https://doi.org/10.1016/J.IJFORECAST.2019.03.029
25. What Is the Telecommunications Sector? https://www.investopedia.com/ask/answers/070815/what-telecommunications-sector.asp. Accessed 1 Aug 2021

26. Reasons for customer churn in telecoms [Survey results]|TechSee. https://techsee.me/resources/surveys/2019-telecom-churn-survey/. Accessed 1 Aug 2021
27. Telecom Churn Dataset|Kaggle. https://www.kaggle.com/mnassrib/telecom-churn-datasets. Accessed 1 Aug 2021
28. Metrics and scoring: quantifying the quality of predictions—scikit-learn 0.24.2 documentation. https://scikit-learn.org/stable/modules/model_evaluation.html#accuracy-score. Accessed 17 Aug 2021
29. Classification: precision and recall|machine learning crash course. https://developers.google.com/machine-learning/crash-course/classification/precision-and-recall. Accessed 14 Aug 2021
30. Internet World Stats (2021) History and growth of the internet from 1995 till today. Retrieved July 14, 2021, from https://www.internetworldstats.com/emarketing.htm
31. Statista (15 July 2020) Number of social network users worldwide from 2017 to 2025 (in billions) [Graph]. In Statista. Retrieved July 14, 2021, from https://0-www-statista-com.seyhan.library.boun.edu.tr/statistics/278414/number-of-worldwide-social-network-users/
32. McKinsey (2012) The social economy: unlocking value and productivity through social technologies. Retrieved July 14, 2021, from www.mckinsey.com/industries/technology-media-and-telecommunications/our-insights/the-social-economy
33. Edosomwan S, Prakasan SK, Kouame D, Watson J, Seymour T (2011) The history of social media and its impact on business. J Appl Manag Entrep 16(3):79–91
34. Paridon T, Carraher SM (2009) Entrepreneurial marketing: customer shopping value and patronage behavior. J Appl Manag Entrep 14(2):3–28
35. Burby J, Brown A, Standards Committee WAA (2007) Web analytics definitions. Web Analytics Association, Washington DC
36. Jansen BJ (2009) Understanding user-web interactions via web analytics. Synth Lectureson Inf Concepts, Retrieval, Serv 1(1):1–102
37. Kotler P, Kelle KL (2006) A Framework for marketing management, 6th edn. Pearson Prentice Hall
38. Lovelock C H, Wirtz J (2011) Services marketing-people, technology, and strategy. 7th edn. Pearson Prentice Hall
39. Kim J, Lim C (2021) Customer complaints monitoring with customer review data analytics: an integrated method of sentiment and statistical process control analyses. Adv Eng Inf 49:101304
40. Kang D, Park Y (2014) Review-based measurement of customer satisfaction in mobile service: sentiment analysis and VIKOR approach. Expert Syst Appl 41(4):1041–1050
41. Hwangbo H, Kim YS, Cha KJ (2018) Recommendation system development for fashion retail e-commerce. Electron Commer Res Appl 28:94–101
42. Jivani AG (2011) A comparative study of stemming algorithms. Int J Comp Tech Appl 2(6):1930–1938
43. Scikit Learn (2021) Sklearn feature extraction, Count vectorizer. https://scikitlearn.org/stable/modules/generated/sklearn.feature_extraction.text.CountVectorizer.html
44. Machine Learning (2021) Text classification, Step 3: prepare your data. https://developers.google.com/machine-learning/guides/text-classification/step-3
45. Goutte C, Gaussier E (2005) A probabilistic interpretation of precision, recall and *F*-Score, with implication for evaluation. In: Losada DE, Fernández-Luna JM (eds) Advances in information retrieval. ECIR 2005. Lecture notes in computer science, vol 3408. Springer, Berlin, Heidelberg. https://doi.org/10.1007/978-3-540-31865-1_25
46. Nick Brooks (2018) Women's e-commerce clothing reviews. https://www.kaggle.com/nicapotato/womens-ecommerce-clothing-reviews
47. Herlocker J, Konstan J (2004) Evaluating collaborative filtering recommender systems. ACM Trans Inf Syst 22(1):5–53. Retrieved from http://portal.acm.org/citation.cfm?doid=963770.963772, http://dl.acm.org/citation.cfm?id=963772. https://doi.org/10.1145/963770.9637722
48. Agrawal R, Srikant R (1994) Fast algorithms for mining association rules, proceedings of the 20th VLDB conference. Santiago, Chile
49. Hegland M (2003) Algorithms for association rules, lecture notes in computer science (2600)

50. Jooa JH, Bangb SW, Parka GD (2016) Implementation of a recommendation system using association rules and collaborative filtering. Procedia Comput Sci (91):944–952
51. Viktoratos I, Tsadiras A, Bassiliades N (2018) Combining community-based knowledge with association rule mining to alleviate the cold start problem in context-aware recommender systems. Expert Syst Appl 101:78–90
52. Chee CH, Jaafar J, Aziz IA, Hasan MH, Yeoh W (2019) Algorithms for frequent itemset mining: a literature review. Artif Intell Rev 52:2603–2621
53. Yazgan P, Kusakci I (2016) A literature survey on association rule mining algorithms. Southeast Eur J Soft Comput 5(1). https://doi.org/10.21533/scjournal.v5i1.102
54. Kaggle (2021) https://www.kaggle.com/c/movielens-100k. Accessed 18 Oct 2021
55. Lin W, Alvarez S, Ruiz C (2001) Efficient adaptive support association rule mining for recommender systems. Kluwer Academic Publishers pp 1–20
56. Smyth B, McCarthy K, Reilly J, O'Sullivan D, McGinty L, Wilson DC (27–30 June 2005) Case-studies in association rule mining for recommender systems, conference: proceedings of the 2005 international conference on artificial intelligence, ICAI 2005. Las Vegas, Nevada, USA
57. Mican D, Tomai N (2010) Association-rules-based recommender system for personalization in adaptive web-based applications. In: Daniel F, Facca FM (eds) Current trends in web engineering. ICWE 2010. Lecture notes in computer science, vol 6385. Springer, Berlin, Heidelberg. https://doi.org/10.1007/978-3-642-16985-4_8
58. Kumara Swamy M, Krishna Reddy P (2015) Improving diversity performance of association rule based recommender systems. In: Chen Q, Hameurlain A, Toumani F, Wagner R, Decker H (eds) Database and expert systems applications. Globe 2015, DEXA 2015. Lecture notes in computer science, vol 9261. Springer, Cham. https://doi.org/10.1007/978-3-319-22849-5_34
59. García MNM, Lucas JP, Batista VFL, Vicente MDM (2010) Semantic based web mining for recommender systems In: de Leon F, de Carvalho AP, Rodríguez-González S, De Paz Santana JF, Rodríguez JMC (eds) Distributed computing and artificial intelligence. Advances in intelligent and soft computing, vol 79. Springer, Berlin, Heidelberg. https://doi.org/10.1007/978-3-642-14883-5_3
60. Paranjape-Voditel P, Deshpande U (2011) An association rule mining based stock market recommender system. Second Int Conf Emerg Appl Inf Technol 2011:21–24. https://doi.org/10.1109/EAIT.2011.90
61. Sandvig JJ, Mobasher B, Burke R (2007) Robustness of collaborative recommendation based on association rule mining. In: Proceedings of the 2007 ACM conference on recommender systems (RecSys '07): 105–112. Association for computing machinery, New York, NY, USA. https://doi.org/10.1145/1297231.1297249
62. Sobhanam H, Mariappan AK (2013) Addressing cold start problem in recommender systems using association rules and clustering technique. Int Conf Comput Commun Inf 2013:1–5. https://doi.org/10.1109/ICCCI.2013.6466121
63. Ali Z, Khusro S, Ullah I (2016) A hybrid book recommender system based on table of contents (ToC) and association rule mining conference: the 10th international conference on informatics and systems (INFOS '16) At: Giza, Egypt Volume: ACM/ICPS
64. Chen M, Lin C (2008) A data mining approach to product assortment and shelf space allocation. Expert Syst Appl 32(2007):976–998

Sultan Ceren Oner has received her M.Sc. degree from Istanbul Technical University while working as a research assistant in Industrial Engineering Department. She currently works at Huawei Turkey R&D Center as an experienced researcher. Her research interests are location-based advertising, recommendation systems, and decision theory.

Yusuf Isik is a M.Sc. student in Industrial Engineering Department of Boğaziçi University. He earned his B.Sc. from Istanbul Technical University from the same department in 2019 and is currently working as a research assistant at Istanbul Technical University.

Abdullah Emin Kazdaloglu is a research assistant and a Ph.D. student in industrial engineering department at Istanbul Technical University (ITU). He received his B.Sc. degree in management engineering from ITU in January of 2018 and also received his M.Sc. degree in industrial engineering from ITU in August of 2021. He has a passion for data science, machine learning, and artificial intelligence, and his current research interests are statistics, data science, and business analytics methods. He is eager to develop his professional skills and curious to discover and implement new manners, technologies, and tools especially in the field of machine learning, statistic and computer science ecosystems, data analytics, and business knowledge.

Mirac Murat received the B.S. degree in Industrial Engineering from Istanbul University (2012) and the M.S. degree in Industrial Engineering from Karadeniz Technical University (2018), where he has been working as a research assistant since 2014. He is a Ph.D. candidate and continues his Ph.D. in Industrial Engineering at Istanbul Technical University in his main research area of intelligent decision making and data analytics. His research interest also includes artificial intelligence and heuristic methods.

Tolga Ahmet Kalayci completed his undergraduate education at Istanbul Technical University Industrial Engineering Department (2008) and his Master's degree from Boğaziçi University Industrial Engineering Department (2011). Since 2011, he has taken various roles in the fields of information technology and data science in finance, insurance, retail, and e-commerce sectors. Currently, he works as the data science manager at Hepsiburada, Turkey's leading e-commerce platform, and continues his Ph.D. thesis focused on artificial neural networks at Istanbul Technical University Industrial Engineering Department. As of 2021, he has published an international article and two papers in the fields of machine learning and econometrics. Throughout his academic and professional life, he has produced and implemented solutions to many practical and theoretical problems in the field of data science.

Kubra Cetin Yildiz is a research assistant and a Master's student in the Department of Industrial Engineering at Istanbul Technical University, Turkey. In 2019, she received her B.Sc. degree in Industrial Engineering from Gazi University. Her current research interests lie in the area of human–computer interaction, usability testing and evaluation, qualitative data analysis, and predictive modeling. She plans to continue her academic career by conducting studies focused on enabling individuals and organizations to interact with technology more effectively.

Aycan Pekpazar is a research assistant at the Industrial Engineering Department of Istanbul Technical University (ITU), Turkey. She received her Ph.D. degree in the Department of Industrial Engineering from ITU. Her research interests are human–computer interaction, user experience, usability, technology acceptance, and optimization.

Mahmut Sami Sivri is currently a lecturer at Industrial Engineering Department in Istanbul Technical University. He is also working on some R&D projects as the director of Data Analytics at ITU's Technopark. He received the B.S. degree in Computer Engineering and the M.Sc. degree in Engineering Management from Istanbul Technical University. He worked in various companies and positions in the IT industry since 2008. His current research interests include machine learning, sentiment analysis, deep learning, big data and its applications, Industry 4.0, financial technologies, data analytics, supply chain, and logistics optimization.

Nevcihan Toraman graduated from Gaziantep University, Industrial Engineering Department in 2011 and Master's Degree in 2015, and she is still carrying out her academic thesis studies at Istanbul Technical University Industrial Engineering Doctorate Program. She continues her professional career as a data scientist. Throughout her business life, she has been involved in analytical projects in the manufacturing, banking and aviation sectors, as well as teaching roles in data science and analytics.

Basar Oztaysi is a full-time professor at Industrial Engineering Department of Istanbul Technical University (ITU). He teaches courses on data management, information systems management and business intelligence, and decision support systems. His research interests include intelligent systems, fuzzy sets, and data mining. He has four edited books and has more than 200 publications. He has been in the editorial board of Journal of Intelligent and Fuzzy Sets, Journal of Fuzzy Logic and Modeling in Engineering, and Recent Patents on Engineering.

Umut Asan is currently an associate professor in the Industrial Engineering Department at Istanbul Technical University, Turkey. He received his B.Sc. and M.Sc. degrees in Industrial Engineering from Istanbul Technical University. He holds a doctoral degree in Industrial Engineering from Technical University of Berlin (with summa cum laude). His present research interests include marketing analytics, data-driven decision making, applied multivariate statistical analysis, and uncertainty modeling.

Cigdem Altin Gumussoy is an associate professor at Industrial Engineering Department of Istanbul Technical University (ITU), Turkey. She holds her M.Sc. and Ph.D. degree in the Department of Industrial Engineering from ITU. Her research interests are human–computer interaction, user experience, usability, and technology acceptance.

Financial Analytics

Mahmut Sami Sivri, Abdullah Emin Kazdaloglu, Emre Ari, Hidayet Beyhan, and Alp Ustundag

1 Introduction

Financial analytics is the application of data science methods and techniques in finance domain. Data analytics has a significant and rising role in helping companies reduce risk and make more efficient financial decisions. AI-based real-time risk detection systems are used to assess risks across the company. Financial statements are automatically examined, including balance sheets, income statements and cash flows, and early warning systems are established. Credit allocation decisions are made using sophisticated machine learning models that estimate how likely a consumer is to repay a loan. Customers receive smart recommendations from financial organizations based on their banking or investment preferences. Financial firms use algorithmic trading and sophisticated mathematical models to develop new trading techniques. Fraudulent transactions and money laundering activities can be detected using anomaly detection algorithms. Robo-advisors provide automated and algorithm-driven financial planning services to their customers without human intervention. In this chapter, four cases are presented regarding financial statement analysis, credit risk management and investment analytics.

M. S. Sivri · A. E. Kazdaloglu · E. Ari · A. Ustundag (✉)
Department of Industrial Engineering, Faculty of Management, Istanbul Technical University, 34367 Macka, Istanbul, Turkey
e-mail: ustundaga@itu.edu.tr

H. Beyhan
Department of Management Engineering, Faculty of Management, Istanbul Technical University, 34367 Macka, Istanbul, Turkey
e-mail: beyhan17@itu.edu.tr

A. Ustundag et al. (eds.), *Business Analytics for Professionals*, Springer Series in Advanced Manufacturing, https://doi.org/10.1007/978-3-030-93823-9_13

2 Financial Statements, Ratio Analytics and Bankruptcy Prediction

In its simplest terms, financial analysis is a process that is generally carried out to generate long-term plans and to evaluate the financial performance, convenience and sustainability of the company in the context of the company's sector, economic conjuncture and financial policies. This process could be adapted not only to the company context but also to projects, investments, budgets and other finance-based transactions. If the financial analysis is done externally, the answer can be sought whether any business is profitable and stable enough to guarantee the return of the monetary investment made in it. If done internally, it enables managers and financial analysts within the company to draw inferences from the company's past financial situation. It helps them make their decisions and recommendations about the future of the company. In this way, financial analysis reveals the strengths and weaknesses of the company that affects its competitive position among other companies, and it is in a principal position for the development of the company [1]. When both ways are evaluated in detail, financial analysis, in short, expresses the company's capacity to carry out its operational processes profitably, to fulfil its obligations and to generate sufficient cash for the opportunities that may arise instantly in the market. At the same time, it should not be forgotten that the company is financially responsible for issues such as paying interest, meeting the principal debt and paying dividends to its investors, who try to maximize their earnings by minimizing their risks [2].

In other words, financial analysis is a general term that expresses the ability to make business and investment-based decisions using financial data. This analysis is performed in two different ways under the headings of technical and fundamental analysis. Technical analysis is statistical trend analysis regarding the company's past data such as share price changes and sales volume. Technical analysis is based on understanding the behaviour pattern of investors in the market rather than looking for the main reasons underlying the share price. In this way, it is tried to make it possible to predict the fluctuations in the shares more easily according to the actions of the market participants. On the contrary, fundamental analysis provides a perspective on the fundamental value of the company by using the financial statements of the company and the ratios derived from these statements. Typically, analysts make this analysis by considering macroeconomic factors such as government policies, financial regulations, microeconomics such as internal policies and management of the company, and environmental factors, including the internal balances of the company's industry [3]. In this way, it creates a foresight to value security. Generally, fundamental analysis is carried out in two ways, and these ways are qualitative and quantitative methods. Qualitative methods include issues such as brand awareness, patents and technology, while quantitative methods are carried out using the three primary expressions used in financial analysis (balance sheet, profit & loss statement and cash flow statement). Thus, companies use these tables both for their internal operations and for the convenience and transparency of reporting to their investors. All three tables are interrelated and shed light on the company's activities, status and financial performance [4]. In the light of all this information, it is obvious that financial statements are of great importance to determine the status of companies.

2.1 Problem Definition

Fundamental financial analysis is one of the most important financial building blocks for all companies and is carried out by examining the financial reports of companies. Fundamental analysis includes many distinct steps in order to present an entire picture of the company's performance. The starting point for this application is the company's financial statements, specifically the income statement (profit & loss statement), balance sheet and cash flow statement. The income statement is used to analyse the company's performance in terms of sales, revenue, business profitability and future cash flow projections over a certain period of time and is usually the starting point of financial statement analysis. The balance sheet shows the financial position of the company at a specific point in time. The balance sheet includes the company's resource (assets) and capital (liability and equity) information. The cash flow statement, on the other hand, shows the operational, financial and investment cash flow, source and use of the company in a certain time period by highlighting the liquidity, solvency and financial flexibility of the company [5]. In the light of all these financial statements, it is possible to calculate the performance or value of any company and to make trend and ratio analysis for each statement with various performance metrics.

However, despite all the benefits that financial statements provide for the company's situation, there are certain limitations as well. First of all, companies' financial statements are mostly unique and the wording in the statements varies from company to company. For instance, a company in the music industry expresses its sales under the name 'Total Revenue', while another company in the retail industry shows this expression as 'Total Net Sales'. Another problem is that different companies collect various financial information differently to elicit cumulative sums. Because of such differences, making comparisons between companies stands out as one of the constraints of financial statements. Another problem is that financial information can be misleading due to the fact that financial statements provide data for a specific time period and effects such as seasonality. The third problem is that there are many financial ratios. Although each of these ratios has a healthy proportional range, it is not possible to evaluate several of them together manually. The fourth problem can be expressed as the fact that the financial statements do not provide enough information about the future legitimacy of the companies, since they contain historical data. The firm's viability and legitimacy are of great importance to the firm's creditors and investors, and therefore, it is critical to assess the firm's probable bankruptcy. In order to eliminate all these problems, it is clear that companies should seek robust and dynamic solutions when evaluating their financial statements.

With this point of view, it is critical to render financial statements with dense information more understandable, to visualize important parameters in all three financial statements and to calculate important ratios. Moreover, it is necessary to develop a system supported by machine learning methods that will evaluate whether the company will go bankrupt in the light of the financial information and ratios in all three financial statements. In this context, answers will be sought to the following main issues:

- What factors and financial information should be considered when analysing the company's historical data?
- What are the changes in the profitability, efficiency, liquidity and solvency of the company over a certain period of time?
- How to visualize important financial information?
- What kind of model should be developed in order to make the comparison of financial ratios of different companies meaningful and to evaluate the health of companies for the future? Which ratios and/or metrics should be included in this model?

2.2 *Case Study: A Technology Company*

Founded in the autumn months of 2000, *X* company is a multinational technology company with its headquarters in *Y* country and more than 1 million employees worldwide, serving both online and offline platforms. In 2020, more than 1 billion unique users visited the website of company *X*, whose focus is electronic commerce. In 2020, more than 1 billion products were sold through the websites of company *X*, and the total annual revenue from these sales (online + offline) reached the $400 billion-mark. On a monthly basis, the website of company *X* receives approximately 15 million clicks and over 50 thousand transactions are made. At the same time, company *X* has 150 offline stores worldwide and more than 100 million products are sold in these stores on an annual basis.

In this case study, the financial statements of Company *X* from the first quarter of 2001 to the last quarter of 2020 will be our primary focus. Although there are fluctuations from time to time in the 20-year period, it can be indicated that Company *X* is in a serious upward trend in terms of financial information, such as sales, net income, assets, capital and net cash flow. Company *X*'s income statement includes financial items for which cumulative sums can be calculated, such as net sales, gross profit, total operating expenses, operating income, profit before tax and net profit. Assets in the balance sheet consist of current and non-current assets. Its liabilities, on the other hand, include current and non-current liabilities and include long-term liabilities among them. In addition, the total equity of the stakeholders is also shown. The cash flow statement of company *X* is basically divided into three as operating, investment and financial cash flow. As a whole, company *X*'s income statement, balance sheet and cash flow statement are described item by item and all financial statements of Company *X* have the following assumptions and operational and financial environment:

- The data in the financial statements of Company *X* was not defected intentionally or unintentionally, and these data were audited by an independent auditor and an audit opinion was presented for the accuracy of the financial statements.
- Financial statements of *X* company online and offline stores are evaluated together.
- It is assumed that government policies, macroeconomic changes and potential competitors do not have an impact on Company *X*'s financial statements and will not have any future effect.
- The companies in the data set to be used to predict bankruptcy are not affiliated with Company *X* in any way, and there is no correlation between their financial information.
- All aspects of technical analysis, such as company X's share price, are not covered by this case study.

The IT department of *X* company wants to develop a system by directly assisting its creditors and investors externally, and its financial analysts and senior management internally. In this context, the balance sheet, income statement and cash flow statement of *X* company will be analysed in terms of 20-financial year period, and this analysis will be supported by dynamic graphics and financial ratio calculations will be made for the financial health of the company. Thus, financial statements will be saved from their complex structure and will be made easier to comprehend. In addition, a bankruptcy prediction model will be developed using machine learning methods and data from more than 6800 companies in country Z. In this way, the data obtained by *X* company using the current financial statement analysis methods will be tested with a more realistic model, and thus, it will be possible to have clearer information about the future and legitimacy of the company. Hence, in the context of this case study, first of all, comprehensive financial statement and ratio analysis of *X* company will be made and then the bankruptcy prediction of *X* company will be made with the help of the data containing financial information of more than 6800 companies used in machine learning methods. A foresight will be created regarding the future financial situation of the company. The data sets containing the financial statements of *X* company and the financial data of 6800 companies are given in the data set folder of the chapter.

2.3 Model

In this case study, the model is examined under two main headings. First of all, a structure will be established in which the general trend of *X* company is analysed, financial ratio analysis is made, and all findings are supported by graphics. Then, a comprehensive bankruptcy prediction model will be developed with machine learning algorithms and the bankruptcy prediction of company *X* will be made thanks to the findings obtained from the first subsection.

2.4 *Fundamental and Ratio Analysis*

In this part of the model, the three financial statements will be carefully examined, and the quarterly data will be expressed as fiscal years, the cumulative totals of the financial statements will be calculated, and thus, the completed financial statements will be visualized. Then, the important financial items in all three financial statements will be graphed. Finally, financial ratios will be calculated by making the necessary calculations and the ratios will be visualized on the basis of categories.

2.4.1 Profit and Loss Statement Analysis

Profit & loss statement is a financial report used by companies for internal control and planning, and externally for stakeholders to evaluate the company's financial performance [6]. This report summarizes the financial information of companies such as revenue, loss, expense, profit/loss for a certain period of time (i.e. quarterly, fiscal year). Profit & loss statement, which is also referred to as a profit statement or a statement of activities in some sources, shows the ability of companies to maximize their profits by increasing their revenue and reducing their expenses. Essentially, it most quickly reflects the financial return of how much profit or loss companies generate with the work they do.

2.4.2 Balance Sheet Analysis

The balance sheet, which is also expressed as the table of the financial position of the company and is the starting point of the fundamental analysis, describes the resources (assets), liabilities and equities (net worth) of the companies. Unlike the profit & loss statement, it shows a snapshot of the company at a certain time. Namely, it is not related to the period spread over the annual process as in the profit & loss statement. Basically, it works in the context of the assets necessary for the company to continue its activities equal to the sum of the company's financial liabilities and equity. One of the main reasons why the balance sheet analysis is so important is that it shows whether the company has a balanced and stable structure enough to pay its current debt. Another critical reason is that analysis and liquidation of assets that can attract potential investors can be performed with this financial statement.

2.4.3 Cash Flow Statement Analysis

The cash flow statement is a financial statement that shows where a company gets its cash from and how it is spent. In its simplest terms, it is a summary of the amount of cash or cash equivalents entering and leaving the company. Essentially, cash flow consists of three key components: cash flow from operating activities, cash flow from investment and cash flow from financing. The cash that the company generates (or spends) as a result of its internal activities is analysed under the heading of operational cash flow. This section includes records of changes in net profit, working capital (e.g. changes in assets and liabilities). Cash flow for transactions on an investment basis includes financial items such as capital expenditures on equipment and property, sales of assets. Financial cash flow is dominated by debt and equity. It includes items such as repayments of short- and long-term debts and dividend payments.

2.4.4 Financial Ratio Analysis

Financial ratios, which can be used for many purposes, are one of the important elements in the process of evaluating the performance and financial condition of an entity. Financial ratios are numerical ratios obtained from three financial statements of companies. They are powerful tools that summarize the financial statements of companies, such as the company's ability to pay its debts and to possess managerial adequacy, and are used to measure the overall financial health of companies [7]. In this way, the ratios create a prediction about whether companies will remain "viable" in future. They allow monitoring of company performance from year to year and observing potential trends (changes). In addition, even though there are many financial ratio values, the ratios help to compare different companies' financial health more easily [8]. Thus, it provides a very effective solution for both internal and external users. Therefore, it is necessary to focus on certain ratios that can play a key role in measuring the financial health of companies. These ratios are liquidity ratios, leverage ratios, effectiveness ratios, profitability ratios, market value ratios, price multiples and valuation ratios.

Liquidity ratios: Liquidity ratios are financial measures that measure a company's ability to meet its short-term liabilities and cash flows without any capital increase. With this ratio, a company's ability to pay its debt obligations is measured by calculating the current ratio, acid-to-test ratio, cash ratio and operating cash flow ratio metrics.

Leverage Ratios: Like liquidity ratios, leverage ratios directly measure a company's financial health, but leverage ratios examine a company's ability to meet its long-term obligations. The main leverage ratios are as follows: debt ratio, debt-to-equity ratio and interest coverage ratio.

Efficiency Ratios: Efficiency ratios measure how effectively a company utilizes its assets. Thanks to these ratios, they can be used as an indicator of the efficiency of how much income is generated by using the assets. Common effectiveness ratios include: asset turnover ratio, inventory turnover ratio and days sales in inventory ratio.

Profitability Ratios: Profitability ratios are used to evaluate the company's ability to generate net profit according to its revenue, costs, assets and equities by looking at the data of the company over a period of time. In other words, profitability ratios are a measure of how effectively profits can be made and value produced for stakeholders.

Commonly used ratios are: gross margin ratio, operating margin ratio, return on assets ratio and return on equity ratio.

Market Value Ratios: Market value ratios are used to evaluate the current share price of a publicly held company's stock. This only helps potential investors understand whether the company's stock is overpriced or underpriced. The most commonly used market value ratios are as follows: book value per share ratio, dividend yield ratio, earnings per share ratio and price-earnings ratio.

Price Multiples: It gives the ratios of the current share prices of a publicly held company in terms of financial items such as earnings, sales and cash flow. Thus, it is possible for investors and analysts to make comparisons between companies and between different financial years of the same company. Commonly used ratios are: price to earnings ratio, price to sales ratio, price to book value and price to free cash flow.

Valuation Ratios: Valuation ratios are used in calculating the company's value. Commonly used ratios are: EV/EBITDA, EV/Sales, EV/FCF and book-to-market value.

2.5 Bankruptcy Analysis

The term bankruptcy is expressed as the inability of a company to pay its debts to its creditors [9]. The bankruptcy of a company and even the possibility of going bankrupt is important for the company's investors and society. Therefore, bankruptcy prediction should be made before the bankruptcy of a company and necessary and appropriate models should be built. In this part of the model, machine learning algorithms are used to predict whether companies will go bankrupt. In this way, it will be possible to predict the bankruptcy of companies with their financial statements and financial ratios.

2.5.1 Oversampling with SMOTE

Unbalanced data set is a problem that is frequently seen in classification problems and occurs when the class distributions are quite far from each other. This problem arises because the majority class dominates the minority class in machine learning algorithms. Owing to this, algorithms often predict the entire data set very poorly for the minority class, showing proximity to the majority class. Even though there are different metric selection and re-sampling methods to solve such problems, the SMOTE sampling method is the easiest and most useful method to apply.

SMOTE oversampling technique starts from the samples of the minority class and generates synthetic new observations in the feature space randomly by interpolation method. Thus, it balances the majority class with the number of observations. However, it does not interfere with the model other than increasing the number of samples and does not provide extra information to the model. According to some sources, the point to be considered while applying SMOTE is that the method should be applied only to the train data set and the original test data should be used while testing the data. Nevertheless, in some models, after all the data is balanced with the SMOTE method, the split of train and test data and the application of algorithms in this way also stand out in practice.

2.6 *Solution and Analysis*

The results about the financial fundamental analysis and ratio analysis of company *X* and its bankruptcy prediction model are given in this section.

2.7 *Fundamental Analysis and Ratio Analysis*

The profit & loss statement, balance sheet, cash flow statement and financial ratios of *X* company mentioned in 7 different categories in the model section will be given in this section.

First of all, the financial items of net sales, gross profit, total operating expenses, operating income, profit before tax, profit before change in accounting and net profit have been calculated in the profit & loss statement. In order not to adversely affect the general flow and visuality of the section, the profit & loss statement of *X* company is not given here. The relevant tables and code outputs can be examined in the Jupyter notebook of the chapter.

Company *X*'s balance sheet is also examined, and its total assets, total current liabilities, total non-current liabilities, total liabilities and total stakeholder equity items are calculated.

Finally, the cash flow statement of company *X* has been examined. Cash used for operating activities, cash used for investment activities and cash used for financial activities are calculated, and the total cash flow is found on a yearly basis.

2.7.1 Profit and Loss Statement Analysis

Profit & loss statement analysis was made, after the necessary cumulative calculations and adjustments in the profit & loss statement, balance sheet and cash flow statement of *X* company. Gross profit, total operating expenses and operating income are visualized with two different graphical displays. Similarly, profit before tax and net profit are compared on a yearly basis. Thus, the trend based on years is followed by analysing the profit & loss statement graphically.

Gross profit, total operating expenses and operating income are shown in Fig. 1. As it can be understood from this figure, it is seen that company *X* makes more sales from year to year and its operating expenses increase accordingly. However,

Fig. 1 Gross profit, total operating expenses and operating income with stacked chart

the noticeable increase in operating profit, especially after 2014, means that the company makes more sales in return for its expenses. In Jupyter notebook, these three financial items are more clearly described and the increase in operating income is more evident. In order not to adversely affect the general flow and visuality of the sections, a symbolic visual is presented in each of the profit & loss, balance sheet and cash flow statements of *X* company. Other related graphics can be examined through Jupyter notebook of the chapter.

Finally, the pre-tax profit and net profit of company *X* are given in Jupyter notebook. There is a significant exponential increase in the net profit of the company after 2014.

2.7.2 Balance Sheet Analysis

Secondly, the balance sheet of company *X* has been analysed. First of all, the distribution of total assets, total liabilities and total stakeholder equity has been examined. Then the distribution of all asset types is visualized. In the third chart, total liabilities and total stakeholder equity are plotted on a multiple axis. In the fourth chart, the distribution of current and non-current liabilities is given and the sub-distributions of both types of liabilities are drawn. The distribution of stakeholder equity types is found in the last chart. However, these graphics are not presented in the text section and should be followed through Jupyter notebook. Symbolically, the chart of assets, total liabilities and stakeholder equity are given here.

Figure 2 shows the graph of assets, liabilities and stakeholders' equity. The equation of the "Asset = Liability + Equity" equation can also be seen graphically. It

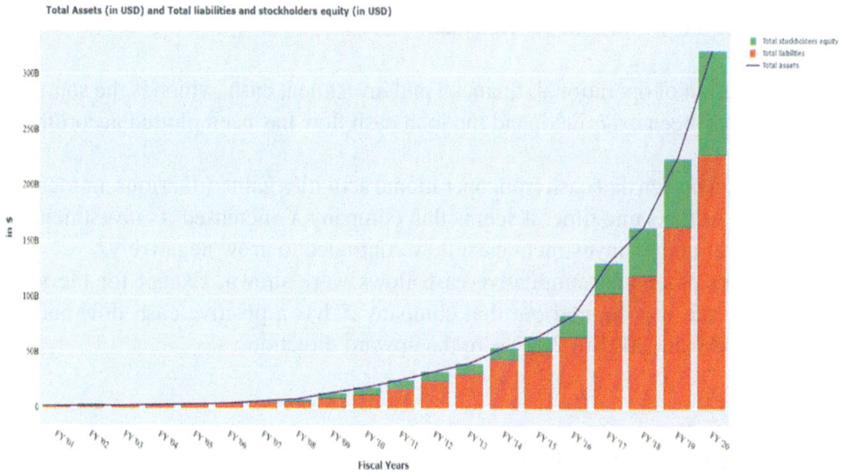

Fig. 2 Assets, total liabilities and stakeholder equity

seems that Company X has increased its assets tremendously over the years, and in turn, there has been a significant increase in its liabilities.

Afterwards, the distribution of asset items has been examined and shown. It is observed that property and equipment assets in particular have increased significantly from year to year. In addition, it is a remarkable case that the amount of market security assets can be easily converted into cash due to the speed of circulation in the market, especially in 2020. Cash and cash equivalents have also increased significantly from year to year.

In Jupyter notebook, the liabilities and stockholder equity of company X are shown on a 2-axis graph. While liabilities are followed on the left axis, equity capital is followed on the right axis. It seems that the stockholder equity tends to fluctuate relatively more over the years and increases more slowly than the liabilities. Generally, this situation is interpreted as a high risk for stakeholders. Yet, it would not be wrong to think that Company X increases its debts by following an aggressive policy in order to be more competitive.

Additionally, the distribution of current and non-current liabilities expressing long-term and short-term liabilities has been given, and the distribution for sub-financial items of both liabilities is also shown. It is realized that the long-term rental debt of X company and its debt to suppliers—as a short-term debt—(Payable accounts) have increased significantly over the years.

The equity distribution of company X is also shown in Jupyter notebook. The increase in retained earnings, which parallels the increase in net income, especially after 2015, is enormous. It can be deduced that Company X did not distribute dividends after 2015.

2.7.3 Cash Flow Statement Analysis

The distribution of operational, financial and investment cash, which is the source of total cash, has been examined, and the total cash flow has been plotted according to years.

In Fig. 3, it is seen that cash from operational activities gained a serious momentum after 2008. At the same time, it seems that company *X* increased its investments in the same year and its investment cash flow continued to grow negatively.

In Jupyter notebook, cumulative cash flows were shown. Except for the years 2001 and 2005, it is figured out that company *X* has a positive cash flow and the general trend from 2001 to 2020 is in the upward direction.

2.8 Financial Ratio Analysis

In this section, the financial ratios of *X* company in seven different categories are drawn.

Liquidity ratios can be seen in Fig. 4. Even though the general trend over the 20-year period is that Company *X* has generally worsened in terms of liquidation, it is clear that the company can continue to exist financially. However, the fact that the cash ratio is even below 0.5 most of the time creates the impression that Company *X* might have had problems in finding the cash it needs to pay its debts in some periods. In particular, the decrease in the current ratio over the years means that the current liabilities have increased against the current assets.

Figure 5 shows the ratios regarding the solvency of company *X*. Particularly, the serious fluctuations in the interest coverage rate can be shown as evidence that the

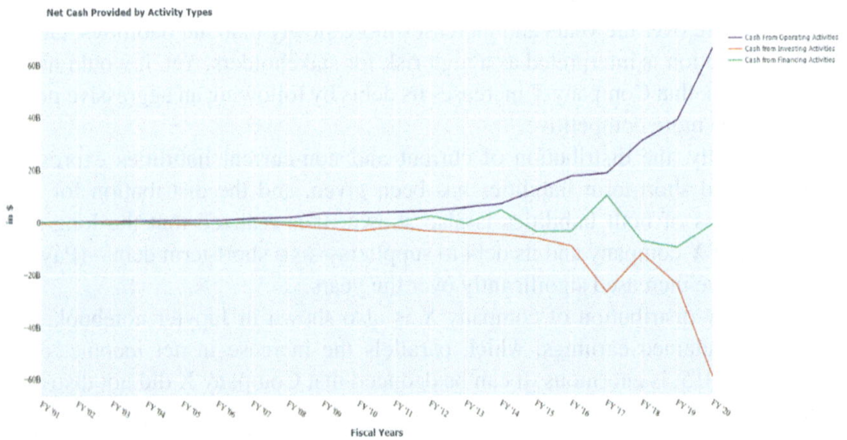

Fig. 3 Net cash provided by activity

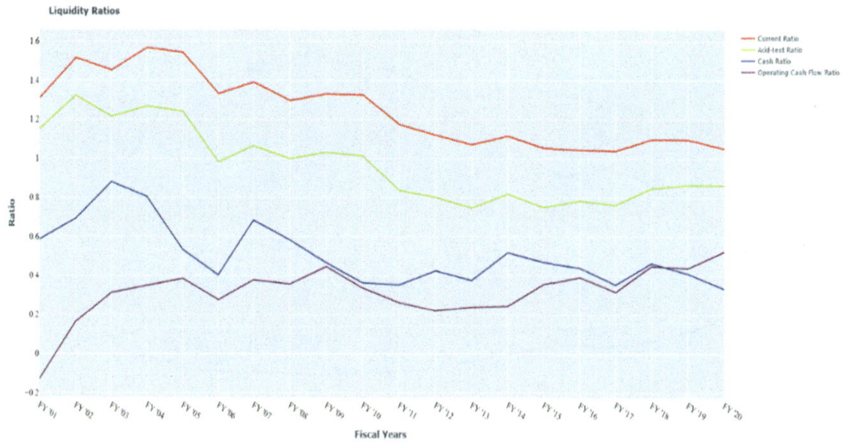

Fig. 4 Liquidity ratios

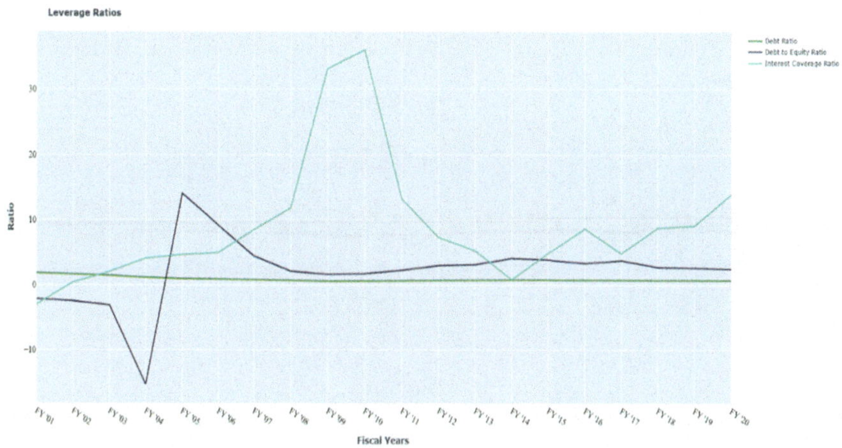

Fig. 5 Leverage ratios

company might have had difficulties in interest payments in some years. Also, a debt-to-asset ratio often above 0.7 means that most of Company *X*'s assets are financed by creditors.

The efficiency ratios are shown in Fig. 6. Both inventory and asset turnover ratios appear to be good overall. Assets and inventory appear to be handed over in relatively short times compared to the industry average. This is a piece of evidence of the efficiency of Company *X*.

In Fig. 7, profitability ratios are shown. Especially the good enough ROA and ROE ratios reveal the efficiency of the company's profit generating capacity.

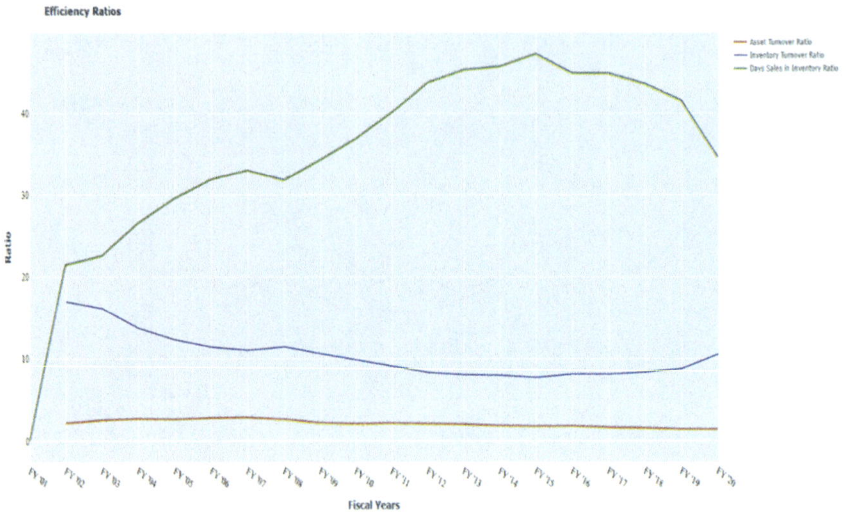

Fig. 6 Efficiency ratios

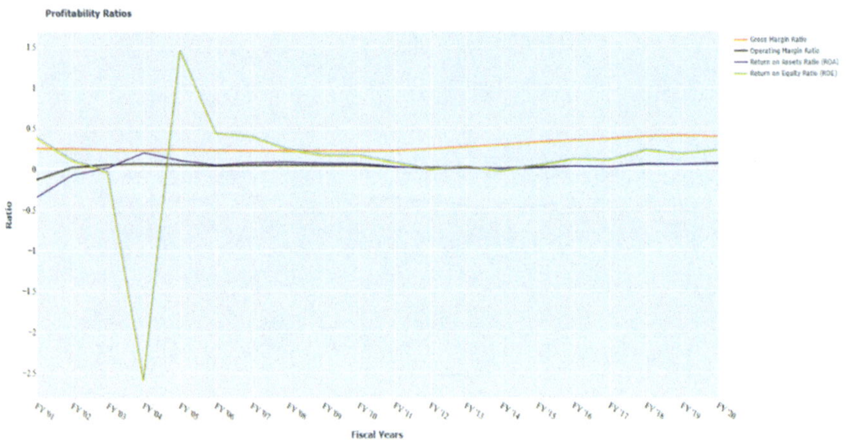

Fig. 7 Profitability ratios

In Fig. 8, there are ratios regarding the market value of Company *X*. Ratios such as earnings per share and sales show that the value per share of Company *X* has increased over the years. The significant increase in the book value, especially after 2015, is a proof of the increase in the net assets (total assets − total liabilities) of Company *X* per share.

Price multiples are given in Fig. 9. These ratios mainly concern investors who want to invest in company *X*.

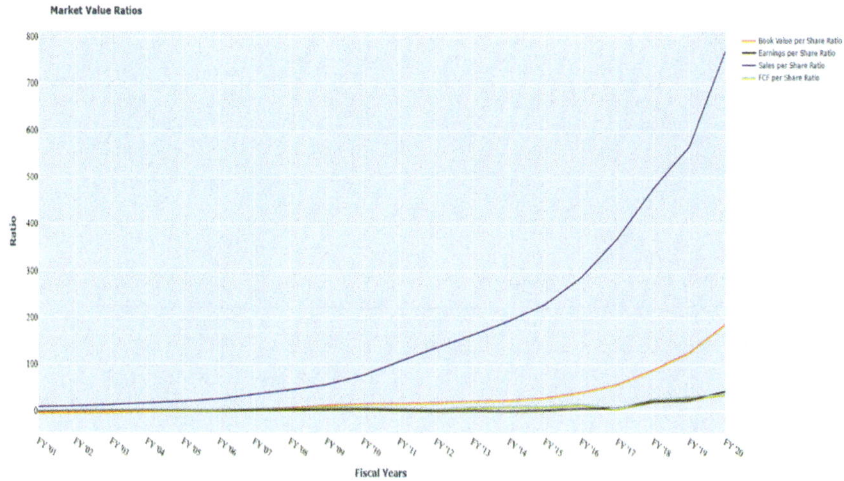

Fig. 8 Market value ratios

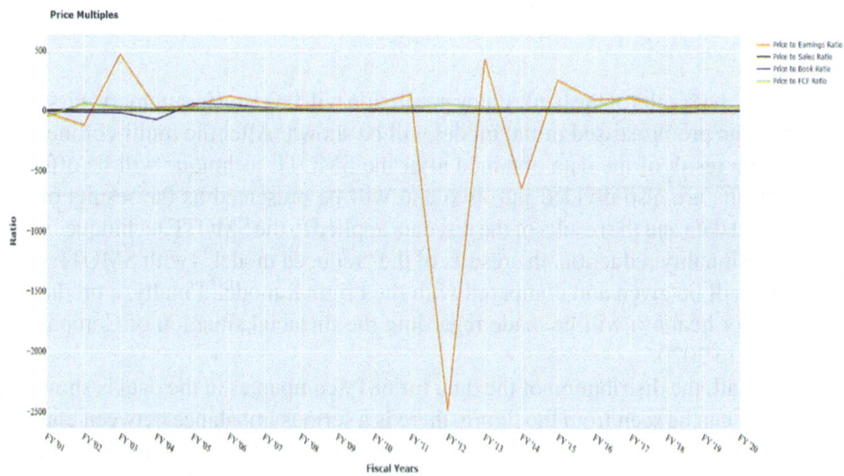

Fig. 9 Price multiples

Ratios presenting the overall value of the company are given in Fig. 10 under the name of valuation metrics.

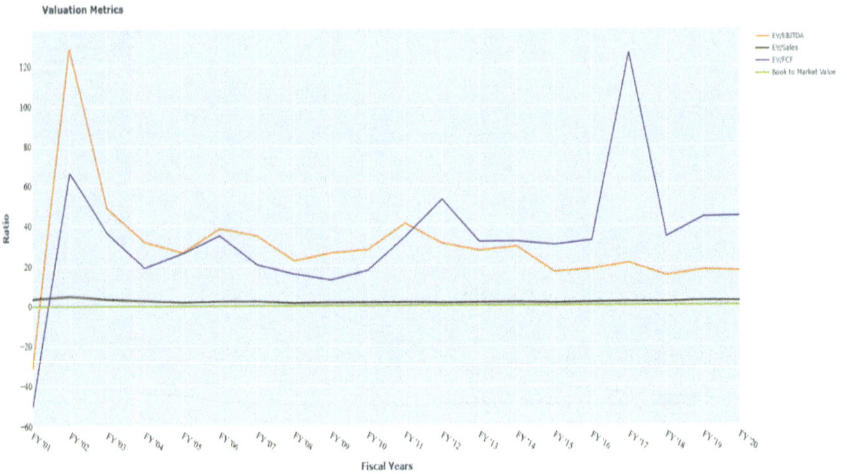

Fig. 10 Valuation metrics

2.9 Bankruptcy Prediction

In this part, the results of the bankruptcy prediction will be given in summary. First, the results of the pre-processed initial model will be shown. After the multi-collinearity control, the result of the data obtained after the SMOTE technique will be offered. These results are also divided into two and will be presented as the results of the original test data and the results of the test data applied to the SMOTE technique. After the dimensionality reduction, the results of the "reduced model" (with SMOTE used test data) will be given and compared with the original model. Finally, a prediction (bankrupt or healthy) will be made regarding the financial situation of Company *X* from 2001 to 2020.

First of all, the distribution of the data for 6819 companies in the data is shown in Fig. 11. As can be seen from the figure, there is a serious imbalance between classes.

In Fig. 12, the confusion matrix values of the initial models divided into the train–test data with 80–20 are shown. As can be seen from the matrices, none of the models can predict "class 1 data" well, since the amount of data with class 1 is very small. Therefore, it is clear that the data should be processed a little more and models should be developed with a different approach.

In the next stage of the analysis, multi-collinearity control was performed and 12 features with high correlation with each other were excluded from the analysis. Afterwards, the sample has been increased with the SMOTE technique and a balance has been achieved between the classes. This is shown in Fig. 13.

Then, all data have been trained with 12 models using the train data. The point to note in this regard is that SMOTE is applied only to the train data in the initial data. That is, the test data is kept as the original data at this stage. The sample has been expanded with SMOTE and then the data has been trained with a larger sample,

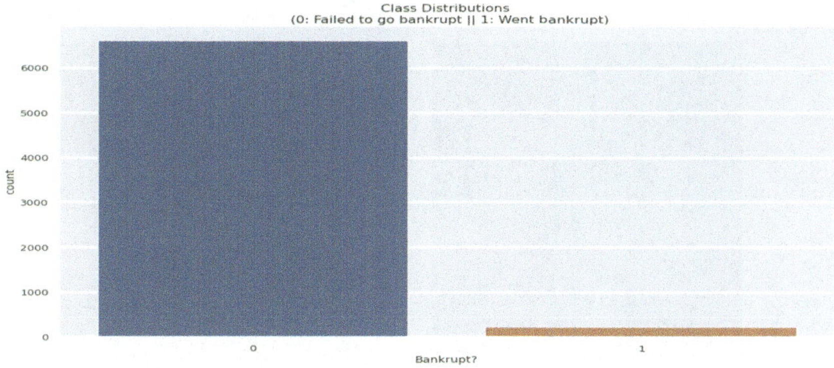

Fig. 11 Class distributions

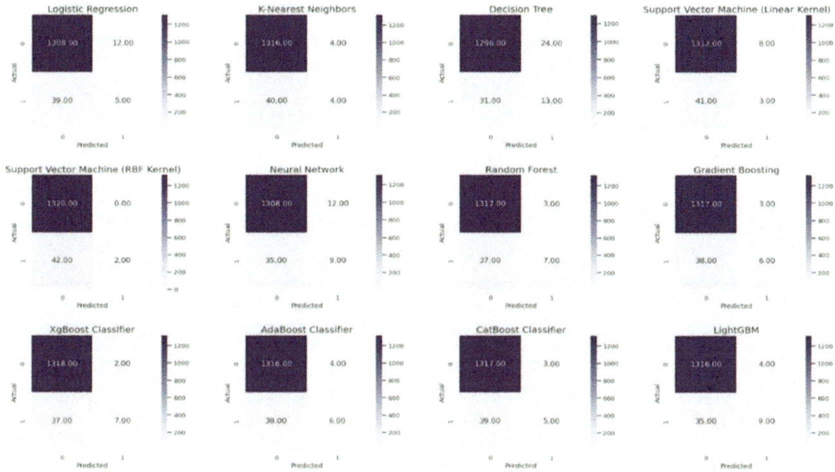

Fig. 12 Confusion matrix of initial models

but the models have been tested with the original test data. Since the models were trained with more data, the test data has been predicted to be better than the initial model. Figure 14 gives the ROC curve of the models, and Fig. 15 gives the confusion matrices. As can be seen from the ROC curve and confusion matrices, while logistic regression, SVM methods and AdaBoost classifier give relatively good results, the results of other models are only mediocre.

In the next model, the SMOTE technique has been applied directly to the raw data (before train–test data split). In this analysis, SMOTE technique has been applied to the raw data. Afterwards, train and test data have been created. Thus, the trained models were validated with the test data, which has a more balanced class distribution. Nevertheless, due to the loss of originality of the test data, there was a problem of overfitting in the models. Figure 16 shows the ROC curve of the model, and Fig. 17

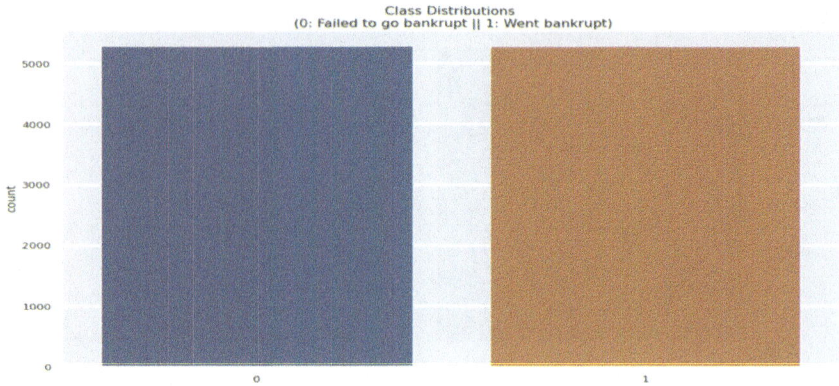

Fig. 13 Class distributions (original test data)

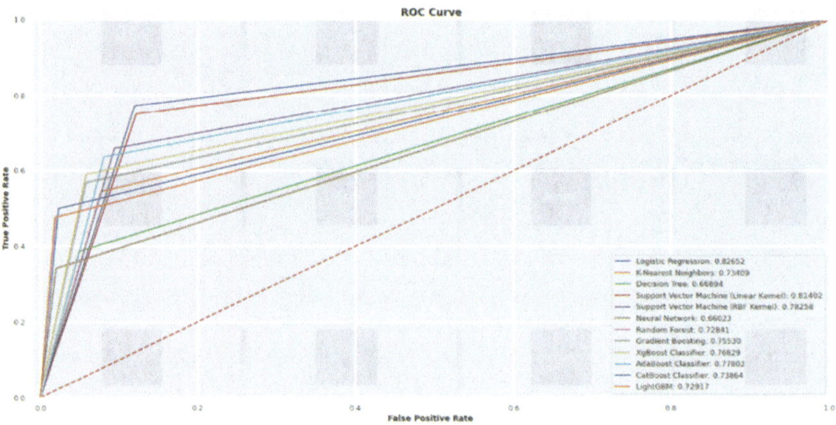

Fig. 14 SMOTE model ROC curve (original test data)

shows the confusion matrices. As can be seen from both performance metrics, many models predicted classes with around 95% success. Especially the success of models such as neural network, CatBoost, LightGBM is very high.

Afterwards, dimension reduction has been applied in order to express the model with fewer attributes. In this way, it is aimed to establish a decent enough model with fewer variables. As a result of the analysis depicted in Fig. 18, the contribution of the variables to the model has been shown and a "reduced model" has been established with 4 attributes according to the graph. Since the marginal contribution of the features after the 4 features to the model has decreased noticeably, it is enough to select 4 features.

After the dimension reduction was performed, the new data with 4 variables was rerun with the SMOTE model (in which the test data was also SMOTE). ROC curve and confusion matrices gave relatively poor results compared to the original model,

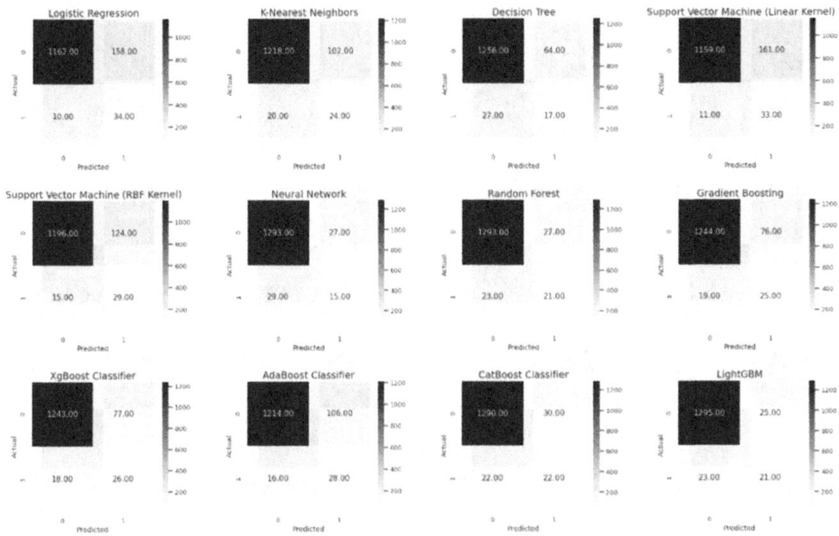

Fig. 15 SMOTE model confusion matrices (original test data)

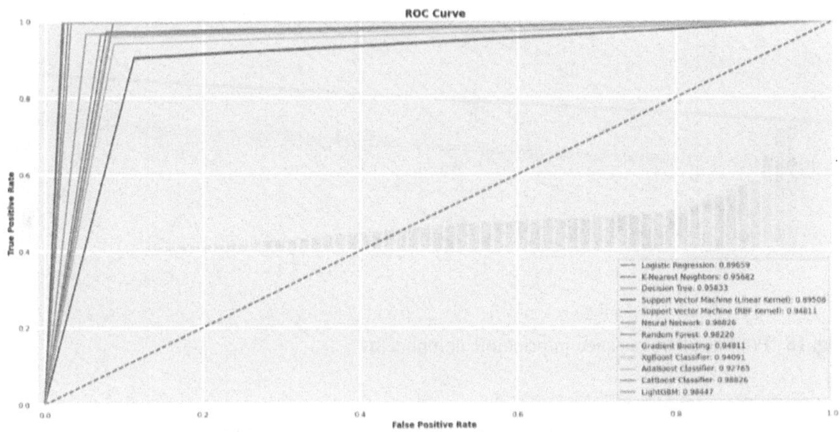

Fig. 16 ROC curve (SMOTE test data)

but despite the fact that the model could be expressed with only 4 variables instead of 62, no tremendous performance degradation was observed. Figure 19 shows how the performance of the 12 models changed after dimension reduction. In spite of the decrease of about 10% in some models, it can be said that it is quite satisfactory to establish models with only 4 features.

In the last part of the analysis, all features in the bankruptcy forecast data were calculated using the 20-year data of company *X*. This data of company *X* was given to both original models (original test data and SMOTE test data models) and the

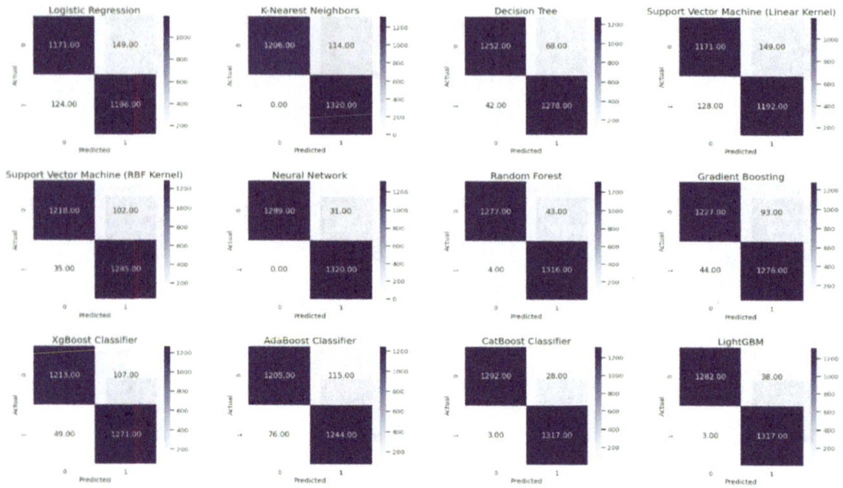

Fig. 17 Confusion matrices (SMOTE test data)

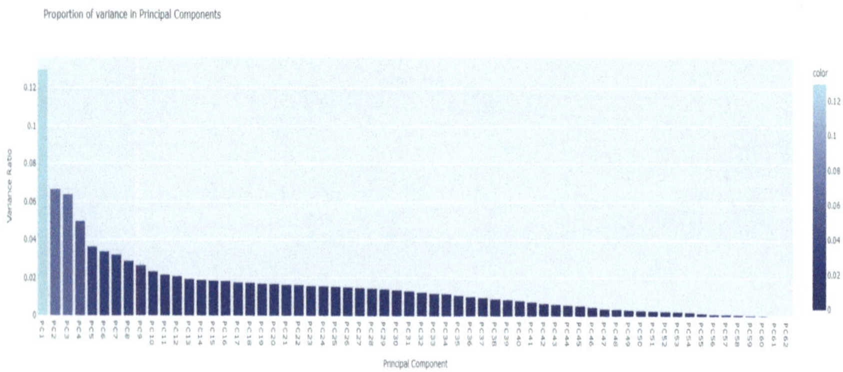

Fig. 18 Proportion of variance in principle components

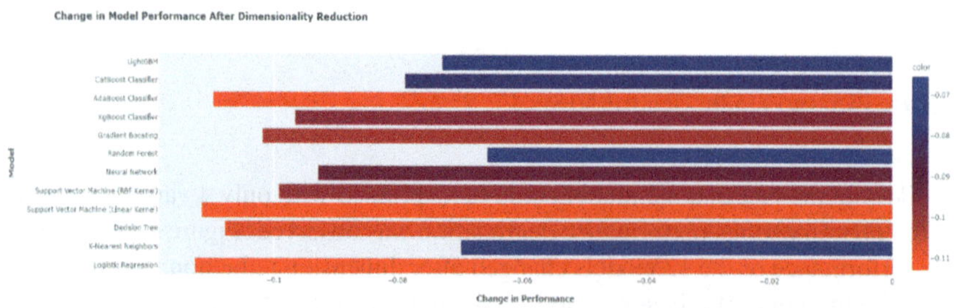

Fig. 19 Change in model performance after dimensionality reduction

prediction was made. Since the tables are difficult to read in the text, references are made to the Jupyter notebook. The results of the models using the original test data are also given there. The majority of algorithms in this model did not show Company *X* as insolvent for any year. However, a significant portion of the algorithms, again analysed using the SMOTE used test data showed that Company *X* went bankrupt in some years.

Finally, the bankruptcy prediction of company *X* was fulfilled by using the ensemble method which combines more than one model and generally gives better results than a single method. Ensemble was applied for both SMOTE and original test data. Accordingly, the financial situation of company *X* in the past years is shown in Jupyter notebook.

3 Conclusion

In this case study, first of all, comprehensive profit & loss statement, balance sheet and cash flow statement of *X* company are analysed. Afterwards, the necessary financial ratios have been calculated and the fundamental analysis has been completed. In the second stage of the case study, a bankruptcy prediction model using 12 machine learning algorithms has been developed as a result of data obtained from more than 6800 companies. This model is also supported by 3 sub-models created within itself. This case study can be developed by selecting the best parameters for the 12 machine learning algorithms and by doing cross-validation.

4 Credit Risk Analysis

4.1 Problem Definition

Credit risk analytics is a type of analytics or techniques that investigate and analyse the financial history of a person, institution or a company whether they can use and apply a credit for some purposes, or they are good at pay their credit back on time. Credit risk analytics can use different models to evaluate the applicant's financial situations. These models vary during analyses according to type of data and number. In the following case study, first the problem is going to be introduced and data will be prepared for analysing the applicants' situations and different models have been used to evaluate the customers' financial situations. After evaluation results, which model is the best model to be used in the credit risk analytics. In the final stage, a brief theory of the best model has been introduced.

4.2 Case Study

The German Bank wants to determine whether customers who apply for credit can be approved to give credit. The bank desires to know information about customers' economic conditions, such as how much their salaries are, for what customers want to take credits (i.e. buying a car, having a house, a tv or a mobile phone, etc.) and so on. If the applicant is in good condition to pay back money, the bank opens a credit to customers who apply the bank credit services. The case is how to make a faster decision that giving a credit to a new customer is risky or not instead of investigating one by one each applicant. According to previous applicants, that is the research question. All it is needed to find the most accurate model to evaluate the applicant credit approval. The German Bank System has applicants to banks to take credit. All data has been obtained from [10]. The credit approval services demand some information about applicants and keep this information to evaluate accepting or rejecting credit applications. Following information is about the applicants' features the German Bank System's requirements:

- Age: This feature is numerical information and legal age for applying the credit.
- Gender: This is string and categorical information that has two categories, namely male or female.
- Job: Job feature is a categorical feature that has four categories which are expressed numerically instead of categorical strings; 0: unskilled and non-resident, 1: unskilled and resident, 2: skilled, 3: high skilled.
- Housing: This feature is string variable which means categorical that these are three categories which are own, rent and free.
- Saving Account: This is a string variable that means categorical variable indicates if applicants have bank account, there are four categories; little, moderate, quite rich and rich.
- Checking Account: This is a categorical variable with three categorical class; little, moderate and rich.
- Credit Amount: This is a numeric variable that how much money applicants apply on Euro.
- Duration: This feature is a numeric variable to show how many months the applicants want to pay back credit.
- Purpose: This feature is a string variable that means categorical variables consist of eight categories. That is the aim of taking credit such as car, furniture/equipment, radio/TV, domestic appliances, repairs, education, business, vacation/others.
- Risk Situation: This feature is value target that string with two categorical variables as good or bad situation.

For the research question, original data sets include 1000 entries and 20 features. But in this case, only few of features are used to analyse to model research questions. Used features have been described in the problem definition section. There are ten features used in the analysis of this case which are age, gender, job, housing, saving account, checking account, credit amount, duration, purpose and risk situation.

The aim of this case study is to find out what algorithm or machine learning method gives the best accuracy or the best performance. The bank credit approving department uses the algorithm to make more fast and efficient method without losing time on approvals.

4.3 Data Information

In this section, first it is necessary to explore the data and its features. Data has ten features columns, and some of them are numerical variables and the rest of them categorical. Since categorical ones are sex, job, housing, saving accounts, checking account, credit amount, purpose and risk. numerical variables are age and duration. From these variables, risk categorical variable has been assigned to the dependent variable as binary classification good referred 1 that takes the credit and bad referred 0 that does not take the credit. This risk variable means that situation of a given customer is good or bad. Since customers can be approved giving credit defined as good which is accepted, otherwise defined as bad condition which is rejected. Apart from the risk feature, rest of the features have been assigned to independent variable (Table 1).

In the case study, there are seven methods that are applied to the data and get the results. These methods are decision tree classifier, random forest, naive Bayes, *k*th nearest neighbour, logistic regression, support vector machine and XGBoost classifier as machine learning methods. All these methods are called classification methods that means labelled as supervised learning methods on machine learning field.

There are missing values in data sets, there are two types of implementations of algorithms. One of these types is to clean missing data and another one is to proceed without cleaning missing values. These approaches are common in the machine

Table 1 Number of features non-null values

#	Column	Non-null count	Dytpe
0	Unnamed: 0	1000	Int64
1	Age	1000	Int64
2	Sex	1000	Object
3	Job	1000	Int64
4	Housing	1000	Object
5	Saving accounts	817	object
6	Checking account	606	Object
7	Credit amount	1000	Int64
8	Duration	1000	Int64
9	Purpose	1000	Object
10	Risk	1000	Object

learning field, and they have some advantages and disadvantages. One of the most popular advantages is all information for modelling is available that algorithms are easily run. One of the most well-known disadvantages is removing missing values from the data will cause loss of the information. Most researchers do not want to lose information to implement algorithms. In this case, all algorithms and methods have been applied with missing values and without missing values.

4.4 Data Cleaning and Preparing

First, all algorithms will be applied to cleaned data. Missing values have been cleaned with their rows. Saving account and checking account features have missing cells. Saving account feature has 183 missing cells, and the checking account has 394 missing cells. After cleaning missing values on both features, there are 522 rows left. Why 522 rows are that there are some missing values intercepted with saving accounts and checking accounts. All data has been cleaned but not prepared yet for modelling and evaluating. For evaluation, it is necessary to group data as a one binary target variable and the rest of the features are going to be independent variables. Risk feature has been selected as target variable that is a categorical variable as defined before. This risk categorical variable constituted from good and bad categories means that the good category means that credit application approved; on the other hand, bad category means that credit application not approved. This risk category defines binary classification that good is 1, bad is 0.

In the second stage, categorical independent variables have been group by numbers to make categories and assigned using dummy variables method. In order to prevent from multi-collinearity problem, first of columns has been deleted for each explanatory variable group. In the beginning data has 10 features that one of the features is assigned the target variable, which is risk, rest of them are explanatory variables which are age, credit amount, duration, sex, job, housing, saving accounts, checking accounts and purposes. There are nine explanatory variables are separated by three of them are numerical variable six of them categorical variables. To implement models to the data, categorical variables must be grouped by dummy variables that make categorical variables to numerical variables. After organizing data with dummy variables, 27 features have been gotten. Normally, nine features have been in data but with dummies with one-columns deleting totally of 21 explanatory variables have been displayed. Once cleaning and preparing data following structures have been founded for target variables and independent variables. These independent variables features are as given in Fig. 21 (Table 2).

The above explanatory variables are after the cleaning and preparing the data. There are 522 entries and explanatory variables. Finally, the data has been ready to be run.

There are a lot of methods to make preparation data to be modelled. One of them has been already mentioned above is to remove missing values. Since removing data is going to cause loss of information, many researchers do not desire to apply this

Table 2 After dropping null cells

#	Column	Non-null count	Dytpe
0	Age	522	Int64
1	Credit amount	522	Int64
2	Duration	522	Int64
3	Sex_female	522	Uint8
4	Job_1	522	Uint8
5	Job_2	522	Uint8
6	Job_3	522	Uint8
7	Housing_own	522	Uint8
8	Housing_rent	522	Uint8
9	Saving accounts_moderate	522	Uint8
10	Saving accounts_quite rich	522	Uint8
11	Saving accounts_rich	522	Uint8
12	Checking account_moderate	522	Uint8
13	Checking account_rich	522	Uint8
14	Purpose_car	522	Uint8
15	Purpose_domestic appliances	522	Uint8
16	Purpose_education	522	Uint8
17	Purpose_furniture/equipment	522	Uint8
18	Purpose_radio/TV	522	Uint8
19	Purpose_repairs	522	Uint8
20	Purpose_vacation/others	522	Uint8

method to data sets. Because, in order to clean missing values, one needs to remove all columns or rows that this approach is going to corrupt lots of other information including other features. Instead of removing the missing values, all missing values can be filled with one of some sensible approaches. This without dropping missing values analysis separated into two sub-methods. One of the methods is to run models that fill empty cells without knowing which model fill cells, the other one is to first fill cells according to the most sensible method in the literature. In the literature, there are many filling techniques that help to avoid data loss [11] which is called imputation. Imputation means filling missing values with some numbers or some categorical properties.

According to the type of data, there are some preferred imputation techniques that can be used to prepare data to be modelling. Data has two categorical variables which are saving account and checking account. To fill missing values cells, maximum repeating sub-feature impute the empty cells. Saving accounts missing cells are filled with the little categorical values because most of the customer approximately more fifty per cent have little accounts. Checking account missing cells are filled with the moderate categorical variable that checking account customer mostly have

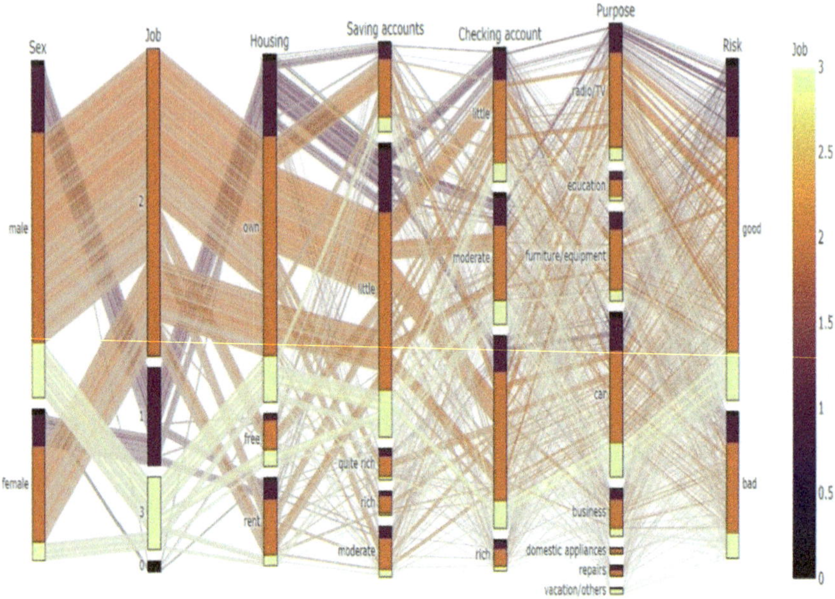

Fig. 20 Parallel diagram of the categorical features

moderate account more than eighty per cent of the applicants. With imputation of the missing values, data have been prepared for the modelling.

According to parallel categorical diagram, most of the applicant's gender is male compared to female. From the diagram, generally people work second job type and mostly have their own house. The applicants tend to have quite little saving accounts in the bank and little and moderate checking accounts. They apply mostly to buy TV/radio, furniture/equipment, car. Risk situations are good even if some of them are bad situation. All information can be gotten from Fig. 20.

4.5 Model

After the data have been cleaned and prepared for the modelling, seven machine learning algorithms have been applied which are naive Bayes (NB), decision tree (DT), random forest (RF), logistic regression (LR), K-nearest neighbour (KNN), support vector machine (SVM) and XGBoost (XGB) classifier models have been run. The aim of the applying seven classification algorithm is to find out what algorithm gives the more accurate result. There are three data types that models have been implemented; first algorithms are applied with dropping missing values, and then second all algorithms are applied without dropping any of the missing values. And lastly, all algorithms are run with imputed cell values.

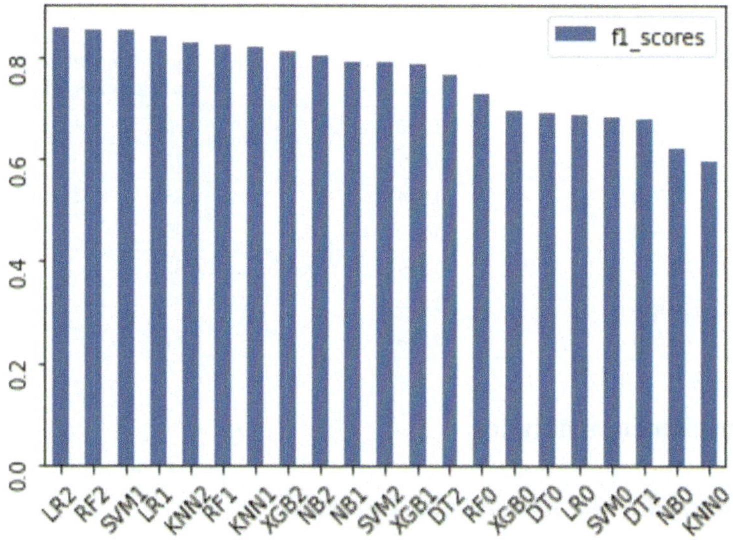

Fig. 21 F1 scores bar chart for all situation on models

For the evaluation of the model performance, classification matrix has been used by a matrix. In the model evaluation, F1 score model evaluation metric has been considered because of the data that give more accurate results for the models. From the model performance analysis, after the missing value imputation applied to data, logistic regression has given the best performance metric among the classification methods that have been already mentioned which these models are LR, DT, RF, NB, SVM, KNN and XGB models.

All measures of F1 scores have been compared in the following table (f1_scores are the approximate values).

It can be understood that from Fig. 21, LR2 model has the highest f1_score among the rest of the other applied models.

As it can be seen from Table 3, the highest f1_score is found from the LR model overall when it is tested with different test and train data sets cover 80 per cent of data and 20 per cent of data is test data with random state is 7. As a result, the LR model has been selected by the German Bank to apply which decision has made new customer applications. There are other models which also have enough f1_score to use but LR is the highest one even if there is no big difference.

Table 3 F1 scores for all situation on models

F1_score	Dropped missing values	Without dropped missing values	Imputation with maximum repeating categorical values
NB	0.622	0.792	0.806
DT	0.692	0.679	0.767
RF	0.730	0.824	0.855
LR	0.690	0.840	0.859
KNN	0.598	0.819	0.827
SVM	0.682	0.854	0.790
XGB	0.696	0.788	0.811

5 Investment Analytics

5.1 Introduction to Portfolio Analytics

One of the most fundamental problems of businesses today is to provide an efficient financial management. One of the main issues that businesses should be interested in the field of finance is the creation and management of an investment portfolio. This issue has become even more important in a complex environment such as a globalizing society, rapidly increasing competition and extensive economic changes at the national and international levels [12].

In 1952, Markowitz (Markowitz) developed the mean variance model that formulates the mathematical relationship between the returns and risks of assets [13]. In the following years, portfolio creation methodologies have diversified, with other models trying to expand it and eliminate its weaknesses. Thanks to these models, investors began to create portfolios that support certain investment styles and preferences.

In this section, besides the basic financial calculations, performance metrics and portfolio construction methods will be presented. In addition, the portfolios created with different models for the five technology stocks traded in the S&P 500 index will be compared with the performance metrics whose definitions are given.

5.2 Return & Risk Calculations

Return can be defined as how much gain or loss at time $t + 1$ from time t. It can be formulated as follows:

$$\text{Return}_t = \frac{\text{Price}_t - \text{Price}_{t-1}}{\text{Price}_{t-1}} \tag{1}$$

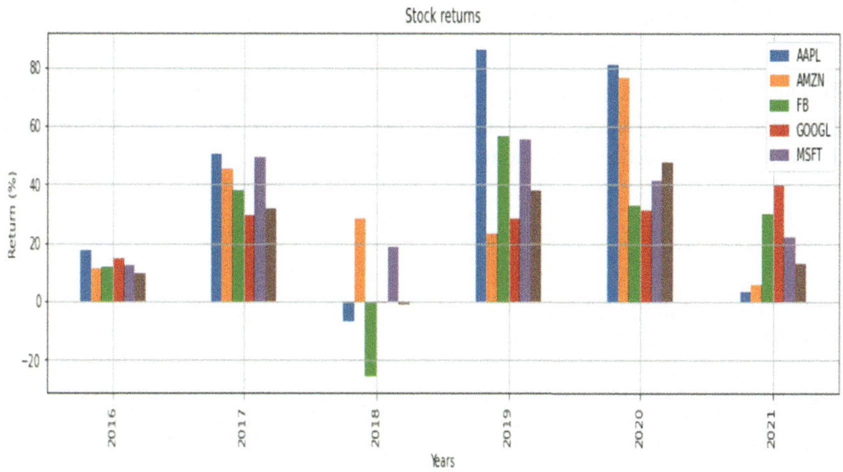

Fig. 22 Yearly stock returns

The cumulative return is an aggregate return or the total change in the investment price over a period of time. Along with the cumulative return, the compounded return is the rate of return that takes into consideration the compounding effect of the investment for each period (Fig. 22).

The risk of an asset is often expressed as its volatility, and it is usually defined as the standard deviation (or variance) of the returns on the asset (Table 4).

Similarly, the volatility of a portfolio can be calculated by the variance, weights and correlations of the assets it contains. Portfolio variance is formulated as follows:

$$\sigma_p^2 = w_1^2\sigma_1^2 + w_2^2\sigma_2^2 + 2w_1w_2\sigma_{1,2} \tag{2}$$

σ_p^2: Variance of the portfolio.
w_n: Weight of the nth asset.
σ_n^2: Variance of the nth asset.
$\sigma_{1,2}$: Covariance between assets 1 and 2.

Table 4 Expected returns and volatility of the stocks

Stock	Expected return (%)	Volatility (%)
AAPL	36.98	30.20
AMZN	37.28	30.32
FB	25.16	32.95
GOOGL	20.24	26.34
MSFT	31.92	27.76

5.3 Performance Metrics

Sharpe Ratio: It is the most widely used measure of risk-adjusted rate of return. It is calculated by subtracting the portfolio return from the risk-free asset's return and dividing it by the portfolio's standard deviation.

$$\text{Sharpe Ratio} = \frac{R_p - R_f}{\sigma_p} \tag{3}$$

R_p: Return of the portfolio.
R_f: Risk-free rate.
σ_p: Standard deviation of the portfolio.

Sortino Ratio: It is the modified version of the Sharpe ratio but uses a different standard deviation, downside deviation.

$$\text{Sharpe Ratio} = \frac{R_p - R_f}{\sigma_d} \tag{4}$$

σ_d: Standard deviation of the downside.

Treynor Ratio: It is also known as the reward ratio for risk taken. The risk expressed in the Treynor ratio refers to the systematic risk measured by the beta of a portfolio, which will be explained in the next section.

$$\text{Treynor Ratio} = \frac{R_p - R_f}{\beta_p} \tag{5}$$

β_p: Beta of the portfolio.

Drawdown: It is defined as percentage loss from the last maximum peak. Maximum drawdown can also be defined as the worst percentage loss from a market peak to the very lowest point. It is a strong indicator that measures the downside risk of the portfolio.

5.4 Factor Models

Capital Asset Pricing Model (CAPM): The model was developed by Sharpe [14], Linter [15] and Mossin [16] to explain why different securities have different expected returns. CAPM explains that every investment carries two different risks, systematic and unsystematic.

The risk called systematic risk or beta is the risk of being in the market and expressed as follows:

$$R_i = R_f + B_i(R_m - R_f) \tag{6}$$

Table 5 Beta of the stocks

Stock	Beta
AAPL	1.14
AMZN	1.04
FB	1.09
GOOGL	1.00
MSFT	1.13

R_i: Expected return of the asset i.
R_f: Risk-free rate.
R_m: Expected return of the market.
β_i: Beta of the asset i (Table 5).

In the CAPM, beta is used to describe the relationship between systematic risk and an asset's expected return. Higher beta indicates that the asset is more volatile than the market. The CAPM is also graphically illustrated as the security market line (SML). As a straight line intersecting the vertical axis at the risk-free rate, SML depicts the risk–return trade-off (Fig. 23).

Fama French Multi-Factor Model: This model is usually expressed as a three-factor model that improves the CAPM (single factor model). These three factors are market, size and value factors [17]. In the formula below, $R_m - R_f$ represents the market factor, SMB represents the size factor and HML represents the value factor.

$$R_i = R_f + \beta_m (R_m - R_f) + B_s SMB + B_h HML \tag{7}$$

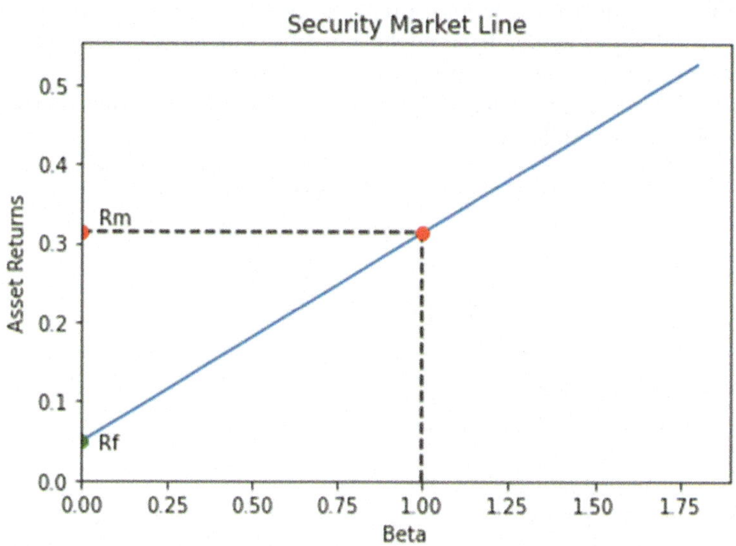

Fig. 23 Security market line of NASDAQ

R_i: Expected return of the asset i.
R_f: Risk-free rate.
R_m: Expected return of the market.
β_m: Market beta of the asset i.
β_s: Size beta of the asset i.
β_h: Value beta of the asset i.
SMB: is the portfolio returns of small stocks minus the portfolio returns of big stocks.
HML: is the portfolio returns for high book-to-market value minus returns of low book-to-market value stock.

5.5 Modern Portfolio Theory

Portfolio selection includes mathematical models designed to best meet the needs of the investor and investors seek the highest possible return against minimum risk. Therefore, they want to maximize risk-adjusted returns. Markowitz [13] firstly introduced the modern portfolio theory (MPT) that helps to determine the minimum risk level that the investor should undertake in order to reach the targeted return level. The Markowitz's MPT is often generically known as the mean–variance optimization framework. In this section, MPT will be used to create optimal portfolios that meet the needs of investors [18].

The efficient frontier is a graph showing the rate of return corresponding to volatility. It includes optimal sets of portfolios that offer the highest expected return for a given level of risk or the lowest risk for a given expected return level. If the points are derived by assuming all possible weights with different combinations of entities, the efficient frontier chart is arrived (Fig. 24).

If a risk-free security is available, then an investor's portfolio optimization problem can be formulated as follows.

$$\text{minimize}\frac{1}{2}w'\sum w \tag{8}$$

$$\text{subject to}\left(1-\sum_{i=1}^{n} w_i\right)r_f + w'\mu \tag{9}$$

w: is the vector of portfolio weights.
$\sum$: is the covariance matrix of the asset returns.
r_f: is the risk-free rate.
μ: is the expected returns (Fig. 25).

Minimum volatility portfolio aims to minimize the volatility of the portfolio, and it ignores expected return and focuses only on the risk. It is formulated as follows.

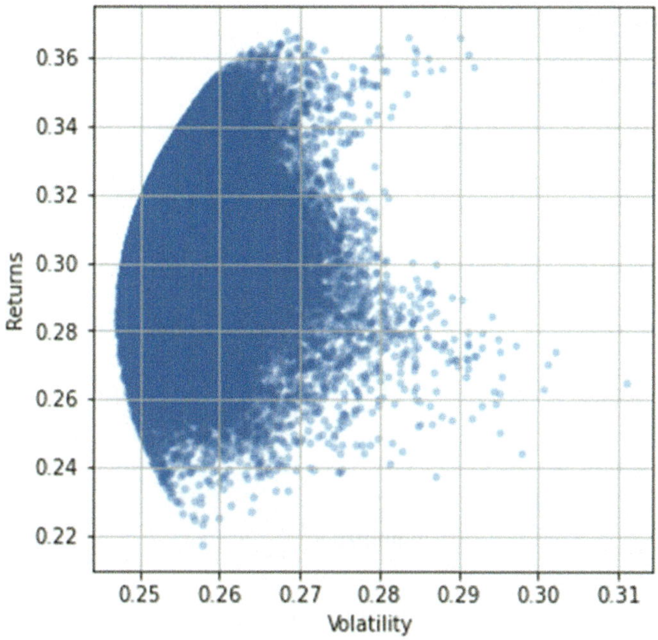

Fig. 24 Efficient frontier chart

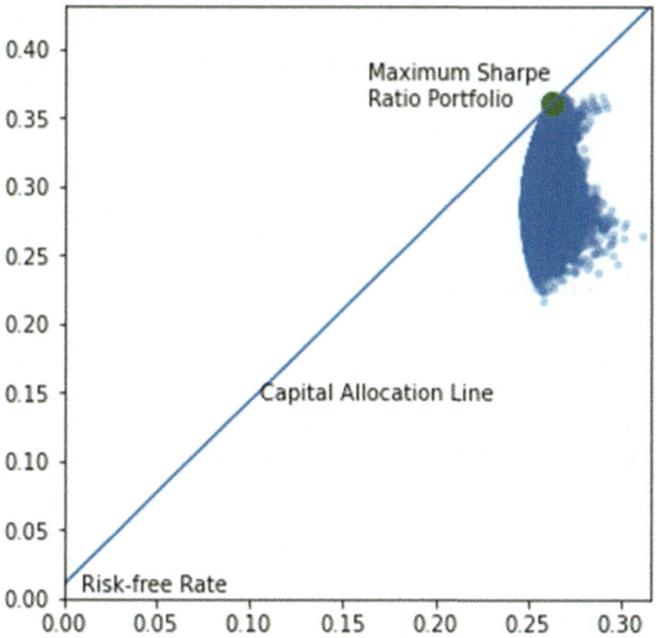

Fig. 25 Capital allocation line

Table 6 Results of portfolio optimization

Objective	Expected return (%)	Volatility (%)	Sharpe ratio	Sortino ratio	Beta	Treynor ratio	Max drawdown (%)
Equal weighted	30.29	25.16	1.16	1.43	1.08	0.27	77.77
Minimum volatility	28.31	24.71	1.11	1.36	1.06	0.26	75.20
Maximize sharpe	36.44	26.45	1.34	1.72	1.09	0.32	84.42
Efficient risk (25%)	31.75	25.0	1.23	1.53	1.08	0.29	79.60
Efficient return (35%)	35.00	25.78	1.32	1.66	1.1	0.31	83.02

$$\text{minimize } w^T \sum w \tag{10}$$

$$\text{subject to } 1^T w = 1 \tag{11}$$

Also, return can be maximized for a given target risk β_p. It is formulated as follows (Table 6).

$$\text{maximize } w^T \mu \tag{12}$$

$$\text{subject to } w^T \sum w \leq \beta_p \tag{13}$$

5.6 *Black and Litterman*

The Black–Litterman (BL) model is a portfolio allocation model that creates portfolios based on an investor's unique insights [19]. This model is also on the basis of mean–variance. In the model, expected returns are calculated as the new expected returns, which are found by the following expression.

$$E(R) = \left[\left(\tau \sum\right)^{-1} + P^T \Omega^{-1} P\right]^{-1} \left[\left(\tau \sum\right)^{-1} \Pi + P^T \Omega^{-1} Q\right] \tag{14}$$

$E(R)$: is the new expected return.
τ: is a scalar constant.
$\sum$: is the covariance matrix of the asset returns.

P: is a matrix that identifies assets involved in the views.
Ω: is the uncertainty matrix of views.
Π: is the prior expected return.
Prior expected returns in the above formula are calculated as follows.

$$\Pi = \lambda \sum w_{mkt} \tag{15}$$

λ: is the risk aversion coefficient or market-implied risk premium.
w_{mkt}: is the market capitalization weight of the assets.
Risk aversion coefficient is found by the following formula:

$$\lambda = \frac{R_m - R_f}{\sigma_m^2} \tag{16}$$

σ_m^2: is the variance of the market.

In the BL model, investors can provide different type of views. They can state "AMZN will return 30%" or "FB will drop 10%" as an absolute view. On the other hand, relative views can be stated like "AAPL will outperform GOOGL by 2%". In addition, investors can provide confidence level in their views.

In our case, absolute views will be provided as an example. Also, short sells will be allowed. Below table shows our views and also weights produced by the model using these views (Tables 7 and 8).

Table 7 Investor views and resulted weights in BL model

Stock	View (%)	BL weight (%)
AAPL	+20	55
AMZN	+30	107
FB	−10	−78
GOOGL	5	−20
MSFT	15	36

Table 8 Result table of BL optimization model

Objective	Expected return	Volatility	Sharpe ratio	Sortino ratio	Beta	Treynor ratio	Max drawdown
BL market risk aversion	48.02%	36.67%	1.28	1.88	1.09	0.43	91.61%

6 Financial Hedging Analysis

6.1 Problem Definition

Corporate risk management refers to all methods that companies use to reduce the risk they may face. Hedging is one of the financial risk management strategies that is defined as taking an offsetting position in derivatives market of related security. There are several ways to hedge current positions by using derivatives such as futures, options and forwards.

Futures contracts are agreements that secure a specified price at a specified time in future for two parties: seller and buyer. In this sense, futures contracts are commonly used for hedging currencies, stock market indexes and commodities. These contracts are traded on exchanges which regulate clearance and pricing of contracts. The price of future contract, which will be referred to as "futures", is the same regardless of which broker is used due to the centralized price mechanism.

Exporting and importing is a way to expand businesses with its risks such as foreign exchange risk. Internationally trading companies are usually exposed to foreign exchange risk, especially if the local currency is unstable and aims to limit the losses they may incur due to currency fluctuation. Companies aim to minimize this risk by taking a position in currency derivative markets that is called hedging. Financial risk manager of a company tries to find out answers for the following questions among available contracts [20]:

- Which futures contract is appropriate to use?
- When to take buy (long) or sell (short) position?
- What is the optimal hedge ratio for the amount exposed to risk?

Risk managers can determine the most appropriate answers to these questions by using hedging strategies. The aim of hedging is to neutralize the risk of exchange rate fluctuations and guarantee a fixed price for a specified future date. Futures are leveraged financial instruments in which investors only put in a margin which is a fraction of total notional value on contract. Futures have standardized specifications which are determined by exchange. These specifications are: (1) the asset: this is underlying product of future contract that can be a foreign currency or a physical commodity like gold; (2) the contract size: this is the amount of the asset that needs to be delivered within one contract; (3) delivery arrangements: the method or place of delivery for underlying asset that is specified by exchange; (4) delivery months: this refers to the month that the delivery occur; 5) price quotes: this refers that how price of contract will be quoted (6) tick size: this is the minimum price movement exchange allows; (7) position limits: this is the limits of daily price movements specified by exchange. As an example, Chicago Mercantile Exchange (CME) currency futures contract specifications are illustrated in Table 9.

Businesses or individuals open accounts with brokers to trade on futures and brokers direct their orders to exchange. This account is generally called margin account, traders are required to deposit an initial amount of cash specified by

Table 9 Contract specifications for foreign currency futures: Eurodollar [21]

EURO FX futures—contract specs	
Contract unit	125,000 Euro
Price quotation	US dollars and cents per Euro increment
Trading hours	CME Globex
	Sunday–Friday 5:00 p.m.–4:00 p.m. (6:00 p.m.–5:00 p.m. ET) with a 60-min break each day beginning at 4:00 p.m. (5:00 p.m. ET)
	CME ClearPort
	Sunday 5:00 p.m.–Friday 5:45 p.m. CT with no reporting Monday–Thursday from 5:45 p.m.–6:00 p.m. CT
Minimum price fluctuation	CME Globex
	0.00005 per Euro increment = \$6.25
	Consecutive month spreads: 0.00001 per Euro increment = \$1.25
	All other spreads: 0.00002 per Euro increment = \$2.50
	CME ClearPort
	0.00001 per Euro increment = \$1.25
Product code	CME Globex: 6E
	CME ClearPort: EC
	Clearing: EC
Listed contracts	Quarterly contracts (Mar, Jun, Sep, Dec) listed for 20 consecutive quarters and serial contracts listed for 3 months
Settlement method	Deliverable
Termination of trading	Trading terminates at 9:16 a.m. CT, 2 business day prior to the third Wednesday of the contract month
Settlement procedures	Physical delivery
	EUR/USD futures settlement procedures
Position limits	*CME position limits*
Exchange rulebook	*CME 261*
Block minimum	*Block minimum thresholds*
Price limit or circuit	*Price limits*
Vendor codes	*Quote vendor symbols listing*

exchange. They are also required to keep a maintenance margin which is the minimum amount of equity investors must hold in margin account. Gains and losses of futures are calculated every trading day with "marking to market" procedure.

An example of currency futures is given in the following case study.

6.2 Case Study

On 1 July 2021, ABC company from USA sells machinery to a German company and will receive €3.75 m on 25 September 2021. They expect that Euro will weaken

Table 10 Contract details of September Euro futures

Specification	Explanation
Contract size	125,000 Euro
Price quotation	US dollars and cents per Euro increment
Settlement process	Mark to market
Termination of trading	2 business day prior to the last Wednesday of the contract month
Delivery month	September
Initial margin	$3000
Maintenance margin	$2200

against US Dollar; therefore, they choose to take a position in the currency futures market to hedge their position. The current spot rate in the market is $1.1764/1€ and futures contracts at $1.170/1€ are available for delivery month. Contract details of Euro Futures are given in Table 10.

6.2.1 Setting up the Hedging

Under given conditions, the company sets up the hedging strategy against losses. The will be received in the end of September, and therefore, a future contract with same month is obtained. After the decision of expiry month, the appropriate position is taken since company is in long equity cash position, so they take put position in futures market. Lastly, company will decide the fraction of equity to be hedged. Investors can do either partial hedge or full hedge their current positions. The fraction of total equity hedged is called hedge ratio. In our case, different scenarios of hedge ratio will be applied to see gain/losses at expiry. If company aim to do full hedging, they need to put 30 futures contracts.[1]

The company can take position under these setups.

6.2.2 Simulating Spot Price

The company takes position in the currency futures market, and the settlement process is marked to market. The aim of this process is to enable investors to see their current financial situation based on current futures price. Therefore, the margin account will be updated based on the current situation. A futures price for Euro/USD futures is simulated and given in Fig. 26.

Based on the generated price, total gain/losses are marked to the margin account daily and first 10 days and last 10 days of contract holding period are given in Table 11.

Here in Table 11, there are 30 contracts and $3000 is put in as initial margin for each contract, so $90,000 in total. Margin account starts with $90,000 and changes daily

[1] Number of contract for full hedging is calculate as: 3,750,000/125,000 = 30.

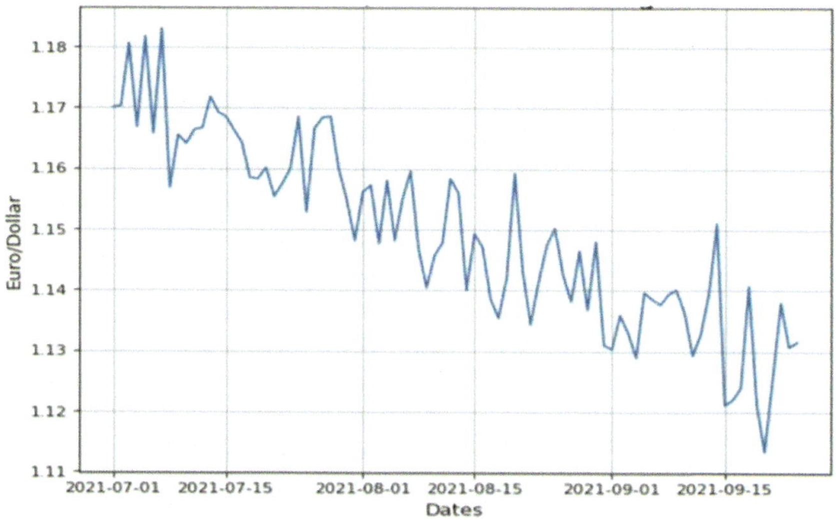

Fig. 26 Simulated euro/US dollar futures price

Table 11 Margin account balance

Dates	Futures price	Change in value	Gain/Loss ($)	Account balance ($)
2021–07–01	1.170000	–	–	90,000
2021–07–02	1.170292	0.0003	−1125	88,875
2021–07–03	1.180560	0.0103	−38,625	50,250
2021–07–04	1.166781	−0.0138	51,750	102,000
2021–07–05	1.181667	0.0149	−55,875	46,125
2021–07–06	1.165743	−0.0159	59,625	105,750
2021–07–07	1.182957	0.0172	−64,500	41,250
2021–07–08	1.156793	−0.0262	98,250	139,500
2021–07–09	1.165531	0.0087	−32,625	106,875
2021–07–10	1.164129	−0.0014	5250	112,125
⋮	⋮	⋮	⋮	⋮
2021–09–15	1.121084	−0.0299	112,125	274,125
2021–09–16	1.121991	0.0009	−3375	270,750
2021–09–17	1.123935	0.0019	−7125	263,625
2021–09–18	1.140673	0.0167	−62,625	201,000
2021–09–19	1.121300	−0.0194	72,750	273,750
2021–09–20	1.113382	−0.0079	29,625	303,375
2021–09–21	1.125451	0.0121	−45,375	258,000
2021–09–22	1.137948	0.0125	−46,875	211,125
2021–09–23	1.130630	−0.0073	27,375	238,500
2021–09–24	1.131484	–	–	–

Table 12 Sensitivity analysis for different spot rate with full hedging at expiry date[2]

Spot price	Cash flow at spot rate ($)	Cash flow with full hedging ($)	Total gain/loss ($)
1.11	4,162,500.0	4,387,500.0	225,000.0
1.12	4,200,000.0	4,387,500.0	187,500.0
1.13	4,237,500.0	4,387,500.0	150,000.0
1.14	4,275,000.0	4,387,500.0	112,500.0
1.15	4,312,500.0	4,387,500.0	75,000.0
1.16	4,350,000.0	4,387,500.0	37,500.0
1.17	4,387,500.0	4,387,500.0	0.0
1.18	4,425,000.0	4,387,500.0	−37,500.0
1.19	4,462,500.0	4,387,500.0	−75,000.0
1.20	4,500,000.0	4,387,500.0	−112,500.0
1.22	4,575,000.0	4,387,500.0	−187,500.0
1.24	4,650,000.0	4,387,500.0	−262,500.0

based on current futures price. Simulated price is 1.170292 on 2021–07-02 which has increased by 0.000292 compared to the day before. The gain/loss on day 2021–07-02 is calculated as: $125000 * (1.170292 - 1.17) * 30 = \$1095$, where 125,000 is contract size and 30 is number of contracts. The amount \$1095 is subtracted from \$90,000 since the company is in put position.

The scenario analysis is for full hedging and different rate of hedge can be obtained. The company may gain or loss at expiry date, and they therefore may not take full hedge position.

6.2.3 Sensitivity Analysis of Spot Rate

On day of expiry, spot rate may take different values than expected. At this point, companies may create a sensitivity analysis table to see possible gains or losses. In Table 12, total gain/losses are given for different spot prices at expiry.

Since the company takes short position in Euro futures, they make gains for any spot price less than 1.17 and vice versa. The total gain/losses of company is illustrated in Fig. 27.

Another sensitivity analysis is made for hedge ratios, and gain/losses is given for different hedge ratios in Table 13.

Total cash flows under different scenarios for hedge ratios are illustrated in Fig. 28.

In this case, a hedging method is used to stabilize the financial position. A company aims to hedge their cash flow on a future date. A margin account is created, and marking to market process is shown in accordance with simulated futures price. Two different sensitivity analyses are also applied for both spot rate and hedge ratio. All

[2] All values given in Table 12 are in US Dollar.

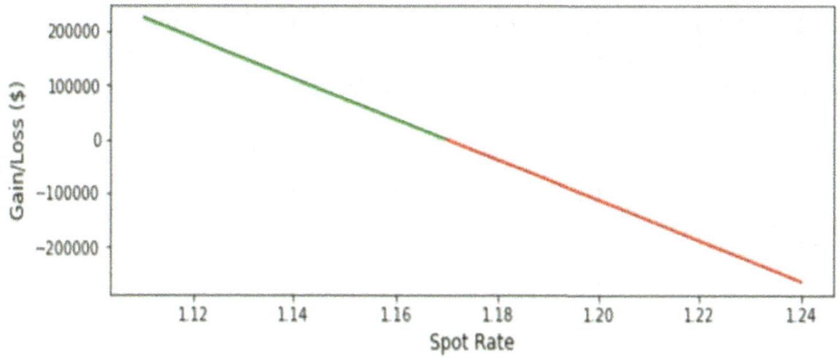

Fig. 27 Gain and losses under different spot rate scenarios

Table 13 Total cash flow under different hedge ratios

Tables hedge ratio (%)	Tables total cash flow ($)
0.0	4,162,500.0
10.0	4,185,000.0
20.0	4,207,500.0
30.0	4,230,000.0
40.0	4,252,500.0
50.0	4,275,000.0
60.0	4,297,500.0
70.0	4,320,000.0
80.0	4,342,500.0
90.0	4,365,000.0

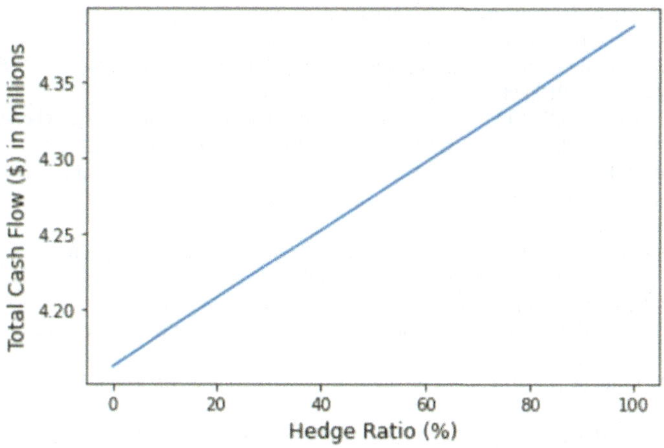

Fig. 28 Total cash flow under different hedge ratios

scenarios on expiry date were summarized in Tables 11, 12 and 13 and Figs. 26, 27 and 28. The optimal hedge ratio can also be determined theoretically but it is preferred to use sensitivity analysis to decide the amount of equity to hedge.

References

1. Tuovila A (2021) Financial analysis. Investopedia. https://www.investopedia.com/terms/f/financial-analysis.asp/. (Access: 23 July 2021)
2. Friedlob GT, Schleifer LL (2003) Essentials of financial analysis. Wiley p 23
3. Kumawat D (2020) An introduction to financial analysis, analytic steps. https://www.analyticssteps.com/blogs/introduction-financial-analysis/. (Access: 23 July 2021)
4. Fridson MS, Alvarez F (2011) Financial statement analysis: a practitioner's guide. Wiley p 597
5. Approved prep provider CFA institute: financial reporting and analysis (2021). https://ift.world/booklets/fra-understanding-cash-flow-statements-part1/. (Access: 23 July 2021)
6. Elliott JW, Uphoff HL (1972) Predicting the near term profit and loss statement with an econometric model: a feasibility study. J Account Res 259–274
7. Barnes P (1987) The analysis and use of financial ratios. J Bus Finance dan Acc 14(4):449
8. Chen KH, Shimerda TA (1981) An empirical analysis of useful financial ratios. Financ Manage 51–60
9. Rao NV, Atmanathan G, Shankar M, Ramesh S (2013) Analysis of bankruptcy prediction models and their effectiveness: an Indian perspective. Gt Lakes Her 7(2)
10. Url: https://archive.ics.uci.edu/ml/datasets/statlog+(german+credit+data)
11. Musil CM, Warner CB, Yobas PK, Jones SL (2002) A comparison of imputation techniques for handling missing data. West J Nurs Res 24(7):815–829
12. Sarmas E, Xidonas P, Doukas H (2020) Multicriteria portfolio construction with python. Springer International Publishing, pp 1–3. https://doi.org/10.1007/978-3-030-53743-2_1
13. Markowitz H (1952) Portfolio selection. J Financ 7:77–91
14. Sharpe WF (1964) Capital asset prices: a theory of market equilibrium under conditions of risk. J Financ 19:425–442. https://doi.org/10.1111/j.1540-6261.1964.tb02865.x
15. Lintner J (1965) Security prices, risk, and maximal gains from diversification. J Financ 20:587–615. https://doi.org/10.1111/j.1540-6261.1965.tb02930.x
16. Mossin J (1966) Equilibrium in a capital asset market. Econometrica 34(4):768–783. https://doi.org/10.2307/1910098
17. Fama EF, French KR (1983) Common risk factors in the returns on stocks and bonds. J Financ Econ 33(1):3–56. https://doi.org/10.1016/0304-405X(93)90023-5
18. Romero PJ, Balch T (2014) Modern portfolio theory: the efficient frontier and portfolio optimization. What hedge funds really do: an introduction to portfolio management. Business Expert Press. pp 71–81
19. Black F, Litterman R (1991) Combining investor views with market equilibrium. J Fixed Income 1(2):7–18. https://doi.org/10.3905/jfi.1991.408013
20. Hull J (2018) Options, futures, and other derivatives, 10th edn. Pearson
21. CME Group (2021) Retrieved 1 September 2021, from https://www.cmegroup.com/markets/products.html#sortAsc&sortField
22. Endel F, Piringer H (2015) Data wrangling: making data useful again. IFAC-PapersOnLine 48(1):111–112
23. Furche T, Gottlob G, Libkin L, Orsi G, Paton NW (2016) Data wrangling for big data: challenges and opportunities. EDBT 16:473–478
24. Field A (2009) Logistic regression. Discovering Stat Using SPSS 264:315
25. Yacouby R, Axman D (2020) Probabilistic extension of precision, recall, and F1 score for more thorough evaluation of classification models. In: Proceedings of the first workshop on evaluation and comparison of NLP systems. pp 79–91

Mahmut Sami Sivri is currently a lecturer at Industrial Engineering Department at Istanbul Technical University. He is also working on some R&D projects as Director of Data Analytics at ITU's Technopark. He received the B.S. degree in Computer Engineering and the M.Sc. degree in Engineering Management from Istanbul Technical University. He worked in various companies and positions in the IT industry since 2008. His current research interests include machine learning, sentiment analysis, deep learning, big data and its applications, Industry 4.0, financial technologies, data analytics, supply chain and logistics optimization.

Abdullah Emin Kazdaloglu is a research assistant and a Ph.D. student in industrial engineering department at Istanbul Technical University (ITU). He received his B.Sc. degree in management engineering from ITU in January of 2018 and also received his M.Sc. degree in industrial engineering from ITU in August of 2021. He has a passion for data science, machine learning and artificial intelligence, and his current research interests are statistics, data science and business analytics methods. He is eager to develop his professional skills and curious to discover and implement new manners, technologies and tools especially in field of machine learning, statistic and computer science ecosystems, data analytics and business knowledge.

Emre Ari is a Research Assistant and a Ph.D. student in the Department of Industrial Engineering at Istanbul Technical University, Turkey. In 2016, he received his M.Sc. Degree in Statistics from Queen Mary University of London in UK and he received his B.Sc. degree in Mathematics from Kahramanmaraş Sütçü İmam University in 2009. His current research interests lie in the area machine learning, deep learning, reinforcement learning, data analysis, financial approaches and statistics. He plans to continue his academic career by trying to find new approaches in his interest areas.

Hidayet Beyhan is a Research Assistant and a Ph.D. Candidate in the Management Program at Istanbul Technical University, Turkey. He received his M.Sc. degree in Accounting and Finance from Swansea University and B.Sc. degree in Mathematics from Eskişehir Osmangazi University. His current research interests are agent-based modelling in finance, quantitative finance and computational finance. He continues his research in multi-agent financial markets and plans to extend his research with the application of machine learning methods in finance.

Alp Ustundag is the Head of Industrial Engineering Department of Istanbul Technical University (ITU) and the coordinator of M.Sc. in Big Data & Business Analytics Program. He is also the CEO of Navimod Business Analytics Solutions located in ITU Technopark (http://navimod.com/). He has worked in IT and finance industry from 2000 to 2004. He continued his research studies at the University of Dortmund between 2007 and 2008 and completed his doctorate at ITU in 2008. He has conducted a lot of research and consulting projects in the finance, retail, manufacturing, energy and logistics sectors. His current research interests include artificial intelligence, data science, machine learning, financial and supply chain analytics. He has published many papers in international journals and presented various studies at national and international conferences.

Human Resources Analytics

Ozgur Akarsu, Cigdem Kadaifci, and Sezi Cevik Onar

1 Introduction

Human resource analytics is a special part of analytics where the main focus is the human resource. In HR analytics, the analytical process is applied to the organization's human resources. The main objective of this process is not only to enhance employee performance but also to improve overall employee satisfaction. HR analytics helps institutions by providing useful insights for the human resource functions by collecting data and transforming it into useful information for empowering human resource-related processes. HR analytics is very useful for improving HR functions such as workforce management, recruitment, performance evaluation, and development management. In this chapter, four different HR problems are focused, and these problems are tried to solve by using HR analytics tools. The rest of the chapter is organized as follows. Chapter two gives the four different HR problems, namely attrition risk, recruitment, performance evaluation, and training planning. These problems are defined via a comprehensive literature review. For each problem, a case study is defined and HR analytics tools are applied. The last section concludes and gives further suggestions.

O. Akarsu
Koc Digital, Camlica Business Center No: 11, 34700 Unalan, Uskudar, Istanbul, Turkey

C. Kadaifci · S. C. Onar (✉)
Department of Industrial Engineering, Faculty of Management, Istanbul Technical University, Macka, 34367 Istanbul, Turkey
e-mail: cevikse@itu.edu.tr

A. Ustundag et al. (eds.), *Business Analytics for Professionals*, Springer Series in Advanced Manufacturing, https://doi.org/10.1007/978-3-030-93823-9_14

2 Employee Turnover (Attrition)

Employee turnover (i.e., attrition) refers to the number (or ratio) of employees leaving the organization either voluntarily or involuntarily. Voluntary turnover is based on the decision of employees, while the involuntary one occurs when an employer decides to end the employment relationship, as a result of the death or retirement of the employee [1, 2]. From an organization's perspective, performance under expectations, improper behavior, or adaptation problems regarding the culture or working environment may cause the termination of the employment agreement [1, 3]. Since the control of the organizations over the employee-related involuntary turnover is low, in the scope of this study, the voluntary turnover will be examined.

Starting from recruitment, organizations invest in each employee through selection, orientation, training, improvement, and positioning. When an employee leaves the organization, then the position should be filled immediately with an appropriate candidate. Accordingly, an employee loss causes a significant cost as well as affecting the organizational value [4], the long-term strategies [5], the competitive advantage [1, 6], the organizational performance [7, 8], and even loss of customers, especially if the leaving employee is a talented and highly capable one. Being an important factor affecting the organization's overall performance, employee turnover deserves considerable attention from both of the organizations and the researchers in order to enable taking precautions to diminish (or manage) the employee loss. Thanks to the developments and improvements in the learning-based data analysis techniques, the employee data representing the reasons for the voluntary employee attrition can be examined, and whether the employee quits or not can be predicted. The common approach in the recent literature on the prediction of employee turnover is applying different methods or algorithms and comparing them based on the performance measures such as accuracy, precision, recall, specificity, the area under the curve (AUC), and F-measure. In Table 1, some studies are provided along with the performance measures and the dataset utilized.

According to a detailed literature review, gender [1, 6, 8–10], state of origin [1], duration of service [1, 4, 8–10], the title in the organization or job level [1, 6, 8, 9], annual or monthly salary [1, 4, 8, 9], job satisfaction [4, 6, 8–10], environment satisfaction [6, 8–10], relationship satisfaction [8, 9], performance of the employee [4, 6, 8, 9], work–life balance [6, 8–10], the number of trainings the employee participated in the previous year [8, 9], promotion [4], feeling appreciated [10] are among the factors considered for the prediction. Even though several algorithms and methods including support vector machines [4, 5, 8, 9, 11–13], AdaBoost [4, 8], neural networks [13], KNN [8, 9, 11, 12], and naïve Bayes [5, 8, 9, 11, 12] are used, the methods preferred considering the performance measures are decision tree [1, 4, 8, 9, 11–13], random forest [4, 5, 8–10, 12, 13], and XGBoost [5, 11, 13, 14].

To achieve a well-managed prevention process to diminish employee attrition, the organizations better consider the most important factors affecting employees' quitting decisions. According to recent studies, the relatively high attrition is mostly related to the lower involvement [10], motivation, and satisfaction of employees [9];

Table 1 Learning-based methods to predict employee attrition

Methods	Accuracy	Precision	Recall/Sensitivity	F-measure	AUC	Specificity	Dataset	N	References
Decision tree (C4.5 (J48) classifier)	0.670	0.613	0.670	0.636	0.784	–	Higher Institutions in Nigeria	309	[1]
Logistic regression	0.871	–	0.871	0.853	0.816	0.454	IBM HR analytics		[8]
Decision tree	0.982	0.950	0.975	0.962	–	–	–	14,999	[4]
Random forest (with feature selection)	0.986	0.993	0.948	0.970	–	–	–		[4]
Logistic regression	0.750	0.356	0.733	–	0.743	0.753	–	4410	[6]
Random forest	0.808	–	–	–	0.851	–	IBM HR analytics	1470	[9]
Voting classifier (with feature selection)	0.960	–	–	0.620	–	–	IBM HR analytics	1470	[13]
Voting classifier	0.980	–	–	0.880	–	–	Kaggle HR dataset	15,000	[13]
Voting classifier (with feature selection)	0.990	–	–	0.910	–	–	Small-sized HR real data	450	[13]
Random forest	0.988	0.983	0.994	0.990	–	–	Real data	1649	[10]
XGBoost	0.960	–	0.950	0.960	0.950	–			[11]
XGBoost	0.891	–	–	–	–	–	IBM HR analytics	1470	[14]
XGBoost	0.880						HR Information Systems (HRIS) database and Bureau of Labor Statistics	73,115	[5]
Random forest	0.990	0.990	0.989	0.993		0.994	IBM HR Analytics	1470	[12]

thus, improving the working environment, balancing the workload, and sustaining a strong relationship between the employee and management may help to deal with this problem [6, 10]. Using existing employee data to extract invaluable information may support organizations to enhance their decision-making capabilities regarding employment strategies.

Case Study: Employee Turnover (Attrition)

In this use case, a machine learning model was built to predict employee attrition by using an employee dataset. The dataset is a modified version of IBM HR Analytics Employee Attrition Data [50]. The main research question is to figure what is the attrition risk of my current employees and what are the factors that affect employee attrition?

The dataset includes 20 variables of 1102 employees in a company. The list of variables and their definitions can be seen in Table 2.

For the next step, unnecessary variables were removed, and a brief exploratory data analysis was conducted. Since the main goal of this analytics model was to predict

Table 2 List of variables_employee attrition

Variable name	Variable description
EmpID	Unique ID of employees
Age	Age of the employee
Daily rate	The daily wage rate of an employee
Distance from home	The distance between the household and company location
Education	Education level of the employee (1: High School,3: Undergraduate, 4: Masters, 5: Ph.D.)
MonthlyIncome	The monthly wage of an employee
NumCompaniesWorked	Number of companies that employee worked before
PerformanceRating	Last performance score of the employee
RelationshipSatisfaction	Satisfaction score of the employee
TrainingTimesLastYear	Number of trainings employee took during last year
YearsInCurrentRole	How many employee spent in the current role
YearsSinceLastPromotion	How many years have passed since the last promotion
YearsWithCurrManager	How many years passed with the current manager
Tenure	Total number of years within the company
TotalExperience	Total number of years in the work-life
WageRate	Comparison of the wage with the company mean
WorkingModel	Working model of the employee (Hybrid, working in the office, working remote)
OrgFunction	Organizational department of the employee
PositionEngagement	Engagement score for the position
Attrition	Active or former employee (0: Active, 1: Former employee)

the attrition risk of employees and identify the reasons behind attrition, a bivariate analysis was proceeded to understand the distribution of attrition. To conduct the bivariate analysis, the data types of the features were modified and checked. The dataset included 4 categorical, 16 numeric, and 1 string features. Since the feature "EmpID" represents the unique ID of each employee, it was removed from the analysis. Additionally, null values were checked for each column. It was seen that the dataset did not include any null value. The attrition ratios of former and current employees for each categorical variable is compared by using bi-variate analysis. The results of this anaysis can be given as follows:

- For the education level = 1, attrition level %17.3 is slightly higher than the overall ratio (%15.6)
- For the education level = 5, attrition level %7.4 is lower than the overall ratio (%15.60)
- The employees whose wages are less than the company mean (Wage rate = 1) have a slightly higher attrition ratio (%17.9) than the other ones
- The employees who are working remote (Working Model = 2) have a higher attrition rate (%25.4) than the other groups
- Employees working in the office (Working Model = 1) have the lowest attrition level when compared with the other two working models.
- Employees from departments "0" and "2" have slightly higher attrition ratios than the expected mean.

As a result of the exploratory data analysis, it is possible to say that all of the features within the dataset could be used for modeling. Possible signs of relations between target column "Attrition" and dependent variables such as "Age," "Monthly Income," "Years in current role," "Education Level," and "Working Model" are possible predictors of attrition of employees. However, since these comparisons were conducted on variable levels separately, the outputs of the model should be analyzed to understand the complex relationships.

For the next step, a classification model was applied to predict the attrition risk of current employees. All of the variables except "EmpID" were used as predictors and the variable "Attrition" was used as the target. In this study, three different machine learning algorithms were applied to the Attrition dataset. The results showed that "LightGBM" classifier outperformed the other two algorithms. Since data is imbalanced with an attrition rate of 15.6% checking, both accuracy and F1 scores are crucial to assess model performance. Even though the decision tree model is a simpler model than random forest and LightGBM, it outperformed the other two with a recall score 0.34 and F1 score 0.32. However, its overall accuracy is worse than the other two (0.76). Random forest model has a higher accuracy rate, but the recall metric is considerably low (0.09). Nonetheless, F1 Score is 0.16. LightGBM has a better F1 score (0.27) than random forest. Additionally, the recall metric for LightGBM is also higher than random forest (0.18). Additionally, in terms of overall accuracy, it outperformed the other two with a score of 0.84. As a result, the LightGBM algorithm was chosen to predict the employee attrition risk with the best performance metrics.

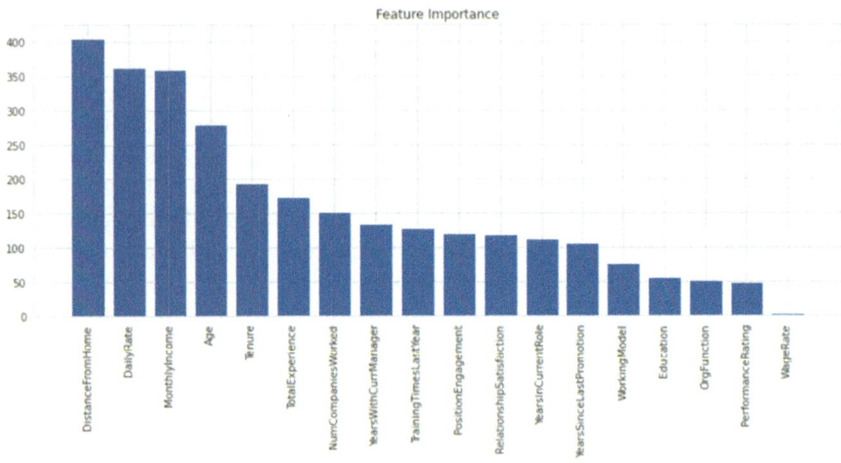

Fig. 1 Future importance_employee attrition

For the next step, the feature importance of the LightGBM model was calculated. The feature importance plot is given in Fig. 1.

Figure 1 shows that "Distance from Home," "DailyRate," and "MonthlyIncome" are the most important three features for the classification model.

Deep dive analysis provided that if the Dailyrate decreases, the attrition risk of employees increases. Lower monthly income increases the risk of attrition. Low levels of distance from home decrease the risk of attrition.

3 Recruitment Analytics

Organizations aim to hire appropriate employees having the appropriate knowledge, skills, and abilities to contribute to the success and sustain the competitive advantage, ensuring the right fit for the vacant positions is in the scope of the human resources recruitment function.

Recruitment, consisting of activities to be performed to find and attract qualified employees, affects the organizational performance through the success of the subsequent HR functions such as selection, compensation, and training [15]. Recruiting requires considerable knowledge and expertise, especially as it is a data-driven and knowledge-sensitive process, to attract the job seekers to apply to the job openings, maintain the interest of job seekers in the position as long as being open, review the curriculum vitae (CV) of the candidates, and match the requirements of the com company with potential applicants [16, 17]. As involving a huge amount of data, automatizing the process is among the primary concerns of both the practitioners and researchers. Turning the existing data into invaluable business intelligence, recommender systems become appealing both to the recruiters and to the job seekers.

In order to offer job seekers positions attracting their attention, job recommender systems are developed. For instance, to match job seekers with available job openings, a career platform classifies both the CVs of the registered users and available job titles using support vector machines and k-nearest neighbor methods [18]. A job recommendation system consisting of random forest and support vector machines is proposed to target the price and location sensitive people separately [19].

To support recruiters, an automated recruitment system imitating the knowledge of the recruiter, extracting data from CVs, and matching the requirements and profiles is built [17]. In a study proposing a new data-based design for electronic human resources management (e-HRM) activities including e-recruitment, e-selection, e-performance management, e-compensation, and e-learning, attrition is used to predict potential positions to be input to the e-recruitment system using decision tree [classification and regression trees (CART)], support vector machines, and k-nearest neighbor methods [20]. Decision tree performs better compared to other methods. The effectiveness and usability of these systems are quite important for their prevalence. Accordingly, information quality, popularity, and security of the source are found to be the factors affecting the effectiveness of e-recruitment [21].

Recruitment sources should be examined carefully due to their effect on employee performance, employee turnover [22], or employee loyalty [23]. Even though there is not any consensus on the sources resulting in the recruitment of employees performing better, several studies are conducted to understand the nature of the relationship. The primary reasons of differences in the prominent sources of recruitment can be summarized as follows: (i) using a different number of sources [24] and even combining them into different groups (e.g., internal and external [25], being controllable or uncontrollable by the organization [16]), (ii) assuming that the job seekers and/or recruiters use single source and neglecting the usage of multiple sources [24], (iii) the differences in the employee demographics or characteristics, (iv) the differences in the type of jobs (i.e., white/blue-collar positions, permanent/temporary jobs, etc.) [24], (v) the differences in the type of organization and industry (i.e., local/global [26]), (vi) the differences in sample sizes [24].

An organization might fill existing or expected vacancies internally by recruiting its own employees or externally [16]. Interdepartmental transfers, promotions, and internship programs are considered in internal sources while external sources of recruiting include recommendations, employment agencies, universities and colleges, referral programs, job fairs, professional associations, and trade unions, rehiring former employees, and advertisements, as well as employee exchange [24–26].

Performance-related studies indicate that internal sources have a significant effect on job performance [25], employees hired using employee referral have higher job performance than those hired through other methods [27], and agencies do better in finding the appropriate employees [28].

Table 3 Sample dataset recruitment

	EmpID	PerformanceScore	OrgEngagement	PositionEngagement	HireSource
0	Emp1	18.5	0.87	0.02	Applied online
1	Emp2	35.3	0.18	0.39	Campus
2	Emp3	67.9	2.75	3.11	Referral
3	Emp4	35	1.7	0.92	Campus
4	Emp5	27.5	0.35	0.38	Referral

Case Study: Assessing Recruitment Source Through Employee Performance and Engagement

In this case, an analytics approach was applied to an HR dataset to assess the best medium for hiring. The dataset used in this case is a modified version of IBM HR Analytics Employee Attrition Data [50]. The main objective is to assess recruitment sources through employee performance and engagement scores and define which recruitment source is better to ensure employee performance.

The dataset includes Performance Score, Organizational Engagement Score, Positional Engagement Score, and source of hiring for 852 employees in a company. A sample dataset is given in Table 3.

Brief statistical analysis of numerical variables showed that the performance score deviates between 0 and 100, whereas engagement scores deviate between 0 and 5. If the correlation between numerical features is checked, it is seen that Organizational and Positional Engagement scores are highly correlated (0.83). Additionally, the Performance score is correlated with Position Engagement with a correlation coefficient 0.65 but less correlated with Organizational Engagement with a coefficient of 0.54.

To select the best hiring source, the target variable using engagement and performance scores is defined. Since organizational engagement is related to employee commitment to the company and positional engagement is related to employee's willingness to stay in the current role, both of them are linked with long-term individual performance. By using these three indicators, employee data is segmented and every employee is classified in a segment that represents their engagement and performance scores. The segmentation is conducted by applying an unsupervised machine learning algorithm and on the next step, hiring sources are evaluated according to these created segments.

By using K-means clustering algorithm and elbow method, the optimum number of employee segments is determined as 4 (Fig. 2).

After the cluster analysis the employee groups are defined as follows: employees with a low level of engagement and performance as "Low," employees with mid-level performance level and mid-level engagement as "Medium," and employees with high performance and high engagement as "High". Since engagement is an important factor for sustainable performance, it is possible to assert that hiring source that provides more employees in class "High" is a better method for recruiting engaged

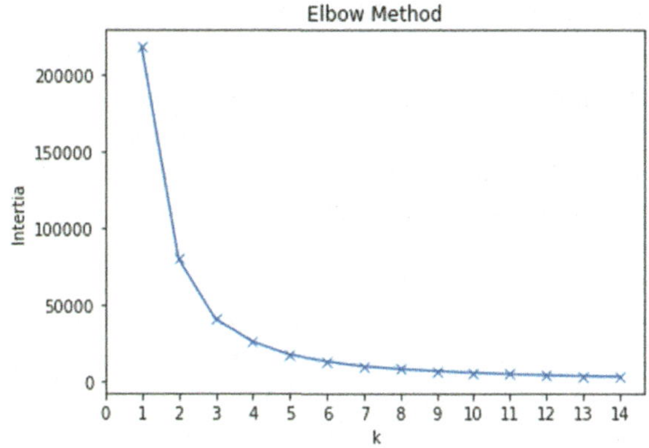

Fig. 2 K-means analysis-recruitment analysis

and high-performer employees. The distribution of hiring sources and the employee clusters are given in Table 4.

When the total distribution of high performers are compared for each hiring source, it is possible to say that "Search Firm" has the highest high performer ratio with %34.4 which is higher than the overall average (%27.5). On the other hand, it is possible to say that the hiring source "Referrals" has the lowest high performer ratio with %22.5 which is lower than the overall average (%27.5). As a result, a clustering approach based on the performance and engagement scores of employees showed us that "Search Firm" is the best hiring source to ensure long-term success within the company.

Table 4 The distribution of hire source and the clusters

HireSource	Perf_Cluster	
Applied online	High	29.1
	Low	26.5
	Mid	44.4
Campus	High	27.6
	Low	30.0
	Mid	42.4
Referral	High	22.5
	Low	26.4
	Mid	51.1
Search Firm	High	34.4
	Low	21.9
	Mid	43.8

4 Performance Analytics

Performance management, an approach for enhancing the overall targets of the organization by improving the performance of the employees (both individuals and teams) to benefit from their potential, ensures that the employees embrace the values and the goals of the organization [29]. An effective performance management approach requires measuring employee performance using appropriate tools and techniques to provide input to other HR functions such as career planning and training, develop strategies, or take necessary actions related to possible risks.

Employee performance directly affects the success, survival, and competitive power of the organizations. So, it is expected that the performance assessments support the managerial decisions by providing the necessary information. Considering the dynamic and rapidly changing nature of the business environment, the costly and time-consuming traditional performance assessment tools and techniques may fail to provide instant real-time information to support the decisions [30, 31]. At this point, data analytics methods and algorithms are used to observe the actual performance and even take precautions for future deteriorations likely to be based on continuous predictions [32].

In order to predict the employee performance, stochastic gradient descent and bagging classifier [33], neural networks [34], decision tree [33, 35–37], naïve Bayes [34, 36, 38], logistic regression [33, 34, 36, 39], random forest [33, 34, 39], and support vector machines [34, 39] are used. As given in Table 5, classification methods and algorithms are preferred. As the employee characteristics and the nature of the work have an impact on the performance, these studies consider various factors including gender, age, education, marital status, industry, organizational support, total experience in years, title of the position (e.g., managerial or non-managerial), leadership type, nature of the task (e.g., complexity), socio-economic status, location, income group, wage system, workspace and environment, task complexity, motivation, and competence, [34–37]. Personality-related factors such as the abilities of creative thinking, conflict resolution, decision-making, and personal relationships are also included in the models [34].

With the rise of information technologies, digitalization, and other related technologies [31], performance management activities are evolving to become more artificial intelligence-oriented (AI). Using AI is expected to result in continuous, real time, flawless data with a robust and unbiased performance management process [40].

Case Study: Who are the best performers? Identifying top manager profile through analytics

In this case, an analytics approach was applied to assess the profile of top-performer managers using an employee dataset. The dataset is a modified version of IBM HR Analytics Employee Attrition Data [50]. The objective of this case is to identify the top performer managers and their common characteristics.

Table 5 Methods to predict employee performance

Methods	Accuracy	Precision	Recall/Sensitivity	F-measure	AUC	*N*	References
Support vector machines	0.853	0.870	0.850	–	0.890	1500	[34]
Decision tree (CART)	0.906	–	–	–	–	120	[37]
Logistic regression	0.756	–	–	–	–	227	[39]
Decision tree	0.708	0.727	0.708	0.708	0.859	206	[35]
Naïve Bayes	0.806	0.804	0.806	0.805	0.894	1037	[38]
Logistic regression	0.834	0.843	1.000	0.915	0.799	858	[36]
Logistic regression	0.690	–	–	–	–	512	[33]
K-means clustering and artificial neural network (ANN)	–	–	–	–	–	220	[41]

The dataset includes 8 variables of 218 managers in a company. The list of variables and their definitions can be seen in Table 6.

To identify top performer managers within the company, evaluation scores of subordinates and business performance outcomes were analyzed. Basic descriptive statistics show that TeamAsessment has a mean of 2.24, and PerformanceScore has

Table 6 List of variables and their definitions

Variable name	Variable Description
Age	Age of the manager
TotalExperience	Total number of years in the work life
Education	Education level of the manager (1: High School,3: Undergraduate, 4: Masters, 5: Ph.D.)
EducationField	Graduation major of the manager
OrgFunction	Organizational department of the manager
Tenure	Total years of employment within the company
TeamAssessment	Team evaluation score of the manager
PerformanceScore	Performance evaluation score of the manager

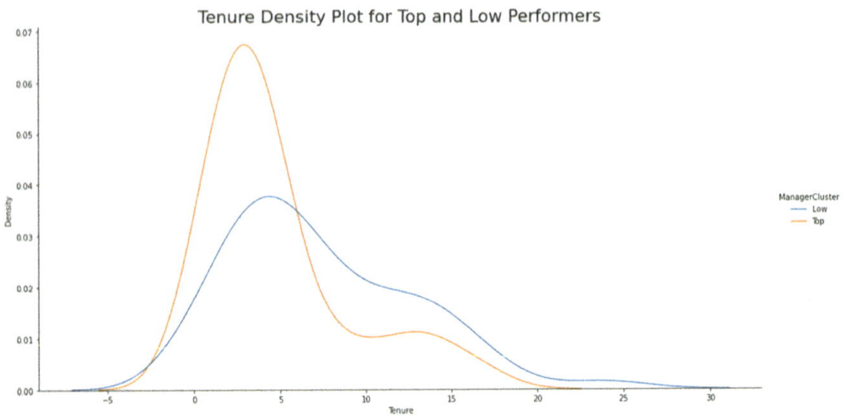

Fig. 3 The tenure density plot for top and low performers

54.2. The correlation coefficient between TeamAssessment and Performance Score is −0.0089 which indicates that these two variables are not correlated. To identify the top performer managers, firstly, managers who have higher performance scores than the company average were identified. Secondly, managers whose team assessment scores were higher than the company average were selected.

Analysis results proved that there are 60 top performers and 55 low-performer managers. Comparison of basic descriptive statistics of top and low performers show that top performers' average age is lower than low performers (41.3–37.3). Secondly, top performers have a higher total experience average and a lower tenure than low performers. Figure 3 illustrates the tenure density plot for top and low performers.

The density plot showed that the number of top performers who have a tenure less than 5 is higher than the low performers. On the contrary, the number of low performers with a tenure of more than 5 is higher than the top performers. Even though distributions of top and low performers in terms of total experience are similar, more top performers have a total experience of more than 15. When the top and low performers' education levels are compared, it is seen that the number of top performers who have a Master's degree is more than the low performers. On the contrary, the number of low performers who have a high school degree is higher than the top performers.

Analysis of the education field marks a considerable difference between technical degree and life sciences graduates. Figure 4 shows the performance difference between education degrees.

The number of top performers with a technical degree is much higher than the low performers. On the contrary, the number of low performers with life sciences degrees is much higher than the top performers. When the number of top and low performers are compared in different organizational departments, it is seen that the number of top performers is slightly higher in information technologies and lower in operations. In summary, the common characteristics of the top performers are they

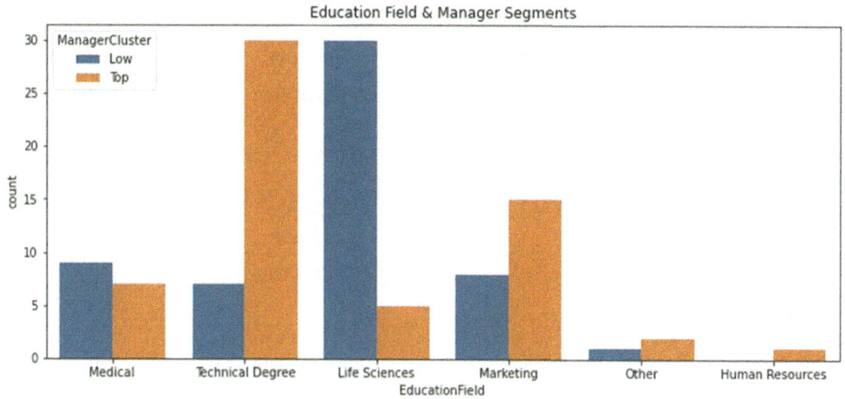

Fig. 4 The performance difference between education degrees

are below 40, have a tenure less than 5, have a total experience of more than 10, have a technical degree in IT and marketing in operations.

5 Training Analytics

Training and development have a significant role in the professional and individual improvement of employees in terms of knowledge, skill, and attitude [42, 43]. As a long-term process aiming to enhance employee performance considering the possible requirements of the job which may arise in the future, the development covers strategic level improvements [43].

Effective training and development support the motivation [44] and self-fulfillment of employees and in turn, reduce employee turnover, ensure successful and sustainable employee retention strategies [45], and contribute to the performance, growth, and competitive power of organizations [46].

Similar to other HR functions, data-based methods and algorithms are used in examining the training and development activities. To automatize the assessment of employees' learning types, the determination of training materials and the development of training strategies, an artificial intelligence-based expert system consisting of a rule-based approach and association rule mining is proposed [47]. This expert system operates using employee information such as department, seniority, and position.

Classifying employees based on their knowledge and expertise also supports HR to determine the training needs. K-means clustering is used to classify the expertise of employees (four classes from excellent to poor) and four classification algorithms including AdaBoost, random forest, support vector machines, and logistic regression are used to identify the development potentials where the random forest performs better with 0.80 accuracy [48]. Employees are classified whether being a part of

Table 7 Training development management-related studies

Purpose	Methods	accuracy	F-measure	Dataset	N	References
A training plan based on performance assessment	k-means clustering	–	–	Real data	100	[49]
Classification	Random forest	0.963	–	Kaggle dataset	54,808	[43]
Development and innovation performance	Logistic regression	–	–	Real data	1384	[46]
Classification	Random forest	0.800	0.796	Real data	1,652,815	[48]

personalized training or not using decision tree, random forest, and support vector machines where the random forest performs better with 0.963 accuracy [43]. This study uses a dataset consisting of variables including department, length of service, the number of previous training, average training score, and awards.

Besides, the department and the industry organizations operate in [43, 46], the tools of training are considered in the studies examining the relationship between training and development and employee performance. Training on the job, coaching, mentoring, using development centers or project teamwork may affect the performance of employees or organizations [46]. Table 7 summarizes development management studies.

Case Study: Data-driven training and development plan through analytics

In this case, an analytics approach was conducted to identify the best training method for the employee to increase performance. The dataset is a modified version of IBM HR Analytics Employee Attrition Data [50]. The main questions of the study are: "Which training method is more effective than others in terms of employee performance? Are there any differences between different departments in terms of training method and performance? Can we predict employee performance through the number of classes and online training?"

The dataset includes performance rating, department information, and training types of 1102 employees of a company. In this case, the main question is to understand which training type is more effective in terms of employee performance. There are multiple factors to evaluate the effectiveness of training. Performance Rating is the observed employee performance. It is the most tangible output of training investments. For this reason, the effect of training on employee performance is explored. Since the training needs of different departments could vary, the relationship between employee performance and training type will be explored on the department level. The list of variables and their definitions can be seen in Table 8.

In the dataset, there are 43 employees from finance, 709 employees from IT, and 350 employees from sales. When correlation analysis is performed, it is seen that

Table 8 List of variables_training and development plan

Variable name	Variable description
OrgFunction	Organizational department of the employee (0: Finance, 1: IT, 2: Sales)
ClassTrainingLastYear	Number of class training assigned to the employee last year
OnlineTrainingsWeeks	Number of online training weeks that employee finished last year
PerformanceRating	Last performance score of the employee

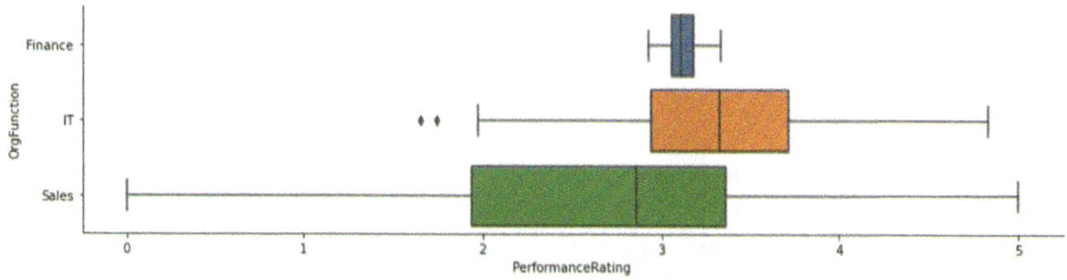

Fig. 5 Boxplot analysis of performance rating

relatively high correlations between performance rating and training types. "Online training weeks" is positively correlated with performance rating with a Pearson coefficient of 0.44. "In Class Trainings" is positively correlated with performance rating with a Pearson coefficient of 0.33.

Boxplot analysis of Class training variable shows that the distributions of three organizations are identical. Secondly, boxplot analysis of Performance Rating given in Fig. 5 shows that IT has a higher average score than Finance and Sales Finance which has a very narrow distribution with low variance. The standard deviation for Finance is 0.09, whereas for IT, it is 0.52 and for Sales it is 0.86.

Exploratory data analysis showed that organizational function is an indicator for performance rating where IT has higher values. To understand the relationship between performance rating and online/class trainings, firstly, the relationship between the target variable performance with other two numeric variables are controlled, and then drilling down to the organizational department level are performed. When the relationship between performance rating and online trainings are explored, a very strong correlation in IT department are occurred. The correlation table on organization department level shows that, Performance scores and online trainings are strongly positively correlated in IT department (0.84). The correlation table between Performance scores and class trainings shows that these two variables are strongly positively correlated in sales department. Since the effect of online and class trainings on employee performance are wanted to be explored, the high correlations in Sales and IT reveal provide a baseline for further analysis. Even though the performance of employees within a company could be related with many factors, the high correlations in Sales and IT enable us to assert that number of class trainings could be a good indicator in Sales department and the number of online trainings that an employee gets could be a good indicator for performance in IT department.

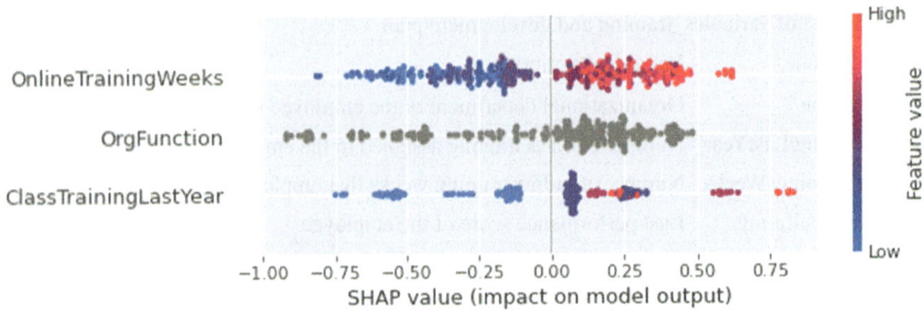

Fig. 6 Shap value_trainin and development management

In the next step, a machine learning algorithm will be applied to predict employee performance through available dependent variables Organizational Function, Online training hours, and the number of class trainings.

To predict employee performance through online and class trainings, a lightgbm regressor model was built. The results show that the model can explain the %78 of target variable (r2_score: 0.78) which is a very good score. Root mean score of the model 0.32 which is also a good error value when it is compared with the mean of target variable performance score (3.1) and standard deviation (0.71). Lastly, shap values of the model were calculated to explain the behavior of the lightgbm model (Fig. 6).

The results of the predictive model show that it is possible to predict employee performance through online and class trainings and organizational department information. Shap values show that higher values of Online trainings also increase the performance score. The exploratory analysis showed this relationship is since Online trainings are highly correlated with performance score in IT department.

In this chapter, the importance of HR analytics is highlighted and showed how HR problems can be solved by using HR analytics tools. HR analytics can be defined as to collect data and transform it into useful information for improving human resource management. HR analytics is very useful for organizations since they create useful insights for managing the organization and help them to reach strategic goals. HR analytics is especially very useful for improving HR functions such as workforce management, recruitment, performance evaluation, and development management. In this chapter, four different HR problems are handled and tried to solve these problems by using HR analytics tools. For future studies, HR analytics tools can be applied to various HR problems.

References

1. Alao D, Adeyemo AB (2013) Analyzing employee attrition using decision tree algorithms. Comput Inf Syst Dev Inf Allied Res J 4:17–28
2. Shaw JD, Delery JE, Jenkins GD Jr, Gupta N (1998) An organization-level analysis of voluntary and involuntary turnover. Acad Manag J 41:511–525
3. Barrick MR, Mount MK, Strauss JP (1994) Antecedents of involuntary turnover due to a reduction in force. Pers Psychol 47:515–535
4. Yadav S, Jain A, Singh D (2018) Early prediction of employee attrition using data mining techniques. In: 2018 IEEE 8th international advance computing conference (IACC). IEEE, pp 349–354
5. Punnoose R, Ajit P (2016) Prediction of employee turnover in organizations using machine learning algorithms. (IJARAI) Int J Adv Res Artif Intell 5:22–26
6. Setiawan I, Suprihanto S, Nugraha AC, Hutahaean J (2020) HR analytics: Employee attrition analysis using logistic regression. In: IOP conference series: materials science and engineering. IOP Publishing, p 032001
7. Bhartiya N, Jannu S, Shukla P, Chapaneri R (2019) Employee attrition prediction using classification models. In: 2019 IEEE 5th international conference for convergence in technology (I2CT). IEEE, pp 1–6
8. Ozdemir F, Coskun M, Gezer C, Gungor VC (2020) Assessing Employee Attrition Using Classifications Algorithms. In: Proceedings of the 2020 the 4th international conference on information system and data mining. Hawaii, USA, pp 118–122
9. Pratt M, Boudhane M, Cakula S (2021) Employee attrition estimation using random forest algorithm. Baltic J Modern Comput 9:49–66
10. Sadana P, Munnuru D (2021) Machine Learning Model to Predict Work Force Attrition. In: 2021 6th international conference for convergence in technology (I2CT). IEEE, pp 1–6
11. Mhatre A, Mahalingam A, Narayanan M, et al (2020) Predicting employee attrition along with identifying high risk employees using big data and machine learning. In: 2020 2nd International conference on advances in computing, communication control and networking (ICACCCN). IEEE, pp 269–276
12. Sisodia DS, Vishwakarma S, Pujahari A (2017) Evaluation of machine learning models for employee churn prediction. In: 2017 International conference on inventive computing and informatics (ICICI). IEEE, pp 1016–1020
13. Yahia NB, Hlel J, Colomo-Palacios R (2021) From big data to deep data to support people analytics for employee attrition prediction. IEEE Access 9:60447–60458
14. Jain R, Nayyar A (2018) Predicting employee attrition using xgboost machine learning approach. In: 2018 International conference on system modeling & advancement in research trends (SMART). IEEE, pp 113–120
15. Barber AE (1998) Recruiting employees: Individual and organizational perspectives. Sage Publications.
16. Breaugh JA (2013) Employee recruitment. Annu Rev Psychol 64:389–416
17. Shahbaz U, Beheshti A, Nobari S, et al (2018) irecruit: Towards automating the recruitment process. In: Service research and innovation. Springer, pp 139–152
18. Javed F, Jacob F (2015) Data science and big data analytics at career builder. In: Big-data analytics and cloud computing. Springer, London, pp 83–96
19. Martinez-Gil J, Freudenthaler B, Natschläger T (2018) Recommendation of job offers using random forests and support vector machines. In: Proceedings of the workshops of the EDBT/ICDT 2018 joint conference (EDBT/ICDT 2018)
20. Nasar N, Ray S, Umer S, Mohan Pandey H (2020) Design and data analytics of electronic human resource management activities through Internet of Things in an organization. Softw Pract Exp 1–17. https://doi.org/10.1002/spe.3006
21. Ikram A, Su Q, Fiaz M, Khadim S (2017) Big data in enterprise management: transformation of traditional recruitment strategy. In: 2017 IEEE 2nd international conference on big data analysis (ICBDA). IEEE, pp 414–419.

22. Blau G (1990) Exploring the mediating mechanisms affecting the relationship of recruitment source to employee performance. J Vocat Behav 37:303–320
23. Jamil R, Neem H (2013) The impact of outsourcing external recruitment process on the employee commitment and loyalty: empirical evidence from the telecommunication sector of Pakistan. J Bus Manag 8:69–75
24. Zottoli MA, Wanous JP (2000) Recruitment source research: current status and future directions. Hum Resour Manag Rev 10:353–382
25. Al-Khasawneh AL, Malkawi NM, AlGarni AA (2018) Sources of recruitment at foreign commercial banks in Jordan and their impact on the job performance proficiency. Banks Bank Syst 13:12–26
26. Peltokorpi V, Jintae Froese F (2016) Recruitment source practices in foreign and local firms: a comparative study in Japan. Asia Pac J Hum Resour 54:421–444
27. Breaugh JA (2014) Predicting voluntary turnover from job applicant biodata and other applicant information. Int J Sel Assess 22:321–332
28. Swamy CJ, Beloor V, Nanjundeswaraswamy TS (2021) Recruitment and selection process in the IT firms. GIS Sci J 8:343–356
29. Armstrong M (2006) Performance management, 3rd ed. Kogan Page Limited, London and Philadelphia
30. Garg S, Sinha S, Kar AK, Mani M (2021) A review of machine learning applications in human resource management. Int J Prod Perf Manage
31. Sahlin J, Angelis J (2019) Performance management systems: reviewing the rise of dynamics and digitalization. Cogent Bus Manage 6:1642293
32. Jerath A (2018) The role of analytics in predicting employee performance. In: SHRM. https://www.shrm.org/shrm-india/pages/the-role-of-analytics-in-predicting-employee-performance.aspx. Accessed 24 July 2021
33. Keya MS, Emon MU, Akter H, et al (2021) Predicting performance analysis of garments women working status in Bangladesh using machine learning approaches. In: 2021 6th International conference on inventive computation technologies (ICICT). IEEE, pp 602–608
34. Lather AS, Malhotra R, Saloni P, et al (2019) Prediction of employee performance using machine learning techniques. In: Proceedings of the international conference on advanced information science and system. Singapore, pp 1–6
35. Kirimi JM, Moturi CA (2016) Application of data mining classification in employee performance prediction. Int J Comput Appl 146:28–35
36. Li MGT, Lazo M, Balan AK, de Goma J (2021) Employee performance prediction using different supervised classifiers. Singapore, pp 6870–6876
37. Pahmi S, Saepudin S, Maesarah N, Solehudin UI (2018) Implementation of CART (classification and regression trees) algorithm for determining factors affecting employee performance. In: 2018 International conference on computing, engineering, and design (ICCED). IEEE, pp 57–62
38. Valle MA, Varas S, Ruz GA (2012) Job performance prediction in a call center using a naive Bayes classifier. Expert Syst Appl 39:9939–9945
39. Tang A, Lu T, Lynch Z, et al (2020) Enhancing promotion decisions using classification and network-based methods. In: 2020 Systems and information engineering design symposium (SIEDS). IEEE, pp 1–6
40. Ray AS (2019) AI in performance management. In: peopleHum. https://www.peoplehum.com/blog/scope-of-ai-in-performance-management Accessed 24 July 2021
41. Aktepe A, Ersoz S (2012) A quantitative performance evaluation model based on a job satisfaction-performance matrix and application in a manufacturing company. Int J Ind Eng 19:264–277
42. Ameen A, Baharom MN (2019) An appraisal of the effect of training on employee performance in an organisation: a theoretical discussion. Asian J Multidisc Stud 7:27–31
43. Kaewwiset T, Temdee P, Yooyativong T (2021) Employee classification for personalized professional training using machine learning techniques and SMOTE. In: 2021 Joint international conference on digital arts, media and technology with ECTI northern section conference on electrical, electronics, computer and telecommunication engineering. IEEE, pp 376–379

44. Ozkeser B (2019) Impact of training on employee motivation in human resources management. Procedia Comput Sci 158:802–810
45. Fletcher L, Alfes K, Robinson D (2018) The relationship between perceived training and development and employee retention: the mediating role of work attitudes. Int J Hum Resour Manage 29:2701–2728
46. Berber N, Lekovic B (2018) The impact of HR development on innovative performances in central and eastern European countries. Empl Relat 40:762–786
47. Chen N, Xu Z, Xia M (2013) Interval-valued hesitant preference relations and their applications to group decision making. Knowl-Based Syst 37:528–540
48. Liu J, Huang J, Wang T et al (2021) A data-driven analysis of employee development based on working expertise. IEEE Trans Comput Soc Syst 8:410–422
49. Escolar-Jimenez CC, Matsuzaki K, Okada K, Gustilo RC (2019) Enhancing organizational performance through employee training and development using k-means cluster analysis. Int J Adv Trends Comput Sci Eng 8:1576–1582
50. https://www.kaggle.com/pavansubhasht/ibm-hr-analytics-attrition-dataset

Ozgur Akarsu is the head of AI & Data Analytics. He graduated from ITU Industrial Engineering and finished his Ph.D. on HR Analytics in 2016. Ozgur has more than 15 years of experience in digital transformation and analytics in various industries. He also taught classes on machine learning, digital transformation and value chain management in several universities. He is specialized in machine learning and led several projects such as demand forecasting, HR analytics, predictive maintenance.

Cigdem Kadaifci is currently working at the Department of Industrial Engineering, Istanbul Technical University. Dr. Kadaifci got her Ph.D. from Istanbul Technical University. Her research areas are multi-criteria decision making, decision making under uncertainty, scenario analysis, and statistics.

Sezi Cevik Onar is a Professor in the Industrial Engineering Department of Istanbul Technical University (ITU) Management Faculty. She earned her BSc in Industrial Engineering and MSc in Engineering Management, both from ITU. She studied her Ph.D. courses at ITU and Eindhoven Technical University. She has completed her PhD thesis at ITU and visited Copenhagen Business School during these studies. Her Ph.D. was on strategic options. Her research interests include intelligent systems in management, innovation management, creative thinking, strategic management, business planning and multiple criteria decision making. She took part as a researcher in many privately and publicly funded national and international projects including the projects funded by European Union such as intelligent system design, organization design, and human resource management system design. She worked as a visiting professor at Technical University of Troyes. Her refereed articles have appeared in a variety of high impact journals including, Energy, Computers & Industrial Engineering, Supply Chain Management: An International Journal and Expert Systems with Applications. Her publications have been cited more than 3000 papers and her h-index is 28.

Manufacturing Analytics

Nursah Alkan, Dogan Oruc, Arif Gulbiter, and Mehmet Ali Ergun

1 Introduction

With the advent of the Industry 4.0 revolution, the manufacturing industry uses analytics enhanced with real-time production data to enable and maintain enterprise-wide automation as well as making better and faster decisions. Today, this situation has taken the name of manufacturing analytics, and the use of new technologies such as the cloud and the Internet of things (IoT) has enabled the possession of self-managed and healing assets. Manufacturing analytics gives decision makers a competitive advantage by optimizing costs, increasing productivity of operations and quality, accelerating innovation and redefining the customer experience. In this area, with the use of machine learning models and data visualization tools, it can reveal insights in data, optimize processes, make accurate and real-time decisions, and maximize performance. Intelligent manufacturing using sensors and edge devices connected to equipment provides manufacturers with complete and final product testing and customer satisfaction, thanks to a data-driven integrated view of all operations from the supply chain.

In this chapter, two cases are presented: statistical quality control and predictive maintenance.

N. Alkan (✉) · D. Oruc · A. Gulbiter · M. A. Ergun
Department of Industrial Engineering, Faculty of Management, Istanbul Technical University, 34367 Macka, Istanbul, Turkey
e-mail: nalkan@itu.edu.tr

A. Ustundag et al. (eds.), *Business Analytics for Professionals*, Springer Series in Advanced Manufacturing, https://doi.org/10.1007/978-3-030-93823-9_15

2 Statistical Quality Control

Problem Definition

For most businesses today, the quality of products and services is an important decision factor [1]. In order to meet and exceed customer expectations, one needs to ensure a high design quality as well as high quality of conformance. While design quality reflects how well a product design meets customer needs, quality of conformance measures the fit between design specifications and the actual output of the manufacturing process [2]. That is, a high quality of conformance level means the final product is meeting the design specifications and it requires systematic elimination of defective units. The use of statistical methods in explaining the variability in the process plays an important role for identifying and minimizing defective units and is a key element for successful quality improvement [3]. Statistical process control is a tool that aims to adhere to standards by considering the compliance of production with predetermined quality specifications, is used to minimize the production of nonconforming products, and provides decision-making based on data [1]. The tools that are widely used in solving quality problems and especially used for process control can be listed as follows:

- Histograms
- Pareto charts
- Cause-and-effect diagrams
- Defect-concentration diagrams
- Scatter diagrams
- Check sheets
- Control charts.

One of the main objectives of statistical quality control is to quickly identify the assignable causes of process shifts so that precautions and corrective actions are taken before many nonconforming products are manufactured [3]. Control charts are one of the most widely used tools for this purpose. A control chart is simply an online process monitoring technique in which a centerline as well as an upper and lower control limit for the monitored quantity. When there are unusual sources of variability, sample means will fall outside the control limits [1]. In more technical terms, a control chart is a series of hypothesis tests with the null hypothesis the process is in a state of statistical control. The selection of the quality control chart first depends on whether the monitored quantity is continuous or discrete. Control charts can be considered in three classes as given in Fig. 1. When examining control charts based on quantitative data, if the sample size is less than 10, process variability can be controlled by considering control charts based on ranges $\left(\overline{X}, R\right)$. In addition, for cases where the sample size is more than 10, the process can be controlled by considering the control charts based on standard deviations $\left(\overline{X}, S\right)$. In cases where the sample size is 1, the control charts are used for individual measurements $\left(\overline{X}, MR\right)$ in which the moving range of two successive observations is considered. When the control charts based on qualitative data are examined, p control chart is used for fraction nonconforming,

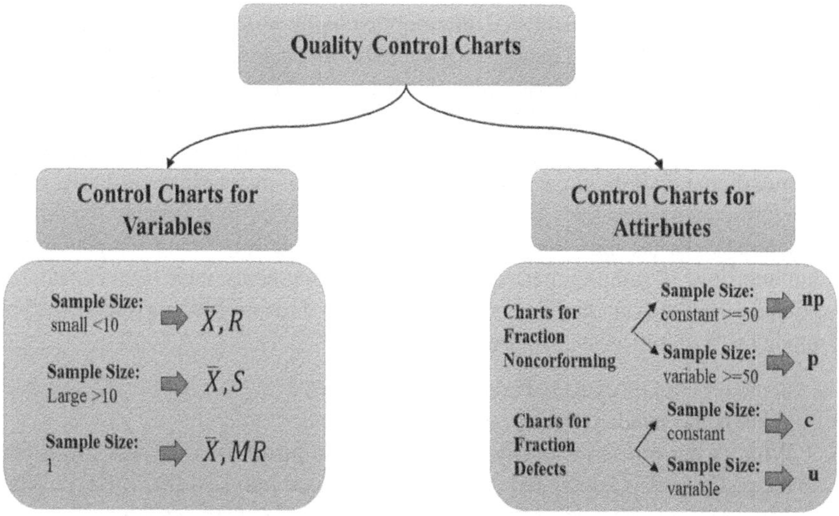

Fig. 1 Types of control charts

np control chart is used for the number nonconforming, *c* control chart is used for nonconformities, and *u* control chart is used for nonconformities per unit. Apart from the control charts given in Fig. 1, there are also different control charts such as Cumulative Sum Control Charts (CUSUM) [4] and Exponentially Weighted Moving Average (EWMA) [5], used for more precise analysis when the monitored quantity is continuous.

In a control chart, the lower and upper specification limits (LSL and USL, respectively) of the process represent the limits within which the process is said to be in control. The ability to interpret the causes of out-of-control situations resulting from a systematic or random error in the control chart requires experience and knowledge of the process. Western electric rules have been proposed to detect nonrandom patterns in control charts, and it has been concluded that the process is out of control in the following cases:

- A single point plots outside 3-sigma control limits.
- 4 of 5 consecutive points plot at a distance of $\mp 1\sigma$ or beyond from the centerline.
- 2 of 3 consecutive points are plot outside the $\mp 2\sigma$ range in the same direction
- 8 consecutive points plot on the same side of the centerline.
- 9 consecutive points plot ranked above or below the mean.
- Points showing an increasing or decreasing trend plot.

Case Study

Piston rings located between the cylinder and the piston constitute important and necessary components that ensure the efficient operation of the automobile engine.

In particular, in the manufacture of automobile engine piston rings, the inner diameter of the rings is a critical quality characteristic for rings to be viable. *X* company, one of the automobile main parts manufacturers, produces the necessary main and side parts of the automobile. The quality characteristics of the products in the manufacturing process have a critical importance. The company prioritizes the quality of the inner diameter of the rings, during the production of piston rings, as it affects the entire subsequent processes. In the case study, it is aimed to determine the quality characteristics of the inner diameter of the rings during the company's piston ring manufacturing. 37 samples, each consisting of five segments, have been considered for this aim and the standard deviation and average of the process are unknown. The company is trying to answer following questions:

1. Is the process in statistical control?
2. What is the standard deviation of the process?
3. What is the process capacity achieved for the piston ring process given that the piston ring is known to be within specification limits 74.000 ∓ 0.05? What percentage of piston rings are manufactured outside of specifications?
4. What are the limits of the process for the 95% confidence level?
5. What are the other relevant analyses used to monitor the adequacy of the process?

Model

First of all, it is necessary to determine which control chart should be used. Since there are 37 samples of size $n = 5$ and the standards are unknown, the process is controlled by using $\overline{X}$ and s control charts. An unbiased estimator of σ^2 is estimated by using the sample variance as given in Eq. (1) for each sample, since σ^2 is the unknown variance of a probability distribution.

$$s_i^2 = \frac{\sum_{i=1}^{n}\left(X_i - \overline{X}\right)^2}{n-1} \tag{1}$$

Since the standard deviation of the ith sample is s_i for m preliminary samples, each of size n, the standard deviation of the process is calculated as given in Eq. (2).

$$\overline{s} = \frac{1}{m}\sum_{i=1}^{m} s_i \tag{2}$$

An unbiased estimator of σ is $\overline{s}/c_4$, and the control chart s for standard deviations is written as Eq. (3).

$$\text{UCL} = B_4\overline{s} \quad \text{CL} = \overline{s} \quad \text{LCL} = B_3\overline{s} \tag{3}$$

where UCL, CL, and LCL are upper control limit, center line, and lower control limit, respectively. Then, the control chart $\overline{X}$ for averages is written as Eq. (4).

$$\text{UCL} = \overline{\overline{X}} + A_3\overline{s} \quad \text{CL} = \overline{\overline{X}} \quad \text{LCL} = \overline{\overline{X}} - A_3\overline{s} \tag{4}$$

where $\overline{\overline{X}}$ is the average of the averages. The constants B_3, B_4, A_3, and c_4 for construction of $\overline{X}$ and s charts from past data are listed in Table 1 for various sample sizes.

After investigating the stability of the process using control cards, the capability of the process should be measured using various tools. The process capability ratio (PCR), also called the process capability index, is a statistical measure of process capability used for process improvement efforts. It also indicates the ability of the process to meet specifications or meet customer requirements. Process capability indices allow different processes to be compared based on how well the process is controlled by organizations (see in Table 2). A PCR is obtained as in Eq. (5).

$$C_p = \frac{\text{UCL} - \text{LCL}}{6s} \tag{5}$$

In some cases, the process mean may not be centered between specification limits, resulting in the C_p index overestimating process capacity. Therefore, if a single specification limit is defined and there is no centralization, the process capability criticality ratio C_{pk} is used. The C_{pk} index is presented as given in Eq. (6).

$$C_{\text{pk}} = \min\left(\hat{C}_{\text{pu}}, \hat{C}_{\text{pl}}\right) \tag{6}$$

where C_{pu} and C_{pl} are the process capability criticality ratio for UCL and LCL, respectively.

$$C_{\text{pu}} = \frac{\text{UCL} - \mu}{3\sigma} \tag{7}$$

$$C_{\text{pl}} = \frac{\mu - \text{LCL}}{3\sigma} \tag{8}$$

The percentage of the specification band used up by the process is calculated as in Eq. (9).

$$P = \frac{1}{C_p} \times 100 \tag{9}$$

In industrial use areas, PCRs are used to calculate and interpret a point estimate of the desired quantity. However, confidence intervals need to be established for any alternative to become standard practice in the industrial area, as capability ratios (C_p and C_{pk}) are subject to statistical fluctuations. The confidence intervals for PCRs are obtained as given in Eq. (10).

Table 1 Factors for constructing control chart [3]

Observation in sample	Chart for averages					Chart for standard deviation				Factors for center line		Chart for ranges				
	Factors for control limits			Factors for center line		Factors for control limits						Factors for control limits				
	A	A2	A3	c4	1/c4	B3	B4	B5	B6	d2	1/d2	d3	D1	D2	D3	D4
2	2.121	1.880	2.659	0.7979	1.2533	0	3.267	0	2.606	1.128	0.8865	0.853	0	3.686	0	3.267
3	1.732	1.023	1.954	0.8862	1.1284	0	2.568	0	2.276	1.693	0.5907	0.888	0	4.358	0	2.574
4	1.500	0.729	1.628	0.9213	1.0854	0	2.266	0	2.088	2.059	0.4857	0.880	0	4.698	0	2.282
5	1.342	0.577	1.427	0.9400	1.0638	0	2.089	0	1.964	2.326	0.4299	0.864	0	4.928	0	2.114
6	1.225	0.483	1.287	0.9515	1.0510	0.030	1.970	0.029	1.874	2.534	0.3946	0.848	0	5.078	0	2.004
7	1.134	0.419	1.182	0.9594	1.0423	0.118	1.882	0.133	1.806	2.704	0.3698	0.833	0.204	5.204	0.076	1.924
8	1.061	0.373	1.099	0.9650	1.0363	0.185	1.815	0.179	1.751	2.847	0.3512	0.820	0.388	5.306	0.136	1.864
9	1.000	0.337	1.032	0.9693	1.0317	0.239	1.761	0.232	1.707	2.970	0.3367	0.808	0.547	5.393	0.184	1.816
10	0.949	0.308	0.975	0.9727	1.0281	0.284	1.716	0.276	1.669	3.078	0.3249	0.797	0.687	5.469	0.223	1.777
11	0.905	0.285	0.927	0.9754	1.0252	0.321	1.679	0.313	1.637	3.173	0.3152	0.787	0.811	5.535	0.256	1.744
12	0.866	0.266	0.886	0.9776	1.0229	0.354	1.646	0.346	1.610	3.258	0.3069	0.778	0.922	5.594	0.283	1.717
13	0.832	0.249	0.850	0.9794	1.0210	0.382	1.618	0.374	1.585	3.336	0.2998	0.770	1.025	5.647	0.307	1.693

Table 2 Recommended minimum values of the PCR [3]

	Two-sided specifications
Existing processes	1.33
New processes	1.5
Safety, strength, or critical parameter, existing process	1.5
Safety, strength, or critical parameter, new process	1.67

$$\hat{C}_p\sqrt{\frac{\chi^2_{1-\alpha/2,n-1}}{n-1}} \le C_p \le \hat{C}_p\sqrt{\frac{\chi^2_{\alpha/2,n-1}}{n-1}} \tag{10}$$

While the spread of the process can be examined with the C_p index, the placement of the process mean can also be examined with the C_{pk} index. Although the C_p index can successfully present information about the spread of the process, it cannot provide information about the placement of the mean. The C_{pk} value is used to estimate how consistent it is with the average performance. With the C_{pk} index, it is tried to interpret how far the process mean is from the specification limits. The higher the C_{pk} value, the better the process will be. Although the placement of the process mean is examined with C_{pk}, since healthy results cannot be obtained in some cases, the C_{pm} index should be used instead. The C_{pm} index is based on the difference between the target value and the process mean and provides more accurate information about the placement of the process mean. The C_{pm} index is calculated as given in Eq. (11).

$$C_{\text{pm}} = \frac{\text{USL} - \text{LSL}}{6\sqrt{s^2 + \left(\overline{X} - T\right)^2}} \tag{11}$$

where, $\overline{X}$ and T indicate the mean value and the target value, respectively.

Solution and Analysis

In the sample data considered [3], firstly, $\overline{X}$ and s values of each sample are calculated to create the $\overline{X}$ and s control charts. Sample data summary including $\overline{X}$ and s values is as presented in Table 3.

Table 3 Sample data summary based on $\overline{X}$ and s

Sample	X_1	X_2	X_3	X_4	X_5	$\overline{X}$	s
0	74.030	74.002	74.019	73.992	74.008	74.0102	0.0148
1	73.995	73.992	74.001	74.011	74.004	74.0006	0.0075
2	73.988	74.024	74.021	74.005	74.002	74.0080	0.0147
3	74.002	73.996	73.993	74.015	74.009	74.0030	0.0091
4	73.992	74.007	74.015	73.989	74.014	74.0034	0.0122

Since the sample size is $n = 4$, the B_3 and B_4 values used in the calculation of the acceptance area of the s control chart are found as 0 and 2.089, respectively, as given in Table 1. The A_3 value used in the calculation of the acceptance area of the $\overline{X}$ control chart is found as 1.427 from Table 1. Then, the control charts obtained for $\overline{X}$ and s are presented in Figs. 2 and 3, respectively.

In Fig. 2, it is seen that the 37th sample value in the $\overline{X}$ control chart falls outside of the control limits. In this case, the 37th sample value is removed from the dataset and the control charts are reconstructed. The resulting charts are given in Figs. 4 and 5.

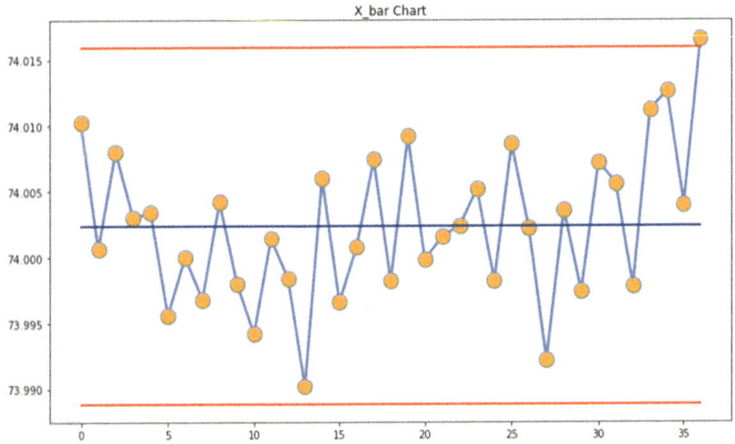

Fig. 2 $\overline{X}$ control chart

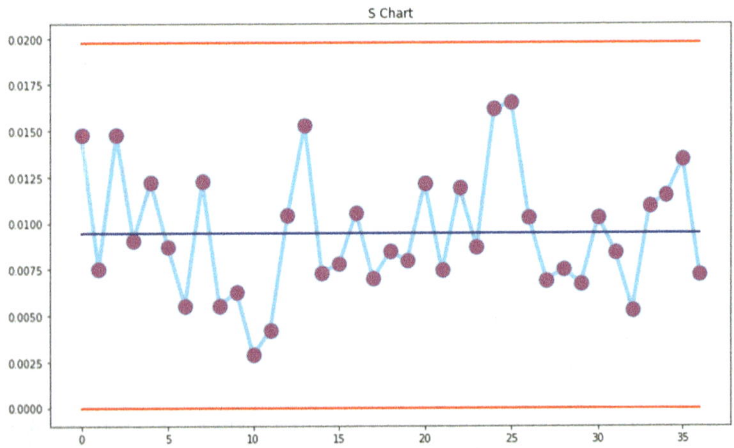

Fig. 3 S control chart

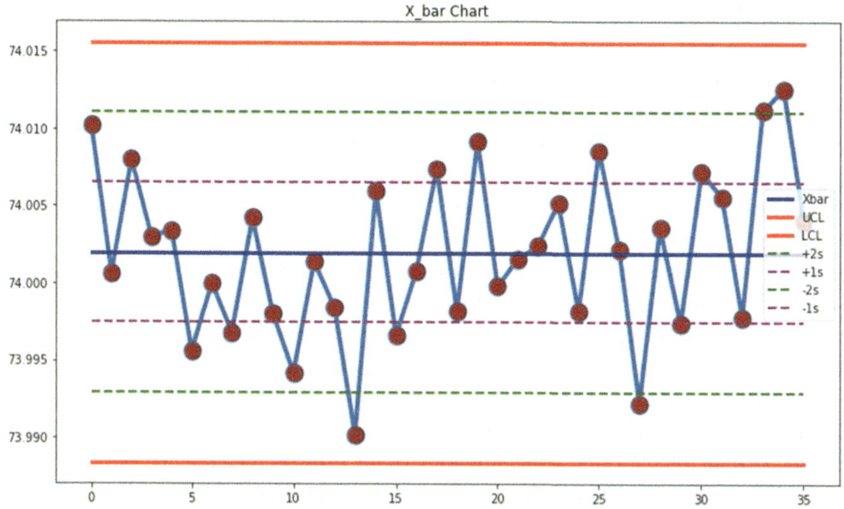

Fig. 4 $\overline{X}$ control chart for new values

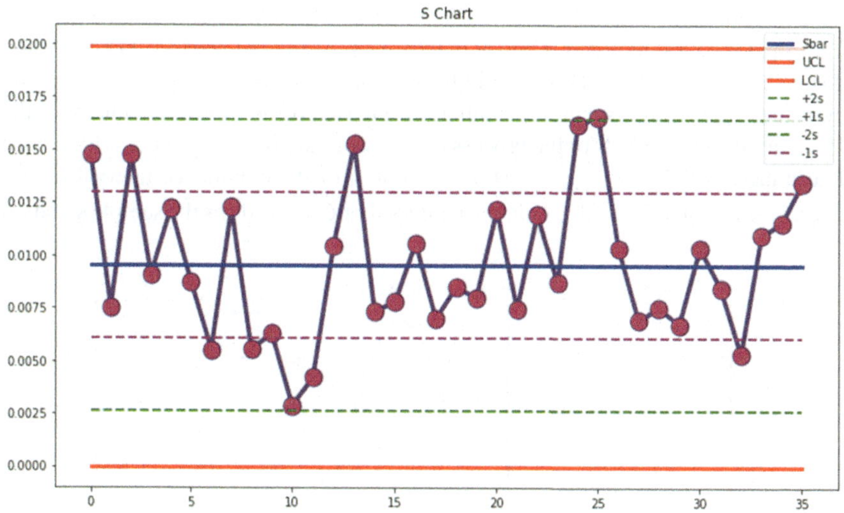

Fig. 5 S control chart for new values

As it is observed in Figs. 4 and 5, none of the observations are out of control. Besides, special conditions have also been met. The standard deviation of the process according to new values is obtained as 0.01. According to the 74.000 ∓ 0.05 specification limits, the process capability C_p is found as 1.64. The C_p value which is greater than 1 indicates that the process is sufficient and reliable. If the C_p index is less than 1, it means that the process mean is far from the target or the variation is too wide.

The C_p index is only a measure of the process spread and it does not consider whether or not the process mean conforms with the specifications. To test the proximity of process mean and the specified value, C_{pk} index can be used. In order to obtain the C_{pk} value, the lower and upper process capability ratios must be first calculated. The upper process capacity C_{pu} and the lower process capacity C_{pl} are calculated to be 1.58 and 1.71 respectively (See Eq. (7) and (8)). Since $C_{pu} < C_{pl}$, it is observed that the mean value is closer to LSL. Moreover, the C_{pk} value, which is the minimum of C_{pu} and C_{pl}, is obtained as 1.58. By looking at the C_{pk} value, it can conclude the process is capable and meets specification limits since it is greater than 1.33 (Please check Table 2 for minimum acceptable values).

In addition, if the C_{pk} value is in close proximity to the C_p value, it can conclude that the process mean is close to the specified value. On the other hand, C_{pk} value being less than the C_p value is an indication of the process mean being a considerable distance from the target value. According to the results, since the C_{pk} value is smaller than the C_p value, the mean is off of the target value, as seen in Fig. 6. Furthermore, the percentage of the specification limits used up by the process is also calculated as 60.81.

As mentioned earlier, the C_{pm} index shows how much the mean value deviates from the target value and is a better indicator of a centered process. For this case, it is calculated as 0.842. This is an indication of the mean of the process deviating from the target value. When the confidence interval of the process is checked, the 95% confidence intervals of the process capacity of the piston ring process have been determined as $0.213 < \hat{C}_p < 2.027$. The width of the confidence interval for $\hat{C}_p$ index is wider than desired. Confidence interval of $\hat{C}_p$ is wide as the sample standard

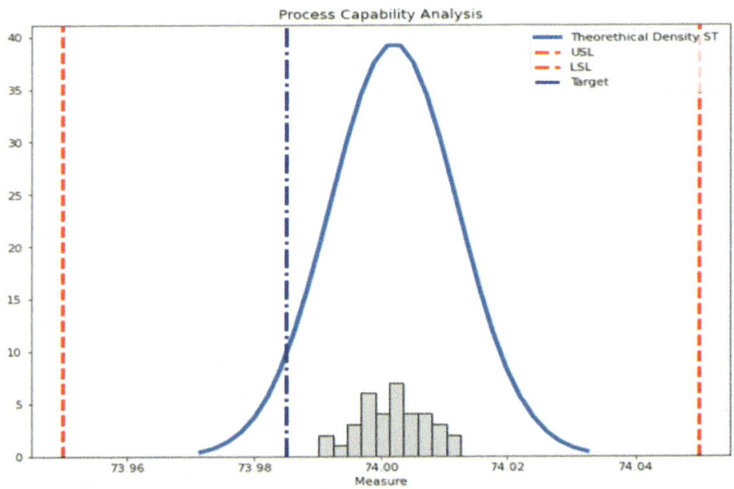

Fig. 6 Histogram of the piston ring process

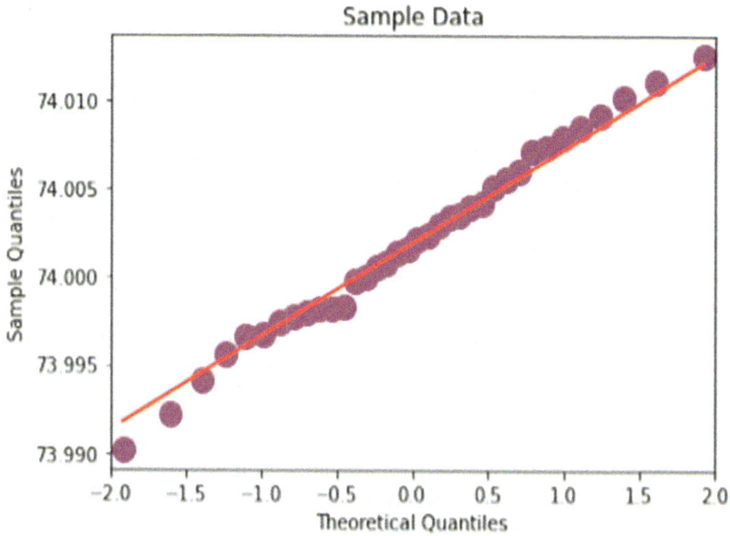

Fig. 7 Q-Q plot for normality test

deviation s showed significant fluctuations in small samples. Since the sample size was small, the confidence interval of the C_p index has given a wide result. For this reason, the confidence interval should be narrowed by increasing the sample size.

All analyzes so far assume that the inner diameter measurements are normally distributed. To make sure that this assumption holds, it is necessary to check if the sample measurements are normally distributed. The normal distribution and histogram plot of the piston ring process are given in Fig. 6. In the histogram obtained from the data, it is observed a shape-resembling a normal distribution. Besides, the histogram is completely within the specification limits also indicates that the process capability is satisfactory.

A Q-Q plot is created and an Anderson–Darling normality test is conducted to check for normality. As seen in Fig. 7, a 45-degree linear trend is observed in the Q-Q plot, which is a good indication of normality. Anderson–Darling normality test also concluded that there is no evidence to reject the null hypothesis of data being normally distributed.

3 Predictive Maintenance

Problem Definition

Predictive maintenance (PdM) is a condition-based maintenance type that monitors the condition of assets using sensor devices, and also, it is known as a proactive maintenance type. To assess the actual condition of a machine, an inspection through infrared, acoustic (partial discharge and ultrasound), vibration analysis, and sound level measurements can be used. These sensor devices are used to predict when the asset will require maintenance and provide real-time data to prevent equipment failure.

Monitoring the status of assets started with using a periodic or offline approach. A continuous or online approach is used to monitor the conditions of assets today. With the IoT, all devices on earth have become able to exchange information and data with each other, and systems have almost started to get smart [6]. Thus, it is possible to remotely monitor the IoT sensor device by connecting it to the maintenance software.

The first step in predictive maintenance practices is to establish baselines. Before installing the sensors, you have to monitor the conditional baselines of the entities and collect data. In this way, when you start conditional data collection, a control portion is created where you can compare anomalies. The hardest challenge in dealing with predictive maintenance is the processing of this big amount of data. This data must be stored, processed, and analyzed using intelligent algorithms. Then, any behavior that a piece of equipment performs outside of normal parameters triggers your predictive maintenance protocol on the sensors. Typically, a related work order is created and assigned to a technician so that he or she can make the necessary repairs to fix the defect.

Anomaly Detection

Big data, which manages the world's rapid change, has started to become a problem due to the necessity of making faster decisions in real time in cases where it cannot be managed. One way to process data faster and more efficiently is connected with detecting anomalous events, changes, or clusters in datasets. As such, anomaly detection, a technology based on artificial intelligence, has become one of the number one tools of IoT to identify these behaviors that are collected in the data pool and characterized as anomalous.

Anomaly detection in data mining means identifying data points, in other words events and/or observations, of the dataset that do not fit into an expected pattern. Such anomalous data are indicative of problems such as a technical glitch, structural defect, or fraud. Anomaly detection when finding out anomalies can be a key to resolving such interferences, as it signals the presence of disruptions of normal behavior, intended or unintended induced attacks, flaws, bugs, and the like.

If anomaly detection algorithms are applied for a dataset, it is expected that three possible situations will arise [7];

- *True Positive*: The anomalies detected in the data fully correspond to the anomalies occurring in the process.
- *False Positives*: The operation continues to be normal, but unexpected data values are detected (for instance: due to noise in the system).
- *False Negatives*: Anomalies occur in the process, but the noises in the system are stronger than the anomaly data.

Defined data consists of a series of values over time. This means that each point is typically a pair of two items, the timestamp of the one being measured, and the value associated with it. The data is not meaningful on their own, but they contain the information necessary to make educated guesses about what can reasonably be expected in the future. Necessary studies such as dynamic analyze in the automatic working principle for automation, observations that cannot be noticed by humans to ensure consistency, skipping false positives/negatives for accuracy are the results of machine learning-based, artificial intelligence-supported self-learning.

Comprehending the type of outliers an anomaly detection algorithm can identify is essential to obtaining the highest accuracy. Because when you do not know what you are up against, you come up against the risk of making wrong decisions when anomaly detection alerts you to a problem or opportunity. The details of anomaly types are:

Point anomaly: These outliers known as global anomalies correspond to data points that deviate greatly from the rest of data points.

Contextual anomaly: The deviation that leads to these anomalies, called conditional outliers, depends on contextual information. These contexts are governed by contextual features and behavioral features. An anomaly in the context of one dataset may not be an anomaly in the context of another dataset.

Collective anomaly: When a subset of data points in a cluster is abnormal to the entire dataset, these values are called collective outliers. Single values in this category are not considered point or contextually abnormal. When different datasets are analyzed together, coincidences may arise where you can make this type of identification. While there is no deviation of a behavior in a single dataset, more important anomalies become clearer when it is combined with another dataset.

Unsupervised techniques, which are among the techniques of anomaly detection, do not require manually labeled training data. They are based on two basic assumptions. First, they assume that most network connections are normal traffic and only a small amount is abnormal. Second, they estimate that malicious traffic is statistically different from normal traffic. Based on these two assumptions, datasets with frequent similar occurrences are considered normal traffic and rarely malicious datasets. The most popular unsupervised algorithms include K-means, autoencoders, Gaussian mixture modeling (GMM), principal component analysis (PCA), a class support vector machine (SVM), and analysis based on hypothesis testing. They generally work well for discovering new patterns in a dataset and clustering data into several categories based on various characteristics.

But only a tiny fraction of the tons of data in the world is labeled, including texts, images, time series, and more. But most of the time, labeled data is needed when supervised machine learning is desired. Suppose you want to train a model to classify text documents, but you want to give your algorithm only one hint on how to create the categories. At this point, there is another class of algorithms called semi-supervised algorithms that can learn from partially controlled datasets that you will use. As you can imagine, semi-supervised learning algorithms are trained from a combination of labeled and unlabeled data. This is useful for several reasons. For example, the process of labeling large amounts of data for supervised learning is often time consuming. Moreover, too much labeling can cause human biases on the model. Labeled data is used in this manner to help identify specific groups of types in the data and what they might be. The algorithm is then trained on unlabeled data to identify the boundaries of these types.

Failure Prediction

Predictive maintenance actually consists of catching or predicting failures and taking certain precautions according to these failures. After the latest developments, with the support of machine learning and cloud servers, the collection and processing of data from factories, buildings, machines, and sensors has become easier. With this ease, not only the status of the assets was observed, but also the possible failures and malfunctions started to be predicted with the data obtained.

The purpose of error estimation is to establish models that can predict errors that may occur in assets and sub-parts with a high accuracy rate. The success of a strong model is mainly based on three principles; accessing the right data, framing the problem appropriately, and evaluating the predictions correctly. Besides, acquiring high accuracy rate and applying failure prediction effectively require the selection and correct adjustment of the most accurate algorithm for specific system setups [8].

To be able to predict errors with predictive maintenance, it is necessary to have sufficient data first. That is why it is best to start collecting historical data on machine performance and service records. Using historical data is the most important step in failure prediction. In addition, static information about the machine such as mechanical properties and typical usage behaviors will also be useful at this stage.

At this point, having enough data is a good starting for us. However, having too much data brings with it a problem such as noise during measurement, data pollution caused by unusual data. At this point, before the model is built, the data should be cleaned correctly and the meanings in the data should not be lost. The next step after the data are collected correctly and ready for use is framing the problem, that is, creating the right model for the system. There are some points to consider at this point;

- What are the expectations from model outputs?
- Is historical data available for the model?
- Are the times apparent when the machine breaks down and works in the dataset? So, the data has the label?
- Is the amount of data that the machine corrupts and the amount of data that it works properly proportional in the dataset to each other?

Once this information is obtained, the strategy of the model needs to be decided. There are many modeling strategies in failure prediction for predictive maintenance. It can basically be explained as follows;

Remaining Useful Life (RUL): In this strategy scenario, the remaining useful life for the part is calculated. Thus, it is tried to estimate how long the usable cycle time determined for the part can be. Both static and historical data are used in the model. In general, data for all types of deterioration are included in the dataset. However, if the response of the system is different for the error types found, at this point, it is necessary to create separate models for each type of error.

Time Interval: Although the use of this strategy is challenging for modeling, there is no need to precisely estimate the remaining time. In general, in this type of modeling, it is tried to be predicted whether the asset will deteriorate in any specified time interval. Likewise, static and historical data are needed. Generally, this strategy is referred to as classification. This method can be used in two different ways. Different time intervals for any type of deterioration can be created, and with these time intervals, it can be predicted in which time period the asset will deteriorate by using the multi-class classification method. In another method, a model is created for many types of deterioration. For example, class zero shows the probability of asset no failure, class 1 indicates the probability of type 1 asset failure in the next X days, and class 2 shows the probability of asset failure in the next X days. Types and number of classes can be increased according to the models. However, the much increase in the number of classes will also create problems in the establishment of the model.

Once models are successfully established, their performance needs to be measured for future datasets before they can be used. In a good estimation, the hyperparameters determined by the training set should show high performance also on the validation data or cross validation data. Performance criteria vary according to the model strategy used.

When the model is created with the classification strategy, the accuracy score performance criterion is checked first. The formula is simply the number of correctly classified data divided by the total amount of data. However, the accuracy score is very sensitive to data distribution and unbalanced data. If the amount of data of any class in a dataset is much higher than the amount of data of the other class, then an unbalanced dataset situation occurs. In this case, other criteria should be examined together with the accuracy score. Another popular criterion is the confusion matrix. It is also known as the error matrix. With the confusion matrix, it can be seen as separated by class how correct and incorrect the estimated values are (Table 4).

Case Study: A turbofan manufacturer

XYZ company is an acclaimed turbofan manufacturer. Besides manufacturing, they are giving maintenance service for turbofan engines. The company, which generally uses the scheduled maintenance method, has recently decided to make all its planning on predictive maintenance. With predictive maintenance, the company wants to create two different forecasting models and provide anomaly detection or failure prediction services according to customer requests. In order to set up the system and create the

Table 4 Confusion matrix

	Actually positive (0)	Actually negative (0)
Predicted negative (0)	True positive (TPs)	False positives (FPs)
Predicted positive (1)	False negative (FNs)	True negatives (TNs)

Table 5 Variable list of data collected from engines

Variable name	Variable description
Time, in cycles	Engine working cycle
os1	Operational settings 1 for engine performance
os2	Operational settings 2 for engine performance
os3	Operational settings 3 for engine performance
sensor_01	Measurement data of Sensor 1
sensor_02	Measurement data of Sensor 2
sensor_03	Measurement data of Sensor 3
.	.
.	.
sensor_25	Measurement data of Sensor 25
sensor_26	Measurement data of Sensor 26

required model, 100 newly produced engines were tested and data were collected from 26 different sensors (temperature, pressure, speed, etc.) for each engine during the test [9]. The list of variables from the collected data is given below (Table 5).

Statistical Analysis of Collected Data

Statistical examination and cleaning of the collected data before training will facilitate the creation phase of models and will also help determine feature importance. To begin with, it would be factual to look at the mean, standard deviation, and min and max values of features.

According to statistical characteristics in Fig. 8, data of sensor 22, 23, 24, 25 and 26 are empty, so no calculations can be made. Things to note here are the calculations made in the columns that include the machine setup os3, sensor 3, sensor 1, sensor 5, sensor 10, sensor 16, sensor 18, and sensor 19. In these calculations, except for sensor 6, the minimum and the maximum values of the data belonging to these columns are equal to each other and the arithmetic mean. Standard deviations were measured as 0. This means that these columns actually read constant values and will continue to be removed from the dataset. As a result of the examinations, columns were removed from the dataset and correlation analysis was performed.

To better understand the data, let us take the data collected by sensors 15, 17, 20, and 21 as an example. Different colored lines in Fig. 9 represent different engine units. The remaining useful lifetime for 100 engines is in the range of 0–350. As can be seen in the graphics in Fig. 9, the data coming in under normal operating conditions of the engines oscillate around certain values. When the engine's lifetime comes to an end, there are stubs in the data. Especially large deviations can be observed after RUL is 100 cycles.

To add a different interpretation to Fig. 10, as can be seen from the scale on the right of the figure, the relation size in white cells reaches from plus one to minus one. Especially between sensor 9 and sensor 14, the correlation size between sensor 4 and sensor 11 and 12, sensor 7 and sensor 11 and 12, sensor 8 and sensor 13, and finally

	unit_num	cycle_time	os1	os2	os3	sensor_01	sensor_02	sensor_03	sensor_04	sensor_05
count	20631.000000	20631.000000	20631.000000	20631.000000	20631.0	20631.00	20631.000000	20631.000000	20631.000000	2.063100e+04
mean	51.506568	108.807862	-0.000009	0.000002	100.0	518.67	642.680934	1590.523119	1408.933782	1.462000e+01
std	29.227633	68.880990	0.002187	0.000293	0.0	0.00	0.500053	6.131150	9.000605	1.776400e-15
min	1.000000	1.000000	-0.008700	-0.000600	100.0	518.67	641.210000	1571.040000	1382.250000	1.462000e+01
25%	26.000000	52.000000	-0.001500	-0.000200	100.0	518.67	642.325000	1586.260000	1402.360000	1.462000e+01
50%	52.000000	104.000000	0.000000	0.000000	100.0	518.67	642.640000	1590.100000	1408.040000	1.462000e+01
75%	77.000000	156.000000	0.001500	0.000300	100.0	518.67	643.000000	1594.380000	1414.555000	1.462000e+01
max	100.000000	362.000000	0.008700	0.000600	100.0	518.67	644.530000	1616.910000	1441.490000	1.462000e+01

...	sensor_17	sensor_18	sensor_19	sensor_20	sensor_21	sensor_22	sensor_23	sensor_24	sensor_25	sensor_26
...	20631.000000	20631.0	20631.0	20631.000000	20631.000000	0.0	0.0	0.0	0.0	0.0
...	393.210654	2388.0	100.0	38.816271	23.289705	NaN	NaN	NaN	NaN	NaN
...	1.548763	0.0	0.0	0.180746	0.108251	NaN	NaN	NaN	NaN	NaN
...	388.000000	2388.0	100.0	38.140000	22.894200	NaN	NaN	NaN	NaN	NaN
...	392.000000	2388.0	100.0	38.700000	23.221800	NaN	NaN	NaN	NaN	NaN
...	393.000000	2388.0	100.0	38.830000	23.297900	NaN	NaN	NaN	NaN	NaN
...	394.000000	2388.0	100.0	38.950000	23.368800	NaN	NaN	NaN	NaN	NaN
...	400.000000	2388.0	100.0	39.430000	23.618400	NaN	NaN	NaN	NaN	NaN

Fig. 8 Statistical characteristics of the collected data

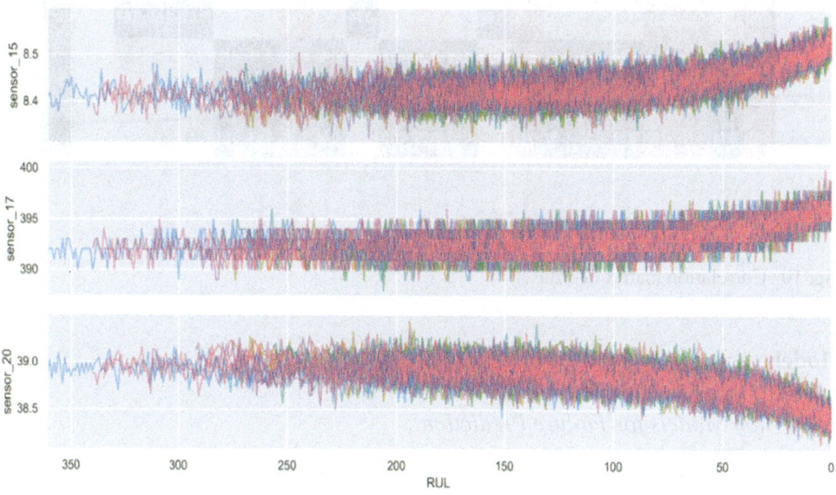

Fig. 9 Sensor data in total cycles of all unit

sensor 11 and sensor 12 is greater than 0.8 or smaller than −0.8. This means they have a high correlation between. For this reason, some cleanings have been made, such as discarding sensor 9 in order to reduce the data size, where some sensors can actually be used in estimation through other sensors.

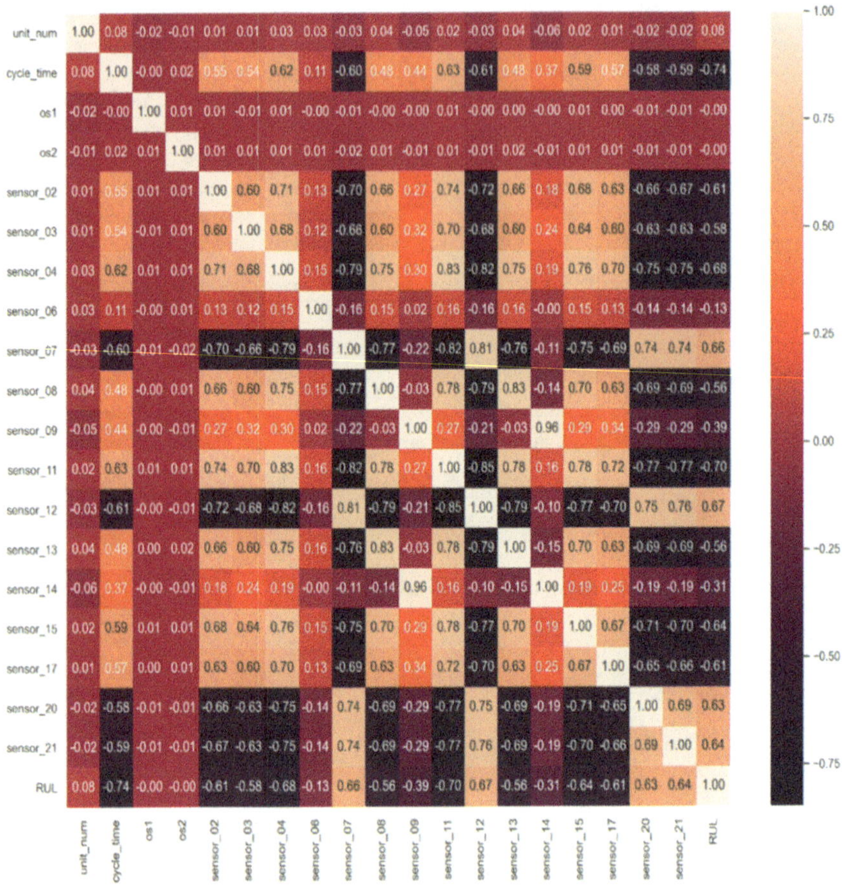

Fig. 10 Correlation matrix of dataset

Models

Regression Models for Failure Prediction

7 different algorithm types were selected to predict the remaining useful lifetime using regression models. The details of the algorithms are;

- *Lasso*: The l1-type regulation of linear regression applied. Alpha value was determined as 0.5 for the application.
- *Support Vector Regressor:* Like lasso, it has a regularization equation in it. The C value was chosen as 5 in the application.
- *K-Nearest Neighbors:* The most suitable number of neighbors for the application was determined as 9. The 'weights' parameter is determined according to the distance. With the P parameter selected as 1, the Manhattan distance is used for the distance calculation.

- *Artificial Neural Networks:* Default settings are used in the artificial neural network and only 1 hidden layer consisting of 100 neurons is used. In practice, the max iter parameter is set to 500 in order to achieve convergence in the loss function.
- *Random Forest:* Number of predictive decision trees is set to 75.
- *AdaBoost Regressor:* Decision trees are used as the basic estimator in its use. The number of trees to be used for estimation was determined as 150, and the learning coefficient was chosen as 0.5.
- *XGBoost Regressor:* Square error is used for loss function. The maximum depth for the decision trees was determined as 5, and the learning coefficient was determined as 0.3.

As seen in Table 6, 7 algorithms were trained with training sets and tested with test sets. As a result, the metrics used for model evaluation were calculated. The most important metrics in the application are primarily RMSE, and the other is R^2 score. The reason why RMSE is more important than MAE is that it is required to penalize major errors within the framework of predictive maintenance. Because, as a result of estimating the remaining life of the machines with a great error, the machines cannot be intervened correctly, and the processes in the enterprise may be disrupted if the deterioration is not prevented. As can be seen, the best case is the XGBoost algorithm and AdaBoost algorithm came after. The equation established in both models can explain the dataset with high accuracy. For the regression modeling, the winner algorithm chosen based on the RMSE and R2 score was chosen as XGBoost (Fig. 11).

Classification Models for Failure Prediction

Two different classes were created according to the remaining useful lifetime in the classification application. In order to perform the classification application, the classes of data with a remaining lifetime below 30 cycles were determined as 1, and those with more than 30 labeled as 0. 7 different algorithm types were used in the classification model. The details of the algorithms are;

Table 6 Model evaluation metrics for regression algorithms

Method	MAE	MSE	RMSE	R^2 score (%)
Lasso	30.24	1520.82	38.99	66.79
SVM	23.72	1228.11	35.04	73.18
Nearest neighbors	24.18	1174.31	34.27	74.36
Neural networks	23.78	1097.24	33.12	76.04
Random forest	11.15	279.45	16.72	93.89
AdaBoost	7.03	131.94	11.49	97.11
XGBoost	7.73	107.86	10.39	97.65

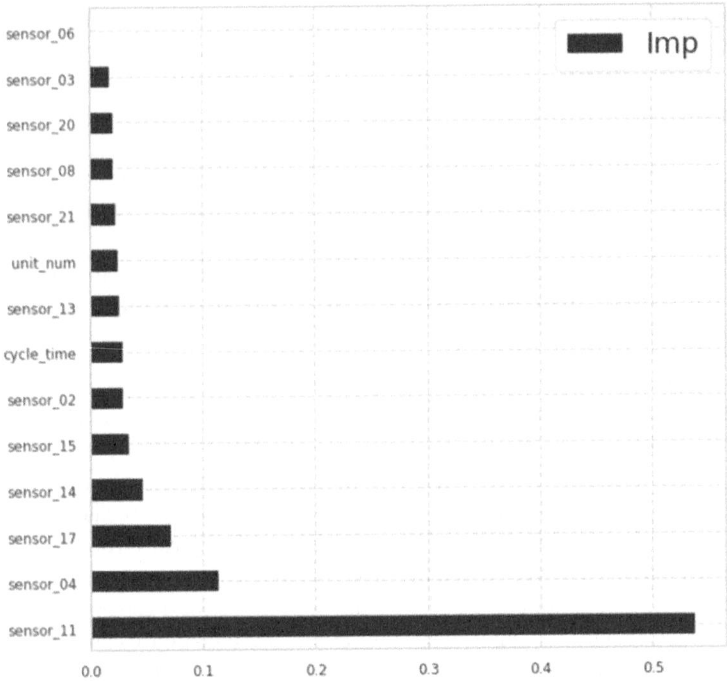

Fig. 11 Feature importance of XGBoost regressor

- *K-Nearest Neighbor*: Classification according to the 9 nearest neighbors has been determined. With the P parameter selected as 1, the Manhattan distance is used for the distance calculation.
- *Logistic Regression*: The default hyperparameters given by scikit-learn are used in the classification application.
- *Support Vector Classifier*: Default hyperparameters given by scikit-learn have been used within the classification application.
- Random Forest: The number of predictive decision trees has been determined as 75. The n_jobs parameter is set to -1 in order to obtain a hardware faster result. Thus, parallel computations were used.
- *Artificial Neural Networks*: A multi-layered perceptron structure with 3 hidden layers and 40 neurons in each layer is used in classification applications. The Broyden–Fletcher–Goldfarb–Shanno algorithm, which is in the quasi-Newton method family, was chosen as the solver parameter. The alpha parameter is also set to 1.
- *AdaBoost Classifier*: Decision trees with a maximum depth of 4 and discriminating according to the entropy criterion are used as the basic estimator in the AdaBoost algorithm. The number of determined estimators is 150.

Table 7 Model evaluation metrics for classification algorithms

	Accuracy score (%)	Precision score (%)	Recall score (%)	F1 score (%)	Auc score (%)
Nearest neighbor	96.30	90.06	84.73	87.31	91.54
Logistic regression	95.99	87.56	85.49	86.51	97.67
RBF SVM	96.49	90.47	85.69	88.01	92.05
Random forest	96.22	88.16	86.45	87.29	92.19
Neural nets	96.28	87.23	88.17	89.23	92.95
AdaBoost	96.80	90.31	88.17	89.23	93.25
XGBoost	97.72	92.74	92.04	92.39	95.39

- *XGBoost Classifier*: Logistic regression has been used as objective function for classification application. The maximum depth of the decision trees was determined as 4, and the learning rate was determined as 0.4.

As seen in Table 7, 7 different algorithms have been trained in the application and tested with test sets. As a result, the metrics explained in the chapter were calculated. The metrics are listed in the table from smallest to largest. Recall score has been chosen as the most important metric since it is desired to correctly predict the machines that are close to failure in the application. Considering the recall score, the best algorithm is the XGBoost classifier, and the second-best ones are AdaBoost and artificial neural networks. As can be seen, the recall score is equal for Adaboost and artificial neural networks. Therefore, by looking at the Auc score, it can be observed how well the model separates the probabilities of the positive class from the probabilities of the negative class. Therefore, AdaBoost can be chosen as the second-best classifier.

In the dataset created for classification, there are 3100 data in class 1 and 17,531 data in class 0 in total. It has been observed that the accuracy score should not be overly relied on because of an unbalanced dataset. At this point, besides looking at the Recall and Auc score, the confusion matrix can be used to see how much these values are. As an example, the confusion matrix obtained for XGBoost in Fig. 12 and Roc-Auc curve in Fig. 13 can be shown. It can be seen that the model is successful in separating the classes from each other.

Clustering Models for Anomaly Detection

5 different algorithms that can learn unsupervised within the application were selected for anomaly detection. The details of the algorithms are;

- Local outlier factor: Contamination rate parameter was set as 0.248. This ratio is the ratio of the elements of the first cluster to the whole data. Novelty parameter was set to true in order to make the prediction with the trained algorithm.

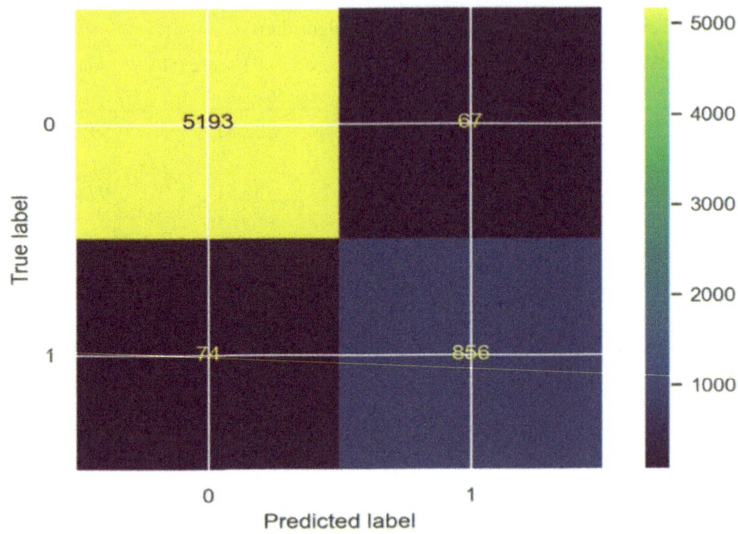

Fig. 12 Confusion matrix of XGBoost classifier

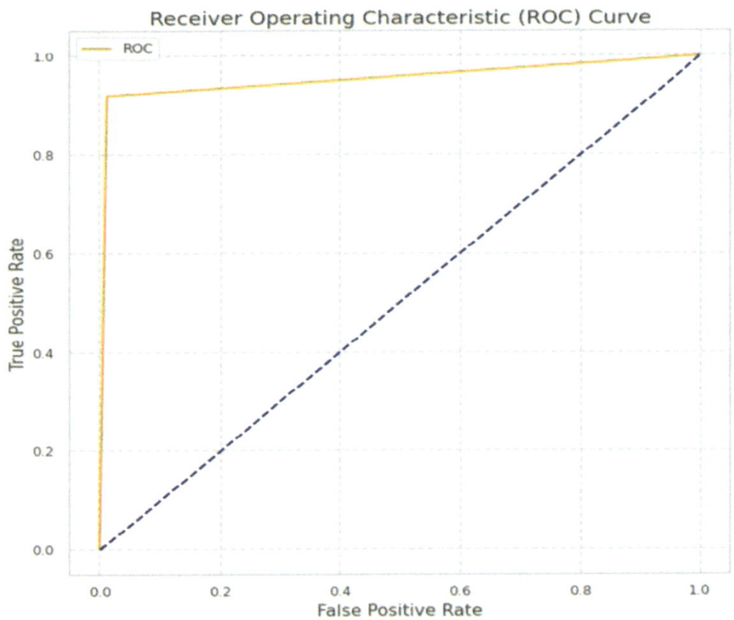

Fig. 13 Roc-Auc curve of XGBoost classifier

- One-class support vector machine: The default hyperparameters given by the scikit-learn library are used within the clustering application.
- *Isolation forests:* The number of predictive decision trees in the clustering application was determined as 100. The contamination rate parameter was determined as 0.248 and the n jobs parameter was set to -1 to use all CPUs.
- *Autoencoder neural networks:* 4 hidden layers having 20, 12, 12, and 20 neurons were used while creating the neural network layer. The contamination rate parameter was also determined as 0.248.
- *K-Means:* The number of clusters was determined as 2 in the clustering application. One of these clusters takes the value 0 and the other the value 1. The set that takes the value of 1 denotes the anomaly set.

A few steps must be followed in order to determine the anomalies in the dataset and to evaluate the model outputs. First of all, the remaining useful lifetime column was created. As a result of the visualization of the dataset in Fig. 14, it was observed that there are large deviations in the sensor data when the useful lifetime is between 25 and 50 cycles. Therefore, data rows with less than 40 cycles are classified as anomalies for the application (class 1). At this point, when the clusters in the dataset are examined, it is observed that there are 4100 elements in the first cluster and 16,531 elements in the 0th cluster.

As can be seen in the table above, 5 different algorithms were trained, and then evaluation metrics were calculated by comparing the predictions with the actual values. Correctly estimating the failure moments of the engines is more important than the wrong estimation of the intact engines. The recall score shows how much

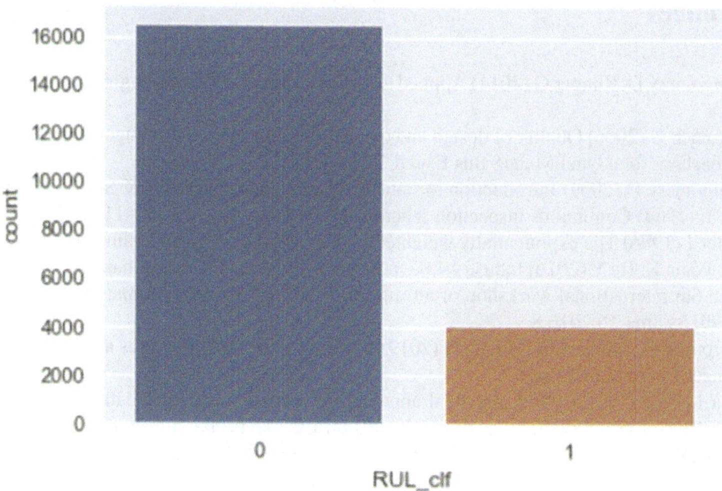

Fig. 14 Distribution of classes in the dataset

Table 8 Model evaluation metrics for clustering algorithms

Method	Adjusted Rand Score	Fowlkes Mallows Score	Accuracy score (%)	Recall score (%)	Auc score (%)
Local outlier factor	0.01	0.67	66.81	23.73	50.61
One-class SVM	0.03	0.60	58.22	70.73	62.92
Isolation forest	0.37	0.78	82.74	69.00	77.58
Autoencoder	0.39	0.79	83.24	70.24	78.36
K-means	0.25	0.70	75.39	96.32	83.26

those predicted as positive clusters actually belong to the positive cluster. The positive set represents the engines that are close to failure within the application. Therefore, recall is the most important metric in application. According to the recall metric, K-means was significantly better than the other models in the application, close values were found for automatic encoders, one-class SVM and isolation forests. At this point, the Auc score was examined in order to observe how much success was achieved in separating the two clusters from each other. Under these conditions, the most suitable first model for anomaly detection application was the K-means, while the second algorithm was determined as neural networks with automatic encoder (Table 8).

References

1. Montgomery D, Runger G (2014) Applied statistics and probability for engineers. Wiley, United States
2. Meirovich G (2006) Quality of design and quality of conformance: contingency and synergistic approaches. Total Qual Manag Bus Excell 17(2):205–219
3. Montgomery D (2009) Introduction to statistical quality control. Wiley & Sons, United States
4. Page E (1954) Continuous inspection schemes. Biometrika 41(1–2):100–115
5. Hunter J (1986) The exponentially weighted moving average. J Qual Technol 18:203–210
6. Li Z, Wang K, He Y (2016) Industry 4.0—potentials for predictive maintenance. In: Proceedings of the 6th international workshop of advanced manufacturing and automation. https://doi.org/10.2991/iwama-16.2016.8
7. Mehrotra KG, Mohan CK, Huang H (2019) Anomaly detection principles and algorithms. S.l.: Springer International Pu
8. Irrera I, Vieira M (2014) A practical approach for generating failure data for assessing and comparing failure prediction algorithms. In: 2014 IEEE 20th Pacific rim international symposium on dependable computing. https://doi.org/10.1109/prdc.2014.19
9. Saxena A, Goebel K (2008) Turbofan engine degradation simulation data set. NASA Ames Prognostics Data Repository. http://ti.arc.nasa.gov/project/prognostic-data-repository, NASA Ames Research Center, Moffett Field, CA

Nursah Alkan is a Ph.D. candidate and a research assistant in the Department of Industrial Engineering at Istanbul Technical University since 2019. She received the M.Sc. degree in Department of Industrial Engineering from Yildiz Technical University, Turkey, in 2019. Her research interests include fuzzy sets, multi-criteria/objective decision-making, data analysis, machine learning. and digital transformation.

Dogan Oruc is a data analyst at Trendyol.com Company. Before working in Trendyol.com, he worked one year at Acrome Robotics as Software Engineer. He completed his undergraduate education at Istanbul Technical University, Department of Industrial Engineering (ITU-IE) and Department of Mechanical Engineering (ITU-ME) with double major program. He studied courses about operational research, data science, probability, and statistics. His work which is *'Humanoid Robot Head Mechatronics Design: MUDA'* has been featured in Turkey Robotics Conference, TORK. His last research explored how to model product design about predictive maintenance in Industry 4.0.

Arif Gulbiter is Master Student at Industrial Engineering Department of Ghent University (UGent-IEOR). Before studying master program in UGent-IEOR, he worked one year at Turkish Economy Banks as a business analyst and one year at Acrome Robotics as a process analyst. He completed his undergraduate education at Istanbul Technical University, Department of Industrial Engineering (ITU-IE). He studies courses about operations research, decision-making under uncertainty and industrial statistics. His last research explored how to model product design about predictive maintenance in Industry 4.0.

Mehmet Ali Ergun is an academician and researcher currently working at Industrial Engineering Department of Istanbul Technical University as an Assistant Professor. He holds a Ph.D. degree in Industrial Engineering from University of Wisconsin—Madison and a B.S. degree in Computer Engineering from Bogazici University. His research focuses on applying stochastic optimization, simulation and machine learning techniques to help making data-driven decisions in healthcare related challenges. In one of his work, he was a major contributor to the National Cancer Institute-funded Cancer Intervention and Surveillance Modeling Network (CISNET) project that was used to develop the breast cancer screening guidelines published in the U.S.

The manufacturer's authorised representative in the EU is Springer Nature Customer Service Centre GmbH, Europaplatz 3, 69115 Heidelberg, Germany. If you have any concerns regarding our products, please contact ProductSafety@springernature.com

Printed and bound by CPI Group (UK) Ltd, Croydon, CR0 4YY
22/09/2026
02220954-0007